W9-BVH-624

Advances in Well Test Analysis

Robert C. Earlougher, Jr.

Senior Research Engineer

Marathon Oil Co.

Second Printing

Henry L. Doherty Memorial Fund of AIME

Society of Petroleum Engineers of AIME

New York 1977 Dallas

DEDICATION

To Evelyn, whose patience, understanding, and encouragement were essential to completion of this monograph.

ISBN 0-89520-204-2

Contents

SPE Monograph Series

The Monograph Series of the Society of Petroleum Engineers of AIME was established in 1965 by action of the SPE Board of Directors. The Series is intended to provide members with an authoritative, up-to-date treatment of the fundamental principles and state of the art in selected fields of technology. The work is directed by the Society's Monograph Committee, one of 40 national committees, through a Committee member designated as Monograph Coordinator. Technical evaluation is provided by the Monograph Review Committee. Below is a listing of those who have been most closely involved with the preparation of this book.

Monograph Coordinator

William C. Miller, Shell Development Co., Houston

Monograph Review Committee

Earl E. Morris, chairman, Amoco Production Co., Houston
John M. Campbell, Continental Oil Co., Houston
Vance J. Driscoll, Amoco Production Co., Houston
Bruce B. McGlothlin, Gulf Research & Development Co., Houston
L. J. Sanders, Amoco Production Co., Houston
Juris Vairogs, Cities Service Oil Co., Tulsa
Jerry L. Zink, Continental Oil Co., Houston

SPE Monograph Staff

Thomas A. Sullivan
Technical Services
Manager-Editor

Ann Gibson
Production Manager

Georgeann Bilich
Project Editor

Acknowledgments

This manuscript exists only because of significant support from many individuals. In particular, I am indebted to three of my Marathon colleagues, H. C. Bixel, Hossein Kazemi, and Shri B. Mathur,* who have heavily influenced the content and philosophy of this monograph. The four of us have worked closely in well test analysis for many years. As a team, we have prepared company handbooks and presented training courses in well test analysis for Marathon engineers. In so doing, we have developed a philosophy for presenting well test analysis material, and have learned which techniques are most successful and which are least successful for conveying information in an understandable and useable fashion. That philosophy and background pervade this monograph — it is impossible to separate my contributions from those of Messrs. Bixel, Kazemi, and Mathur. Additionally, Messrs. Bixel, Kazemi, and Mathur have contributed significantly to the presentation of much of the material in the monograph, if not in detail at least in substance. I very much appreciate all they have contributed toward making this a useful book.

The material presented in this monograph has been reviewed by members of the SPE-AIME Monograph Review Committee. They have spent hundreds of hours reading, checking, and critically commenting on all aspects of the material and its presentation. There is no doubt that the monograph is a much better volume than it would have been without their aid.

Marie E. LeBlanc typed the many versions of the manuscript required to reach the final form. Her secretarial skills and command of the English language have enabled preparation of this volume to proceed smoothly and on schedule.

Sally M. Andrews illustrated the monograph. Besides preparing all the original illustrations, she redrew many illustrations taken from the references to provide a consistent nomenclature and format. Her artist's viewpoint, her skill, and her highly accurate work have added substantially to this monograph.

*Presently with Shell Oil Co.

Preface

By about 1973, recent publications and advances in well test analysis were numerous enough to justify some type of update to C. S. Matthews' and D. G. Russell's monograph, *Pressure Buildup and Flow Tests in Wells*. In 1974, the Monograph Committee asked me to prepare an updated monograph that would include enough information to stand alone, rather than to just be a supplement or an update to the Matthews-Russell monograph. Although this book draws heavily on information in *Pressure Buildup and Flow Tests in Wells*, it is my belief that it can be used for most well test analysis situations without requiring other material.

In the mid-1960's, when the Matthews-Russell monograph was being prepared, reservoir simulation, and particularly the application of reservoir simulation to well test problems, was in its infancy. Subsequently, there has been a significant expansion of knowledge about well testing, much of it a result of the application of reservoir simulators. Publication of this monograph does not imply the belief that such advances will not continue, for they certainly will. I expect that there will be updated well test analysis monographs at regular intervals for many years to come.

The subject of this monograph is a broad and general one that is hard to define completely and concisely. I have attempted to present a valid and useful range of information rather than a completely comprehensive treatment (which would require many times the present length). As a result, there are many compromises. A high degree of technical accuracy (not always available) is sometimes sacrificed to provide methods with practical utility. Many available testing and analysis techniques are just referenced, while only the essence of others is presented, without consideration of minor qualifications and special cases that often appear in the original articles. In writing this book, I have made many value judgments, not only as to the accuracy and validity of a particular technique, but also as to its practical application and utility to the engineer. In general, I have attempted to provide information that can be used readily for practical and real problems. Most parts of the monograph give guidelines for applicability of various analysis techniques. One set of nomenclature is applied to all types of well testing. To show the interrelation and the minor differences between various tests, I include in Appendix E a comparison table that should be useful to the frequent monograph user.

The monograph is not written as a textbook or to provide theoretical background. There are no derivations, although the method for deriving some of the equations is suggested. The reader must return to the original reference material cited to find derivations. Worked examples are an important part of this monograph — more than 50 are included to illustrate analysis techniques presented and, frequently, to emphasize practical problems that arise in well test analysis.

I believe that most users of this monograph will find it logically organized and readily applicable to many well testing problems. Nevertheless, there will arise many situations for which the answer does not appear in this monograph and which will require further research on the part of the reader. As years pass, many currently unanswered questions will be answered; it is hoped that the next volume covering this subject will include many of those new answers.

Littleton, Colorado — ROBERT C. EARLOUGHER, JR.

February, 1976

Chapter 1

Introduction

1.1 Purpose

In 1967, Matthews and Russell published the first complete, cohesive treatment of well testing and analysis.[1] The Matthews-Russell monograph has become a standard reference for many petroleum engineers. Since the publication of that monograph, more than 150 additional well test analysis technical papers have been published. Those papers have extended the scope of well test analysis, publicized many new problems, provided solutions for previously unsolved problems, and changed the approach to some phases of well test analysis. Thus, it is appropriate to provide an updated monograph dealing with advances in well test analysis in a manner that presents an up-to-date treatment of the state of the art.

Enough material is presented so that this book can be used alone rather than solely as a supplement to the Matthews-Russell monograph. Matthews and Russell have presented the applicable history, the theoretical background of fluid flow, and the derivation of most of the equations used in well test analysis. Therefore, this monograph does not treat those subjects in detail, but refers to more rigorous treatment. The theory is brief and simple and derivations are minimized, since a detailed understanding of the mathematics involved in developing well test analysis equations is not necessary for correct engineering application. However, an understanding is often required of what a given method physically represents for appropriate engineering application. Thus, an attempt is made to be conceptually clear about different analysis techniques and to present estimates of the range of applicability. Examples illustrate most analysis techniques.

1.2 Use of Pressure Transient Testing in Petroleum Engineering

Reliable information about in-situ reservoir conditions is important in many phases of petroleum engineering. The reservoir engineer must have sufficient information about the reservoir to adequately analyze reservoir performance and predict future production under various modes of operation. The production engineer must know the condition of production and injection wells to coax the best possible performance from the reservoir. Much of that information can be obtained from pressure transient tests.

Pressure transient testing techniques, such as pressure buildup, drawdown, injectivity, falloff, and interference, are an important part of reservoir and production engineering. As the term is used in this monograph, pressure transient testing includes generating and measuring pressure variations with time in wells and, subsequently, estimating rock, fluid, and well properties. Practical information obtainable from transient testing includes wellbore volume, damage, and improvement; reservoir pressure; permeability; porosity; reserves; reservoir and fluid discontinuities; and other related data. All this information can be used to help analyze, improve, and forecast reservoir performance.

It would be a mistake to either oversell or undersell pressure transient testing and analysis. It is one of the most important in a spectrum of diagnostic tools. In certain situations it is indispensable for correct well or reservoir analysis; for example, in definition of near-wellbore and interwell conditions as opposed to composite properties that would be indicated by steady-state productivity index data. In other cases, a simpler approach is adequate, or a different or combined approach is needed to solve a problem.

Consider the case of a pumping oil well with substantial production decline. It usually would be inappropriate to run a pressure buildup test without first determining whether the problem was merely a worn pump and high working fluid level or some other mechanical problem. If a simple approach fails to identify the problem, a pressure buildup test could be indispensable in pinpointing that the specific problem is related to damage at or near the formation face rather than to rapid reservoir depletion.

On the other hand, even with the most complex and thorough transient analysis, a unique solution often is not possible without considering other information. Pressure interference or pulse testing could establish the possible existence and orientation of vertical fractures in a reservoir. However, other information (such as profile surveys, production logs, stimulation history, well production tests, borehole televiewer surveys or impression packer tests, core descriptions, and other geological data about reservoir lithology and continuity) would be useful in distinguishing between directional permeability and fractures or estimating whether the fractures were induced or natural.

In practice, engineering application of pressure transient analysis is often limited by (1) insufficient data collection; (2) inappropriate application of analysis techniques; or (3) failure to integrate other available or potentially available information. Most practicing engineers are aware of instances where a definitive analysis has been precluded by a lack of accurate early pressure and withdrawal information or prior base data for comparative purposes.

It is generally good practice to run a base pressure transient test on a producing well shortly after completion or on an injection well after a suitable period of injection. This can lead to early recognition and correction of many problems, of which insufficient stimulation is only the most obvious. Such tests also provide in-situ data for reservoir simulation and a base for comparison with reservoir or well problems as they arise.

1.3 Organization, Scope, and Objectives

The data in this monograph should enable the petroleum engineer to design, conduct, and analyze pressure transient tests to obtain reliable information about reservoir and well conditions. Each chapter is, as nearly as possible, an independent unit. For completeness, Chapter 1 includes a short discussion of unit conversion factors and the SI (metric) unit system. Appendix A provides a list of conversion factors and a tabulation of some of the more important equations in oilfield units, groundwater units, and three sets of metric units.

Chapter 2 is a summary of transient fluid flow behavior and sets the stage for all transient test analysis procedures in the text. The approach is a pragmatic one that provides the reader with material to derive methods of test analysis and to calculate expected transient response in wells. Since recent advances have modified some older methods, we attempt to integrate and present what appears to be best current engineering practice. Many research studies, invaluable in themselves for providing insight or cross-checks on the validity of other work, are not necessarily suitable for direct field application. Others, although complex, provide ways to estimate important reservoir properties. In situations where only minor differences in accuracy would result by using simpler methods, preference has been given to the simpler method. Nevertheless, test analysis procedures in the monograph may be used without complete understanding of Chapter 2. Appendix B presents a detailed theoretical treatment of the use of superposition to generate new solutions that may be useful to some readers. Appendix C presents a wide range of dimensionless solutions incorporating various geometries and boundary conditions.

The chapters describing basic testing and analysis techniques utilize the flow theory of Chapter 2, but otherwise stand alone. Since the primary thrust of the monograph is toward the practicing engineer, an effort has been made to set bounds and define the range of applicability of various solutions or techniques. Chapter 3 covers pressure drawdown testing, the most theoretically simple form of pressure transient testing. It also introduces type-curve matching. That relatively new (to the petroleum industry) approach allows the engineer to effectively use more sophisticated transient solutions incorporating wellbore storage effects, deep fracturing, complex boundary conditions, etc., when simpler analysis techniques are not applicable.

Chapter 4 covers multiple-rate testing and discusses how superposition may be used where variable rates are involved. Chapter 5 treats pressure buildup test analysis; and Chapter 6 presents methods for estimating average pressure for well drainage areas and the entire reservoir. Chapter 7 deals with injection well testing, a matter of ever-increasing importance. Chapter 8 discusses drillstem test analysis.

Chapter 9 gives transient testing techniques utilizing more than one well. Chapter 10 covers the effects of reservoir heterogeneities on pressure behavior. Chapter 11 provides more detailed information on the effects of wellbore storage and induced fractures on pressure transient behavior. Chapter 12 briefly discusses computer methods; and Chapter 13 considers design and instrumentation of pressure transient tests.

Appendix D presents methods and correlations for estimating many reservoir rock and fluid properties; and Appendix E summarizes well test analysis equations.

1.4 Nomenclature and Units

As much as possible, the standard symbols adopted by the Society of Petroleum Engineers of AIME[2-4] are used throughout this monograph. "Oilfield units" are used in equations *consistently:* flow rate, q, is in stock-tank barrels per day; permeability, k, is in millidarcies; time, t, is in hours; viscosity, μ, is in centipoise; compressibility, c, is in volume/volume/pounds per square inch; and porosity, ϕ, is always used as a fraction. Units are included in the nomenclature list. Occasionally, different units are used to be consistent with industry usage; such cases are clearly identified.

Throughout the monograph, a positive flow rate, $q > 0$, signifies production, while a negative flow rate, $q < 0$, designates injection. The sign convention requires that the correct sign be given to slopes of various data plots. That results in some equations that are slightly different from forms commonly seen in the literature. However, this is a practical way to approach transient test analysis.

We expect that metric units eventually will be the only accepted units in engineering. For that reason, Appendix A provides information outlining the definition of the "SI" units of weights and measures, along with factors for converting to SI from customary units. SI is the official abbreviation, *in all languages,* for the International System of Units (les Systéme International d'Unités). The International System is neither the centimetre-gram-second (cgs) system nor the metre-kilogram-second (mks) system, but is a modernized version of mks. A complete description of SI is presented by Hopkins.[5] The American Petroleum Institute has proposed a set of metric standards for use in the petroleum industry.[6] Most nations are gravitating toward exclusive use of SI, so SI units are given top billing in the conversion tables in Appendix A.

Tables A.1 and A.2 provide general information about the SI system. Table A.3 gives values for physical constants useful in petroleum engineering. Table A.4 gives general conversion factors. Table A.5 presents conversion factors that include permeability. Table A.6 deals with temperature scales and conversions. Finally, Table A.7 compares units and equations for well testing from five unit systems. Oilfield units are used *exclusively* throughout the remainder of this monograph.

In this monograph, the term permeability (k) is sometimes used even though the terms mobility (k/μ) or mobility-thickness product (kh/μ) may be more appropriate. This is done because permeability is a property of the rock rather than a combined property of rock and fluid. Even though this convention is used, it is important to recognize that the mobility-thickness product almost always appears as a unit in the flow and transient test analysis equations. Similarly, porosity (ϕ) is sometimes used rather than the commonly associated porosity-thickness product (ϕh) or porosity-compressibility-thickness ($\phi c_t h$) product.

References

1. Matthews, C. S. and Russell, D. G.: *Pressure Buildup and Flow Tests in Wells,* Monograph Series, Society of Petroleum Engineers of AIME, Dallas (1967) **1.**
2. "Letter Symbols for Petroleum Reservoir Engineering, Natural Gas Engineering, and Well Logging Quantities," Society of Petroleum Engineers of AIME, Dallas (1965).
3. "Supplements to Letter Symbols and Computer Symbols for Petroleum Reservoir Engineering, Natural Gas Engineering, and Well Logging Quantities," Society of Petroleum Engineers of AIME, Dallas (1972).
4. "Supplements to Letter Symbols and Computer Symbols for Petroleum Reservoir Engineering, Natural Gas Engineering, and Well Logging Quantities," Society of Petroleum Engineers of AIME, Dallas (1975).
5. Hopkins, Robert A.: *The International (SI) Metric System and How It Works,* Polymetric Services, Inc., Tarzana, Calif. (1974).
6. "Conversion of Operational and Process Measurement Units to the Metric (SI) System," *Manual of Petroleum Measurement Standards,* Pub. API 2564, American Petroleum Institute (March 1974) Chap. 15, Sec. 2.

Chapter 2

Principles of Transient Test Analysis

2.1 Introduction

This chapter summarizes the basic transient flow theory for the well testing and analysis techniques presented in this monograph. An understanding of the following material should clarify the techniques presented later, as well as allow the reader to devise additional testing and analysis techniques. Nevertheless, it is possible to use the material in Chapters 3 through 13 without a thorough reading and understanding of this chapter.

All basic theory needed in the monograph is summarized here. We neither derive the basic flow equations nor show how to solve them. Rather, a general equation is used for transient pressure behavior with dimensionless pressure solutions for the specific conditions desired. Some important dimensionless pressure functions are presented in this chapter and in Appendix C, and references to others are provided. The dimensionless pressure approach provides a way to calculate pressure response and to devise techniques for analyzing transient tests in a variety of systems.

Sections covering wellbore storage effects and wellbore damage and improvement are included, since those effects have a significant influence on transient well response. The reader is encouraged to study those sections, even if he only scans the rest of the chapter. Chapter 11 provides additional information about the effects of those two quantities.

2.2 Basic Fluid-Flow Equation

The differential equation for fluid flow in a porous medium, the diffusivity equation, is a combination of the law of conservation of matter, an equation of state, and Darcy's law.[1-4] When expressed in radial coordinates, the diffusivity equation is*

$$\frac{\partial^2 p}{\partial r^2} + \frac{1}{r}\frac{\partial p}{\partial r} = \frac{1}{0.0002637}\,\frac{\phi \mu c_t}{k}\,\frac{\partial p}{\partial t} \quad \text{. . . . . (2.1)}$$

Matthews and Russell[1] present a derivation of Eq. 2.1 and point out that it assumes horizontal flow, negligible gravity effects, a homogeneous and isotropic porous medium, a single fluid of small and constant compressibility, and applicability of Darcy's law, and that μ, c_t, k, and ϕ are independent of pressure. As a result of those assumptions, and since the common boundary conditions are linear, Eq. 2.1 is linear and readily solved. Therefore, solutions (dimensionless pressures) may be added together to form new solutions, as indicated in Section 2.9. If ϕ, μ, c_t, or k are strong functions of pressure, or if varying multiple fluid saturations exist, Eq. 2.1 must be replaced by a nonlinear form. That equation usually must be solved using computer analysis methods (numerical reservoir simulation) beyond the scope of this monograph.

*Symbols and units are defined in the Nomenclature. Normally, only deviations from that list are discussed in the text.

Boundary conditions are an important factor in solutions to Eq. 2.1. Most transient-test analysis techniques assume a single well operating at a constant flow rate in an infinite reservoir. That boundary condition is useful because every well transient is like that of a single well in an infinite reservoir — at early time. At later times the effects of other wells, of reservoir boundaries, and of aquifers influence well behavior and cause it to deviate from the "infinite-acting" behavior. Thus, different solutions to Eq. 2.1 are required for longer time periods. Superposition or other solutions are needed to include other factors, such as gradually changing rate at the formation face (wellbore storage), hydraulic fractures, layered systems, or the presence of multiple fluids or boundaries. Many of those solutions are presented in Appendix C and Chapters 10 and 11; Matthews and Russell[1] present others. The solution for a constant-pressure well is given in Chapter 4.

Although Eq. 2.1 appears to be severely restricted by its basic assumptions, under certain circumstances it can be applied to both multiple-phase flow and gas flow, as indicated in Sections 2.10 and 2.11.

2.3 Solutions to the Flow Equation — Dimensionless Quantities

Comprehensive treatments of transient well testing normally use a general approach for providing solutions to the diffusivity equation, Eq. 2.1. Such an approach provides a convenient way of summarizing the increasing number of solutions being developed to more accurately depict well or reservoir pressure behavior over a broad range of time, boundary, and geometry conditions. The general solutions rely on the concepts of dimensionless pressure and dimensionless time, explained later in this section. Some solutions are identical to others in certain time ranges, but are significantly different in others. Thus, throughout the monograph, guidelines indicate where complex solutions are needed and where the simpler solutions normally give adequate results.

An unfortunate consequence of the generalized dimensionless-solution approach is that the dimensionless parameters do not provide the engineer with the physical feel available when normal dimensional parameters are used. For example, a real time of 24 hours may correspond to a dimensionless time range from about 300 for a tight gas reservoir to more than 10^7 for a highly permeable oil reservoir. The pressure corresponding to a 24-hour time in those two situations might vary by hundreds of pounds per square inch. Fortunately, after one works long enough with dimensionless variables, one does begin to get a feel for them. Nevertheless, it is always good practice to calculate physical quantities from dimensionless ones; that is easily done because physical quantities are *directly proportional* to dimensionless quantities.

The dimensionless-solution approach can be illustrated by starting with the familiar steady-state radial flow equation:

$$q = 0.007082\ \frac{kh(p_e - p_w)}{B\mu\ \ln(r_e/r_w)}\ .$$

This equation may be solved for the pressure difference,

$$p_e - p_w = 141.2\ \frac{qB\mu}{kh}\ \ln(r_e/r_w),$$

Changing to dimensionless form, the radial flow equation becomes

$$p_e - p_w = 141.2\ \frac{qB\mu}{kh}\ p_D,$$

where

$$p_D = \ln(r_e/r_w).$$

Thus, the physical pressure drop in the steady-state radial-flow situation is equal to a dimensionless pressure drop, which in this case is simply $\ln(r_e/r_w)$, times a scaling factor. The scaling factor depends on flow rate and reservoir properties only. The same concept applies to transient flow and to more complex situations — only the dimensionless pressure is different. It is this generality that makes the dimensionless-solution approach useful.

The advantages of the dimensionless form occur, as indicated previously, when situations get more complex. In general terms, the pressure at any point in a single-well reservoir being produced at constant rate, q, is described with the generalized solution of Eq. 2.1:

$$p_i - p(t,r) = 141.2\ \frac{qB\mu}{kh}\ \left[p_D\,(t_D, r_D, C_D, \text{geometry}, \ldots) + s\right], \quad \ldots\ldots\ldots (2.2)$$

where p_i is the initial, *uniform* pressure existing in the reservoir before production or injection; q is the *constant* surface flow rate; k, h, and μ are *constant* reservoir properties; p_D is the dimensionless-pressure solution to Eq. 2.1 for the appropriate boundary conditions; and s is the skin effect, a dimensionless pressure drop assumed to occur at the wellbore face as a result of wellbore damage or improvement.[5,6] Skin effect, s, only appears in Eq. 2.2 when $r_D = 1$. (See Section 2.5.)

In *transient* flow, p_D is *always* a function of dimensionless time,

$$t_D = \frac{0.0002637\,kt}{\phi\mu c_t r_w^2}\ , \quad \ldots\ldots\ldots\ldots\ldots\ldots\ldots (2.3a)$$

when based on wellbore radius, or

$$t_{DA} = \frac{0.0002637\,kt}{\phi\mu c_t A} = t_D \left(\frac{r_w^2}{A}\right), \quad \ldots\ldots\ldots (2.3b)$$

when based on total drainage area. Dimensionless pressure also varies with location in the reservoir, as indicated in Eq. 2.2 by the dimensionless radial distance from the operating well,

$$r_D = r/r_w. \quad \ldots\ldots\ldots\ldots\ldots\ldots\ldots\ldots\ldots\ldots (2.4)$$

The point location also may be expressed in Cartesian coordinates. Dimensionless pressure is also affected by system geometry, other system wells, the wellbore storage coefficient of the producing well, anisotropic reservoir characteristics, fractures, radial discontinuities, and other physical features.

Dimensionless pressure, p_D, is a solution to Eq. 2.1 for specific boundary conditions and reservoir geometry. Practically speaking, dimensionless pressure is just a number given by an equation, a table, or a graph. Some expressions for p_D are given in Sections 2.4, 2.7, and 2.8, and Appendix C. The dimensionless-pressure approach is used throughout this monograph because of its simplicity and general applicability in well-test development and analysis. The approach, which is easy to apply, results in simple, general equations that apply to any set of reservoir properties. It is easily adapted to mathematical manipulation and superposition (Section 2.9), so more complex systems can be considered. For simplicity, the following conventions apply throughout this monograph:

1. Although dimensionless pressure is generally a function of time, location, system geometry, and other variables, we commonly write $p_D(t_D, \ldots)$, $p_D(t_D)$, or just p_D. Dimensionless pressure, p_D, is a number that may be obtained from an equation, figure or table; it scales linearly to real pressure.

2. The symbol t_D always refers to dimensionless time calculated from Eq. 2.3a using the *wellbore radius*. It is clearly indicated when dimensionless time is based on some other dimension. Dimensionless time is just real time multiplied by a scale factor that depends on reservoir properties.

3. Eq. 2.2 includes the van Everdingen-Hurst[5,6] skin factor. That factor appears only when calculating Δp for a producing or injecting well. In general, s is not shown in equations unless it is specifically used. The reader should recognize that adding s is necessary under the appropriate circumstances.

The following example illustrates the use of Eq. 2.2 to estimate flowing well pressure in a closed system.

Example 2.1 Estimating Well Pressure

Estimate the pressure at a well located in the center of a closed-square reservoir after it has produced 135 STB/D of dry oil for 15 days. Other data are*

$p_i = 3{,}265$ psi $\quad \phi = 0.17$
$k_o = 90$ md $\quad c_t = 2.00 \times 10^{-5}$ psi^{-1}
$\mu_o = 13.2$ cp $\quad r_w = 0.50$ ft
$B_o = 1.02$ RB/STB $\quad A = 40$ acres $= 1{,}742{,}400$ sq ft
$h = 47$ ft $\quad s = 0.$

Curve A of Fig. C.13 is used for p_D since it applies to a closed-square system with a well at the center.

Using Eq. 2.3b,

$$t_{DA} = \frac{(0.0002637)(90)(15 \times 24)}{(0.17)(13.2)(2.00 \times 10^{-5})(1{,}742{,}400)}$$

$$= 0.109.$$

From Curve A of Fig. C.13, $p_D(t_{DA} = 0.109) = 6.95$. Rearranging Eq. 2.2 and substituting values,

$$p_{wf}(t, r_w) = 3{,}265 - \frac{(141.2)(135)(1.02)(13.2)}{(90)(47)}(6.95)$$

$$= 2{,}843 \text{ psi.}$$

Fig. 2.1 schematically illustrates three transient flow regimes for a closed drainage system. Dimensionless pressure is shown as a function of both t_{DA} and log (t_{DA}). The portion marked A is the early transient or infinite-acting flow regime; we prefer the term "infinite-acting", since all wells act as if they were alone in an infinite system at short flow times. The infinite-acting period is characterized by a straight line on the semilog plot, Fig. 2.1b. The portion of the curves labeled C in Fig. 2.1 is the pseudosteady-state flow regime that occurs in all closed systems. During pseudosteady-state flow, pressure changes linearly with time, as shown in Fig. 2.1a. The B portion of the curves is the transition period between infinite-acting and pseudosteady-state flow.

In Fig. 2.1, flow is transient *at all times*. Some systems exhibit true steady-state behavior with p_D constant. Those systems are most commonly observed in laboratory core flooding and permeability measurement experiments; they also may exist in fluid injection projects with balanced production and injection and in reservoirs with a strong natural water drive.

*In the examples in this monograph, data values are not always stated to their full number of significant digits. In such cases, values are assumed to have three significant digits with significant zeroes omitted. Computations are usually done using the intermediate values shown. When intermediate values are not shown, all computed digits have been used and the final result has been rounded off.

2.4 Dimensionless Pressure During the Infinite-Acting Flow Period

Fig. 2.2 is a schematic representation of a single well producing at constant rate q in an infinite, horizontal, thin reservoir containing a single-phase, slightly compressible fluid. When the assumptions of Eq. 2.1 are satisfied, Eq. 2.2, with p_D from Fig. 2.3, describes the pressure behavior at any point in the system. Fig. 2.3 shows p_D is a function of t_D and of r_D, the dimensionless radial distance from the well, for the infinite-acting system. (Fig. C.1 is a full-scale gridded version of Fig. 2.3.) When $r_D \geq 20$ and $t_D/r_D^2 \geq 0.5$, or when $t_D/r_D^2 \geq 25$, the $r_D = 20$ and the "exponential-integral solution" lines on Figs. 2.3 and C.1 are essentially the same, so p_D depends only on t_D/r_D^2 under those conditions.[8] The exponential-integral solution[1,4] (also called the line-source or the Theis[9] solution) to the flow equation is

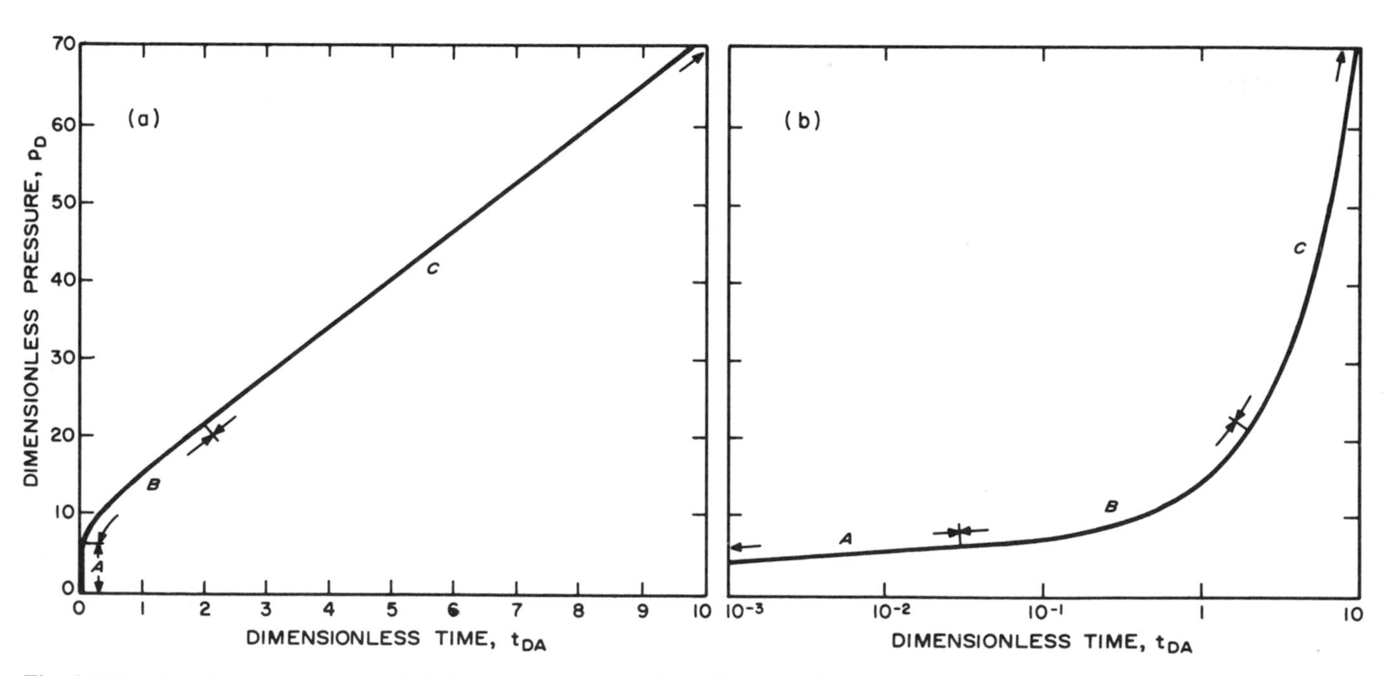

Fig. 2.1 Transient flow regimes: A — infinite acting; B — transition; C — pseudosteady state. Data from Earlougher and Ramey[7] for a 4:1 rectangle with the well at $x/L = 0.75$, $y/W = 0.5$.

$$p_D(t_D,r_D) = -\frac{1}{2}\,\mathrm{Ei}\left(\frac{-r_D{}^2}{4t_D}\right), \quad \ldots\ldots\ldots\ldots\ldots (2.5a)$$

$$\simeq \frac{1}{2}\left[\ln(t_D/r_D{}^2) + 0.80907\right]. \quad \ldots\ldots\ldots (2.5b)$$

Eq. 2.5b may be used when

$$t_D/r_D{}^2 > 100, \quad \ldots\ldots\ldots\ldots\ldots\ldots\ldots\ldots\ldots (2.6)$$

but the difference between Eq. 2.5a and Eq. 2.5b is only about 2 percent when $t_D/r_D{}^2 > 5$. Thus, for *practical* purposes, the log approximation to the exponential integral is satisfactory when the exponential integral is satisfactory. Nevertheless, the more accurate limit of Eq. 2.6 is used in this monograph.

The exponential integral is defined by

$$\mathrm{Ei}(-x) = -\int_x^\infty \frac{e^{-u}}{u}\,du. \quad \ldots\ldots\ldots\ldots\ldots\ldots (2.7a)$$

Values may be taken from tables[10] or may be approximated from

$$\mathrm{Ei}(-x) \simeq \ln(x) + 0.5772 \quad \text{for } x < 0.0025. \quad \ldots\ldots (2.7b)$$

At the operating well $r_D = 1$, so $t_D/r_D{}^2 = t_D$. Since $t_D > 100$ after only a few minutes for most systems, there is practically no difference between the two forms of Eq. 2.5, as illustrated by the following example.

Example 2.2 Estimating Pressure vs Time History of a Well

Use the exponential-integral solution and the data of Example 2.1 to estimate the pressure vs time relationship for a well in an infinite-acting system.

We calculate p_{wf} at 1 minute and at 10 hours to illustrate the procedure; final results are shown in Fig. 2.4. From Eq. 2.3a,

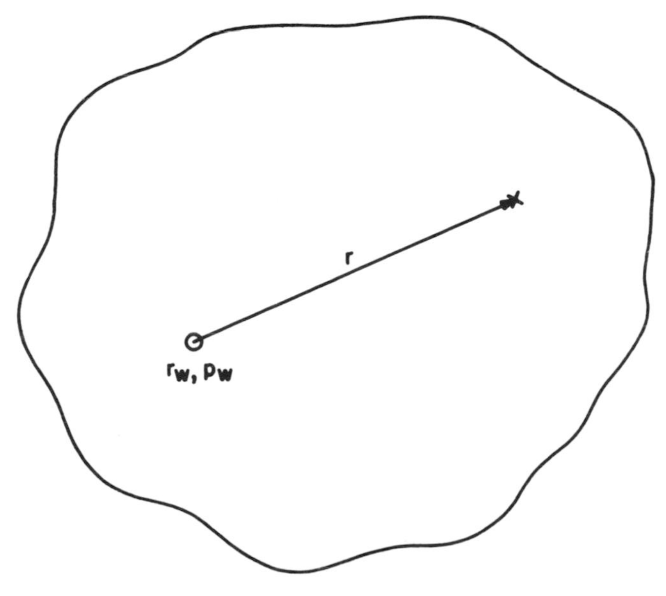

Fig. 2.2 Infinite system with a single well.

$$t_D = \frac{(0.0002637)(90)\,t}{(0.17)(13.2)(2.00\times10^{-5})(0.50)^2}$$

$$= 2{,}115\,t.$$

At 1 minute, $t_D = (2{,}115)(1/60) = 35.25$. The exponential-integral solution, Eq. 2.5a, applies because $t_D/r_D{}^2 = 35.25/1 > 25$. However, since $t_D/r_D{}^2 = 35.25 < 100$, the log approximation, Eq. 2.5b, should not be used. Using Eq. 2.5a and evaluating p_D from Fig. C.1, we get $p_D(t_D = 35.25) = 2.18$. Then, rearranging Eq. 2.2,

$$p_{wf}(t = 1 \text{ minute})$$

$$= 3{,}265 - \frac{(141.2)(135)(1.02)(13.2)}{(90)(47)}(2.18)$$

$$= 3{,}265 - (60.67)(2.18) = 3{,}133 \text{ psi}.$$

At 10 hours, $t_D = (2{,}115)(10) = 21{,}150$ and the log approximation, Eq. 2.5b, can be used:

$$p_D = \frac{1}{2}\left[\ln(21{,}150/1) + 0.80907\right]$$

$$= 5.384,$$

so

$$p_{wf}(t = 10 \text{ hours}) = 3{,}265 - (60.67)(5.384)$$

$$= 2{,}938 \text{ psi}.$$

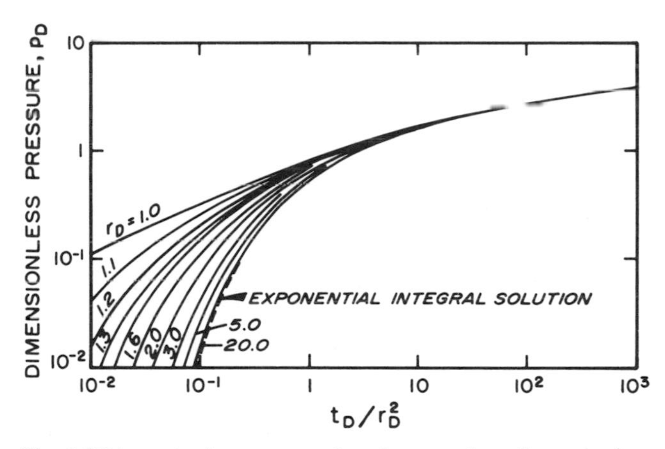

Fig. 2.3 Dimensionless pressure function at various dimensionless distances from a well located in an infinite system. After Mueller and Witherspoon.[8]

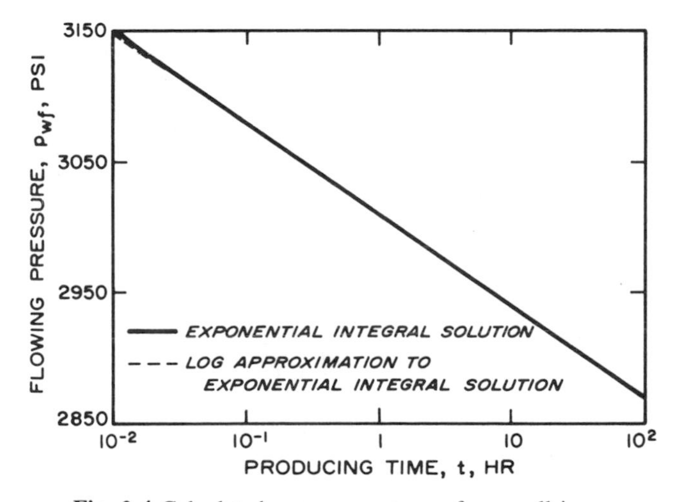

Fig. 2.4 Calculated pressure response for a well in an infinite-acting system. Example 2.2.

The response from 0.01 hour (0.6 minute) to 100 hours is shown in Fig. 2.4. The dashed line is the log approximation for $t_D/r_D^2 < 100$; after that time the two lines coincide. Note how well the two solutions (Eqs. 2.5a and 2.5b) agree, even for $t_D/r_D^2 = (2{,}115)(0.01)/1 = 21$.

As mentioned previously, all wells are infinite-acting for some time after a change in rate. For drawdown, the duration of the infinite-acting period may be estimated from

$$t_{eia} = \frac{\phi\mu c_t A}{0.0002637k}(t_{DA})_{eia}, \quad \text{(2.8a)}$$

where t_{DA} at the end of the infinite-acting period is given in the "Use Infinite System Solution With Less Than 1% Error for t_{DA} <" column of Table C.1. For a well in the center of a closed circular reservoir, $(t_{DA})_{eia} = 0.1$ and

$$t_{eia} \simeq \frac{380\ \phi\mu c_t A}{k}. \quad \text{(2.8b)}$$

Equations for buildup are given in Section 5.3.

2.5 Wellbore Damage and Improvement Effects

There are several ways to quantify damage or improvement in operating (producing or injecting) wells. A favored method represents the wellbore condition by a steady-state pressure drop at the wellface in addition to the normal transient pressure drop in the reservoir. The additional pressure drop, called the "skin effect," occurs in an infinitesimally thin "skin zone."[5,6] In the flow equation, Eq. 2.2, the degree of damage (or improvement) is expressed in terms of a "skin factor," s, which is positive for damage and negative for improvement. It can vary from about −5 for a hydraulically fractured well to +∞ for a well that is too badly damaged to produce. Eq. 2.2 indicates the pressure drop at a damaged (or improved) well differs from that at an undamaged well by the additive amount

$$\Delta p_s = \frac{141.2\ qB\mu}{kh}\ s \quad \text{(2.9)}$$

Fig. 2.5A illustrates the idealized pressure profile for a damaged well ($s > 0$). Since the damage-zone thickness is considered to be infinitesimal, the entire pressure drop caused by the skin occurs at the wellface. The thin-skin approximation results in a pressure gradient reversal for wellbore improvement ($s < 0$), shown in Fig. 2.5B. Although this situation is physically unrealistic, the skin-factor concept is valuable as a measure of wellbore improvement. A more physically realistic pressure profile for the negative skin situation is also shown in Fig. 2.5B.

If the skin is viewed as a zone of finite thickness with permeability k_s, as shown in Fig. 2.6, then[11]

$$s = \left(\frac{k}{k_s} - 1\right)\ln\left(\frac{r_s}{r_w}\right). \quad \text{(2.10)}$$

Either s, k_s, or r_s may be estimated from Eq. 2.10 if the other two parameters are known.

It is also possible to define an apparent wellbore radius for use in Eqs. 2.3 and 2.4 so the correct pressure drop at the well results when $s = 0$ is used in Eq. 2.2:[12]

$$r_{wa} = r_w e^{-s}. \quad \text{(2.11)}$$

For positive s, $r_{wa} < r_w$; for negative s, $r_{wa} > r_w$.

Eqs. 2.2 and 2.9 show that the skin factor simply increases or decreases the pressure change at a well proportional to the flow rate of that well. When dimensionless pressure functions include the skin factor (for example,

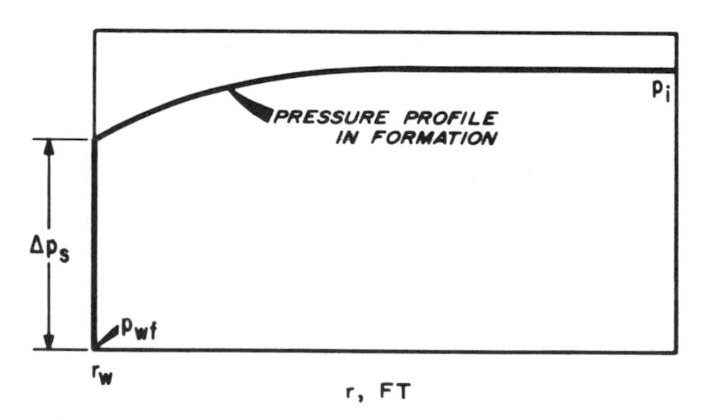

Fig. 2.5A Pressure distribution around a well with a positive skin factor.

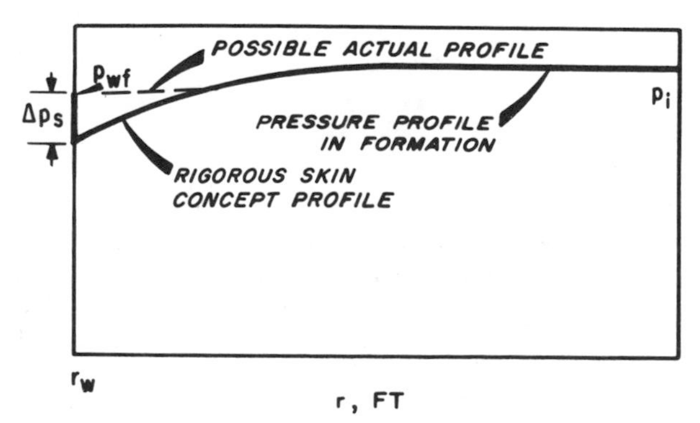

Fig. 2.5B Pressure distribution around a well with a negative skin factor.

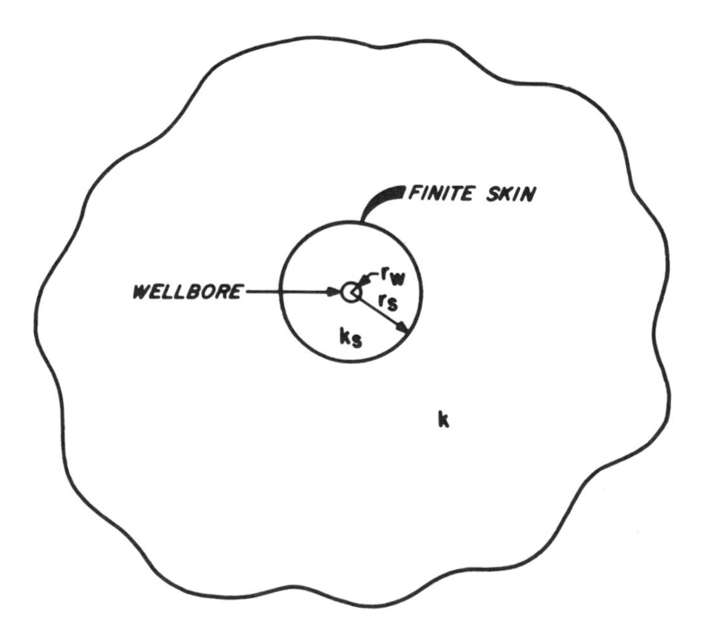

Fig. 2.6 Skin zone of finite thickness.

Figs. C.6 and C.7), it should not be included explicitly in Eq. 2.2. Dimensionless pressure functions that include the skin factor are particularly useful when the skin zone extends for some distance rather than being concentrated within a few feet of the wellbore. The skin-factor concept is used only for pressures at the well. When the skin affects some r_s around the well, the pressure profile in that region will be modified (for example, see Fig. 2.5B). In such situations, special p_D functions that include s must be used to determine pressures near the well. When using Eq. 2.2 to estimate pressure at distances greater than r_w or r_{wa}, the s term is omitted, and normal p_D's are used.

The *flow efficiency* (also called the condition ratio) indicates the approximate fraction of a well's undamaged producing capacity. It is defined as the ratio of the well's actual productivity index to its productivity index if there were no skin.[1] For closed systems, the *flow efficiency* is

$$\frac{J_{\text{actual}}}{J_{\text{ideal}}} = \frac{\bar{p} - p_{wf} - \Delta p_s}{\bar{p} - p_{wf}} \quad . \quad \ldots \ldots \ldots \ldots \ldots (2.12)$$

Although the drainage-area average pressure, $\bar{p}$, should be used in Eq. 2.12, it is frequently permissible to use the extrapolated buildup pressure,[1] p^*. Flow efficiency depends on the flowing pressure, p_{wf}, and thus it depends on how long the well has been operating unless the well is at pseudosteady-state conditions. (At pseudosteady state, $\bar{p} - p_{wf}$ is constant.) For wells operating at true steady state, $\bar{p}$ should be replaced by p_e, the pressure that the area will reach after extended shut-in.

The *damage ratio* and *damage factor* are also relative indicators of wellbore condition. The inverse of the flow efficiency is the *damage ratio*:

$$\frac{J_{\text{ideal}}}{J_{\text{actual}}} = \frac{\bar{p} - p_{wf}}{\bar{p} - p_{wf} - \Delta p_s} \quad . \quad \ldots \ldots \ldots \ldots \ldots (2.13)$$

By subtracting the flow efficiency from 1, we obtain the *damage factor*:

$$1 - \frac{J_{\text{actual}}}{J_{\text{ideal}}} = \frac{\Delta p_s}{\bar{p} - p_{wf}} \quad . \quad \ldots \ldots \ldots \ldots \ldots \ldots (2.14)$$

The following example illustrates the wellbore damage indicators.

Example 2.3 Wellbore Damage Indicators

A pressure buildup-test analysis for a well with q = 83 STB/D, B = 1.12 RB/STB, μ = 3.15 cp, h = 12 ft, r_w = 0.265 ft, and $\bar{p} - p_{wf}$ = 265 psi gave k = 155 md and s = 2.2. Find the pressure drop across the skin, the flow efficiency, the damage ratio, the damage factor, and the apparent wellbore radius.

Using Eq. 2.9,

$$\Delta p_s = \frac{(141.2)(83)(1.12)(3.15)}{(155)(12)} (2.2) = 49 \text{ psi.}$$

The flow efficiency is estimated from Eq. 2.12:

$$\frac{265 - 49}{265} = 0.82 = 82 \text{ percent.}$$

Using Eq. 2.13, the damage ratio is

$$\frac{1}{0.82} = 1.22,$$

and using Eq. 2.14, the damage factor is

$$1 - 0.82 = 0.18.$$

The apparent wellbore radius is estimated from Eq. 2.11:

$$r_{wa} = 0.265 \, e^{-2.2} = 0.03 \text{ ft.}$$

Damage at this well is reducing production to about 82 percent of the value that could be expected without damage or stimulation.

Wells completed with only a part of the formation thickness open to the wellbore can appear to be damaged. Partial penetration (wells not drilled completely through the productive interval) and partial completion (entire productive interval not perforated) are examples.[13] (See Section 11.4 for more information and references.) Fig. 2.7 shows theoretical "pseudoskin" factors for partially penetrating wells. The skin factor estimated from a transient test would be that given by Fig. 2.7 if there were no true *physical* damage (or improvement) at the well. If there is physical damage, the calculated skin factor is higher than indicated by Fig. 2.7.

Skin factors estimated from transient tests on hydrauli-

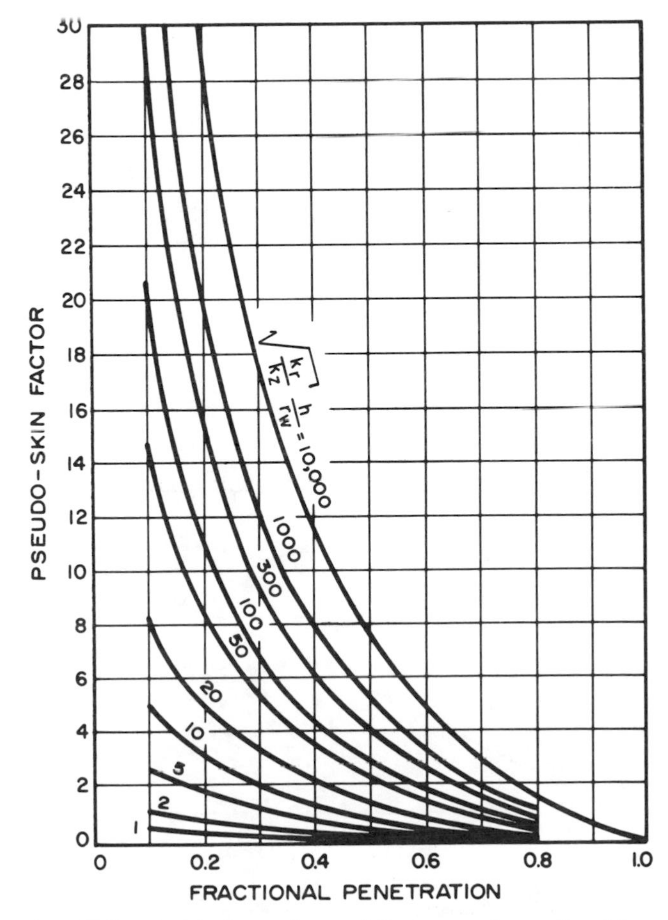

Fig. 2.7 Pseudoskin factor for partially penetrating wells. After Brons and Marting.[13]

cally fractured wells are generally negative. Fig. 2.8 compares the dimensionless pressure for an ideal, undamaged, unstimulated well with that for a hydraulically fractured well with a half-fracture length of 31.63 r_w. At small t_D, the difference between the two dimensionless pressure curves, which is s (Eq. 2.2), varies; at larger t_D that difference is constant. This indicates that reasonable skin values can be estimated from transient tests for many hydraulically fractured wells. However, when large fracture jobs are known to have been performed, the fracture should be accounted for by analyzing well tests using the type-curve matching method (Section 3.3) with Figs. C.3, C.4, C.5, C.17, C.18, or C.19. An important feature of all the log-log plots of fractured-well p_D data is the slope of ½ at small t_D. This slope also will be observed on a log-log plot of transient *pressure difference* data from fractured wells, unless it is obscured by wellbore storage. Section 11.3 provides additional details.

2.6 Wellbore Storage

Wellbore storage, also called afterflow, afterproduction, afterinjection, and wellbore unloading or loading, has long been recognized as affecting short-time transient pressure behavior.[2,14] More recently, several authors[15-26] have considered wellbore storage in detail. It is easy to see that liquid is stored in the wellbore when the liquid level rises. That situation occurs when a pumping well without a packer is shut in; indeed, bottom-hole pressure is often deduced by measuring liquid level. When wellbore storage is significant, it must be considered in transient test design and analysis. If it is not considered, the result may be an analysis of the wrong portion of the transient test data, the deduction of nonexistent reservoir conditions (faults, boundaries, etc.), or an analysis of meaningless data. Fortunately, the effects of wellbore storage usually can be accounted for in test analysis — or can be avoided by careful test design.

The wellbore storage constant (coefficient, factor) is defined[16] by

$$C = \frac{\Delta V}{\Delta p}, \quad \ldots \ldots (2.15)$$

where C = wellbore storage constant (coefficient, factor), bbl/psi,
ΔV = change in volume of fluid in the wellbore, *at wellbore conditions,* bbl, and
Δp = change in bottom-hole pressure, psi.

Applying Eq. 2.15 to a wellbore with a changing liquid level,[16]

$$C = \frac{V_u}{\left(\frac{\rho}{144}\frac{g}{g_c}\right)}, \quad \ldots \ldots (2.16)$$

where V_u is the wellbore volume per unit length in barrels per foot. Eq. 2.16 is valid for both rising and falling liquid levels. When the wellbore is completely full of a single-phase fluid, Eq. 2.15 becomes[16]

$$C = V_w c, \quad \ldots \ldots (2.17)$$

where V_w is the total wellbore volume in barrels and c is the compressibility of the fluid *in the wellbore* at wellbore conditions. Throughout this monograph, the wellbore storage coefficient, C, has units of barrels per psi; some authors prefer cubic feet per psi. The compressibility in Eq. 2.17 is for the fluid in the wellbore; it is *not* c_t for the reservoir. Since the wellbore fluid compressibility is pressure dependent (Appendix D), the wellbore storage coefficient may vary with pressure. Fortunately, such variation in wellbore storage coefficient is generally important only in wells containing gas or in wells that change to a falling or rising liquid level during the test. Those conditions are considered in Section 11.2.

Some dimensionless pressure functions (Appendix C) for systems with wellbore storage use a dimensionless wellbore storage coefficient,

$$C_D = \frac{5.6146\,C}{2\pi\,\phi c_t h r_w^2}. \quad \ldots \ldots (2.18)$$

Note that the total compressibility for the reservoir system, c_t, is used in this definition.

Wellbore storage causes the sand-face flow rate to change

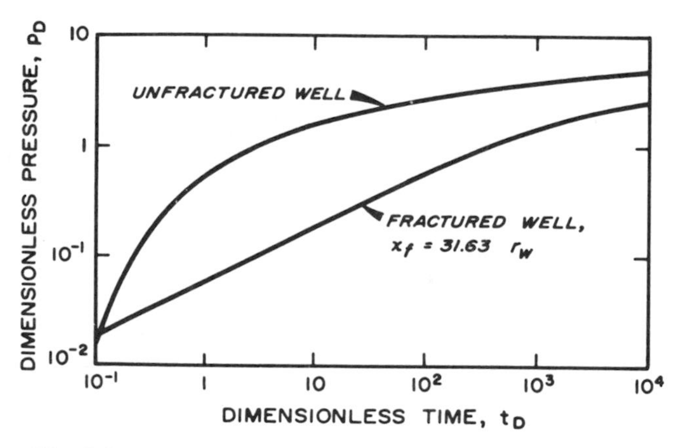

Fig. 2.8 Comparison of dimensionless pressures for an ideal well and for a well with a single vertical fracture. Infinite-acting system.

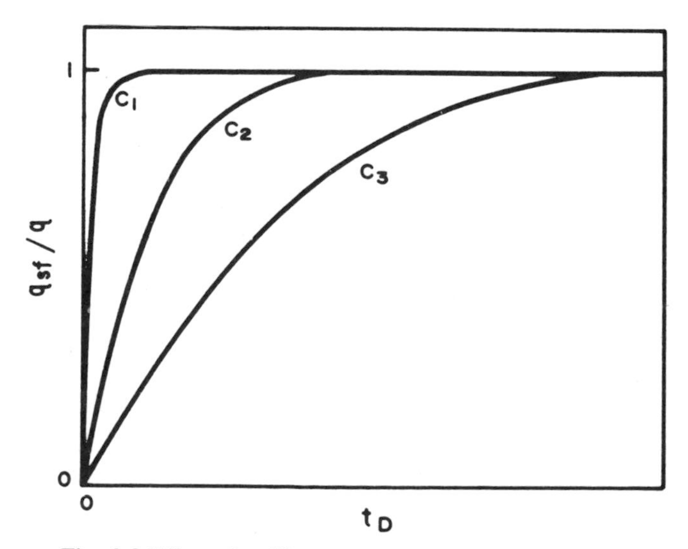

Fig. 2.9 Effect of wellbore storage on sand-face flow rate, $C_3 > C_2 > C_1$.

more slowly than the surface flow rate. Fig. 2.9 schematically shows the ratio of sand-face to surface rate when the surface rate is changed from 0 to q at time 0. When $C = 0$, $q_{sf}/q = 1$ at all times. For $C > 0$, the flow-rate ratio changes gradually from 0 to 1. The larger C, the longer the transition, as indicated in Fig. 2.9. The sand-face flow rate may be calculated from

$$q_{sf} = q + \frac{24C}{B}\frac{dp}{dt}$$

$$= q\left[1 - C_D \frac{d}{dt_D} p_D(t_D, C_D, \ldots)\right]. \quad \ldots\ldots (2.19)$$

Eq. 2.2 relates flowing well pressure to time for a constant flow rate, q. Since Eq. 2.19 indicates that q varies with t and p, it appears that Eq. 2.2 may not be useable. Fortunately, the problem is avoided by using a dimensionless pressure that accounts for wellbore storage and, thus, for the change in flow rate. Such $p_D(t_D, C_D, \ldots)$ are shown in Fig. 2.10, a simplified version of Figs. C.6 and C.7. The effect of wellbore storage on p_D is clear in those figures.

Fig. 2.10 has a characteristic that is diagnostic of wellbore storage effects: the slope of the p_D vs t_D graph on log-log paper is 1.0 during wellbore storage domination. Since p_D is proportional to Δp and t_D is proportional to time (see Eqs. 2.2 and 2.3), Fig. 2.10 indicates a way to estimate when wellbore storage is dominant during a transient test. On log-log paper, plot the pressure change during the test, $p_w - p_w(\Delta t = 0)$, (as a positive number) against test time, Δt, and observe where that plot has a slope of one cycle in pressure change per cycle in time. (Note that the nomenclature has been generalized here: p_w is the bottom-hole pressure during the test, be it flowing or static; $p_w(\Delta t = 0)$ is the pressure at the instant before the start of the test, be it static or flowing; and Δt is running test time, with the test starting at $\Delta t = 0$.) Well test data falling on the unit slope of the log-log plot reveal *nothing* about formation properties, since essentially all production is from the wellbore during that time. The location of the log-log unit slope can be used to estimate the apparent wellbore storage coefficient from[16]

$$C = \frac{qB\Delta t}{24\Delta p}, \quad \ldots\ldots\ldots\ldots (2.20)$$

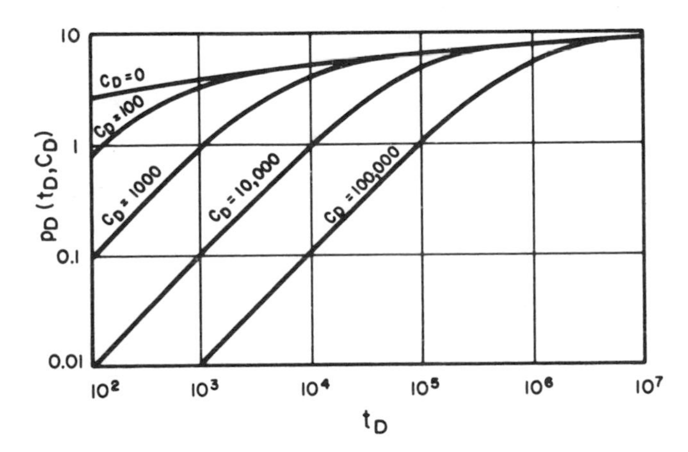

Fig. 2.10 Dimensionless pressure including wellbore storage, $s = 0$. After Wattenbarger and Ramey.[21]

where Δt and Δp are values read from a *point* on the log-log *unit-slope straight line*. C calculated from Eq. 2.20 should agree fairly well with C calculated from Eq. 2.16 or Eq. 2.17. If it does not, a reason should be sought. We have observed that wells producing at high gas-liquid ratios, highly stimulated wells, and wells used for viscous fluid injection commonly indicate wellbore storage coefficients from the log-log data plot that are much higher than those predicted from Eqs. 2.16 and 2.17.

The log-log data plot is a valuable aid for recognizing wellbore storage effects in transient tests when early-time pressure data are available. Thus, throughout this monograph, it is recommended that this plot be made a part of transient test analysis. It often helps the engineer avoid serious analysis mistakes by delineating the period of wellbore storage dominance as a unit-slope straight line. As wellbore storage effects become less severe, the formation begins to influence the bottom-hole pressure more and more, and the data points on the log-log plot fall below the unit-slope straight line and finally approach the slowly curving line for zero wellbore storage. Such behavior is illustrated by Fig. 2.10 for varying degrees of wellbore storage. Sometimes pressure data between the unit-slope line and the zero wellbore-storage line can be analyzed for formation properties, but the analysis may be tedious. The Gladfelter-Tracy-Wilsey[15] and the type-curve matching techniques[18-24] (Sections 3.3 and 8.3) apply in this region. Once the final portion of the log-log plot is reached ($C_D = 0$ line), wellbore storage is no longer important and standard semilog data-plotting analysis techniques apply. As a rule of thumb, that time usually occurs about 1 to 1½ cycles in time after the log-log data plot starts deviating significantly from the unit slope. The time may be estimated from

$$t_D > (60 + 3.5\,s)\,C_D, \quad \ldots\ldots\ldots\ldots (2.21a)$$

or approximately,

$$t > \frac{(200{,}000 + 12{,}000\,s)\,C}{(kh/\mu)}, \quad \ldots\ldots\ldots (2.21b)$$

for drawdown and injection tests. For pressure buildup and falloff tests, Chen and Brigham[25] state that a reasonably accurate analysis is possible when

$$t_D > 50\,C_D\,e^{0.14s}, \quad \ldots\ldots\ldots\ldots (2.22a)$$

or approximately when

$$t > \frac{170{,}000\,C\,e^{0.14s}}{(kh/\mu)}. \quad \ldots\ldots\ldots\ldots (2.22b)$$

Note that skin factor influences pressure buildup (falloff) much more than drawdown (injection).

Fig. C.5 for a horizontally fractured well *without* wellbore storage shows unit-slope straight lines for small h_D. Thus, factors other than wellbore storage can cause the unit-slope straight line on the log-log plot. Fortunately, horizontal fractures are thought to occur rarely.

Example 2.4 Computing Wellbore Storage Coefficients From Well Data

Water is injected into a sand at 2,120 ft through 4.75-in.,

16.00-lb_m casing. Estimate the wellbore storage coefficient for (1) a wellhead injection pressure of 400 psi, and (2) a wellhead vacuum. Use $\phi = 0.15$, $h = 30$ ft, and $r_w = 3.5$ in. to calculate the dimensionless wellbore storage coefficient.

1. When the wellhead pressure is greater than zero, Eq. 2.17 is used to estimate C. Water compressibility is estimated to be 3.25×10^{-6} psi^{-1} from data in Appendix D. For 4.75-in., 16-lb_m casing, V_u = 0.0161 bbl/ft, so V_w = (0.0161)(2,120) = 34.1 bbl. Using Eq. 2.17,

$$C = (34.1)(3.25 \times 10^{-6}) = 1.11 \times 10^{-4} \text{ bbl/psi}.$$

From Eq. 2.18,

$$C_D = \frac{(5.6146)(1.11 \times 10^{-4})}{2\pi(0.15)(3.25 \times 10^{-6})(30)(3.5/12)^2} = 80.$$

2. In this second case there is a changing liquid level in the well, so Eq. 2.16 is used:

$$C = \frac{0.0161}{\left(\frac{62.4}{144}\right)\left(\frac{32.17}{32.17}\right)} = 0.0372 \text{ bbl/psi}.$$

$$C_D = \frac{(5.6146)(0.0372)}{2\pi(0.15)(3.25 \times 10^{-6})(30)(3.5/12)^2}$$

$$= 2.7 \times 10^4.$$

In this case, there is a factor of about 340 difference between compressive and changing-liquid-level storage. Note, we assumed $c_t \simeq c_w$, which is not always a valid assumption.

Example 2.5 Computing Wellbore Storage Coefficient From Well Test Data

Use the log-log data plot shown in Fig. 2.11 to estimate a wellbore storage coefficient. That well has 2.5-in. tubing in 8⅝-in., 35.5-lb_m casing.

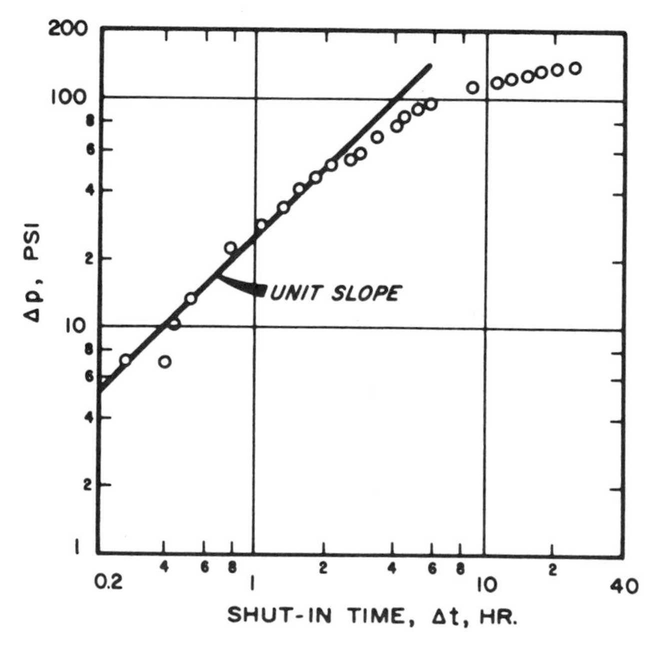

Fig. 2.11 Log-log data plot for Example 2.5. Pressure buildup test in a sandstone reservoir with B = 1.0 RB/STB; q = 66 STB/D; h = 20 ft; and depth = 980 ft.

At Δt = 1 hour, Δp = 26 psi on the unit-slope straight line. Using Eq. 2.20 and the data values in the figure caption, the wellbore storage coefficient is estimated as

$$C = \frac{(66)(1)}{(24)} \frac{1}{26} = 0.106 \text{ bbl/psi}.$$

Using Eq. 2.16 and $\rho = 62.4$ lb_m/cu ft,

$$V_u = C \left(\frac{\rho}{144} \frac{g}{g_c}\right) = (0.106) \left(\frac{62.4}{144} \frac{32.17}{32.17}\right)$$

$$= 0.0459 \text{ bbl/ft}.$$

This indicates that fluid is rising in the annulus between the tubing and casing (V_u for that annulus is 0.0435 bbl/ft). The well is completed without a packer, so the result is reasonable.

For this well, $c_t \simeq 10^{-5}$ psi^{-1}, $\phi = 0.20$, and $r_w = 0.36$ ft, so from Eq. 2.18,

$$C_D = \frac{(5.6146)(0.106)}{2\pi(0.20)(10^{-5})(20)(0.36)^2} = 18{,}300.$$

The wellbore storage coefficient can change during transient testing. For example, consider a falloff test in a water injection well with a high wellhead pressure during injection. When the well is shut in, surface pressure is high initially but could decrease to atmospheric and go on vacuum if the static formation pressure is below hydrostatic. The liquid level must start falling as soon as the wellhead pressure drops below atmospheric. As a result, the wellbore storage coefficient increases from one for fluid compression (Eq. 2.17) to one for a falling liquid level (Eq. 2.16); the second storage coefficient easily could be a hundred to a thousand times the first. The reverse situation can occur as well, with a high, rising-liquid-level storage at the beginning of injection changing to fluid-compression storage as the wellhead pressure begins to increase. Fig. 2.12 schematically illustrates dimensionless pressure behavior when the wellbore storage coefficient changes. When the wellbore storage coefficient increases (from C_1 to C_2 in Fig. 2.12), p_D (or Δp) flattens, begins to increase again, and finally approaches the response curve for the larger storage coefficient. When the wellbore storage coefficient decreases,

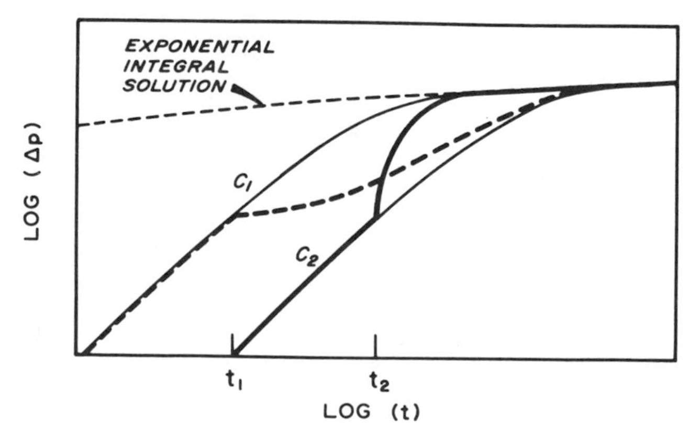

Fig. 2.12 Theoretical pressure response for both increasing and decreasing wellbore storage; $C_2 > C_1$. Adaptation of data from Earlougher, Kersch, and Ramey.[26]

there is a rapid p_D (or Δp) increase as the pressure response approaches the low wellbore storage curve. More discussion about this is given in Ref. 26 and in Section 11.2

Stegemeier and Matthews[27] showed that gas-liquid (phase) redistribution in the wellbore causes anomalous pressure-buildup curves of the form shown in Fig. 2.13. Fig. 2.14 indicates that phase redistribution is similar to wellbore storage, although it is probably more complex than anything else presented in this section. It is important to understand that the behavior illustrated in Fig. 2.13 is wellbore, not formation, dominated. Pitzer, Rice, and Thomas[28] demonstrated this by testing a well once with surface shut in and a second time with bottom-hole shut in.

In summary, wellbore storage effects always should be considered in transient test design and analysis and in computing the expected pressure response of wells. In some cases, transient tests must be designed to minimize or alleviate wellbore storage, or no useful information will be obtained. Wellbore storage effects can be recognized on log-log data plots if sufficient short-time pressure data are available. No information about the formation can be determined from transient test data falling on the unit slope of such a data plot.

2.7 Dimensionless Pressure During the Pseudosteady-State Flow Period

Fig. 2.1 indicates that in closed systems a transition period follows the infinite-acting transient response. That is followed by the pseudosteady-state flow period, a transient flow regime when the pressure change with time, dp/dt, is constant at all points in the reservoir. (That is equivalent to the right-hand side of Eq. 2.1 being constant.) This flow period occasionally has been mistakenly called steady state, although at true steady state pressure is constant with time everywhere in the reservoir (Section 2.8).

Fig. 2.15 schematically shows a single producing well in the center of a closed-square drainage area. Fig. 2.16, a plot of p_D vs t_D at Points A, B, C, and D of Fig. 2.15, illustrates two properties common to all closed systems. First, at small dimensionless times, p_D at the well is given by Eq. 2.5 if $\sqrt{A}/r_w > 50$; the well behaves as if it were alone in an infinite system. Second, at large dimensionless times, p_D *at any point* in the system varies linearly with t_{DA}. This is the pseudosteady-state flow period, which can occur only in bounded systems. During pseudosteady state, dimensionless pressure is given by[29]

$$p_D = 2\pi t_{DA} + \frac{1}{2}\ln\left(\frac{A}{r_w^2}\right) + \frac{1}{2}\ln\left(\frac{2.2458}{C_A}\right) . \qquad (2.23)$$

C_A, the shape factor, is a geometric factor characteristic of the system shape and the well location. Values are given by Brons and Miller,[12] Dietz,[30] and others.[7,31] Both C_A and the final term in Eq. 2.23 are given in Table C.1. Eq. 2.23 may be used for any closed system with known shape factor. If 31.62, the C_A value for a well in the center of a circular system, is used in Eq. 2.23, the last two terms become the familiar $\ln(r_e/r_w) - 0.75$.

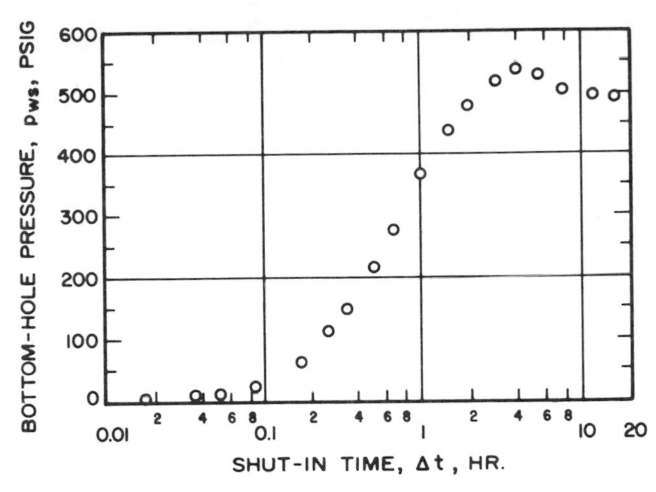

Fig. 2.13 Pressure buildup behavior showing the effect of fluid segregation in the wellbore. After Matthews and Russell.[1]

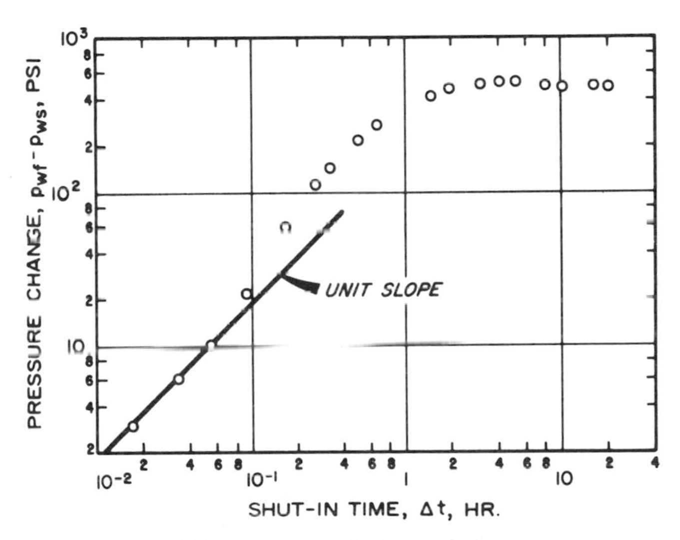

Fig. 2.14 Log-log plot of Fig. 2.13 data.

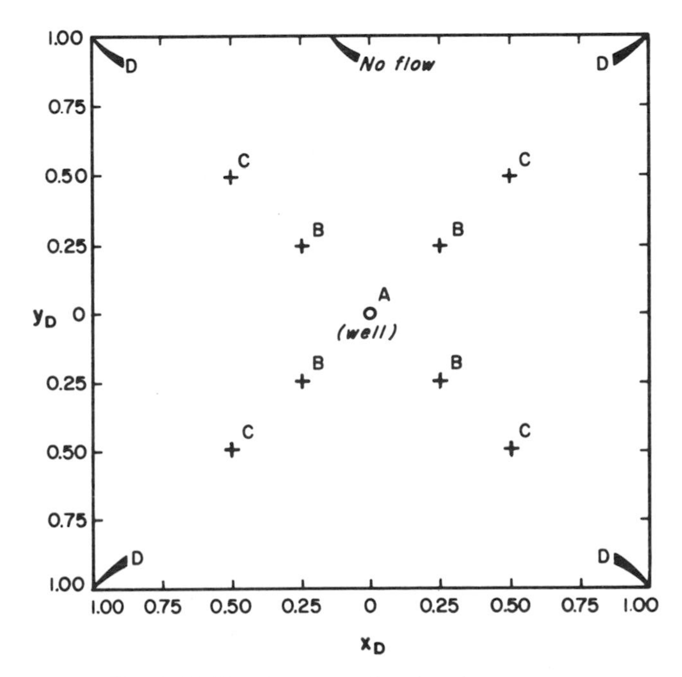

Fig. 2.15 Well in center of a closed square.

Eq. 2.23 applies any time after pseudosteady-state flow begins; that time may be estimated from

$$t_{pss} = \frac{\phi \mu c_t A}{0.0002637\,k} (t_{DA})_{pss} \,, \quad \ldots\ldots\ldots\ldots\ (2.24)$$

where $(t_{DA})_{pss}$ is given in the "Exact for $t_{DA} >$" column of Table C.1. Both C_A and $(t_{DA})_{pss}$ depend on reservoir shape and well location.

Dimensionless pressure data at the well and at several other points in closed rectangular systems are given in Ref. 7.

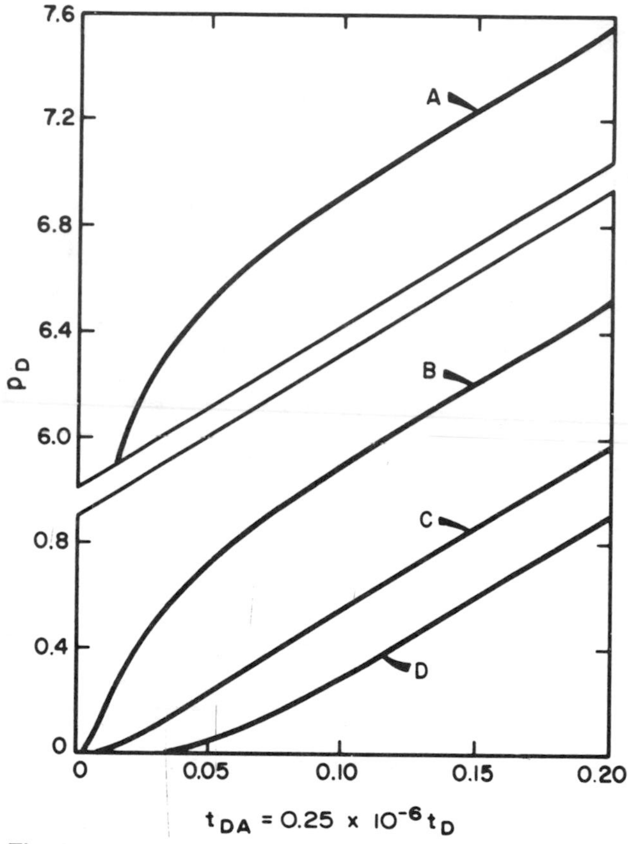

Fig. 2.16 Dimensionless pressure at various points in a closed square caused by a producing well in the center. A, B, C, and D identified in Fig. 2.15; $\sqrt{A}/r_w = 2{,}000$. After Earlougher, Ramey, Miller and Mueller.[31]

Example 2.6 Estimating Well Pressure During Pseudosteady-State Flow

Using the data of Example 2.1, estimate flowing well pressure after 60 days. From Eq. 2.3b,

$$t_{DA} = \frac{(0.0002637)(90)(60 \times 24)}{(0.17)(13.2)(2.00 \times 10^{-5})(1{,}742{,}400)}$$

$$= 0.437.$$

Table C.1 indicates pseudosteady state exists after $t_{DA} = 0.1$ for a well in the center of a square, so Eq. 2.23 applies. From Table C.1, $C_A = 30.8828$. Using Eq. 2.23,

$$p_D = 2\pi(0.437) + \frac{1}{2}\ln\left(\frac{1{,}742{,}400}{(0.50)^2}\right)$$

$$+ \frac{1}{2}\ln\left(\frac{2.2458}{30.8828}\right) = 9.31.$$

Then, from Eq. 2.2,

$$p_{wf}(t, r_w) = 3{,}265 - \frac{(141.2)(135)(1.02)(13.2)}{(90)(47)}(9.31)$$

$$= 2{,}700 \text{ psi}.$$

2.8 Steady-State Flow

When the pressure at every point in a system does not vary with time (that is, when the right-hand side of Eq. 2.1 is zero), flow is said to be steady state. Linear and radial steady-state flow, whose dimensionless pressure distributions are illustrated in Fig. 2.17, usually occur only in laboratory situations. The dimensionless-pressure functions are

steady state, linear

$$(p_D)_{ssL} = 2\pi \frac{Lh}{A} \,, \quad \ldots\ldots\ldots\ldots\ldots\ldots\ldots\ldots\ (2.25)$$

steady state, radial

$$(p_D)_{ssr} = \ln\left(\frac{r_e}{r_w}\right). \quad \ldots\ldots\ldots\ldots\ldots\ldots\ldots\ (2.26a)$$

When Eq. 2.26a is used in Eq. 2.2, we obtain after rearrangement

$$q = \frac{0.007082\,kh\,(p_e - p_w)}{B\mu\,\ln(r_e/r_w)} \,, \quad \ldots\ldots\ldots\ldots\ (2.26b)$$

the familiar radial form of Darcy's law.[32]

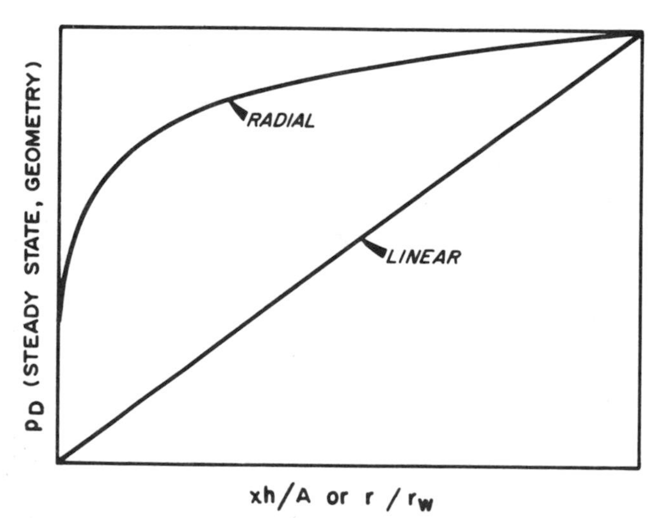

Fig. 2.17 Dimensionless pressure distribution in linear and radial steady-state flow.

In reservoirs, steady-state flow can occur only when the reservoir is completely recharged by a strong aquifer or when injection and production are balanced. Muskat[33] relates flow rate to interwell pressure drop for several flooding patterns. His equations are easily converted to the dimensionless pressure approach used in this monograph. Perhaps the most useful is the dimensionless pressure expression for a *five-spot flooding pattern at steady state* with unit mobility ratio and with r_w the same in all wells:

$$(p_D)_{ss5} = \ln\left(\frac{A}{r_w^{\,2}}\right) - 1.9311. \quad \ldots\ldots\ldots\ldots\ (2.27)$$

Here, A is the five-spot pattern area, not the area per well.

It is useful to recognize that Higgins-Leighton[34] geometric factors are dimensionless pressures for cells within streamtubes operating at steady state. By appropriate addition of available Higgins-Leighton geometric factors, such as those in Ref. 35, one can calculate dimensionless pressures for many irregular steady-state systems.

$$(p_D)_{ss} = \sum_i \left(\frac{2\pi}{\sum_j F_{HLi,j}} \right), \quad \ldots\ldots\ldots\ldots (2.28)$$

where $F_{HLi,j}$ is the Higgins-Leighton geometric factor for cell j in streamchannel i of the pattern. The sums are taken over all cells and all channels. If the pattern is symmetric, the right-hand side of Eq. 2.28 must be multiplied by the number of symmetry units. For the confined five-spot pattern flood of Ref. 35, we calculate $p_D = 10.498$; here, $\sqrt{A}/r_w = 500$. Eq. 2.27 gives the same result to three decimal places.

2.9 The Principle of Superposition

So far, only systems with a single well operating at a constant rate from time zero onward have been considered. Since real reservoir systems usually have several wells operating at varying rates, a more general approach is needed to study the problems associated with transient well testing. Fortunately, because Eq. 2.1 is linear, multiple-rate, multiple-well problems can be considered by applying the principle of superposition. The mathematical basis for this technique is explained by van Everdingen and Hurst,[2] Collins,[36] and others.[1,4,31]

As used here, the superposition principle states that adding solutions to a linear differential equation results in a new solution to that differential equation, but for different boundary conditions. Eq. 2.2 is a solution to Eq. 2.1 for a single well producing at constant rate q. Superposition can be applied to include more than one well, to change rates, and to impose physical boundaries. Superposition is easily applied to infinite systems; but for bounded systems it must be used with more care — not because the principle is different, but because p_D solutions frequently do not give the necessary information for correct superposition.

To illustrate the principle of superposition in space, consider the three-well infinite system in Fig. 2.18. At $t = 0$, Well 1 starts producing at rate q_1, and Well 2 starts producing at rate q_2. We wish to estimate the pressure at the shut-in observation point, Well 3. To do this, we add the *pressure change* at Well 3 caused by Well 1 to the *pressure change* at Well 3 caused by Well 2:

$$\Delta p_3 = \Delta p_{3,1} + \Delta p_{3,2} \; . \quad \ldots\ldots\ldots\ldots\ldots\ldots (2.29)$$

To use Eq. 2.29 we must substitute Eq. 2.2 for Δp. Then, extending to an arbitrary number of wells, $j = 1, 2, \ldots, n$,

$$\Delta p(t, r) = \frac{141.2\mu}{kh} \sum_{j=1}^{n} q_j B_j p_D(t_D, r_{Dj}) \; , \quad \ldots\ldots (2.30)$$

where r_{Dj} is the dimensionless distance from Well j to the point of interest. Note that Eqs. 2.29 and 2.30 add *pressure changes* (or dimensionless pressures), not pressures. If the point of interest is an operating well, the skin factor must be added to the dimensionless pressure for *that well only*.

Fig. 2.19 graphically illustrates the use of Eqs. 2.29 and

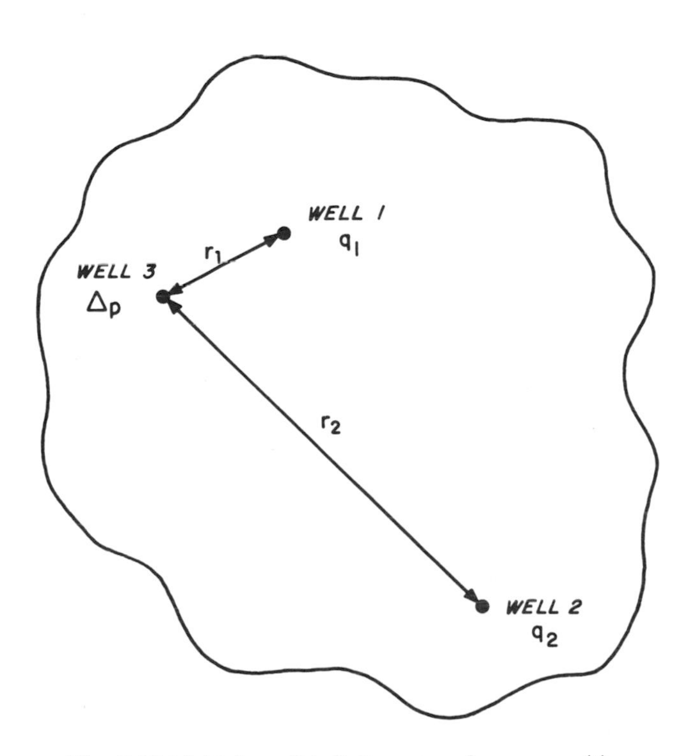

Fig. 2.18 Multiple-well infinite system for superposition explanation.

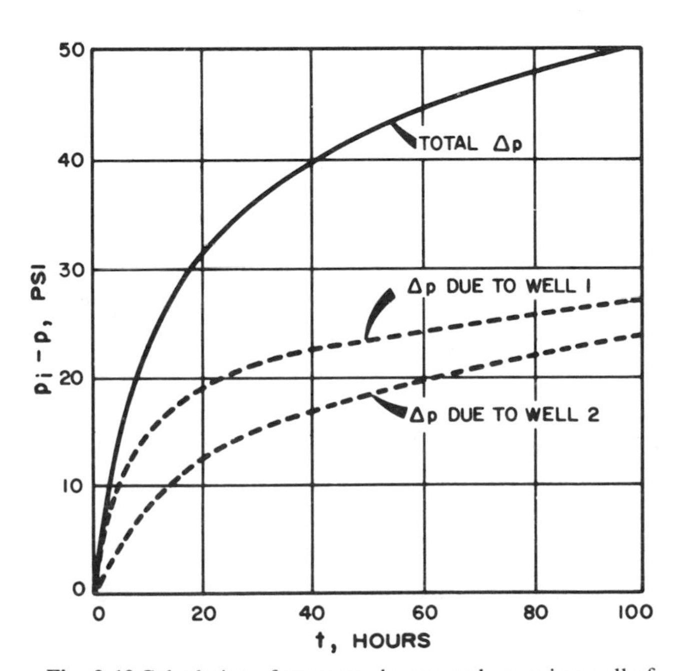

Fig. 2.19 Calculation of pressure change at observation well of Fig. 2.18. $q_1 = 100$ STB/D; $r_1 = 100$ ft; $q_2 = 150$ STB/D; $r_2 = 316$ ft; $k = 76$ md; $\mu = 1.0$ cp; $\phi = 0.20$; $c_t = 10 \times 10^{-6}$ psi^{-1}; $B = 1.08$ RB/STB; $h = 20$ ft; and $s = 0$.

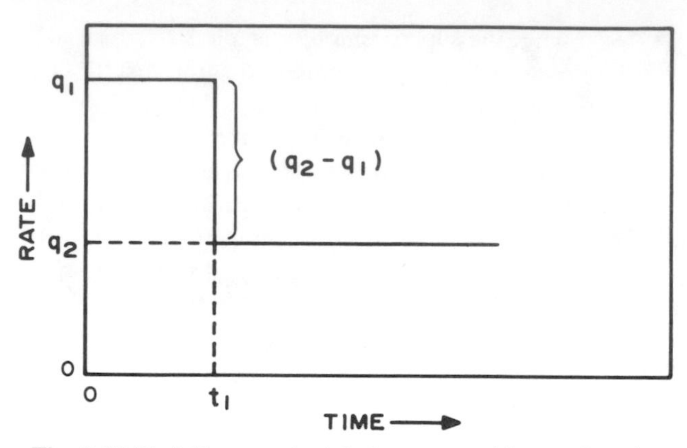

Fig. 2.20 Variable rate schedule for superposition explanation.

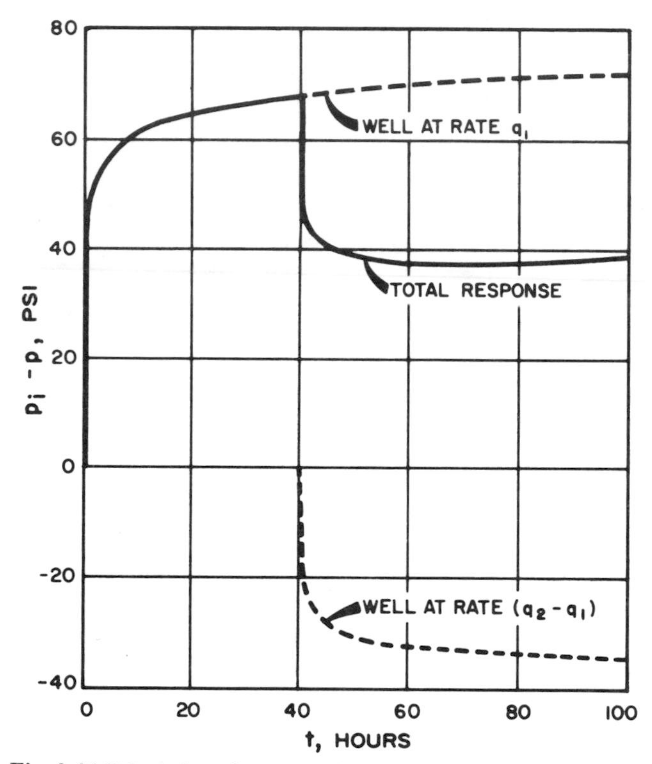

Fig. 2.21 Calculation of pressure change at producing well with the rate history of Fig. 2.20. q_1 = 100 STB/D; q_2 = 50 STB/D; k = 76 md; μ = 1.0 cp; ϕ = 0.20; $c_t = 10 \times 10^{-6}$ psi^{-1}; B = 1.08 RB/STB; h = 20 ft; and s = 0.

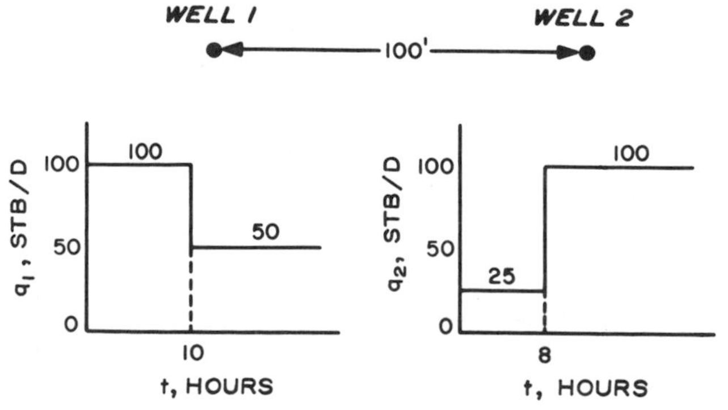

Fig. 2.22 Example superposition calculation for two wells, each produced at two rates. Well 1: $s = 5$, r_w = 1 ft. Well 2: $s = 1.7$, r_w = 1 ft. k = 76 md; μ = 1.0 cp; ϕ = 0.20; $c_t = 10 \times 10^{-6}$ psi^{-1}; B = 1.08 RB/STB; h = 20 ft; and p_i = 2,200 psi.

2.30 for the system of Fig. 2.18 and the exponential integral p_D (Eq. 2.5). The lowest curve in Fig. 2.19 is the pressure change at Well 3 caused by Well 2. The upper dashed curve is the pressure change at Well 3 caused by Well 1. Using Eq. 2.29, we add the two pressure changes to get the solid curve, the total pressure change observed at Well 3.

To illustrate an application of the principle of superposition to varying flow rates, consider a single-well system with the production-rate schedule shown in Fig. 2.20. The production rate is q_1 from $t = 0$ to $t = t_1$ and q_2 thereafter. To perform the superposition calculation the single well may be visualized as two wells located at the same point, with one well producing at rate q_1 from $t = 0$ to t and the second (imaginary) well producing at rate $(q_2 - q_1)$, starting at t_1 and continuing for a time period $(t - t_1)$. The net rate after time t_1 would be $q_1 + (q_2 - q_1) = q_2$. As in the previous illustration, Δp's are added for these conditions. The general form of the equation for N rates, with changes at t_j, j = 1, 2, ... N, is

$$\Delta p = \frac{141.2\,\mu}{kh} \sum_{j=1}^{N} \Big\{ (q_j B_j - q_{j-1} B_{j-1}) \Big[p_D\big([t - t_{j-1}]_D\big) + s \Big] \Big\}. \qquad (2.31)$$

In Eq. 2.31, $[t - t_1]_D$ is the dimensionless time calculated at time $(t - t_1)$. For Fig. 2.20, $N = 2$ and only two terms of the summation are needed. Fig. 2.21 illustrates the calculation. The upper dashed curve (including the first portion of the solid curve) is the pressure change caused by rate q_1. The lower dashed curve is the pressure change caused by rate $q_2 - q_1$ after t = 40 hours; that Δp is negative because $(q_2 - q_1) < 0$. The sum of the two dashed curves, the solid curve, is the pressure response for the two-rate schedule.

To combine varying rates and multiple wells, first apply Eq. 2.31 for each well in the system to estimate Δp caused by that well at the point desired. Then add the Δp for all wells, as in Eq. 2.30, to get the total Δp caused by all wells and all rates. The double summation process is conceptually simple, but can be tedious in application, as illustrated by the following example.

Example 2.7 Principle of Superposition

For the conditions shown in Fig. 2.22, estimate the pressure at Well 1 after 7 hours and at Well 2 after 11 hours. Assume that the system behaves as an infinite one at these short times.

Start by computing the coefficients in the Δp and t_D equations, Eqs. 2.2 and 2.3. Then, at a given time, estimate Δp at the desired well caused by both Well 1 and Well 2; that calculation of Δp may require use of Eq. 2.31 to account for varying rates.

From Eq. 2.3a,

$$t_D = \frac{0.0002637\,kt}{\phi \mu c_t r_w^2} = \frac{(0.0002637)(76)t}{(0.2)(1)(10 \times 10^{-6})(1)^2}$$

$$= 10{,}000\,t.$$

From Eq. 2.2,

$$\Delta p = \frac{141.2\, qB\mu}{kh} \left[p_D(t_D, \ldots)\right]$$

$$= \frac{(141.2)(1.08)(1)q}{(76)(20)} p_D(t_D, r_D, \ldots)$$

$$= 0.1\, q\, p_D(t_D, r_D, \ldots).$$

Recall that s_j must be added to p_D to get Δp at Well j (unless p_D is a function of s, as are some that are given in Appendix C). The appropriate r also must be used to calculate r_D, depending on the p_D function we use.

At t = 7 hours, there is a Δp contribution at Well 1 from a single rate at Well 1 and a single rate at Well 2; so the over-all pressure change would be (Eq. 2.30)

$$\Delta p\ (7 \text{ hours}, r_D = 1)$$
$$= \frac{141.2\, q_1 B_1 \mu}{kh} \left[p_D(t_D, r_D = 1) + s\right]$$
$$+ \frac{141.2\, q_2 B_2 \mu}{kh} p_D(t_D, r_D = 100/1).$$

For the contribution of Well 1, $t_D = (10{,}000)(7) = 70{,}000$. Since $t_D > 100$ Eq. 2.5b is used:

$$p_D(t_D = 70{,}000, r_D = 1)$$
$$= \frac{1}{2}\left[\ln(70{,}000) + 0.80907\right]$$
$$= 5.98.$$

For the contribution of Well 2, at a distance of 100 ft,

$$\frac{t_D}{r_D^2} = \frac{(70{,}000)}{(100/1)^2} = 7.$$

Since $r_D > 20$, we can use the line-source solution, Eq. 2.5a, but we should not use Eq. 2.5b unless $t_D/r_D^2 > 100$. From Fig. C.1 or Fig. C.2 for $t_D/r_D^2 = 7$ and $r_D = 100$,

$$p_D(t_D = 7, r_D = 100) = 1.40.$$

Calculating Δp at Well 1,

$$\Delta p(\text{Well 1, 7 hours}) = (0.1)(100)(5.98 + 5) + (0.1)(25)(1.40) = 113.3 \text{ psi}.$$

The pressure at Well 1 at 7 hours is

$$p_w\ (7 \text{ hours}, r_D = 1) = p_i - \Delta p = 2{,}200 - 113.3 = 2{,}086.7 \text{ psi}.$$

At t = 11 hours, we wish to estimate p at Well 2. We must consider two rates at each well:

$$\Delta p(11 \text{ hours}, r_D = 1)$$
$$= (0.1)(100)\, p_D(\text{Well 1}, t = 11 \text{ hours}, r_D = 100)$$
$$+ (0.1)(50-100)\, p_D(\text{Well 1}, t = [11-10] \text{ hours}, r_D = 100)$$
$$+ (0.1)(25)\left[p_D(\text{Well 2}, t = 11 \text{ hours}, r_D = 1) + s\right]$$
$$+ (0.1)(100-25)\left\{p_D(\text{Well 2}, t = [11-8] \text{ hours}, r_D = 1) + s\right\}.$$

For Well 1, use Fig. C.2:

$$t_D(11 \text{ hours})/r_D^2 = \frac{(10{,}000)(11)}{(100)^2} = 11.$$

$$p_D(\text{Well 1}, t_D = 11, r_D = 100) = 1.61.$$

$$t_D(11-10 \text{ hours})/r_D^2 = \frac{(10{,}000)(1)}{(100)^2} = 1.$$

$$p_D(\text{Well 1}, t_D = 1, r_D = 100) = 0.522.$$

For Well 2, $r_D = 1$:

$$t_D(11 \text{ hours}) = (10{,}000)(11) = 110{,}000.$$

Since $t_D > 100$, we use Eq. 2.5b:

$$p_D(\text{Well 2}, t_D = 110{,}000, r_D = 1)$$
$$= \frac{1}{2}\left[\ln(t_D) + 0.80907\right]$$
$$= 6.21.$$

$$t_D(11-8 \text{ hours}) = (10{,}000)(3) = 30{,}000.$$

$$p_D(t_D = 30{,}000, r_D = 1)$$
$$= \frac{1}{2}\left[\ln(30{,}000) + 0.80907\right]$$
$$= 5.56.$$

Estimating Δp at Well 2,

$$\Delta p(\text{Well 2, 11 hours}) =$$
$$(0.1)(100)(1.61) + (0.1)(50-100)(0.522)$$
$$+ (0.1)(25)(6.21 + 1.7)$$
$$+ (0.1)(100-25)(5.56 + 1.7)$$
$$= 87.7 \text{ psi}.$$

$$p_w(\text{Well 2, 11 hours}) = 2{,}200 - 87.7 = 2{,}112.3 \text{ psi}.$$

Additional applications of the principle of superposition and the method of images are shown in Appendix B.

2.10 Application of Flow Equations to Gas Systems

Although this monograph is concerned only with liquid systems, much of the material presented can be applied to dry gas systems if modified slightly.

Gas viscosity and density vary significantly with pressure, so the assumptions of Eq. 2.1 are not satisfied for gas systems and the equation does not apply directly to gas flow in porous media. That difficulty is avoided by defining a "real gas potential" (also commonly referred to as the real gas pseudopressure or just pseudopressure):[37,38]

$$m(p) = 2 \int_{p_b}^{p} \frac{p}{\mu(p)z(p)} dp, \quad \ldots\ldots\ldots\ldots\ (2.32)$$

where p_b is an arbitrary base pressure. When the real gas potential is used, Eq. 2.1 essentially retains its form but with $m(p)$ replacing p. That equation can be solved and an analog to Eq. 2.2 can be written with $m_D(t_D)$ in place of $p_D(t_D)$. For radial gas flow it has been shown[37,39,40] that when $t_D < (t_D)_{pss}$,

$$m_D(t_D) = p_D(t_D), \quad \ldots\ldots\ldots\ldots\ (2.33)$$

where $p_D(t_D)$ is the liquid flow dimensionless pressure.

Using Eq. 2.33, the gas analog of Eq. 2.2, and substituting the appropriate gas properties, the flow equation for a real gas is

$$m(p_{wf}) = m(p_i) - 50{,}300 \frac{p_{sc}}{T_{sc}} \frac{qT}{kh} \left[p_D(t_D) + s + D|q|\right], \quad (2.34)$$

where q is in Mscf/D. In Eq. 2.34, the term $D|q|$ accounts for non-Darcy flow around the wellbore. Otherwise, the form is like the liquid flow equation. To use Eq. 2.34 it is necessary to construct a high-resolution graph of $m(p)$ vs p from the viscosity and z factor for the gas. If μ and z are not known, information presented by Zana and Thomas[41] may be used to estimate $m(p)$ vs p.

As a result of the characteristics of the real gas potential, Eq. 2.34 can be simplified for certain pressure ranges. Fig. 2.23 shows μz as a function of pressure for a typical gas. At low pressures μz is essentially constant, while at high pressures it is essentially directly proportional to pressure. When this behavior is used in Eq. 2.32, Eq. 2.34 can be simplified to

$$p_{wf} = p_i - 50{,}300 \frac{z_i \mu_{gi}}{2p_i} \frac{p_{sc}}{T_{sc}} \frac{qT}{kh} \left[p_D(t_D) + s + D|q|\right], \quad (2.35)$$

at high pressures, while at low pressures it becomes

$$p_{wf}^2 = p_i^2 - 50{,}300 (z_i \mu_{gi}) \frac{p_{sc}}{T_{sc}} \frac{qT}{kh} \left[p_D(t_D) + s + D|q|\right]. \quad (2.36)$$

As a rule of thumb,[40] Eq. 2.36 generally applies when $p < 2{,}000$ psi, while Eq. 2.35 generally applies for $p > 3{,}000$ psi; for $2{,}000 < p < 3{,}000$ use Eq. 2.34. We suggest that the μz vs p plot be made for the particular gas flowing before choosing between the equations. If neither situation prevails at the pressure level observed or expected, then the real gas potential, $m(p)$, must be used. Eqs. 2.34, 2.35, or 2.36 may be used with liquid dimensionless pressure for most gas systems.

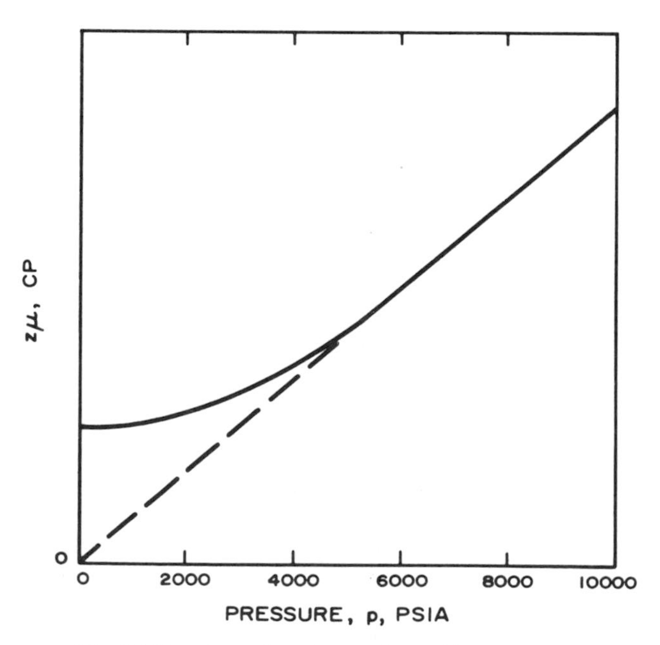

Fig. 2.23 Isothermal variation of μz with pressure.

2.11 Application of Flow Equations to Multiple-Phase Flow

Eqs. 2.1 and 2.2 and the dimensionless pressure information in this monograph are derived for single-phase flow. However, they may be used for certain multiple-phase flow situations with some modifications.[42-45] The basic approach is to replace the mobility terms in Eqs. 2.1 through 2.3 by the total flowing mobility,

$$\lambda_t = \lambda_o + \lambda_g + \lambda_w, \quad (2.37a)$$

or

$$\left(\frac{k}{\mu}\right)_t = k\left(\frac{k_{ro}}{\mu_o} + \frac{k_{rg}}{\mu_g} + \frac{k_{rw}}{\mu_w}\right), \quad (2.37b)$$

and to replace the total system compressibility by

$$c_t = S_o c_{oa} + S_w c_{wa} + S_g c_g + c_f. \quad (2.38a)$$

$$c_t = S_o\left[\frac{B_g}{B_o}\left(\frac{\partial R_s}{\partial p}\right) - \frac{1}{B_o}\left(\frac{\partial B_o}{\partial p}\right)\right] + S_w\left[\frac{B_g}{B_w}\left(\frac{\partial R_{sw}}{\partial p}\right) - \frac{1}{B_w}\left(\frac{\partial B_w}{\partial p}\right)\right] + S_g\left[-\frac{1}{B_g}\left(\frac{\partial B_g}{\partial p}\right)\right] + c_f. \quad (2.38b)$$

With these modifications, single-phase liquid dimensionless pressures may be used to describe multiple-phase systems containing immiscible fluids with fairly uniform saturation distribution. The same modifications are made when analyzing transient well test data. For example, mobility estimating equations take the form

$$\left(\frac{k}{\mu}\right)_o = \pm \frac{162.6\, q_o B_o}{mh} \quad (2.39a)$$

$$\left(\frac{k}{\mu}\right)_g = \pm \left\{162.6\,(1{,}000)\left[q_g - 0.001 (q_o R_s + q_w R_{sw})\right] B_g\right\}\Big/ mh \quad (2.39b)$$

$$\left(\frac{k}{\mu}\right)_w = \pm \frac{162.6 q_w B_w}{mh}. \quad (2.39c)$$

The $\pm$ sign in Eq. 2.39 indicates application to any of the various test analysis techniques.

2.12 Radius of Drainage and Stabilization Time

The concepts of stabilization time and radius of drainage are commonly used in petroleum engineering and in transient testing. These quantities are frequently used without appropriate understanding of their actual meaning and limitations. It is beyond the scope of this monograph to investigate the problems associated with radius of drainage and stabilization time. However, because of their wide use, the quantities are defined and equations are given for them in this section.

Stabilization time has been defined in many ways by various authors.[46,47] Most definitions correspond to the beginning of the pseudosteady-state flow period. Using that as the definition of stabilization time, we can estimate the stabilization time for any shape given in Table C.1 from Eq. 2.24. For a well in the center of most symmetrical shapes, drawdown stabilization time is estimated from

$$t_s \simeq 380 \frac{\phi \mu c_t A}{k}, \quad \text{(2.40a)}$$

where t_s is in hours. If we assume the system is radial,

$$t_s \simeq 1{,}200 \frac{\phi \mu c_t r_e^2}{k}. \quad \text{(2.40b)}$$

For stabilization time in days the coefficient in Eq. 2.40b is 50 and the equation takes the form given by van Poollen.[46] It is important to recognize that stabilization time can be considerably longer than indicated by Eq. 2.40 when the shape is not a symmetrical one with the well in the center, or if two or more noncommunicating layers intersect the wellbore.[48,49]

Radius of drainage is also defined in several ways. Many definitions are presented by van Poollen,[46] Kazemi,[50] and Gibson and Campbell.[51] In most definitions, the radius of drainage defines a circular system with a pseudosteady-state pressure distribution from the well to the "drainage radius." As time increases, more of the reservoir is influenced by the well and the radius of drainage increases, as given by

$$r_d = 0.029 \sqrt{\frac{kt}{\phi \mu c_t}}, \quad \text{(2.41)}$$

where r_d is the radius of drainage in feet and t is in hours. If t is expressed in days, the constant 0.029 rounds to 0.14, and Eq. 2.41 corresponds to equations given in Refs. 46, 50, and 51. Eventually, r_d must stop increasing — either when reservoir boundaries or drainage regions of adjacent wells are encountered, so Eq. 2.41 can only apply until t_{pss}.

Example 2.8 Radius of Drainage

Estimate the radius of drainage created during a 72-hour test on a well in a reservoir with k/μ = 172 md/cp and $\phi c_t = 0.232 \times 10^{-5}$ psi^{-1}. Using Eq. 2.41,

$$r_d = 0.029 \sqrt{\frac{(172)(72)}{0.232 \times 10^{-5}}}$$

$$= 2{,}100 \text{ ft.}$$

This estimate is valid only if no boundaries are within about 2,100 ft of the test well, and if no other operating wells are within about 4,200 ft.

In systems completely recharged by an aquifer or when production and injection are balanced, the concepts of stabilization time and radius of drainage are meaningless. However, Ramey, Kumar, and Gulati[52] define a readjustment time, the time required for a short-lived transient to die out, for such systems. For a single well in the center of a constant-pressure square, which is equivalent to a balanced five-spot water injection pattern with unit mobility ratio, the readjustment time is[52]

$$t_R = 946 \frac{\phi \mu c_t A}{k}. \quad \text{(2.42)}$$

In this equation, A would be approximately one-half the five-spot pattern area.

2.13 Numerical Solution of the Diffusivity Equation

It is possible to obtain analytical solutions to Eq. 2.1 only for the simplest systems. Most dimensionless-pressure functions are from numerical solutions of Eq. 2.1 or its analogs for gas and multiple-phase flow. Computer solution is the only practical method for obtaining dimensionless pressures for extremely heterogeneous systems, layered systems, systems with two or three phases flowing, systems with water or gas coning, or systems with significant gravity effects, for example. During the past several years, many papers that discuss various kinds of reservoir simulators have appeared in the petroleum literature. Three of the classics are by Aronofsky and Jenkins,[53] Bruce, Peaceman, Rachford, and Rice[54], and West, Garvin, and Sheldon.[55] Many facets of reservoir simulation were summarized by van Poollen, Bixel, and Jargon[56] in a series of articles appearing in the *Oil and Gas Journal*. The SPE-AIME Reprint Series booklet[57] on numerical simulation contains many useful papers.

Chapter 12 of this monograph presents some information about the application of computers to transient well testing.

2.14 Summary — A Physical Viewpoint

After presenting the dimensionless-parameter approach to solution of transient flow problems and explaining some of the factors that influence those solutions, it seems worthwhile to summarize the situation from a physical viewpoint.

Fluid withdrawal from a well penetrating a pressurized petroleum reservoir containing a compressible fluid results in a pressure disturbance. Although we might expect that disturbance to move with the speed of sound, it is quickly attenuated, so for any given length of production time there is some distance, the radius of drainage, beyond which no appreciable pressure change can be observed. As fluid withdrawal continues, the disturbance moves farther into the reservoir, with pressure continuing to decline at all points that have started to experience pressure decline. When a closed boundary is encountered, the pressure within the boundary continues to decline, but at a more rapid rate than if the boundary had not been encountered. If, on the other hand, the transient pressure response reaches a replenishable outcrop that maintains constant pressure at some point, pressures nearer the withdrawal well will decline more slowly than if a no-flow boundary had been encountered. Rate changes or additional wells will cause additional pressure transients that affect both pressure decline and pressure distribution. Each well will establish a drainage area that supplies all fluid removed from that well — if there is no fluid injection into the system.

When boundaries are encountered (either no-flow or constant-pressure), the pressure gradient — *not the pressure level* — tends to stabilize after a sufficiently long production time, the stabilization time. For the closed-boundary case, the pressure behavior reaches pseudosteady state with a constant gradient and an over-all pressure decline everywhere that is linear with time. For reservoirs with constant-pressure boundaries, a steady state may be approached. In that case, both pressure gradient and absolute pressure values become constant with time. The pseudosteady-state and steady-state solutions to Eq. 2.1 have a simple form and represent the simplest approach to future performance predictions, when they are applicable.

References

1. Matthews, C. S. and Russell, D. G.: *Pressure Buildup and Flow Tests in Wells,* Monograph Series, Society of Petroleum Engineers of AIME, Dallas (1967) **1,** Chap. 2.
2. van Everdingen, A. F. and Hurst, W.: "The Application of the Laplace Transformation to Flow Problems in Reservoirs," *Trans.,* AIME (1949) **186,** 305-324.
3. Hubbert, M. King: "The Theory of Ground-Water Motion," *J. of Geol.* (Nov.-Dec. 1940) **XLVIII,** 785-944.
4. Horner, D. R.: "Pressure Build-Up in Wells," *Proc.,* Third World Pet. Cong., The Hague (1951) Sec. II, 503-523. Also *Reprint Series, No. 9 — Pressure Analysis Methods,* Society of Petroleum Engineers of AIME, Dallas (1967) 25-43.
5. van Everdingen, A. F.: "The Skin Effect and Its Influence on the Productive Capacity of a Well," *Trans.,* AIME (1953) **198,** 171-176. Also *Reprint Series, No. 9 — Pressure Analysis Methods,* Society of Petroleum Engineers of AIME, Dallas (1967) 45-50.
6. Hurst, William: "Establishment of the Skin Effect and Its Impediment to Fluid Flow Into a Well Bore," *Pet. Eng.* (Oct. 1953) B-6 through B-16.
7. Earlougher, R. C., Jr., and Ramey, H. J., Jr.: "Interference Analysis in Bounded Systems," *J. Cdn. Pet. Tech.* (Oct.-Dec. 1973) 33-45.
8. Mueller, Thomas D. and Witherspoon, Paul A.: "Pressure Interference Effects Within Reservoirs and Aquifers," *J. Pet. Tech.* (April 1965) 471-474; *Trans.,* AIME, **234.**
9. Theis, Charles V.: "The Relation Between the Lowering of the Piezometric Surface and the Rate and Duration of Discharge of a Well Using Ground-Water Storage," *Trans.,* AGU (1935) 519-524.
10. Abramowitz, Milton and Stegun, Irene A. (ed.): *Handbook of Mathematical Functions With Formulas, Graphs and Mathematical Tables,* National Bureau of Standards Applied Mathematics Series-55 (June 1964) 227-253.
11. Hawkins, Murray F., Jr.: "A Note on the Skin Effect," *Trans.,* AIME (1956) **207,** 356-357.
12. Brons, F. and Miller, W. C.: "A Simple Method for Correcting Spot Pressure Readings," *J. Pet. Tech.* (Aug. 1961) 803-805; *Trans.,* AIME, **222.**
13. Brons, F. and Marting, V. E.: "The Effect of Restricted Fluid Entry on Well Productivity," *J. Pet. Tech.* (Feb. 1961) 172-174; *Trans.,* AIME, **222.** Also *Reprint Series, No. 9 — Pressure Analysis Methods,* Society of Petroleum Engineers of AIME, Dallas (1967) 101-103.
14. Chatas, Angelos T.: "A Practical Treatment of Non-Steady State Flow Problems in Reservoir Systems," *Pet. Eng.,* Part 1 (May 1953) B-42 through B-50; Part 2 (June 1953) B-38 through B-50; Part 3 (Aug. 1953) B-44 through B-56.
15. Gladfelter, R. E., Tracy, G. W., and Wilsey, L. E.: "Selecting Wells Which Will Respond to Production-Stimulation Treatment," *Drill. and Prod. Prac.,* API (1955) 117-129.
16. Ramey, H. J., Jr.: "Non-Darcy Flow and Wellbore Storage Effects in Pressure Build-Up and Drawdown of Gas Wells," *J. Pet. Tech.* (Feb. 1965) 223-233; *Trans.,* AIME, **234.** Also *Reprint Series, No. 9 — Pressure Analysis Methods,* Society of Petroleum Engineers of AIME, Dallas (1967) 233-243.
17. Papadopulos, Istavros S. and Cooper, Hilton H., Jr.: "Drawdown in a Well of Large Diameter," *Water Resources Res.* (1967) **3,** No. 1, 241-244.
18. Cooper, Hilton H., Jr., Bredehoeft, John D., and Papadopulos, Istavros S.: "Response of a Finite-Diameter Well to an Instantaneous Charge of Water," *Water Resources Res.* (1967) **3,** No. 1, 263-269.
19. Ramey, H. J., Jr.: "Short-Time Well Test Data Interpretation in the Presence of Skin Effect and Wellbore Storage," *J. Pet. Tech.* (Jan. 1970) 97-104; *Trans.,* AIME, **249.**
20. Agarwal, Ram G., Al-Hussainy, Rafi, and Ramey, H. J., Jr.: "An Investigation of Wellbore Storage and Skin Effect in Unsteady Liquid Flow: I. Analytical Treatment," *Soc. Pet. Eng. J.* (Sept. 1970) 279-290; *Trans.,* AIME, **249.**
21. Wattenbarger, Robert A. and Ramey, H. J., Jr.: "An Investigation of Wellbore Storage and Skin Effect in Unsteady Liquid Flow: II. Finite Difference Treatment," *Soc. Pet. Eng. J.* (Sept. 1970) 291-297; *Trans.,* AIME, **249.**
22. McKinley, R. M.: "Wellbore Transmissibility From Afterflow-Dominated Pressure Buildup Data," *J. Pet. Tech.* (July 1971) 863-872; *Trans.,* AIME, **251.**
23. Barbe, J. A. and Boyd, B. L.: "Short-Term Buildup Testing," *J. Pet. Tech.* (July 1971) 800-804.
24. Earlougher, Robert C., Jr., and Kersch, Keith M.: "Analysis of Short-Time Transient Test Data by Type-Curve Matching," *J. Pet. Tech.* (July 1974) 793-800; *Trans.,* AIME, **257.**
25. Chen, Hsiu-Kuo and Brigham, W. E.: "Pressure Buildup for a Well With Storage and Skin in a Closed Square," paper SPE 4890 presented at the SPE-AIME 44th Annual California Regional Meeting, San Francisco, April 4-5, 1974.
26. Earlougher, Robert C., Jr., Kersch, K. M., and Ramey, H. J., Jr.: "Wellbore Effects in Injection Well Testing," *J. Pet. Tech.* (Nov. 1973) 1244-1250.
27. Stegemeier, G. L. and Matthews, C. S.: "A Study of Anomalous Pressure Build-Up Behavior," *Trans.,* AIME (1958) **213,** 44-50. Also *Reprint Series, No. 9 — Pressure Analysis Methods,* Society of Petroleum Engineers of AIME, Dallas (1967) 75-81.
28. Pitzer, Sidney C., Rice, John D., and Thomas, Clifford E.: "A Comparison of Theoretical Pressure Build-Up Curves With Field Curves Obtained From Bottom-Hole Shut-In Tests," *Trans.,* AIME (1959) **216,** 416-419. Also *Reprint Series, No. 9 — Pressure Analysis Methods,* Society of Petroleum Engineers of AIME, Dallas (1967) 83-86.
29. Ramey, H. J., Jr., and Cobb, William M.: "A General Buildup Theory for a Well in a Closed Drainage Area," *J. Pet. Tech.* (Dec., 1971) 1493-1505.
30. Dietz, D. N.: "Determination of Average Reservoir Pressure From Build-Up Surveys," *J. Pet. Tech.* (Aug. 1965) 955-959; *Trans.,* AIME, **234.**
31. Earlougher, Robert C., Jr., Ramey, H. J., Jr., Miller, F. G., and Mueller, T. D.: "Pressure Distributions in Rectangular Reservoirs," *J. Pet. Tech.* (Feb. 1968) 199-208; *Trans.,* AIME, **243.**
32. Amyx, James W., Bass, Daniel M., Jr., and Whiting, Robert L.: *Petroleum Reservoir Engineering: Physical Properties,* McGraw-Hill Book Co., Inc. New York (1960) 78-79.

33. Muskat, Morris: *Physical Principles of Oil Production,* McGraw-Hill Book Co., Inc., New York (1949) Ch. 12.

34. Higgins, R. V. and Leighton, A. J.: "A Method of Predicting Performance of Five-Spot Waterfloods in Stratified Reservoirs Using Streamlines," *Report of Investigations 5921,* USBM (1962).

35. Higgins, R. V., Boley, D. W., and Leighton, A. J.: "Aids to Forecasting the Performance of Water Floods," *J. Pet. Tech.* (Sept. 1964) 1076-1082; *Trans.,* AIME, **231.**

36. Collins, Royal Eugene: *Flow of Fluids Through Porous Materials,* Reinhold Publishing Corp., New York (1961) 108-123.

37. Al-Hussainy, R., Ramey, H. J., Jr., and Crawford, P. B.: "The Flow of Real Gases Through Porous Media," *J. Pet. Tech.* (May 1966) 624-636; *Trans.,* AIME, **237.**

38. Russell, D. G., Goodrich, J. H., Perry, G. E., and Bruskotter, J. F.: "Methods for Predicting Gas Well Performance," *J. Pet. Tech.* (Jan. 1966) 99-108; *Trans.,* AIME, **237.**

39. Al-Hussainy, R. and Ramey, H. J., Jr.: "Application of Real Gas Flow Theory to Well Testing and Deliverability Forecasting," *J. Pet. Tech.* (May 1966) 637-642; *Trans.,* AIME, **237.** Also *Reprint Series, No. 9 — Pressure Analysis Methods,* Society of Petroleum Engineers of AIME, Dallas (1967) 245-250.

40. Wattenbarger, Robert A. and Ramey, H. J., Jr.: "Gas Well Testing With Turbulence, Damage and Wellbore Storage," *J. Pet. Tech.* (Aug. 1968) 877-887; *Trans.,* AIME, **243.**

41. Zana, E. T. and Thomas, G. W.: "Some Effects of Contaminants on Real Gas Flow," *J. Pet. Tech.* (Sept. 1970) 1157-1168; *Trans.,* AIME, **249.**

42. Martin, John C.: "Simplified Equations of Flow in Gas Drive Reservoirs and the Theoretical Foundation of Multiphase Pressure Buildup Analyses," *Trans.,* AIME (1959) **216,** 309-311.

43. Miller, C. C., Dyes, A. B., and Hutchinson, C. A., Jr.: "The Estimation of Permeability and Reservoir Pressure From Bottom Hole Pressure Build-Up Characteristics," *Trans.,* AIME (1950) **189,** 91-104. Also *Reprint Series, No. 9 — Pressure Analysis Methods,* Society of Petroleum Engineers of AIME, Dallas (1967) 11-24.

44. Perrine, R. L.: "Analysis of Pressure Buildup Curves," *Drill. and Prod. Prac.,* API (1956) 482-509.

45. Earlougher, R. C., Jr., Miller, F. G., and Mueller, T. D.: "Pressure Buildup Behavior in a Two-Well Gas-Oil System," *Soc. Pet. Eng. J.* (June 1967) 195-204; *Trans.,* AIME, **240.**

46. van Poollen, H. K.: "Radius-of-Drainage and Stabilization-Time Equations," *Oil and Gas J.* (Sept. 14, 1964) 138-146.

47. Mathur, Shri B.: "Determination of Gas Well Stabilization Factors in the Hugoton Field," *J. Pet. Tech.* (Sept. 1969) 1101-1106.

48. Cobb, William M., Ramey, H. J., Jr., and Miller, Frank G.: "Well-Test Analysis for Wells Producing Commingled Zones," *J. Pet. Tech.* (Jan. 1972) 27-37; *Trans.,* AIME, **253.**

49. Earlougher, Robert C., Jr., Kersch, K. M., and Kunzman, W. J.: "Some Characteristics of Pressure Buildup Behavior in Bounded Multiple-Layer Reservoirs Without Crossflow," *J. Pet. Tech.* (Oct. 1974) 1178-1186; *Trans.,* AIME, **257.**

50. Kazemi, Hossein: "Pressure Buildup in Reservoir Limit Testing of Stratified Systems," *J. Pet. Tech.* (April 1970) 503-511.

51. Gibson, J. A. and Campbell, A. T., Jr.: "Calculating the Distance to a Discontinuity From D.S.T. Data," paper SPE 3016 presented at the SPE-AIME 45th Annual Fall Meeting, Houston, Oct. 4-7, 1970.

52. Ramey, Henry J., Jr., Kumar, Anil, and Gulati, Mohinder S.: *Gas Well Test Analysis Under Water-Drive Conditions,* AGA, Arlington, Va. (1973).

53. Aronofsky, J. A. and Jenkins, R.: "A Simplified Analysis of Unsteady Radial Gas Flow," *Trans.,* AIME (1954) **201,** 149-154. Also *Reprint Series, No. 9 — Pressure Analysis Methods,* Society of Petroleum Engineers of AIME, Dallas (1967) 197-202.

54. Bruce, G. H., Peaceman, D. W., Rachford, H. H., Jr., and Rice, J. D.: "Calculations of Unsteady-State Gas Flow Through Porous Media," *Trans.,* AIME (1953) **198,** 79-92.

55. West, W. J., Garvin, W. W., and Sheldon, J. W.: "Solution of the Equations of Unsteady State Two-Phase Flow in Oil Reservoirs," *Trans.,* AIME (1954) **201,** 217-229.

56. van Poollen, H. K., Bixel, H. C., and Jargon, J. R.: "Reservoir Modeling — 1: What It Is, What It Does," *Oil and Gas J.* (July 28, 1969) 158-160. (See bibliography for complete series.)

57. *Reprint Series, No. 11 — Numerical Simulation,* Society of Petroleum Engineers of AIME, Dallas (1973).

Chapter 3

Pressure Drawdown Testing

3.1 Introduction

Often, the first significant transient event at a production well is the initial production period that results in a *pressure drawdown* at the formation face. Thus, it seems logical to investigate what can be learned about the well and reservoir from pressure drawdown data. Matthews and Russell[1] state that infinite-acting, transition, and pseudosteady-state pressure drawdown data all may be analyzed for reservoir information. This chapter considers drawdown test analysis for the infinite-acting and pseudosteady-state periods; the transition period (late-transient period) analysis is given in Section 5.2 of Ref. 1. This chapter treats constant-rate drawdown testing only; variable-rate drawdown testing is covered in Chapter 4.

Although drawdown testing is not limited to the initial productive period of a well, that may be an ideal time to obtain drawdown data. Properly run drawdown tests may provide information about formation permeability, skin factor, and the reservoir volume communicating with the well.

Fig. 3.1 schematically illustrates the production and pressure history during a drawdown test. Ideally, the well is shut in until it reaches static reservoir pressure before the test. That requirement is met in new reservoirs; it is less often met in old reservoirs. Fortunately, when the requirement is not satisfied, data may be analyzed by the techniques of Chapter 4. The drawdown test is run by producing the well at a constant flow rate while continuously recording bottom-hole pressure. In this type of test, well-completion data details must be known so the effect and duration of wellbore storage may be estimated.

While most reservoir information obtained from a drawdown test also can be obtained from a pressure buildup test (Chapter 5), there is an economic advantage to drawdown testing since the well is produced during the test. The main technical advantage of drawdown testing is the possibility for estimating reservoir volume. The major disadvantage is the difficulty of maintaining a constant production rate.

3.2 Pressure Drawdown Analysis in Infinite-Acting Reservoirs

The pressure at a well producing at a constant rate in an infinite-acting reservoir is given by Eq. 2.2:

$$p_i - p_{wf} = 141.2 \frac{qB\mu}{kh} \left[p_D(t_D, \ldots) + s \right], \quad \ldots\ldots (3.1)$$

if the reservoir is at p_i initially. The dimensionless pressure at the well ($r_D = 1$) is given by Eq. 2.5b:

$$p_D = \frac{1}{2} \left[\ln (t_D) + 0.80907 \right], \quad \ldots\ldots (3.2)$$

when $t_D/r_D^2 > 100$ and after wellbore storage effects have diminished. Dimensionless time is given by Eq. 2.3a:

$$t_D = \frac{0.0002637\, kt}{\phi \mu c_t r_w^2}. \quad \ldots\ldots (3.3)$$

Eqs. 3.1 through 3.3 may be combined and rearranged to a familiar form of the pressure drawdown equation:[1]

$$p_{wf} = p_i - \frac{162.6\, qB\mu}{kh} \left[\log t + \log \left(\frac{k}{\phi \mu c_t r_w^2} \right) - 3.2275 + 0.86859\, s \right]. \quad \ldots\ldots (3.4)$$

Eq. 3.4 describes a straight-line relationship between p_{wf} and log t. By grouping the intercept and slope terms together, it may be written as

$$p_{wf} = m \log t + p_{1hr}. \quad \ldots\ldots (3.5)$$

Theoretically, a plot of flowing bottom-hole pressure data vs the logarithm of flowing time (commonly called the "semilog plot") should be a straight line with slope m and intercept p_{1hr}. Fig. 3.2 indicates that the straight-line portion (the "semilog straight line") does appear after wellbore damage and storage effects have diminished; no data are shown after the end of the infinite-acting period. The slope of the semilog straight line in Fig. 3.2 and Eq. 3.5 may be determined from Eq. 3.4 to be

$$m = -\frac{162.6\, qB\mu}{kh}. \quad \ldots\ldots (3.6)$$

The intercept at log $t = 0$, which occurs at $t = 1$, is also determined from Eq. 3.4:

$$p_{1hr} = p_i + m \left[\log \left(\frac{k}{\phi \mu c_t r_w^2} \right) - 3.2275 + 0.86859\, s \right]. \quad \ldots\ldots (3.7)$$

Two graphs of pressure drawdown data are required for test analysis. The log-log data plot $[\log(p_i - p_{wf})$ vs $\log t]$ is used to estimate when wellbore storage effects are no longer important (Section 2.6). When the slope of that plot is one cycle in Δp per cycle in t, wellbore storage dominates and test data give no information about the formation. The wellbore storage coefficient may be estimated from the unit-slope straight line with Eq. 2.20. The semilog straight line should begin about 1 to 1.5 cycles in t after the data start deviating from the unit slope. That corresponds to a low-slope, slightly curving line on the log-log plot. Alternatively, the beginning time of the semilog straight line may be estimated from Eq. 2.21b:

$$t > \frac{(200{,}000 + 12{,}000s)C}{(kh/\mu)} \quad \text{. . . . . (3.8)}$$

The second required graph is the semilog plot, p_{wf} vs $\log t$. The slope, m, of the correct straight line is measured from this graph, and formation permeability is estimated from

$$k = -\frac{162.6\, qB\mu}{mh} \quad \text{. . . . . (3.9)}$$

Clearly, kh/μ, kh, or k/μ also may be estimated.

The skin factor is estimated from a rearranged form of Eq. 3.7:

$$s = 1.1513\left[\frac{p_{1hr} - p_i}{m} - \log\left(\frac{k}{\phi\mu c_t r_w^2}\right) + 3.2275\right].$$

$$\text{. . . . . (3.10)}$$

In Eq. 3.10, p_{1hr} must be from the semilog straight line. If pressure data measured at 1 hour do not fall on that line, the line *must be extrapolated* to 1 hour and the extrapolated value of p_{1hr} must be used in Eq. 3.10. This procedure is necessary to avoid calculating an incorrect skin by using a wellbore-storage-influenced pressure. Fig. 3.2 illustrates the extrapolation to p_{1hr}.

If the drawdown test is long enough, bottom-hole pressure will deviate from the semilog straight line and make the transition from infinite-acting to pseudosteady state. If reservoir geometry and properties are known, the end of the semilog straight line may be estimated from Eq. 2.8a.

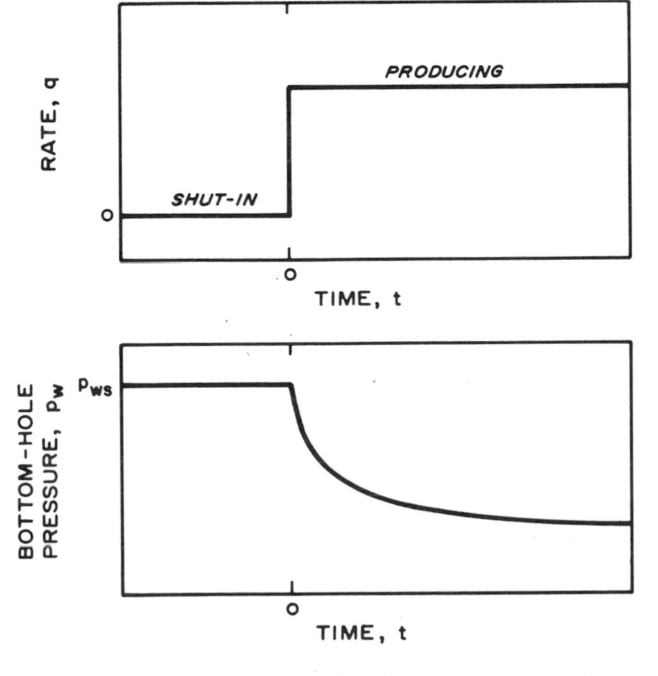

Fig. 3.1 Idealized rate schedule and pressure response for drawdown testing.

Example 3.1 Drawdown Testing in an Infinite-Acting Reservoir

Estimate oil permeability and skin factor from the drawdown data of Figs. 3.3 and 3.4. (Data are from Figs. 5.4 and 5.5 of Ref. 1.) Fig. 3.3, the log-log plot, indicates that wellbore storage effects are not significant for $t > 1$ hr.

Known reservoir data are

$h = 130$ ft	$\phi = 20$ percent
$r_w = 0.25$ ft	$p_i = 1{,}154$ psi
$q_o = 348$ STB/D	$m = -22$ psi/cycle (Fig. 3.4)
$B_o = 1.14$ RB/STB	$p_{1hr} = 954$ psi (Fig. 3.4).
$\mu_o = 3.93$ cp	
$c_t = 8.74 \times 10^{-6}$ psi^{-1}	

Using Eq. 3.9,

$$k_o = -\frac{(162.6)(348)(1.14)(3.93)}{(-22)(130)} = 89 \text{ md}.$$

Eq. 3.10 gives

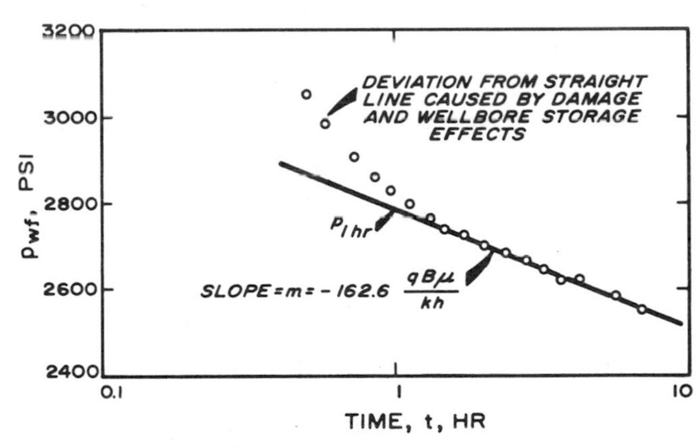

Fig. 3.2 Semilog plot of pressure drawdown data for a well with wellbore storage and skin effect.

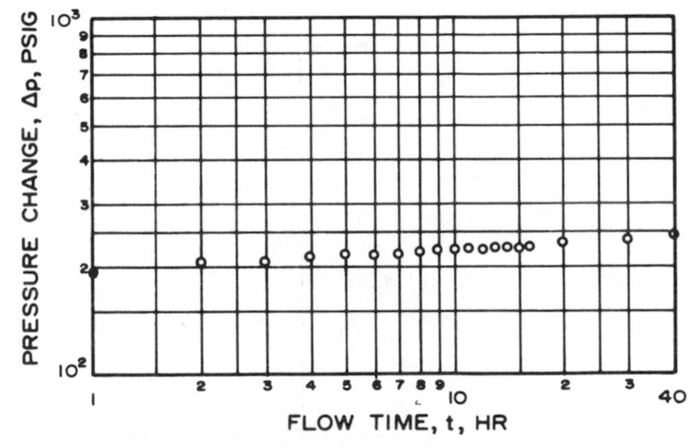

Fig. 3.3 Log-log data plot for Example 3.1.

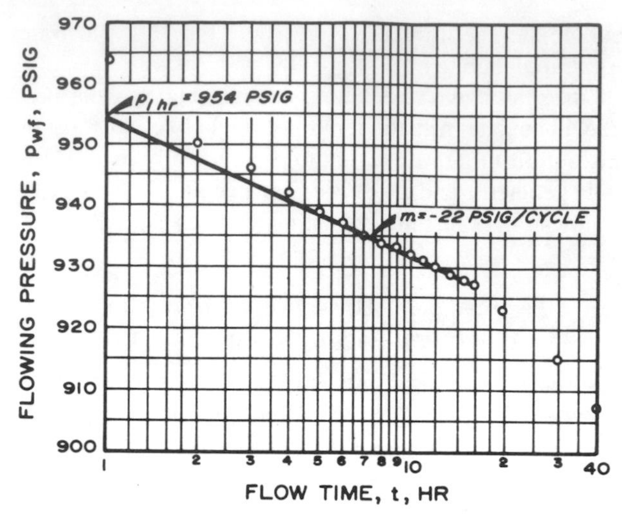

Fig. 3.4 Semilog data plot for the drawdown test of Example 3.1.

$$s = 1.1513\left\{\left(\frac{954 - 1{,}154}{-22}\right) - \log\left[\frac{89}{(0.2)(3.93)(8.74\times10^{-6})(0.25)^2}\right] + 3.2275\right\} = 4.6.$$

3.3 Pressure Drawdown Analysis by Type-Curve Matching

When a drawdown test is too short for the semilog straight line to develop, the data cannot be analyzed with the methods of Section 3.2. *Type-curve matching* techniques[2-9] provide a method for analyzing such data. The general method applies to many kinds of transient well tests for any system with known p_D vs t_D. Type-curve matching may be used for drawdown, buildup, interference, and constant-pressure testing. For single-well testing, type-curve matching should be used only when conventional analysis techniques, such as those illustrated in Section 3.2, cannot be used. In such cases, type-curve analysis can provide approximate results even though normal analysis techniques would fail. The type-curve matching technique has been described[2-9] in many ways; the method outlined here is specifically for use with Figs. C.6 through C.8 for drawdown testing in a well with wellbore storage and skin. The material presented is detailed enough so the reader can devise specific curve-matching techniques for other type curves.* Although the type-curve matching process appears awkward and difficult when described in writing, it is really quite straightforward. The reader is urged to try the method using the step-by-step description and data of Example 3.2.

First, we outline a general type-curve analysis approach for p_D vs t_D type curves similar to those in Figs. C.6 and C.7. Then, we give an explanation and an example of type-curve analysis using Fig. C.8. The general approach to type-curve analysis follows. Fig. 3.5 photographically illustrates the steps.

1. Choose the type curve, usually a log-log plot of p_D vs t_D. To provide specific details, the method is illustrated using Fig. C.6, the type curve for a single well with wellbore storage and skin effect in an infinite system. We must plot observed test data as Δp vs test time, t, on the same size scale as the base type curve. For drawdown tests, the pressure difference is

$$\Delta p = p_i - p_{wf}(t)\ . \qquad (3.11)$$

In general, for any kind of test,

$$\Delta p = \left|p_w(\Delta t = 0) - p_w(\Delta t)\right|\ . \qquad (3.12)$$

Note that Δp is always calculated as a positive number. The time parameter is the running test time, Δt. To plot the data, use tracing paper placed over the desired type curve (Fig. 3.5b); first trace the major grid lines from the type curve for reference (Fig. 3.5c) and mark the Δp (psi) and Δt (hours) scales (Fig. 3.5d). Use the type-curve grid showing through the tracing paper as a guide for plotting the Δp vs Δt data (Fig. 3.5e). This process guarantees that the data plot and the type curve have the same scale. Ignore the curves and the scale on the type curve during the plotting stage; use only the base grid.

2. Slide the tracing paper with the plotted data, keeping the grids parallel, until the data points match one of the type curves (Fig. 3.5f). The type curves are usually similarly shaped, so the matching process can be difficult. After the match is completed, trace the matched curve (Fig. 3.5g) and pick a convenient "match point" on the data plot, such as an intersection of major grid lines. Record values at that point on the data plot $[(\Delta p)_M$ and $(\Delta t)_M]$ and the corresponding values lying beneath that point on the type-curve grid $[(p_D)_M$ and $(t_D)_M]$ (Fig. 3.5h). The match-point data are used to estimate formation properties.

3. In Fig. C.6 (most other Appendix C figures also could be used), the ordinate of the type curve is dimensionless pressure,

$$p_D = \frac{\Delta pkh}{141.2\,qB\mu}\ . \qquad (3.13)$$

By substituting match-point values from Step 2 and rearranging Eq. 3.13, we estimate formation permeability:

$$k = 141.2\ \frac{qB\mu}{h}\ \frac{(p_D)_M}{(\Delta p)_M}\ . \qquad (3.14)$$

4. Similarly, use the definition on the abscissa on the type curve, the dimensionless time in Fig. C.6,

$$t_D = \frac{0.0002637\,kt}{\phi\mu c_t r_w^2}\ , \qquad (3.15)$$

with the time-scale match-point data and the permeability just determined, to estimate the reservoir porosity-compressibility product:

$$\phi c_t = \left[\frac{0.0002637\,k}{\mu r_w^2}\right]\frac{(\Delta t)_M}{(t_D)_M}\ . \qquad (3.16)$$

5. If the type curve is one of several on the graph and is

*Large-scale copies of Figs. 4.12, 8.8A through 8.8C, C.2, C.3, C.5 through C.9, and C.17 through C.19 are available. Information about ordering a packet containing these figures can be obtained by writing to Order Dept. SPE-AIME.

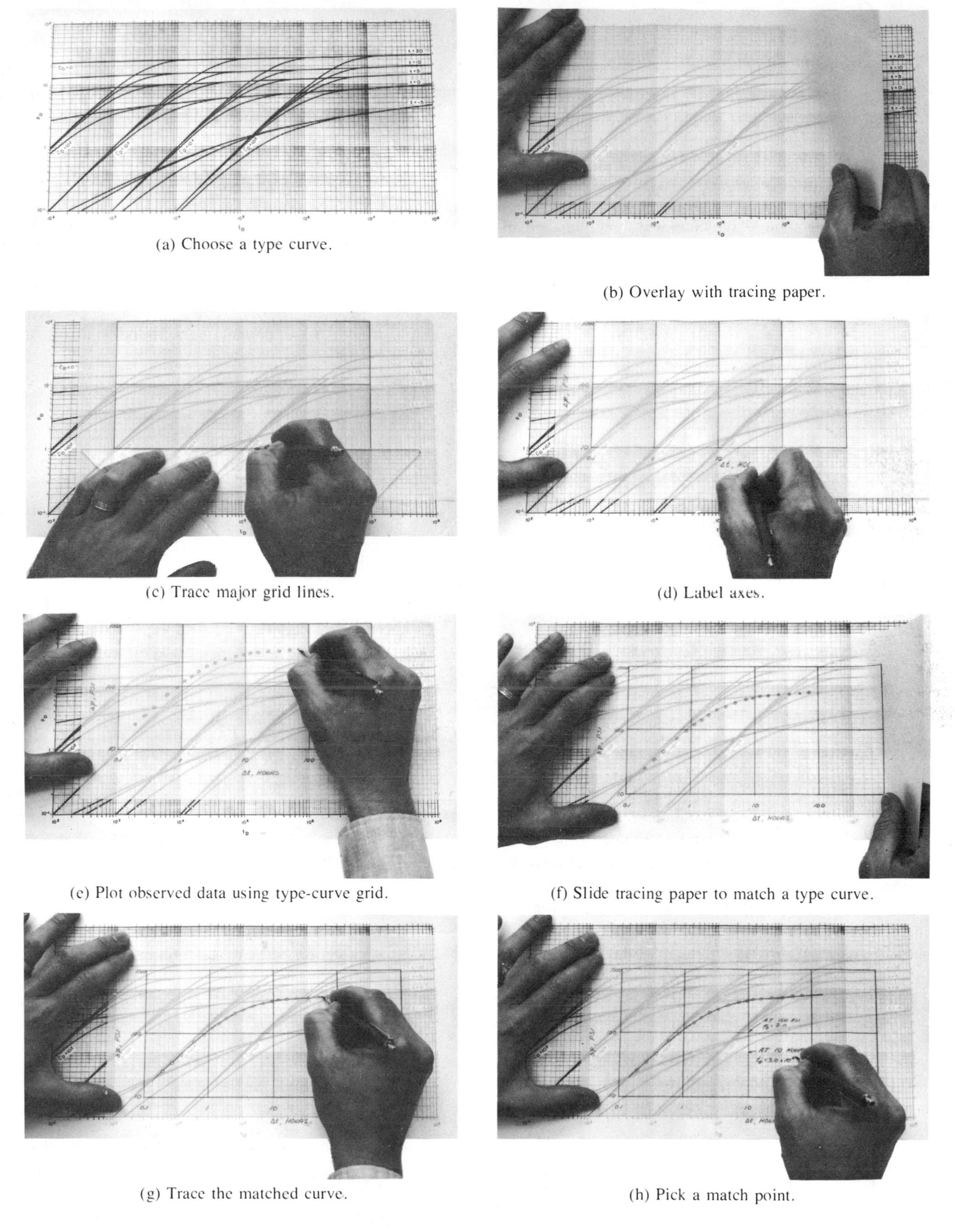

(a) Choose a type curve.

(b) Overlay with tracing paper.

(c) Trace major grid lines.

(d) Label axes.

(e) Plot observed data using type-curve grid.

(f) Slide tracing paper to match a type curve.

(g) Trace the matched curve.

(h) Pick a match point.

Fig. 3.5 Steps in type-curve matching.

identified by a parameter, such as the dimensionless storage coefficient and the skin factor in Fig. C.6, that parameter may be used to estimate additional wellbore or reservoir properties.

Fig. C.6 is one of many useful type curves. Fig. C.8 is useful for drawdown and pressure buildup test analysis for wells with wellbore storage and skin — if the semilog straight line does not develop. The analysis procedure for this type curve, which does not use p_D and t_D, follows. The explanation becomes more understandable if the data of Example 3.2 are used to perform the process.

1a. Plot observed test data as $\Delta p/\Delta t$ (psi/hour) on the ordinate vs Δt (hours) on the abscissa on tracing paper over the grid of Fig. C.8. (Setup is similar to that illustrated in Fig. 3.5.)

1b. Estimate the wellbore storage coefficient expected from completion details (this step can be skipped and C can be estimated from Step 2b):

$$C = V_w c, \qquad (3.17)$$

for a wellbore without a gas-liquid interface or,

$$C = \frac{V_u}{\left(\dfrac{\rho}{144}\dfrac{g}{g_c}\right)}, \qquad (3.18)$$

for a wellbore with changing liquid level.

1c. Estimate the $\Delta p/\Delta t$ value where

$$\left(\frac{\Delta p}{\Delta t}\frac{24C}{qB}\right)_{\text{Fig. C.8}} = 1.0 . \qquad (3.19)$$

This estimate is made from

$$\left(\frac{\Delta p}{\Delta t}\right)_{1.0} = \frac{qB}{24C} . \qquad (3.20)$$

Align the tracing-paper data plot so the value calculated in Eq. 3.20 overlies 1.0 on the ordinate of Fig. C.8 (Eq. 3.19).

2a. Keeping the two grids parallel, slide the tracing-paper data plot *horizontally* until the best match is obtained with one of the curves in Fig. C.8. Slight vertical movement may improve the match. Trace the matched curve onto the data plot and read the value of $(C_D e_{2s})_{\text{Fig. C.8}, M}$ for the matched curve of Fig. C.8. Pick a convenient *match point* with coordinates of $(\Delta p/\Delta t)_M$, $(\Delta t)_M$ from the tracing-paper data plot; read the coordinate values lying directly under this point from Fig. C.8,

$$\left(\frac{\Delta p}{\Delta t}\frac{24C}{qB}\right)_{\text{Fig. C.8}, M}, \quad \left(\frac{kh}{\mu}\frac{\Delta t}{C}\right)_{\text{Fig. C.8}, M}$$

2b. If any vertical movement was made during the curve-matching process, recompute the wellbore storage coefficient from the definition of the ordinate in Fig. C.8:

$$C = \frac{qB\left(\dfrac{\Delta p}{\Delta t}\dfrac{24C}{qB}\right)_{\text{Fig. C.8}, M}}{24\left(\dfrac{\Delta p}{\Delta t}\right)_M}, \qquad (3.21)$$

where q and B are observed data for the test. This value of the wellbore storage coefficient should be essentially the same as the value estimated from Eq. 3.17 or 3.18. If it is not the same, search for a reason, such as washed-out sections of hole, voids connecting with the wellbore, leaking packers, etc.

3. Estimate formation permeability from the definition of the abscissa in Fig. C.8:

$$k = \frac{C\mu\left(\dfrac{kh}{\mu}\dfrac{\Delta t}{C}\right)_{\text{Fig. C.8}, M}}{h(\Delta t)_M} . \qquad (3.22)$$

4. Estimate the skin factor from the parameter on the matched curve:

$$s = \frac{1}{2}\ln\left[\frac{\phi c_t h r_w^2 (C_D e^{2s})_{\text{Fig. C.8}, M}}{0.89359\, C}\right] . \qquad (3.23)$$

This completes the analysis using Fig. C.8.

Other type curves may be used with similar analysis procedures.

Type-curve matching provides a way to analyze transient test data *when insufficient data are available for semilog analysis methods*. If sufficient data are available, semilog methods should be used because they are more accurate than type-curve matching. Nevertheless, when there is no other way to analyze data, when there is insufficient data, or when a fractured well situation is encountered, type-curve matching can provide useful, although approximate, results.

TABLE 3.1—PRESSURE DATA FOR EXAMPLE 3.2.
After Earlougher and Kersch.[8]

Time, Δt (hours)	Pressure Change, Δp (psi)	$\Delta p/\Delta t$ (psi/hr)
0.2	19.7	98.50
0.3	28.1	93.67
0.5	43.1	86.20
0.7	58.3	83.29
1.0	75.1	75.10
2.0	114.5	57.25
3.0	135.5	45.17
5.0	152.2	30.44
7.0	163.2	23.31
10.0	166.7	16.67
20.0	171.2	8.56
30.0	173.9	5.80
50.0	175.2	3.50
70.0	177.1	2.53

Example 3.2 Drawdown Test Analysis by Type-Curve Matching[8]

A pressure drawdown test on a new oil well is strongly influenced by wellbore storage. Nevertheless, enough data exist to use the semilog plot to estimate

$$\frac{kh}{\mu} = 3{,}500 \text{ md ft/cp, and}$$

$$s = 12.$$

Table 3.1 gives Δp and $\Delta p/\Delta t$ data. Other known data are

$q_o = 179$ STB/D $c_t = 8.2 \times 10^{-6}$ psi^{-1}
$B_o = 1.2$ RB/STB $r_w = 0.276$ ft
$h = 35$ ft $\phi = 18$ percent.

Analyze this test using type-curve matching with Fig. C.8 and compare the results with the semilog analysis results.

Since completion details are unknown, the wellbore storage coefficient cannot be estimated. Thus, we must match without this aid (Steps 1b and 1c). We plot $(\Delta p/\Delta t)$ vs Δt on tracing paper laid over the Fig. C.8 grid. Fig. 3.6 shows the resulting data plot. We slide the tracing-paper data plot on Fig. C.8 until a good match results. Fig. 3.7 schematically shows the data plot matched to Fig. C.8. (For clarity in printing, the grid is omitted.) Match-point data are given in Fig. 3.7.

We estimate the wellbore storage coefficient using Eq. 3.21 and the match data from Fig. 3.7:

$$C = \frac{(179)(1.2)(0.1053)}{(24)(10)} = 0.0942 \text{ RB/psi}.$$

We use a modified form of Eq. 3.22 to compute

$$\frac{kh}{\mu} = \frac{(0.0942)(49{,}000)}{(1.0)} = 4{,}620 \text{ md ft/cp}.$$

The skin factor is estimated from Eq. 3.23:

$$s = \frac{1}{2}\ln\left[\frac{(0.18)(8.2\times 10^{-6})(35)(0.276)^2(10^{20})}{(0.89359)(0.0942)}\right]$$

$$= 18.$$

These results are approximate; the technique normally should be used *only when other analysis methods fail*. This example illustrates the analysis method and gives an indication of its accuracy. Thus, we used a test with sufficient data for a conventional, semilog analysis that allows a comparison of the two analysis methods. The kh/μ from type-curve matching is within 32 percent of the value from the semilog plot; but the skin factor is off by 50 percent. In spite of the approximate nature of the analysis technique, useful results are obtained.

The wellbore storage coefficient, $C = 0.0942$ RB/psi, appears to be within reason. Assuming an oil gravity of 30 °API ($\rho = 54.7$ lb_m/cu ft) and a changing liquid level, $V_u = 0.0358$ bbl/ft from Eq. 3.18. That corresponds to about a 6-in.-ID pipe ($r \simeq 0.25$ ft), and is not inconsistent with what little is known about the completion.

3.4 Pressure Drawdown Testing in Developed Systems

Slider[10,11] suggests a technique for analyzing transient tests when conditions are not constant before testing. Fig. 3.8 schematically illustrates a situation with shut-in pressure declining (solid line) before a drawdown test starts at time t_1. The dashed line, an extrapolation of that pressure behavior into the future, represents expected pressure behavior for continued shut-in. Production starts at time t_1, and pressure behaves as shown by the solid line beyond t_1.

Three steps are required to *correctly* analyze such drawdown behavior: (1) determine the correct shut-in pressure extrapolation; (2) estimate the difference between observed flowing pressure and the extrapolated pressure ($\Delta p_{\Delta t}$ in Fig. 3.8); and (3) plot $\Delta p_{\Delta t}$ vs log Δt. A semilog straight line that can be analyzed with Eqs. 3.9 and 3.10 should result.

The preceding analysis procedure usually may be modified. Consider a shut-in well in a developed reservoir with other operating wells. There is a pressure decline at the shut-in test well owing to production at the other wells. After the test well is put on production at time t_1, its pressure is

$$p_{wf} = p_i - \frac{141.2\, qB\mu}{kh}\left[p_D(\Delta t_D, r_{D1} = 1, \ldots) + s\right]$$

$$- \Delta p_{ow}(t), \quad \ldots\ldots\ldots\ldots\ldots\ldots\ldots\ldots \quad (3.24)$$

where $\Delta p_{ow}(t)$ is the pressure drop from p_i at time $t = t_1 + \Delta t$ caused by all other wells in the reservoir. That pressure drop

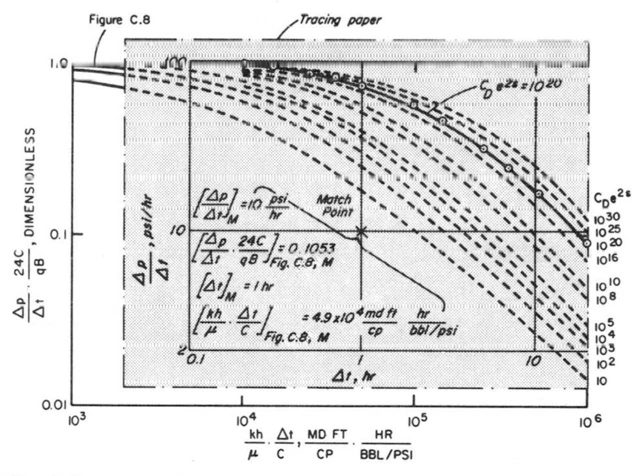

Fig. 3.7 Match of Fig. 3.6 to type curve of Fig. C.8 (part of Fig. 3.6 omitted); Example 3.2, drawdown test on a new oil well. After Earlougher and Kersch.[8]

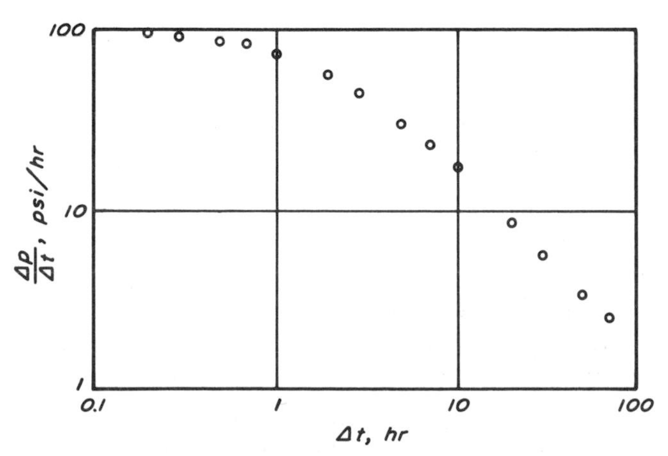

Fig. 3.6 Data plot for Example 3.2 before matching to type curve. Data plotted on tracing paper using grid of Fig. C.8.

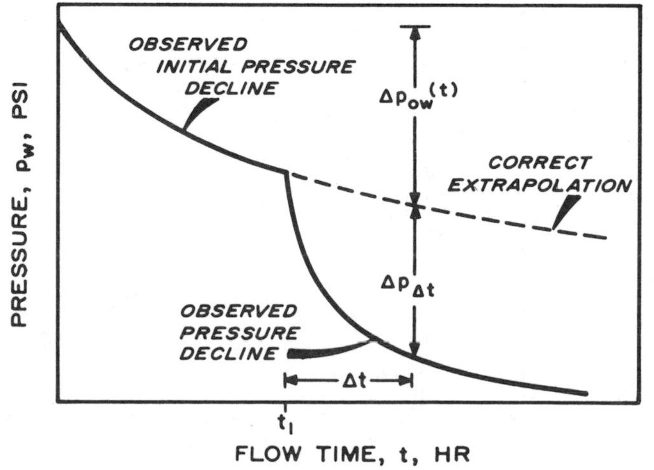

Fig. 3.8 Drawdown testing in a developed reservoir, definition of terms. After Slider.[10]

is indicated schematically in Figs. 3.8 and 3.9 and may be estimated by superposition from

$$\Delta p_{ow}(t) = p_i - p_w(t)$$

$$= \frac{141.2\mu}{kh} \sum_{j=2}^{n} q_j B_j p_D(t_D, r_{Dj}, \ldots). \quad \ldots . (3.25)$$

Eq. 3.25 assumes that all wells start producing at a constant rate at $t = 0$. This assumption may be eliminated by a more complex superposition expression.

If the other wells in the reservoir ($j = 2, 3, \ldots n$) are operating at pseudosteady state, Eq. 3.25 takes the form

$$\Delta p_{ow}(t) = b - m^* t \ , \quad \ldots \ldots (3.26)$$

a straight line with slope $-m^*$ on a plot of Δp_{ow} vs t; or a straight line with slope $+m^*$ on a plot of p_w vs t. (On a reservoir-wide basis, this corresponds to the individual-well, linear, pseudosteady-state Region C of Fig. 2.1a.) The quantity m^* in Eq. 3.26 is determined before the test well starts producing from

$$m^* = \frac{dp_{ws}}{dt} = \frac{(p_{ws})_2 - (p_{ws})_1}{t_2 - t_1} \ . \quad \ldots \ldots (3.27)$$

For declining well pressure, m^* is negative. If pressure data are available before the drawdown test, m^* is easily estimated. It also may be estimated by using Eq. 2.23 in Eq. 3.25:

$$m^* = \frac{-0.23395}{\phi c_t h A} \sum_{j=2}^{n} q_j B_j \ , \quad \ldots \ldots (3.28)$$

where ϕhA is the total reservoir pore volume in cubic feet.

By appropriate combination of Eqs. 3.2, 3.3, 3.24, and 3.26, and by using $t_1 + \Delta t$ in place of t, it can be shown that

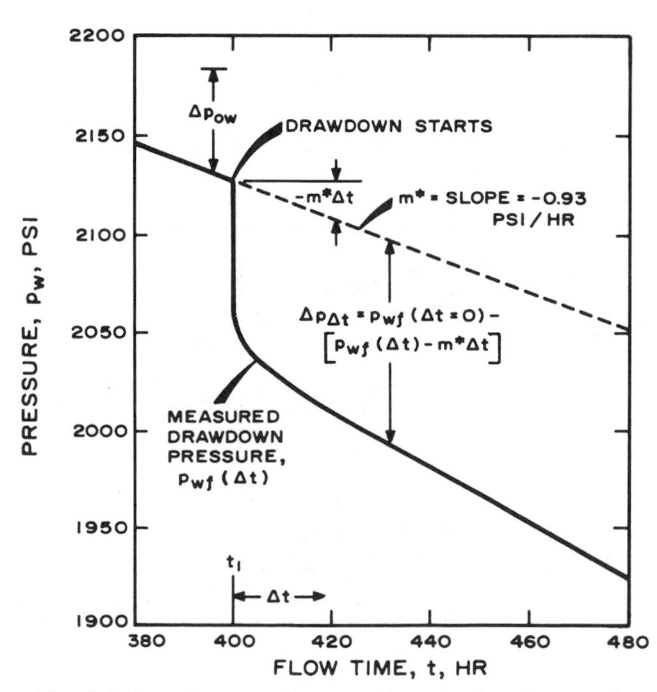

Fig. 3.9 Drawdown test for a well in a developed reservoir, data of Example 3.3.

$$p_{wf} - m^* \Delta t = m \log \Delta t + \Delta p_{1hr} \ , \quad \ldots \ldots (3.29)$$

where m is given by Eq. 3.6 and Δp_{1hr} is given by the right-hand side of Eq. 3.7, with $p_{ws}(\Delta t = 0)$ in place of p_i. Eq. 3.29 indicates that a plot of $(p_{wf} - m^*\Delta t)$ vs log Δt should have a straight-line portion with slope m and intercept Δp_{1hr} at $\Delta t = 1$ hour. The slope, m, can be used with Eq. 3.9 to estimate formation permeability. Skin factor is estimated from

$$s = 1.1513 \left[\frac{\Delta p_{1hr} - p_{ws}(\Delta t{=}0)}{m} - \log\left(\frac{k}{\phi \mu c_t r_w^2}\right) + 3.2275 \right] . \quad \ldots \ldots (3.30)$$

In Eq. 3.30, $p_{ws}(\Delta t{=}0)$ is the shut-in pressure at the beginning of the test.

Example 3.3 Drawdown Test Analysis in a Developed Reservoir

Fig. 3.9 shows simulated drawdown test data for a well in a relatively small multiple-well reservoir. Before the drawdown test, the pressure at the shut-in test well declines linearly, with $m^* = -0.93$ psi/hr, so the analysis method of Eq. 3.29 should apply. At $t_1 = 400$ hours, the test well started producing at 20 STB/D. The pressure response during the flow period is shown in Figs. 3.9 and 3.10. The properties used to simulate the test data are

$k = 20$ md
$h = 10$ ft
$\mu = 1.0$ cp
$\phi = 20$ percent
$r_w = 0.3535$ ft
$q = 20$ STB/D
$B = 1.0$ RB/STB
$c_t = 10^{-5}$ psi^{-1}
$s = 0$
$p_i = 2{,}500$ psi
$p_{ws}(\Delta t{=}0) = 2{,}127$ psi.

To illustrate the importance of a correct analysis, we first estimate k and s from the normal semilog plot of p_{wf} vs log Δt. From Fig. 3.10, $m = -17.1$ psi/cycle and $p_{1hr} = 2{,}051$ psi.

Using Eqs. 3.9 and 3.10,

$$k = \frac{(162.6)(20)(1.0)(1.0)}{(-17.1)(10)} = 19.0 \text{ md},$$

and

$$s = -\ 1.1513 \left\{ \frac{2{,}051 - 2{,}127}{-17.1} - \log\left[\frac{19}{(0.2)(10^{-5})(1)(0.3535)^2}\right] + 3.2275 \right\}$$

$$= -\ 0.24.$$

The calculated permeability is 5 percent too low and the skin factor shows a slight improvement rather than $s = 0$.

We next estimate k and s from the plot of $(p_{wf} - m^*\Delta t)$ vs log Δt shown in Fig. 3.11. Data are

$m^* = -0.93$ psi/hour from Fig. 3.9,
$m = -16.3$ psi/cycle from Fig. 3.11, and
$\Delta p_{1hr} = 2{,}052$ psi from Fig. 3.11.

Using Eqs. 3.9 and 3.30,

$$k = -\frac{(162.6)(20)(1)(1)}{(-16.3)(10)} = 20.0 \text{ md.}$$

$$s = 1.1513\left\{\frac{2{,}052 - 2{,}127}{-16.3} - \log\left[\frac{20}{0.2(10^{-5})(1)(0.3535)^2}\right] + 3.2275\right\}$$

$$= -0.09.$$

A slight improvement in the estimated values of k and s is obtained by including $m^*\Delta t$. The main advantage of including $m^*\Delta t$ is the extended length of the straight line; compare Figs. 3.10 and 3.11. In many cases, the standard analysis techniques given in Section 3.2 can be used with good accuracy. The only difference is that the shut-in pressure just before starting the test rather than p_i should be used to calculate s.

As indicated in Example 3.3, the major advantage of using this kind of analysis is that it extends the length of the semilog straight line and simplifies the analysis. The method may be extended to more complicated systems, although that is beyond the scope of this monograph. The reader is referred to Slider[10,11] for additional technical details.

3.5 Reservoir Limit Testing

A drawdown test run specifically to determine the reservoir volume communicating with the well is called a *reservoir limit test*. Such a test, introduced by Jones,[12,13] uses the pseudosteady-state part of the drawdown data when

$$p_D(t_D, \ldots) = 2\pi t_{DA} + \frac{1}{2}\ln\left(\frac{A}{r_w^2}\right) + \frac{1}{2}\ln\left(\frac{2.2458}{C_A}\right). \quad \ldots\ldots\ldots\ldots (3.31)$$

The dimensionless pressure during pseudosteady-state flow is a linear function of dimensionless time. Eq. 3.31 may be combined with Eqs. 2.2 and 2.3b and simplified to

$$p_{wf} = m^*t + p_{\text{int}}, \quad \ldots\ldots\ldots\ldots\ldots\ldots\ldots\ldots (3.32)$$

where

$$m^* = -\frac{0.23395\,qB}{\phi c_t hA}, \quad \ldots\ldots\ldots\ldots\ldots\ldots (3.33)$$

and

$$p_{\text{int}} = p_i - \frac{70.60\,qB\mu}{kh}\left[\ln\left(\frac{A}{r_w^2}\right) + \ln\left(\frac{2.2458}{C_A}\right) + 2s\right]. \quad \ldots\ldots\ldots\ldots\ldots\ldots (3.34)$$

Eq. 3.32 indicates that a Cartesian plot of bottom-hole flowing pressure vs time should be a straight line during pseudosteady-state flow, with slope m^* given by Eq. 3.33 and intercept p_{int} given by Eq. 3.34. The slope may be used to estimate the connected reservoir drainage volume:

$$\phi hA = -\frac{0.23395\,qB}{c_t m^*}, \quad \ldots\ldots\ldots\ldots\ldots\ldots (3.35)$$

where the volume is in cubic feet. If ϕh is known, the drainage area may be estimated. Other techniques have been proposed[12-14] for analyzing pseudosteady-state data, but this one appears to be the simplest and least prone to errors.

If pressure data are available during both the infinite-acting period and the pseudosteady-state period, it is possible to estimate the drainage shape for the test well. The semilog plot is used to determine m and $p_{1\text{hr}}$; the Cartesian plot is used to get m^* and p_{int}. The system shape factor is estimated from[15]

$$C_A = 5.456\,\frac{m}{m^*}\exp\left[2.303\,(p_{1\text{hr}} - p_{\text{int}})/m\right]. \quad \ldots\ldots (3.36)$$

Knowing the shape factor, use Table C.1 to determine the reservoir configuration with the shape factor closest to that calculated. This process may be refined[15] by computing

$$(t_{DA})_{pss} = 0.1833\,\frac{m^*}{m}\,t_{pss}, \quad \ldots\ldots\ldots\ldots\ldots\ldots (3.37)$$

and using the "Exact for t_{DA} >" column of Table C.1. The time t_{pss} is when the Cartesian straight line starts.

Example 3.4 Reservoir Limit Test[15]

Use the long-time drawdown data of Example 3.1 to estimate the drainage area for that well. Combine the long- and short-time data to estimate the reservoir shape.

Pressure data are shown in Figs. 3.4 and 3.12. From Fig. 3.4, $m = -22$ psi/cycle and $p_{1\text{hr}} = 954$ psi. From Fig. 3.12, $m^* = -0.8$ psi/hour, $p_{\text{int}} = 940$ psi, and $t_{pss} = 11$ hours. Using Eq. 3.35,

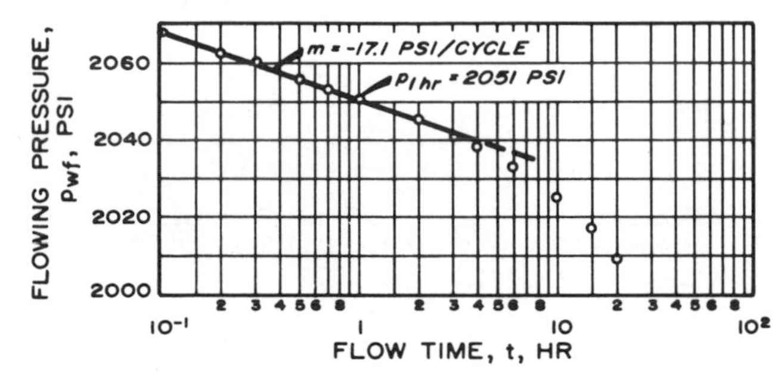

Fig. 3.10 Semilog drawdown curve for the well of Example 3.3.

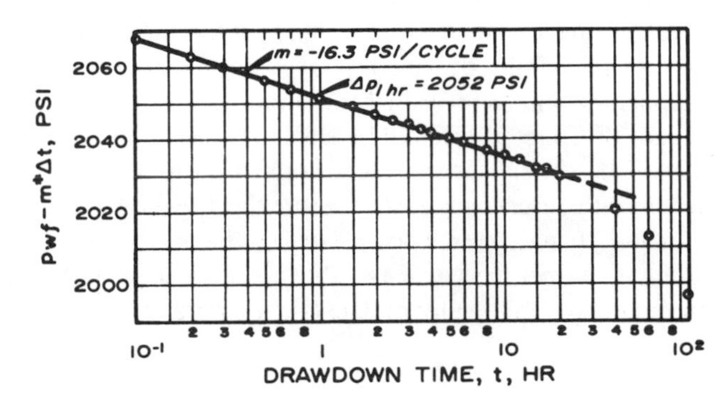

Fig. 3.11 $p_{wf} - m^*\Delta t$ vs log Δt for the well of Example 3.3.

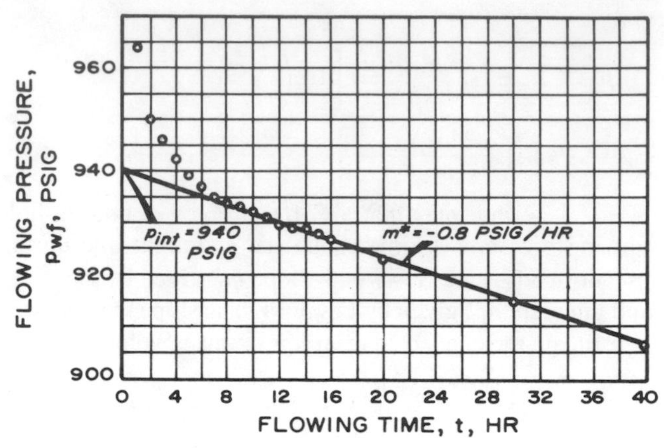

Fig. 3.12 Cartesian plot of the drawdown test data of Examples 3.1 and 3.4.

$$\phi hA = \frac{(-0.23395)(348)(1.14)}{(8.74 \times 10^{-6})(-0.8)}$$

$$= 1.33 \times 10^7 \text{ cu ft}$$

$$= \frac{(1.33 \times 10^7)}{5.6146} = 2.37 \times 10^6 \text{ bbl.}$$

$$A = \frac{1.33 \times 10^7 \text{ cu ft}}{(0.2)(130 \text{ ft})} \times \frac{1 \text{ acre}}{(43{,}560 \text{ sq ft})}$$

$$= 11.7 \text{ acres.}$$

Using Eq. 3.36,

$$C_A = \frac{(5.456)(-22)}{(-0.8)} \exp\left[(2.303)(954 - 940)/(-22)\right]$$

$$= 150.0\, e^{-1.466} = 34.6.$$

In Table C.1, $C_A = 34.6$ corresponds most closely to a well in the center of a circle, square, or hexagon.

For a circle, $C_A = 31.62$.

For a square, $C_A = 30.88$.

For a hexagon, $C_A = 31.6$.

For verification, use Eq. 3.37:

$$(t_{DA})_{pss} = (0.1833)\, \frac{(-0.8)}{(-22)}\, (11) = 0.07.$$

This agrees well with $(t_{DA})_{pss} = 0.1$ for the three shapes.

3.6 Factors Complicating Drawdown Testing

Although a properly run drawdown test yields considerable information about the reservoir, the test may be hard to control since it is a flowing test. If a constant rate cannot be maintained within a reasonable tolerance, the analysis techniques of Chapter 4 should be used. Those techniques also should be used if the well was not shut in long enough to reach static reservoir pressure before the drawdown starts.

The early part of drawdown data is influenced by wellbore storage. Sometimes it is possible to draw a straight line through the semilog plot of data taken during this time. The slope of that line gives incorrect values of permeability and skin. As discussed in Sections 2.6 and 3.2, a log-log data plot of the drawdown data must be made to select the correct semilog straight line.

References

1. Matthews, C. S. and Russell, D. G.: *Pressure Buildup and Flow Tests in Wells,* Monograph Series, Society of Petroleum Engineers of AIME, Dallas (1967) **1,** Chap. 5.
2. Papadopulos, Istavros S. and Cooper, Hilton H., Jr.: "Drawdown in a Well of Large Diameter," *Water Resources Res.* (1967) **3,** No. 1, 241-244.
3. Cooper, Hilton H., Jr., Bredehoeft, John D., and Papadopulos, Istavros S.: "Response of a Finite-Diameter Well to an Instantaneous Charge of Water," *Water Resources Res.* (1967) **3,** No. 1, 263-269.
4. Ramey, H. J., Jr.: "Short-Time Well Test Data Interpretation in the Presence of Skin Effect and Wellbore Storage," *J. Pet. Tech.* (Jan. 1970) 97-104; *Trans.,* AIME, **249.**
5. Agarwal, Ram G., Al-Hussainy, Rafi, and Ramey, H. J., Jr.: "An Investigation of Wellbore Storage and Skin Effect in Unsteady Liquid Flow: I. Analytical Treatment," *Soc. Pet. Eng. J.* (Sept. 1970) 279-290; *Trans.,* AIME, **249.**
6. Wattenbarger, Robert A. and Ramey, H. J., Jr.: "An Investigation of Wellbore Storage and Skin Effect in Unsteady Liquid Flow: II. Finite Difference Treatment," *Soc. Pet. Eng. J.* (Sept. 1970) 291-297; *Trans.,* AIME, **249.**
7. McKinley, R. M.: "Wellbore Transmissibility From Afterflow-Dominated Pressure Buildup Data," *J. Pet. Tech.* (July 1971) 863-872; *Trans.,* AIME, **251.**
8. Earlougher, Robert C., Jr., and Kersch, Keith M.: "Analysis of Short-Time Transient Test Data by Type-Curve Matching," *J. Pet. Tech.* (July 1974) 793-800; *Trans.,* AIME, **257.**
9. Gringarten, Alain C., Ramey, Henry J., Jr., and Raghavan, R.: "Pressure Analysis for Fractured Wells," paper SPE 4051 presented at the SPE-AIME 47th Annual Fall Meeting, San Antonio, Tex., Oct. 8-11, 1972.
10. Slider, H. C.: "A Simplified Method of Pressure Buildup Analysis for a Stabilized Well," *J. Pet. Tech.* (Sept. 1971) 1155-1160; *Trans.,* AIME, **251.**
11. Slider, H. C.: "Application of Pseudo-Steady-State Flow to Pressure-Buildup Analysis," paper SPE 1403 presented at the SPE-AIME Regional Symposium, Amarillo, Tex., Oct. 27-28, 1966.
12. Jones, Park: "Reservoir Limit Test," *Oil and Gas J.* (June 18, 1956) 184-196.
13. Jones, Park: "Drawdown Exploration Reservoir Limit, Well and Formation Evaluation," paper 824-G presented at the SPE-AIME Permian Basin Oil Recovery Conference, Midland, Tex., April 18-19, 1957.
14. Jones, L. G.: "Reservoir Reserve Tests," *J. Pet. Tech.* (March 1963) 333-337; *Trans.,* AIME, **228.** Also *Reprint Series, No. 9—Pressure Analysis Methods,* Society of Petroleum Engineers of AIME, Dallas (1967) 126-130.
15. Earlougher, R. C., Jr.: "Estimating Drainage Shapes From Reservoir Limit Tests," *J. Pet. Tech.* (Oct. 1971) 1266-1268; *Trans.,* AIME, **251.**

Chapter 4

Multiple-Rate Testing

4.1 Introduction

The drawdown testing and analysis methods in Chapter 3 require a constant flow rate; however, it is often impractical or impossible to maintain a constant rate long enough to complete a drawdown test. In such a situation, multiple- (variable) rate testing and analysis techniques are applicable. A multiple-rate test may range from one with an uncontrolled, variable rate,[1,2] to one with a series of constant rates,[3,4] to testing at constant bottom-hole pressure with a continuously changing flow rate.[5] Pressure buildup testing[1] (Chapter 5) is a special kind of multiple-rate well test. Almost any flow-rate change can be analyzed as a well test by using the concepts presented in this chapter.

Accurate flow rate and pressure measurements are essential for the successful analysis of any transient well test. Rate measurements are much more critical in multiple-rate testing, however, than in conventional, constant-rate well tests. Without good flow-rate data, good analysis of multiple-rate tests is impossible.

Multiple-rate testing has the advantage of providing transient test data while production continues. It tends to minimize changes in wellbore storage coefficient and phase segregation (humping) effects and, thus, may provide good results when drawdown or buildup testing would not.

4.2 A General Multiple-Rate Test Analysis Technique

Fig. 4.1 schematically shows a variable production-rate schedule. Although flow rate may change continuously, it is treated as a series of discrete constant rates for analysis purposes. The step-wise approximation improves as the time intervals become smaller. Section B.7 presents the derivation of a general equation for pressure behavior caused by a variable flow rate. The approach presented here requires that the log approximation to the line source (Eq. 2.5b) applies. Then,

$$\frac{p_i - p_{wf}}{q_N} = m' \sum_{j=1}^{N} \left[\frac{(q_j - q_{j-1})}{q_N} \log(t - t_{j-1})\right] + b'. \quad (4.1)$$

Eq. 4.1 is the equation of a straight line with slope

$$m' = \frac{162.6 B\mu}{kh}, \quad (4.2)$$

and intercept

$$b' = m'\left[\log\left(\frac{k}{\phi\mu c_t r_w^2}\right) - 3.2275 + 0.86859\, s\right]. \quad (4.3)$$

Multiple rate transient test data should appear as a straight line when plotted as

$$\frac{p_i - p_{wf}}{q_N} \text{ vs } \sum_{j=1}^{N} \left[\frac{(q_j - q_{j-1})}{q_N} \log(t - t_{j-1})\right].$$

To make that plot correctly, it is important to understand that the rate corresponding to each plotted pressure point is q_N — the last rate that can affect that pressure. As time increases, the number of rates may increase and the last rate may change; but each pressure point is identified with the rate occurring when that pressure was measured. There may be several pressure points associated with a given rate.

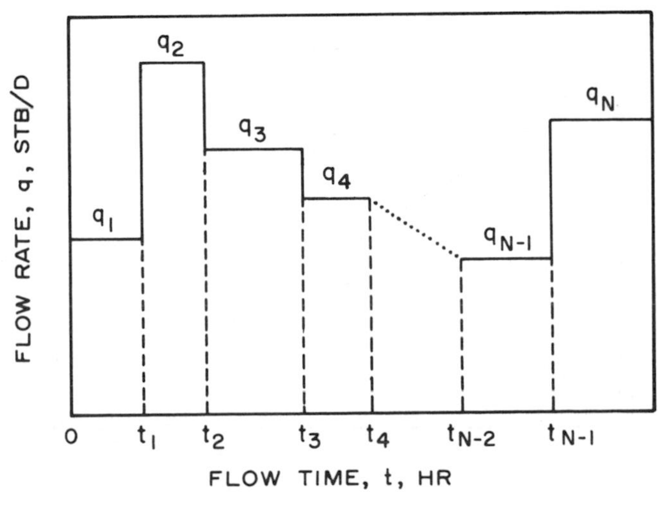

Fig. 4.1 Schematic representation of a variable production-rate schedule.

Examples 4.1 and 4.3 illustrate how the summation term in this plotting technique is calculated.

Once the data plot is made, the straight-line slope and intercept are measured. Permeability and skin factor are estimated from the slope and intercept data using Eqs. 4.2 and 4.3, rewritten as

$$k=\frac{162.6\,B\mu}{m'h}, \quad \text{(4.4)}$$

and

$$s=1.1513\left[\frac{b'}{m'}-\log\left(\frac{k}{\phi\mu c_t r_w^{\,2}}\right)+3.2275\right]. \quad \text{(4.5)}$$

The analysis procedure is direct and simple, but the computations required to make the data plot can be tedious. The analysis has the disadvantage that the initial reservoir pressure, p_i, and the entire flow-rate history must be known; frequently, they are not. As discussed in Section 4.5, the analysis technique may be modified in some situations so that p_i is not used. If the pressure is constant during a test and the rate declines, Eqs. 4.1 through 4.5 generally are not used; instead, the techniques in Section 4.6 are preferred.

When flow-rate variation is a result of wellbore storage, a simplified plotting method, which does not require use of superposition, may be used.[6-8] In this case, one plots $(p_i - p_{wf})/q_{sf}$ vs log t. The result should be a straight line with slope m' given by Eq. 4.2 and intercept b' given by Eq. 4.3. Permeability is estimated from Eq. 4.4 and skin factor is estimated from Eq. 4.5. Ramey[8] points out that the skin factor so calculated may be low by about 0.4. We do not recommend using this analysis technique for variable-rate tests unless the variable rate results only from wellbore storage, in which case the surface rate is constant.

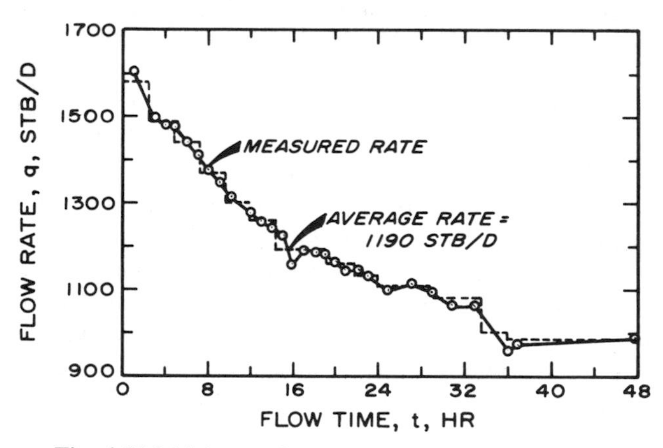

Fig. 4.2 Multiple-rate drawdown test rate history and its approximation. Data for Example 4.1.

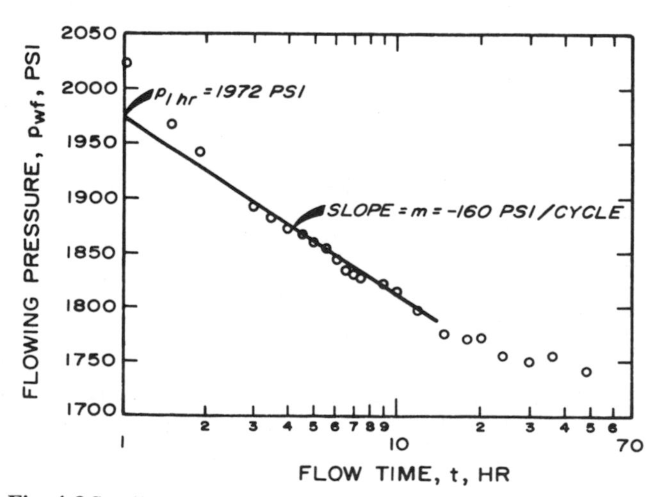

Fig. 4.3 Semilog plot of bottom-hole pressure for the multiple-rate drawdown test of Example 4.1.

Example 4.1 Multiple-Rate Drawdown Test Analysis

Production rate during a 48-hour drawdown test declined from 1,580 to 983 STB/D (Fig. 4.2). Rate and pressure data appear in Table 4.1. Reservoir data are

$p_i = 2{,}906$ psi $\quad \mu = 0.6$ cp
$B = 1.27$ RB/STB $\quad h = 40$ ft.

Fig. 4.3 shows flowing bottom-hole pressure vs log of flow time. That plot, which is normally used for a constant-rate drawdown test, neglects rate variations. Nevertheless, we use the straight line from 3 to 12 hours to estimate permeability. Using an average rate of 1,450 STB/D for the *first 12 hours* of the test, a slope $m = -160$ psi/cycle, and Eq. 3.9,

$$k=-\frac{(162.6)(1{,}450)(1.27)(0.6)}{(-160)(40)}=28.1 \text{ md}.$$

Fig. 4.4 is a plot of $(p_i - p_{wf})/q_N$ vs

$$\frac{1}{q_N}\sum_{j=1}^{N}(q_j-q_{j-1})\log(t-t_{j-1})$$

for this test, using the rate breakdown shown in Table 4.1. Table 4.1 summarizes the calculations of the quantities plotted in Fig. 4.4.

To illustrate the method of computing the time summation, we calculate it at 6.05 and 12.0 hours. At 6.05 hours, $q = 1{,}440$ STB/D is the third rate observed (although the point is the ninth pressure point), so $N = 3$. Computing the summation term,

$$\frac{1}{q_N}\sum_{j=1}^{N}(q_j-q_{j-1})\log(t-t_{j-1})$$

$$=\frac{1}{1{,}440}\{[(1{,}580-0)\log(6.05-0)]_{j=1}$$
$$+[(1{,}490-1{,}580)\log(6.05-2.40)]_{j=2}$$
$$+[(1{,}440-1{,}490)\log(6.05-4.80)]_{j=3}\}$$

$$=\frac{1}{1{,}440}\{[1{,}580\log(6.05)]_{j=1}$$
$$+[-90\log(3.65)]_{j=2}$$
$$+[-50\log(1.25)]_{j=3}\}$$

$$=\frac{1}{1{,}440}\{1{,}235.17-50.61-4.85\}=0.819.$$

Thus, the point for 6.05 hours plots at coordinates (0.819, 0.738) in Fig. 4.4. At 12.0 hours, $q = 1{,}300$ STB/D and $N = 5$. Thus,

$$\frac{1}{q_N}\sum_{j=1}^{N}(q_j - q_{j-1})\log(t - t_{j-1})$$

$$= \frac{1}{1{,}300}\Big[(1{,}580 - 0)\log(12.0 - 0)$$

$$+ (1{,}490 - 1{,}580)\log(12.0 - 2.40)$$
$$+ (1{,}440 - 1{,}490)\log(12.0 - 4.80)$$
$$+ (1{,}370 - 1{,}440)\log(12.0 - 7.20)$$
$$+ (1{,}300 - 1{,}370)\log(12.0 - 9.60)\Big]$$

$$= \frac{1}{1{,}300}\Big[1{,}705.11 - 88.40 - 42.87 - 47.69$$
$$- 26.61\Big]$$
$$= 1.154.$$

The point at 12.0 hours plots at coordinates (1.154, 0.853) in Fig. 4.4.

Two straight lines can be drawn through the data of Fig. 4.4. The slope of the second line is greater than that of the first, possibly indicating transition to pseudosteady state, faulting, or a decrease in permeability away from the well (see Sections 10.2 and 10.4). The incorrect semilog data plot, Fig. 4.3, has a *reduction* in slope for $t > 12$ hours that might be interpreted as increasing permeability away from the well. That is an incorrect conclusion, however, since the slope change in Fig. 4.3 is caused by the declining production rate.

Using the slope of the first straight line in Fig. 4.4 and Eq. 4.4,

$$k = \frac{(162.6)(1.27)(0.6)}{(0.227)(40)} = 13.6 \text{ md}.$$

Thus, the permeability computed from Fig. 4.3 is about 107 percent too high.

TABLE 4.1—VARIABLE FLOW RATE DRAWDOWN DATA FOR EXAMPLE 4.1.

Time, t (hours)	Rate, q (STB/D)	N	p_{wf} (psi)	$p_i - p_{wf}$ (psi)	$\frac{p_i - p_{wf}}{q_N}$ (psi/STB/D)	Σ
1.00	1,580	1	2,023	883	0.5589	0.000
1.50	1,580	1	1,968	938	0.5937	0.176
1.89	1,580	1	1,941	965	0.6108	0.277
2.40	1,580	1	—	—	—	—
3.00	1,490	2	1,892	1,014	0.6805	0.519
3.45	1,490	2	1,882	1,024	0.6872	0.569
3.98	1,490	2	1,873	1,033	0.6933	0.624
4.50	1,490	2	1,867	1,039	0.6973	0.673
4.80	1,490	2	—	—	—	—
5.50	1,440	3	1,853	1,053	0.7313	0.787
6.05	1,440	3	1,843	1,063	0.7382	0.819
6.55	1,440	3	1,834	1,072	0.7444	0.849
7.00	1,440	3	1,830	1,076	0.7472	0.874
7.20	1,440	3	—	—	—	—
7.50	1,370	4	1,827	1,079	0.7876	0.974
8.95	1,370	4	1,821	1,085	0.7920	1.009
9.6	1,370	4	—	—	—	—
10.0	1,300	5	1,815	1,091	0.8392	1.124
12.0	1,300	5	1,797	1,109	0.8531	1.154
14.4	1,260	6	—	—	—	—
15.0	1,190	7	1,775	1,131	0.9504	1.337
18.0	1,190	7	1,771	1,135	0.9538	1.355
19.2	1,190	7	—	—	—	—
20.0	1,160	8	1,772	1,134	0.9776	1.423
21.6	1,160	8	—	—	—	—
24.0	1,137	9	1,756	1,150	1.0114	1.485
28.8	1,106	10	—	—	—	—
30.0	1,080	11	1,751	1,155	1.0694	1.607
33.6	1,080	11	—	—	—	—
36.0	1,000	12	—	—	—	—
36.2	983	13	1,756	1,150	1.1699	1.788
48.0	983	13	1,743	1,163	1.1831	1.800

4.3 Two-Rate Testing

When a multiple-rate test consists of only two flow rates, both testing and analysis are simplified. The two-rate test provides information about k and s while production continues. Wellbore storage effects are often thought to be minimized or eliminated by two-rate tests. In fact, wellbore storage effects last just about the same amount of time in a

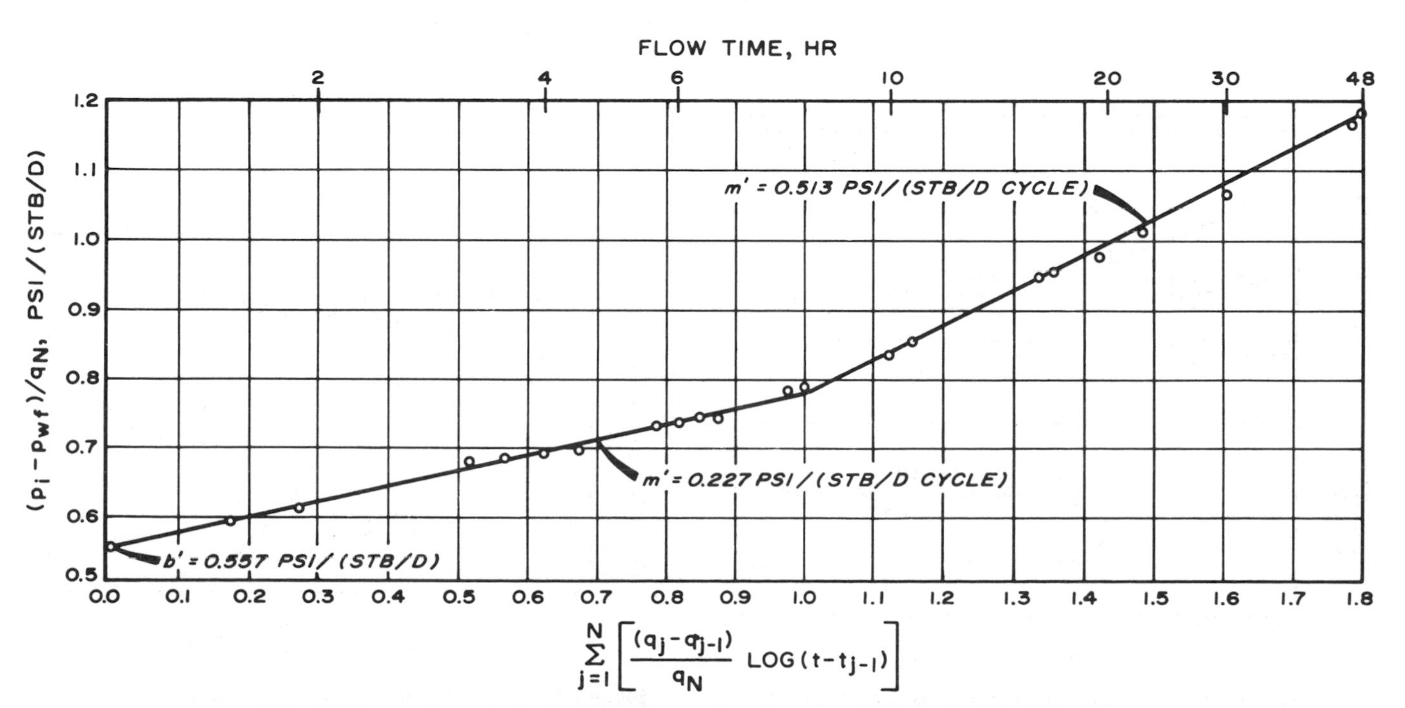

Fig. 4.4 Correct data plot for the multiple-rate drawdown test of Example 4.1; t is in hours.

two-rate test as in a normal buildup, drawdown, falloff, etc., test. However, a two-rate test often can be used to prevent a wellbore storage increase, thus providing an analyzable test when one otherwise might not be possible (see Section 11.2). The main advantage of a two-rate test over a buildup test is that deferred production is minimized.

Fig. 4.5 schematically illustrates the rate and pressure behavior for a two-rate flow test;[3] either a decreasing or increasing rate sequence may be used. Eq. 4.1 may be modified to the form presented by Russell[3] for a two-rate test:

$$p_{wf} = m_1' \left[\log \left(\frac{t_1 + \Delta t}{\Delta t} \right) + \frac{q_2}{q_1} \log \Delta t \right] + p_{\text{int}} . \qquad (4.6)$$

Eq. 4.6 assumes a constant flow rate, q_1, from time 0 to time t_1, at the start of the test. If, instead, the well was stabilized at rate q_1, then Eq. 4.6 is still a good approximation if t_1 is calculated from

$$t_1 = 24 \frac{(V_P)}{q_1} , \qquad (4.7)$$

where V_P is the cumulative volume produced since the last rate stabilization.

Eq. 4.6 implies that a graph of p_{wf} vs

$$\left[\log \left(\frac{t_1 + \Delta t}{\Delta t} \right) + \frac{q_2}{q_1} \log \Delta t \right]$$

should be a straight line with slope

$$m_1' = - \frac{162.6 q_1 B \mu}{kh} , \qquad (4.8)$$

and intercept

$$p_{\text{int}} = p_i + m_1' \frac{q_2}{q_1} \left[\log \left(\frac{k}{\phi \mu c_t r_w^2} \right) - 3.2275 + 0.86859 s \right] . \qquad (4.9)$$

Fig. 4.6 schematically shows such a data plot. Because of the choice of the abscissa variable, time increases from right to left. At long times, the data deviate from the straight line as a result of boundary and interference effects. At short times they have not yet reached the straight line because of rate restabilization and wellbore storage effects. Field test results indicate that rate restabilization is faster for a rate reduction than for a rate increase.[1,3]

Once the slope of the straight line is determined from the data plot, reservoir permeability may be estimated from

$$k = \frac{-162.6 q_1 B \mu}{m_1' h} . \qquad (4.10)$$

The skin factor is estimated from

$$s = 1.1513 \left[\frac{q_1}{q_1 - q_2} \left(\frac{p_{wf}(\Delta t{=}0) - p_{1\text{hr}}}{m_1'} \right) - \log \left(\frac{k}{\phi \mu c_t r_w^2} \right) + 3.2275 \right] . \qquad (4.11)$$

The intercept of the data plot may be used to estimate the false pressure,[3]

$$p^* = p_{\text{int}} - \frac{q_2}{q_1 - q_2} \left[p_{wf}(\Delta t{=}0) - p_{1\text{hr}} \right] , \qquad (4.12)$$

which is used to estimate average reservoir pressure using methods in Chapter 6.

Example 4.2 Two-Rate Flow Test Analysis[3]

Well A is a flowing producer in a low-permeability limestone reservoir in the Permian Basin. Pressure buildup tests in this reservoir usually do not provide interpretable data because of long, low-rate afterflow periods.[3]

A two-rate flow test was run by stabilizing the flow rate at 107 STB/D for several days and then reducing the flow rate to 46 STB/D. The pressure data during the second rate are shown in Fig. 4.7. Other pertinent data are

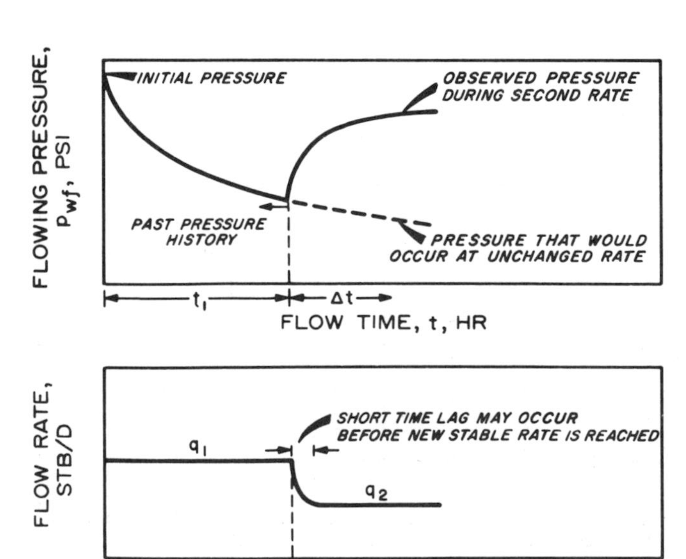

Fig. 4.5 Schematic rate and pressure history for a two-rate flow test, $q_1 > q_2$.

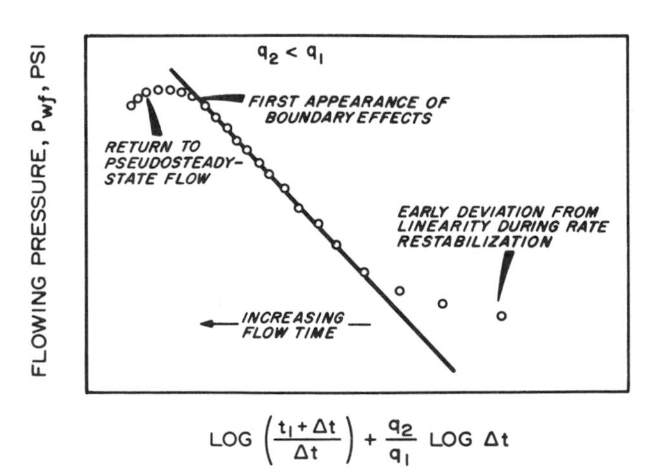

Fig. 4.6 Schematic data plot for a two-rate flow test, $q_1 > q_2$. After Russell.[3]

$$B = 1.5 \text{ RB/STB}$$
$$p_{wf}(\Delta t{=}0) = 3{,}118 \text{ psi}$$
$$h = 59 \text{ ft}$$
$$c_t = 9.32 \times 10^{-5} \text{ psi}^{-1}$$
$$V_p = 26{,}400 \text{ STB}$$
$$t_1 = (26{,}400)(24)/107 = 5{,}921 \text{ hours (Eq. 4.7)}$$
$$\mu = 0.6 \text{ cp}$$
$$\phi = 0.06$$
$$r_w = 0.2 \text{ ft}$$
$$m_1' = -90 \text{ psi/cycle (Fig. 4.7)}$$
$$p_{1hr} = 3{,}169 \text{ psi (Fig. 4.7)}$$
$$p_{int} = 3{,}510 \text{ psi (Fig. 4.7).}$$

Permeability is estimated from Eq. 4.10:

$$k = -\frac{(162.6)(107)(1.5)(0.6)}{(-90)(59)} = 2.9 \text{ md.}$$

The skin factor is estimated from Eq. 4.11:

$$s = 1.1513\left[\left(\frac{107}{107-46}\right)\left(\frac{3{,}118-3{,}169}{-90}\right) - \log\left(\frac{2.9}{(0.06)(0.6)(9.32\times10^{-5})(0.2)^2}\right) + 3.2275\right]$$
$$= -3.6.$$

The false pressure, p^*, is estimated from Eq. 4.12:

$$p^* = 3{,}510 - \frac{46\,(3{,}118-3{,}169)}{(107-46)}$$
$$= 3{,}548 \text{ psi.}$$

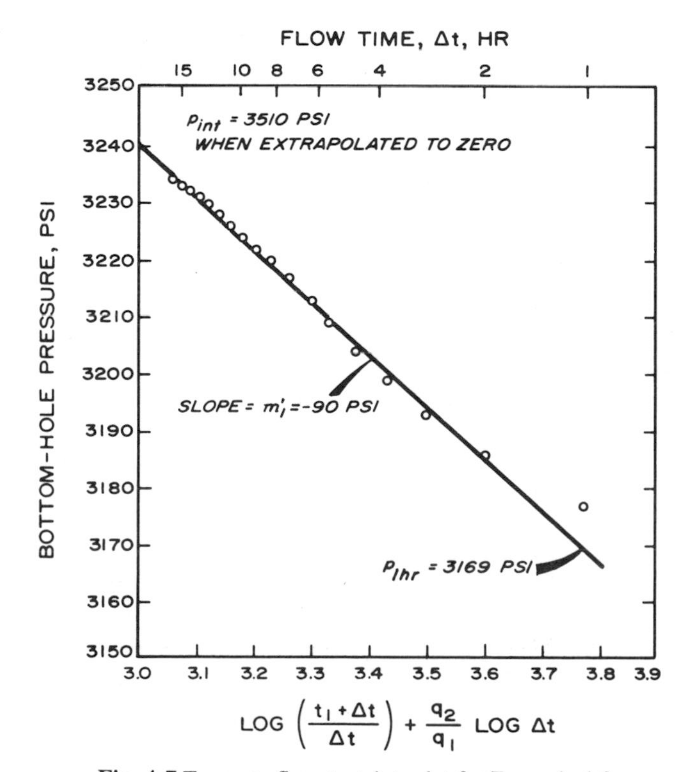

Fig. 4.7 Two-rate flow-test data plot for Example 4.2. After Russell.[3]

The p^* value may be used with material in Chapter 6 to estimate average drainage-region pressure.

In certain circumstances, the plotting technique suggested by Eq. 4.6 may be simplified.[9] When $t_1 >> \Delta t$, $\log(t_1 + \Delta t) \simeq \log t_1$ and $\log[(t_1 + \Delta t)/\Delta t] \simeq \log t_1 - \log \Delta t$. Making these approximations in Eq. 4.6 and rearranging gives

$$p_{wf} = m'' \log \Delta t + p_{int} \; . \qquad (4.13)$$

Thus, for a two-rate test with $t_1 >> \Delta t$, a plot of p_{wf} vs $\log \Delta t$ should be a straight line with slope

$$m'' = \frac{-162.6\,(q_2 - q_1)\,B\mu}{kh}, \qquad (4.14)$$

and intercept

$$p_{int} = p_i + \frac{m'' q_2}{(q_2 - q_1)}\left[\log\left(\frac{k}{\phi\mu c_t r_w^2}\right) - 3.2275 + 0.86859\,s + \frac{q_1}{q_2}\log t_1\right]. \qquad (4.15)$$

The p_{wf} vs $\log \Delta t$ data plot may be used to estimate reservoir permeability from

$$k = -\frac{162.6\,(q_2 - q_1)\,B\mu}{m''h}, \qquad (4.16)$$

and to estimate skin factor from[9]

$$s = 1.1513\left[\frac{p_{1hr} - p_{wf}(\Delta t = 0)}{m''} - \log\left(\frac{k}{\phi\mu c_t r_w^2}\right) + 3.2275\right]. \qquad (4.17)$$

The analysis suggested by Eq. 4.13 is much faster and simpler than that suggested by Eq. 4.6. However, the approximations in Eq. 4.13 cause errors in the results. The error in the permeability is[10]

$$E_k = \frac{k_E - k_{actual}}{k_{actual}}$$
$$= \frac{q_1}{q_1(T^* - 1) - q_2 T^*}, \qquad (4.18)$$

while the error in skin factor is

$$E_s = s_E - s_{actual}$$
$$= \frac{-1.1513\,[p_{1hr} - p_{wf}(\Delta t = 0)]}{m'}\left(\frac{q_1}{q_1 - q_2}\right)\frac{1}{T^*}. \qquad (4.19)$$

In Eqs. 4.18 and 4.19,

$$T^* = \frac{\log \Delta t}{\log\left(1 + \dfrac{\Delta t}{t_1}\right)}. \qquad (4.20)$$

If the second rate of a two-rate test varies significantly, an adaptation of the multiple-rate analysis technique should be used. Odeh and Jones[11] indicate that pressure behavior for a two-rate test with a varying second rate is described by

$$\frac{p_{wf}(\Delta t = 0) - p_{wf}(\Delta t)}{q_N - q_0} = m' \sum_{j=1}^{N} \left[\frac{q_j - q_{j-1}}{q_N - q_0} \times \log(\Delta t - \Delta t_{j-1}) \right] + b', \quad \ldots\ldots (4.21)$$

where Δt is the time from the start of the varying second flow rate. This equation is much like Eq. 4.1 [if $q_0 = 0$ and $p_{wf}(\Delta t = 0) = p_i$, it is identical to Eq. 4.1]; the slope, m', and intercept, b', are given by Eqs. 4.2 and 4.3, respectively. If the data are plotted as the quantity on the left side of Eq. 4.21 vs the summation on the right side, permeability and skin factor may be estimated from Eqs. 4.4 and 4.5. The analysis technique suggested by Eq. 4.21 also applies to variable-rate test analysis if the well has been stabilized for a substantial time at rate q_0.

If pseudosteady state is achieved in the first flow period of a two-rate flow test, the analysis techniques of Section 4.5, Eqs. 4.30 through 4.32, should be used.

Example 4.3 Two-Rate Flow Test, Variable-Rate Case[11]

The data in Table 4.2 are for Well X of Odeh and Jones.[11] Well X was stabilized for several days at $q_0 = 1{,}103$ STB/D. Other data are

$B = 1.0$ RB/STB $h = 18$ ft
$\mu = 1.0$ cp $A = 28.3 \times 10^6$ sq ft
$r_w = 0.26$ ft $\simeq 650$ acres
$c_t = 1.4 \times 10^{-5}$ psi^{-1} $\phi = 11$ percent.

Parameters to be plotted are calculated in Table 4.2. The last value in the right-hand column is calculated as follows. (In this case, $N = 3$ since there are three rates, but five pressure data points.)

$$\sum_{j=1}^{3} \frac{q_j - q_{j-1}}{q_N - q_0} \log(\Delta t - \Delta t_{j-1})$$

$$= \left[\frac{-303}{-223} \log(10 - 0)\right]_{j=1} + \left[\frac{40}{-223} \log(10 - 4)\right]_{j=2} + \left[\frac{40}{-223} \log(10 - 8)\right]_{j=3}$$

$$= [(1.3587)(1)] + [(-0.1794)(0.7782)] + [(-0.1794)(0.3010)]$$

$$= 1.165.$$

The data plot (Fig. 4.8) indicates the slope, m', is 0.148 psi/(STB/D cycle) and the extrapolated intercept, b', is -0.00991 psi/(STB/D). Using Eq. 4.4,

$$k = \frac{(162.6)(1)(1)}{(0.148)(18)} = 61.0 \text{ md},$$

and

$$kh = 1{,}100 \text{ md ft}.$$

Using Eq. 4.5,

$$s = 1.1513 \left[\frac{-0.00991}{0.148} - \log\left(\frac{61.0}{(0.11)(1.0)(1.4 \times 10^{-5})(0.26)^2}\right) + 3.2275\right]$$

$$= -6.5.$$

Odeh and Jones[11] state that pressure drawdown and buildup tests on the well indicated $kh \simeq 1{,}100$ md ft and $s \simeq -7$. They also state that they used the simplified two-rate flow-test analysis method (Eq. 4.13) with an average rate of 832 STB/D. The result was $kh \simeq 1{,}360$ md ft and $s \simeq -6.3$. Thus, the 10-percent rate variation *may have* resulted in a 23-percent error in the value calculated for kh.

Although the production time at the stabilized rate is not known for this example, the range of errors caused by using the simplified two-rate analysis may be estimated by assuming values of the stabilized production time. (This requires assuming that analysis by a two-rate technique with a constant second rate is adequate for the data of this example.) From Eq. 4.20, for $t_1 = 10$ days, and for the maximum test time of $\Delta t = 10$ hours,

$$T^* = \frac{\log 10}{\log\left(1 + \frac{10}{(10)(24)}\right)} = 56.4.$$

The error in kh is estimated from Eq. 4.18 with thicknesses included in the numerator and denominator:

$$E_{kh} = \frac{1{,}103}{1{,}103\,(56.4 - 1) - 832\,(56.4)} = 7.8 \text{ percent}.$$

Thus, if t_1 is indeed 10 days or more, the 23-percent error in kh when using the simplified two-rate analysis method is largely caused by the rate variation, not by the inherent errors in the simplified two-rate analysis technique. How-

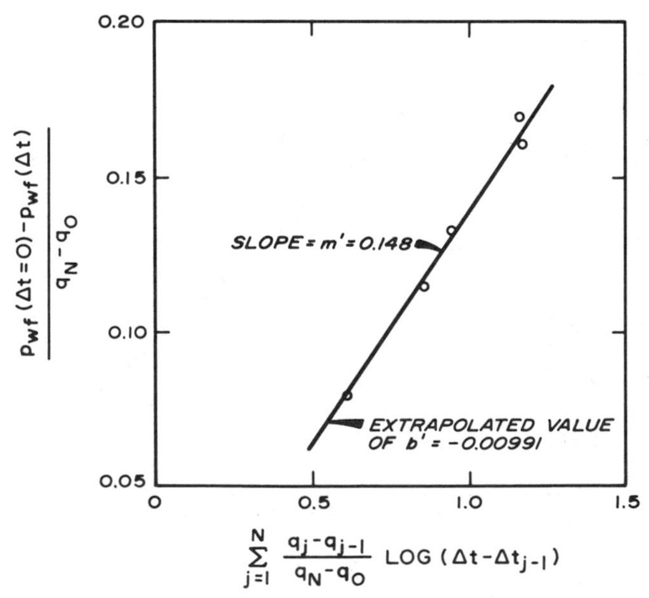

Fig. 4.8 Two-rate flow test, variable-rate case, data plot for Example 4.3. After Odeh and Jones.[11]

TABLE 4.2—SUMMARY OF CALCULATIONS FOR EXAMPLE 4.3.
After Odeh and Jones.[11]

N	j	Δt (hours)	q_j (STB/D)	$\Delta q_N = q_N - q_0$ (STB/D)	$q_j - q_{j-1}$ (STB/D)	$p_{wf}(\Delta t)$ (psi)	$\Delta p = p_{wf}(\Delta t{=}0) - p_{wf}(\Delta t)$ (psi)	$\dfrac{\Delta p}{\Delta q_N}$ (psi/STB)	$\sum_{j=1}^{N} \dfrac{q_j - q_{j-1}}{q_N - q_0} \log(\Delta t - \Delta t_{j-1})$
0	0	0	1,103	0	—	3,630	0	—	—
1	1	4	800	−303	−303	3,654	−24	0.0792	0.602
		6	840	−263	40	3,660	−30	0.114	0.851
2	2	8	840	−263	40	3,665	−35	0.133	0.949
		9	880	−223	40	3,666	−36	0.161	1.171
3	3	10	880	−223	40	3,668	−38	0.170	1.165

ever, the following table indicates that the error could be entirely caused by the analysis technique rather than the variable rate — if production time at the first rate is less than about 4 days.

Production Time at First Rate (days)	Error in Two-Rate Analysis (percent)
1	160
2	50
3	30
4	21
5	16
6	13
8	10
10	8

4.4 Drawdown Testing After a Short Shut-In

It is common practice to run a drawdown test after a shut-in period (pressure buildup test). If the shut-in is too short for the well pressure to stabilize at average reservoir pressure, the drawdown-test analysis techniques of Sections 3.2 and 3.4 should not be used; instead, a multiple-rate-type analysis is applicable. Fig. 4.9 schematically shows a rate history for a drawdown test after a short shut-in. Writing Eq. 4.1 for such a test and rearranging gives

$$p_{wf} = m_3' \left[\frac{q_1}{q_3} \log\left(\frac{t_1 + \Delta t_{si} + \Delta t}{\Delta t_{si} + \Delta t} \right) + \log \Delta t \right] + p_{\text{int}} . \quad \ldots (4.22)$$

Thus, a plot of p_{wf} vs $\{(q_1/q_3) \log[(t_1 + \Delta t_{si} + \Delta t)/(\Delta t_{si} + \Delta t)] + \log \Delta t\}$ should yield a straight line with slope

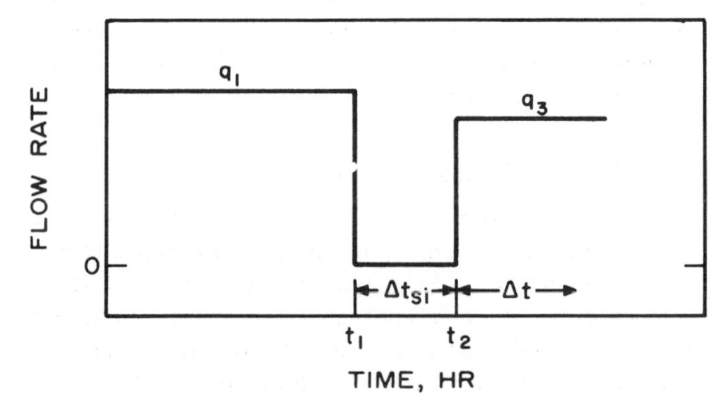

Fig. 4.9 Schematic rate history for a drawdown test after a short shut-in.

$$m_3' = \frac{-162.6\, q_3 B \mu}{kh} , \quad \ldots (4.23)$$

and intercept

$$p_{\text{int}} = p_i + m_3' \left[\log\left(\frac{k}{\phi \mu c_t r_w^2} \right) - 3.2275 + 0.86859\, s \right] . \quad \ldots (4.24)$$

Reservoir permeability can be estimated by solving Eq. 4.23. The equation is the same as the normal drawdown-test equation (Eq. 3.9), but the data plot is different. The skin factor may be estimated from the data plot with

$$s = 1.1513 \left[\frac{p_{\text{int}} - p_{wf}(\Delta t{=}0)}{m_3'} + \frac{q_1}{q_3} \log\left(\frac{t_1 + \Delta t_{si}}{\Delta t_{si}} \right) - \log\left(\frac{k}{\phi \mu c_t r_w^2} \right) + 3.2275 \right] . \quad \ldots (4.25)$$

Eq. 4.25 is similar to Eq. 3.10 for estimating skin factor from a normal drawdown test, with the exception of the additional logarithmic term. The skin factor computed from a drawdown test after a buildup test often does not agree with that computed from the buildup test because the correct data plot and the correct equation (Eq. 4.25) were not used in the analysis. That is particularly true in drillstem-test data analysis.

4.5 Developed Reservoir Effects

When bottom-hole pressure is declining as a result of withdrawals from the test well or from other wells in the reservoir, the analysis methods presented earlier in this chapter must be modified.[12,13] Such modified analysis techniques become increasingly important as the depletion rate increases and as test duration increases. The modifications presented in this section apply when pressure decline at the test well is caused by production from other wells or from the test well itself; the cause of the pressure decline is not important.

Fig. 4.10 illustrates bottom-hole pressure from a tested well in a developed reservoir. The solid line is the observed pressure behavior, while the dashed line represents the pressure that would have been observed had there been no flow-rate change at the well at time t_1. The pressure along the dashed line for the "no test" case is called p_{wext}. Using the approaches outlined by Slider[12,13] (similar to those in Section 3.4), Eq. 4.1 may be modified to

$$\frac{p_{wext}(t_1+\Delta t)-p_{wf}(t_1+\Delta t)}{q_N} = m' \sum_{j=1}^{N} \left[\frac{q_j - q_{j-1}}{q_N} \log(\Delta t - \Delta t_{j-1})\right] + b', \quad \ldots (4.26)$$

where m' and b' are given by Eqs. 4.2 and 4.3, respectively. The analysis is the same as outlined in Section 4.2 except that the pressure quantity used is $p_{wext} - p_{wf}$, the vertical distance between the solid and dashed curves in Fig. 4.10. If the test starts at time zero from p_i in a reservoir with no other operating wells, Eq. 4.26 is identical to Eq. 4.1. A similar equation can be used in place of Eq. 4.21 with p_{wext} replacing $p_{wf}(\Delta t{=}0)$.

For the two-rate test of Section 4.3, the applicable equation is

$$p_{wext}(t_1+\Delta t) - p_{wf}(t_1+\Delta t) = -m_1' \left[\log\left(\frac{t_1+\Delta t}{\Delta t}\right) + \frac{q_2}{q_1}\log \Delta t\right] - \Delta p_{int} . \quad \ldots (4.27)$$

In this case the data plot has slope m_1', given by Eq. 4.8, and an intercept

$$\Delta p_{int} = m_1' \frac{q_2}{q_1}\left[\log\left(\frac{k}{\phi\mu c_t r_w^2}\right) - 3.2275 + 0.86859\, s\right] . \quad \ldots (4.28)$$

Reservoir permeability is estimated from Eq. 4.10 and skin factor from

$$s = 1.1513\left[\frac{\Delta p_{int}}{m_1'}\frac{q_1}{q_2} - \log\left(\frac{k}{\phi\mu c_t r_w^2}\right) + 3.2275\right]. \quad \ldots (4.29)$$

If pseudosteady-state conditions exist during the first flow period in a two-rate flow test, then Eq. 4.27 becomes

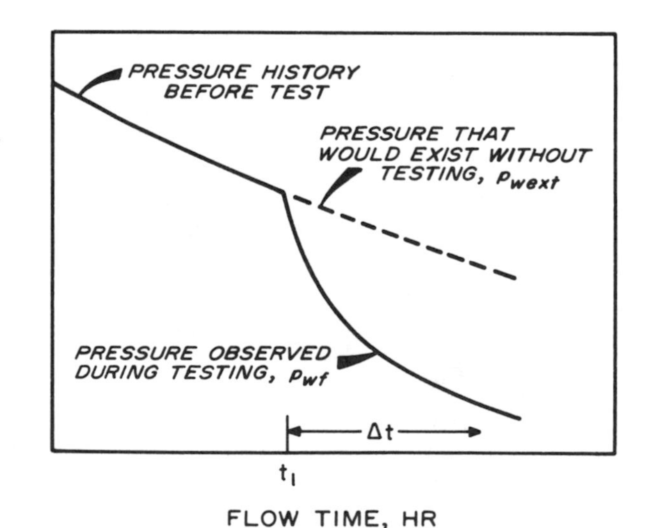

Fig. 4.10 Schematic pressure history for variable-rate testing in a developed reservoir.

$$p_{wext}(t_1+\Delta t) - p_{wf}(t_1+\Delta t) = \Delta p_{1hr} - m'' \log \Delta t. \quad \ldots (4.30)$$

This indicates that a plot of the extrapolated pressure difference vs log Δt should be a straight line with slope $= -m''$, where m'' is given by Eq. 4.14. The intercept is

$$\Delta p_{1hr} = -m''\left[\log\left(\frac{k}{\phi\mu c_t r_w^2}\right) - 3.2275 + 0.86859\, s\right] . \quad \ldots (4.31)$$

Permeability would be estimated from Eq. 4.16 and skin factor would be estimated from

$$s = 1.1513\left[\frac{-\Delta p_{1hr}}{m''} - \log\left(\frac{k}{\phi\mu c_t r_w^2}\right) + 3.2275\right]. \quad \ldots (4.32)$$

This analysis for pseudosteady-state flow during the first rate can be simplified even further, since

$$p_{wext}(t_1+\Delta t) = p_{wf}(\Delta t = 0) + m^* \Delta t, \quad \ldots (4.33)$$

where

$$m^* = \frac{dp_{wf}}{dt}, \quad \ldots (4.34)$$

is estimated from the slope of a Cartesian plot of p_{wf} vs t before the second rate starts. By using Eq. 4.33, Eq. 4.30 may be rewritten:

$$p_{wf}(t_1+\Delta t) - m^* \Delta t = \Delta p_{1hr} + m'' \log \Delta t, \quad \ldots (4.35)$$

where m^* is given by Eq. 4.34 and m'' is given by Eq. 4.14. A plot of $(p_{wf} - m^* \Delta t)$ vs log Δt would be a straight line with slope m'' and intercept

$$\Delta p_{1hr} = \left[p_{wf}(t_1+\Delta t) - m^* \Delta t\right]_{1hr} = p_{wf}(\Delta t = 0) + m''\left[\log\left(\frac{k}{\phi\mu c_t r_w^2}\right) - 3.2275 + 0.86859\, s\right] . \quad \ldots (4.36)$$

The skin factor may be estimated from Eq. 4.17 with Δp_{1hr} in place of p_{1hr}.

The main difference between equations in this section and in Sections 4.2 and 4.3 is that this section does not assume the system pressure is infinite-acting at the time the two-rate or variable-rate test *begins*. In particular, if pseudosteady-state conditions exist at the start of the test, the analysis technique can be expected to give results different from those of the techniques in Sections 4.2 and 4.3.

4.6 Constant-Pressure Flow Testing

The transient behavior of a well operating at constant sand-face pressure is analogous to that of a well operating at a constant flow rate. In a constant-pressure flow test, the well produces at a constant bottom-hole pressure and flow rate is recorded with time. Constant bottom-hole pressure test data are not influenced by wellbore storage. However, if the *surface* pressure is maintained constant, the frictional pressure drop in the flow tubing may act in a manner similar

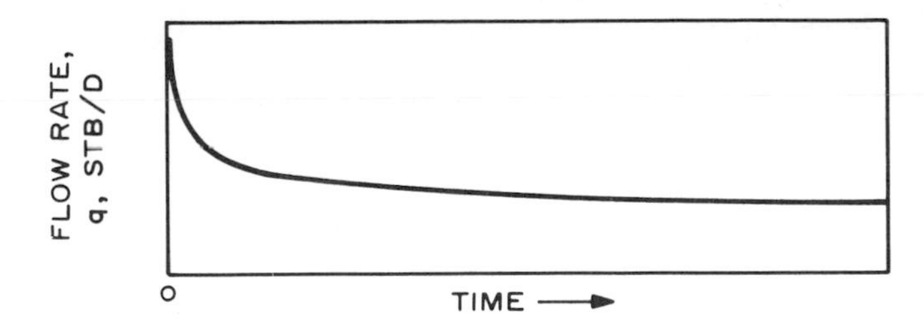

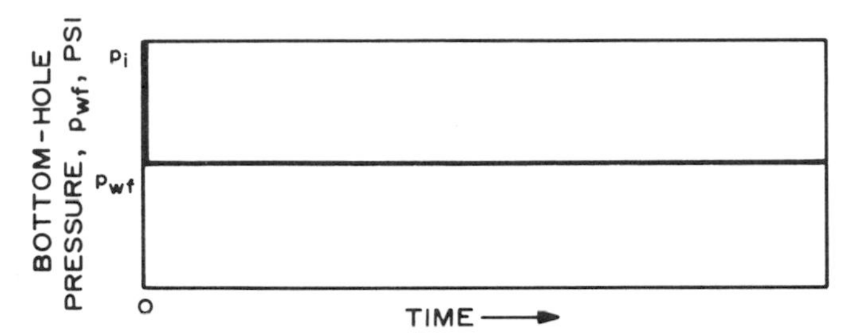

Fig. 4.11 Schematic representation of rate and pressure histories during a constant-pressure test.

to wellbore storage, causing bottom-hole pressure to vary during the test.

Fig. 4.11 schematically represents pressure and rate behavior in a constant-pressure drawdown test. Such a test is seldom performed since it is much easier to measure pressure accurately than it is to measure flow rate accurately. However, constant-rate tests may inadvertently become constant-pressure tests, so it is desirable to have a method for analyzing such tests.

In a manner similar to that used to express pressure as a function of flow rate and time, we may express flow rate as a function of pressure drop and time by

$$q = \frac{kh\,(p_i - p_{wf})}{141.2\,B\mu}\,q_D(t_D)\ . \quad \ldots \ldots \ldots \ldots \ldots (4.37)$$

Dimensionless time has its usual definition:

$$t_D = 0.0002637\ \frac{kt}{\phi\mu c_t r_w^{\,2}}\ . \quad \ldots \ldots \ldots \ldots \ldots (4.38)$$

The principle of superposition may be used with Eq. 4.37 to compute the flow rate resulting from a series of pressure drops during pressure-controlled flow. The calculations are analogous to those given in Section 2.9.

Fig. 4.12,* showing dimensionless rate as a function of dimensionless time for an infinite-acting system,[5] is useful for test analysis by type-curve matching. The technique is similar to that outlined in Section 3.3. Briefly, test data are plotted on tracing paper laid over the grid of Fig. 4.12, with flow rate on the ordinate and the time on the abscissa. The tracing-paper data plot is moved horizontally and vertically until the data match the curve in Fig. 4.12. Then, q_M and t_M are read from an arbitrary match point on the tracing paper and $(q_D)_M$ and $(t_D)_M$ are read from the corresponding point on Fig. 4.12. Reservoir permeability is estimated from

*See footnote on Page 24.

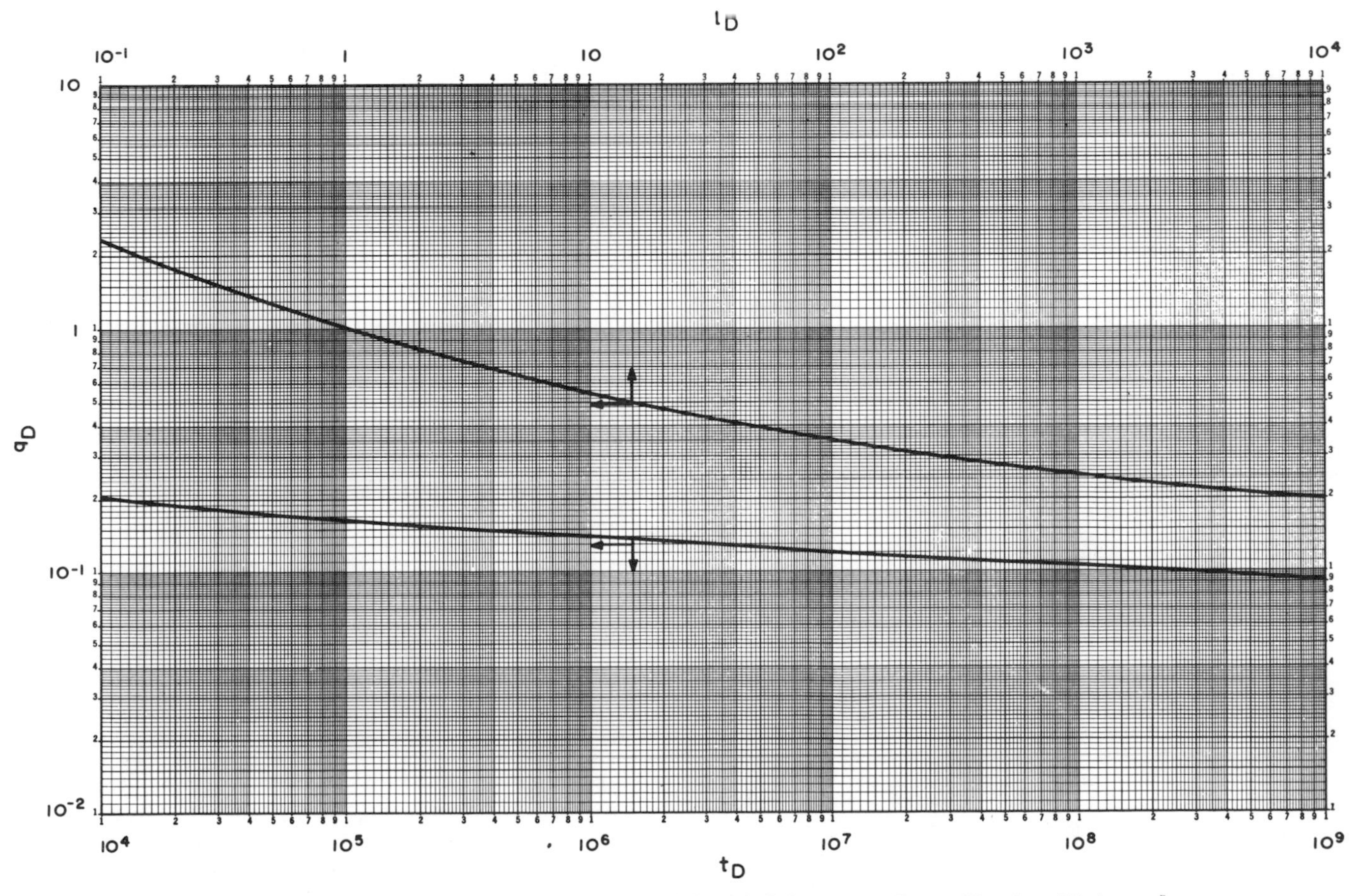

Fig. 4.12 Dimensionless rate for a single well in an infinite system. Data of Jacob and Lohman.[5]

$$k=\frac{141.2B\mu}{(p_i-p_{wf})h}\left[\frac{q_M}{(q_D)_M}\right], \quad (4.39)$$

and reservoir porosity-compressibility-thickness is estimated from

$$\phi c_t h=0.0002637\frac{kh}{\mu r_w^2}\left[\frac{t_M}{(t_D)_M}\right]. \quad (4.40)$$

The curve-matching process *assumes a zero skin factor*.

Jacob and Lohman[5] show that, at long times, q_D may be approximated by

$$q_D=\frac{2}{\ln(t_D)+0.80907} \quad (4.41)$$

Eq. 4.41, which applies only to infinite-acting systems, is correct within 0.1 percent for $t_D \geqslant 5\times10^{11}$. The error is only 1 percent when $t_D \geqslant 8\times10^4$ and is 2 percent when $t_D \geqslant 5\times10^3$. If Eq. 4.41 is used in Eq. 4.37, and if the skin factor is included in the pressure-drop calculation, the result is

$$\frac{1}{q}=m_q\log t+(1/q)_{1\text{hr}}. \quad (4.42)$$

This equation indicates that a graph of $(1/q)$ vs log t should be a straight line with slope

$$m_q=\frac{162.6B\mu}{kh(p_i-p_{wf})}, \quad (4.43)$$

and intercept (at t = 1 hour)

$$(1/q)_{1\text{hr}}=m_q\left[\log\left(\frac{k}{\phi\mu c_t r_w^2}\right)-3.2275+0.86859s\right]. \quad (4.44)$$

Permeability may be estimated from the slope of a $(1/q)$ vs log t data plot with

$$k=\frac{162.6B\mu}{m_q(p_i-p_{wf})h} \quad (4.45)$$

Skin factor may be estimated from a rearranged form of Eq. 4.44:

$$s=1.1513\left[\frac{(1/q)_{1\text{hr}}}{m_q}-\log\left(\frac{k}{\phi\mu c_t r_w^2}\right)+3.2275\right]. \quad (4.46)$$

Example 4.4 Constant-Pressure Testing in an Infinite-Acting Reservoir

The flow-rate data shown in Fig. 4.13 are from a simulated constant-pressure drawdown test. Data used to simulate the test are

$k=6.5$ md	$\phi c_t=2.05\times10^{-6}$ psi^{-1}
$\mu=1.35$ cp	$p_i-p_{wf}=1{,}000$ psi
$h=190$ ft	$r_w=1$ ft
$B=1.0$ RB/STB	$s=0$.

The data of Fig. 4.13 were type-curve matched to Fig. 4.12. Match-point data are

$q_M=1{,}720$ STB/D	$(q_D)_M=0.27$
$t_M=1.0$ hour	$(t_D)_M=600$.

Using Eq. 4.39,

$$k=\frac{(141.2)(1.0)(1{,}720)(1.35)}{(1{,}000)(0.27)(190)}=6.4\text{ md},$$

an error of 1.5 percent. Reservoir porosity-compressibility is estimated from Eq. 4.40:

$$\phi c_t h=\frac{0.0002637\,(6.4)(190)(1)}{(1.35)(1)^2(600)}$$

$$=3.96\times10^{-4}\text{ ft/psi},$$

or

$$\phi c_t=\frac{3.96\times10^{-4}}{190}=2.08\times10^{-6}\text{ psi}^{-1},$$

an error of 1.5 percent.

An alternate analysis uses Fig. 4.14, a plot of $(1/q)$ vs log t. The slope of the straight line drawn through the data for $0.1<t<10$ hours is

$$m_q=1.7\times10^{-4}\text{ (D/STB)/cycle}.$$

Using Eq. 4.45,

$$k=\frac{(162.6)(1.0)(1.35)}{(1{,}000)(1.7\times10^{-4})(190)}=6.8\text{ md},$$

an error of 4.6 percent. The skin factor is estimated from Fig. 4.14 and Eq. 4.46. From Fig. 4.14, $(1/q)_{1\text{hr}}$ = 0.000578 D/STB.

$$s=1.1513\left[\frac{0.000578}{1.7\times10^{-4}}-\log\left(\frac{6.8}{(1.35)(2.08\times10^{-6})(1)^2}\right)+3.2275\right]$$

$$=0.28.$$

This compares with the actual value of 0.

As opposed to constant-rate testing, the analog to pseudosteady-state flow does not develop during constant-pressure testing. When boundary effects influence behavior during the constant-pressure test, there is a rapid decline of flow rate caused by declining reservoir pressure. Flow rate goes to zero as reservoir pressure approaches the wellbore pressure.

4.7 Reservoir Limit Testing When Rate Varies

Section 3.5 discusses reservoir limit testing for constant-rate production; however, it may be difficult to maintain a constant flow rate during long production periods. If flow rate varies in a cyclic or oscillatory manner, reservoir limit testing techniques still can be used.[14,15] The analysis technique is similar to that for constant-rate reservoir limit tests, but the results are less accurate.

To analyze a variable-rate reservoir limit test, one plots observed flowing pressure vs time on Cartesian paper. Pressure points must be segregated by the rate occurring when the pressure measurement was made. Fig. 4.15 is such a plot for a waste-water injection well; the three sets of pressure points occur at three different injection rates. Such a data

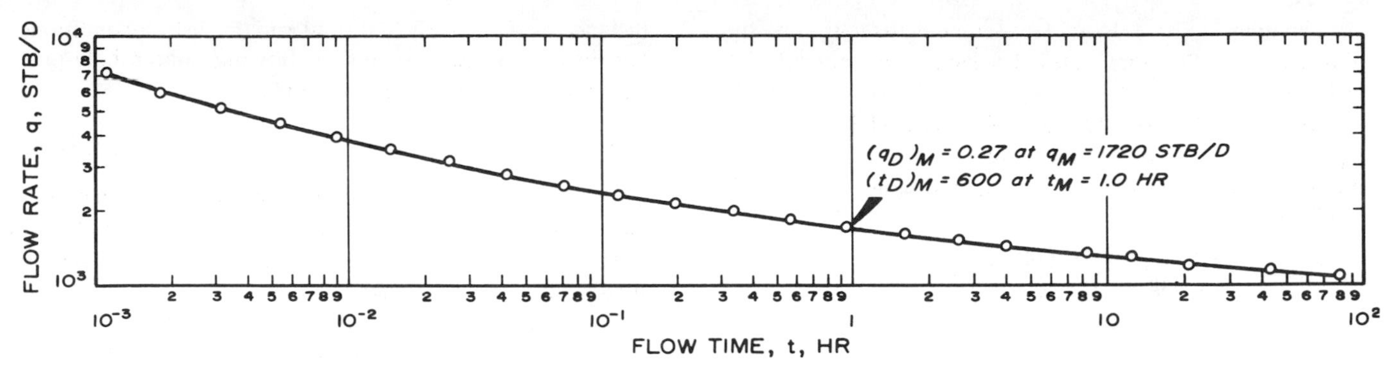

Fig. 4.13 Flow rate-time data for constant-pressure drawdown test of Example 4.4.

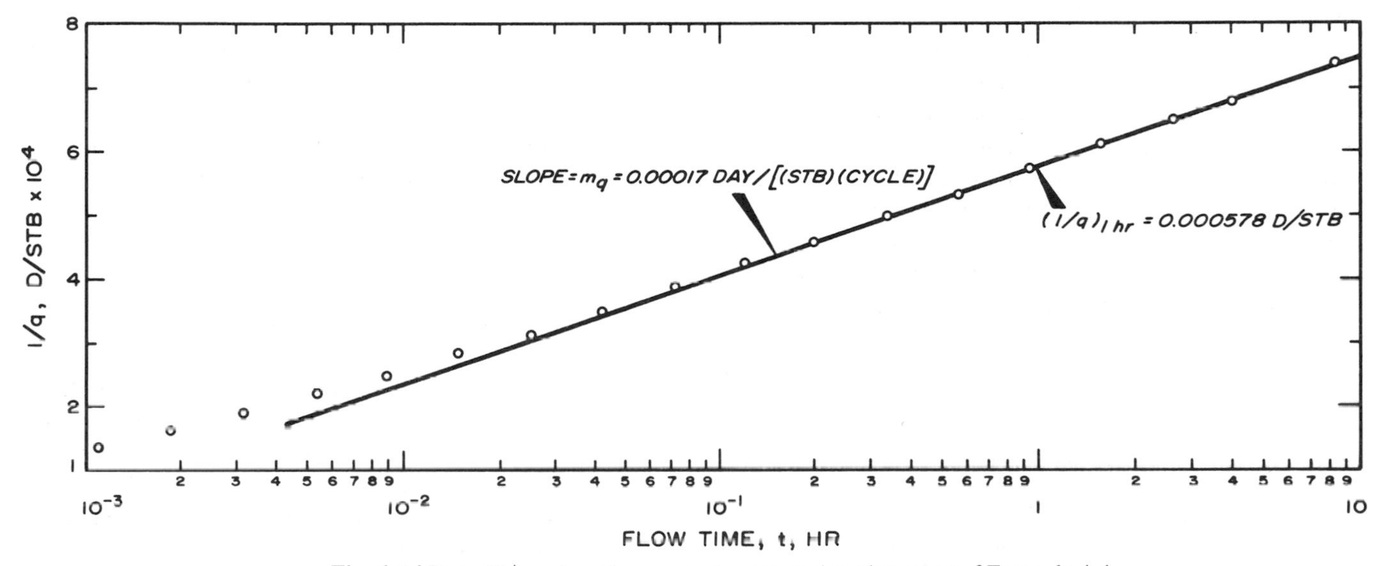

Fig. 4.14 Plot of $1/q$ vs $\log t$ for constant-pressure drawdown test of Example 4.4.

plot should have a straight-line section. Because of flow-rate variation, there is actually never a definite straight line, but a least-squares straight line usually can be fit to the pressure points observed at one of the rates. The slope, m^*, of the straight line may be used to estimate reservoir drainage volume:

$$A\phi h = \frac{-0.23395\,\bar{q}B}{m^* c_t} \quad \text{. . . . . (4.47)}$$

Eq. 4.47 is identical to Eq. 3.35 except that the over-all average flow rate is used. Even though the pressure points are segregated by the rate in effect at the time the pressure was measured, the *total average flow rate* is used to estimate reservoir volume.

Example 4.5 Variable-Rate Reservoir Limit Test

Fig. 4.15 shows pressure data from Ref. 14 for the last 5 years of the 11-year life of an industrial waste-disposal well. Although the data are for injection, the methods of this section can be applied by using a negative rate (see Chapter 7).

The well, with casing set at 1,800 ft, is completed open hole to about 3,000 ft, just into basement. Only a part of the 1,200-ft section is porous and permeable, but porosity and net interval are not known. The injection horizon is known to be of large extent.

Injection is with one, two, or three pumps, so the rate is −5,140, −10,280, or −15,420 STB/D. Pressure and rate data are reported by month only without indication of how long the rate had applied when the pressure was measured. Cumulative fluid injected is known accurately. The average injection rate is −9,660 STB/D for the period shown in Fig. 4.15. The formation volume factor, B, is 1.0; system total compressibility, c_t, is unknown, but is estimated to be about 5×10^{-6} psi^{-1}.

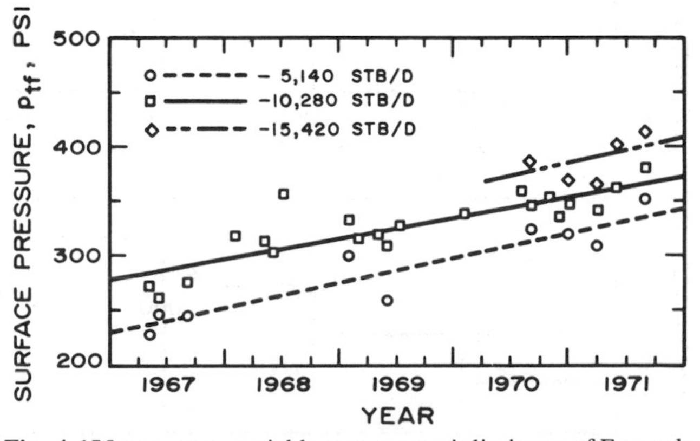

Fig. 4.15 Long-term, variable-rate reservoir limit test of Example 4.5. Five years of injection history for a disposal well with three injection rates; $\bar{q} = -9{,}660$ STB/D during period shown. After Earlougher.[14]

The lines in Fig. 4.15 are least-squares fits to the pressure data for each of the three rates. The slopes, m^*, are $+2.62 \times 10^{-3}$ psi/hr, $+2.15 \times 10^{-3}$ psi/hr, and $+3.16 \times 10^{-3}$ psi/hr for the $-5,140$, $-10,280$, and $-15,420$ STB/D rates, respectively. We estimate the pore volume from Eq. 4.47, being certain to use the average injection rate, $\bar{q} = -9,660$ STB/D.

$$A\phi h = \frac{-0.23395\,(-9,660)(1)}{(2.62 \times 10^{-3})(5 \times 10^{-6})}$$

$$= 1.73 \times 10^{11} \text{ cu ft}$$

$$= 30.7 \times 10^{9} \text{ STB},$$

for the circular data points. Similarly, the estimated pore volume is 37.4×10^9 STB for the square data points and 25.5×10^9 STB for the diamond-shaped data points. These volumes do not contradict the known geology of the formation. The results for the diamond-shaped points ($q = -15,420$ STB/D) should be considered the least reliable since there are only five data points, and they are scattered. The value of 37.4×10^9 STB may be the most reliable, since most data points are for $q = -10,280$ STB/D.

These calculations require that flow be pseudosteady state. A long period of apparently linear pressure increase with time is a good indicator of pseudosteady-state conditions. If this assumption is incorrect, estimated pore volume is too small.

4.8 Deliverability Testing of Oil Wells

Deliverability testing has long been used to predict the capability of a gas well to deliver against a specific flowing bottom-hole pressure.[1,16-19] Fetkovich[20] demonstrates that such testing can be used for oil wells. It is particularly useful for reservoir systems operating below the bubble point, when fluid properties and relative permeabilities vary with distance from the well. Oil flow rate (at surface conditions) has been empirically related to flowing bottom-hole pressure and average reservoir pressure by

$$q_o = J_o' \,(\bar{p}^2 - p_{wf}^2)^n, \quad \ldots \ldots \ldots \ldots \ldots (4.48)$$

where J_o' is a form of productivity index and n is an empirically determined exponent. Fetkovich[20] states that field tests indicate $0.5 \leq n \leq 1.0$. Eq. 4.48 is similar to the deliverability equation used in gas well testing.

Two important deliverability tests are the flow-after-flow test and the modified isochronal test. Fig. 4.16 schematically demonstrates the rate and pressure behavior of a flow-after-flow test. The well is produced at rate q_1 until the pressure stabilizes at p_{wf1}. Then the rate is changed to q_2 until the pressure stabilizes at p_{wf2}, and so on. Normally, four rates are run but any number greater than three may be used. Flow rate may be either increased or decreased. The major disadvantage of the flow-after-flow test is that each rate must remain constant until pressure stabilizes. The time required may be estimated from Eq. 2.40,

$$t_s \simeq 380 \,\frac{\phi \mu c_t A}{k} \quad \ldots \ldots \ldots \ldots \ldots (4.49)$$

For systems that are large or have low permeability, stabilization time can be very long.

To avoid problems with long stabilization times, Cullender[18] proposed the isochronal flow test for gas wells. A shortened version, the modified isochronal flow test,[19] was later suggested and is generally preferred. Fig. 4.17 schematically illustrates flow-rate and pressure histories for a modified isochronal flow test. The well is produced at rate q_1 for time t_1 and the final flowing pressure, p_{wf1}, is observed. Then the well is shut in for time t_1 and the final shut-in pressure, p_{ws2}, is observed. The procedure is repeated at rates q_2, q_3, q_4, etc. The well is usually produced to a stabilized pressure at the final rate, so one stabilized pressure point, $(p_{wf})_{pss}$, is available.

Fig. 4.18 illustrates the analysis method for a modified isochronal deliverability test; $\log(\bar{p}^2 - p_{wf}^2)$ is plotted vs $\log q$. The points usually fall on a straight line with slope $1/n$. The location of the line depends on the flow-period duration. Thus, in normal analysis, the points for the four rates define the straight line and the single stabilized point defines location of the stabilized deliverability curve. The stabilized deliverability curve may be entered at set values of $(\bar{p}^2 - p_{wf}^2)$ to estimate the well's deliverability (flow rate) at a given drawdown. Alternatively, the data plot (Fig. 4.18) may be used to estimate J_o' and n and the flow rate may be estimated from Eq. 4.48. Fig. 4.18 and Eq. 4.48 are written with average reservoir pressure, the pressure used for flow-after-flow and normal isochronal flow tests. *The data plot for a modified isochronal flow test uses the shut-in pressure occurring immediately before the flow rate instead of the average reservoir pressure.*

Fetkovich[20] provides data for many flow-after-flow and isochronal flow tests in several oil wells. Fig. 4.19 demonstrates that the isochronal and flow-after-flow tests can give the same results in oil wells producing from a saturated (both oil and free gas present) reservoir.

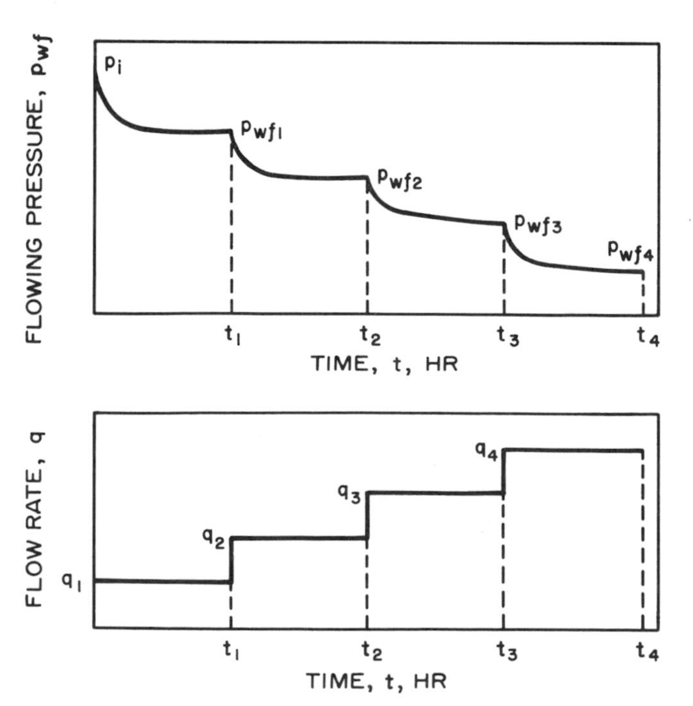

Fig. 4.16 Pressure-rate history for a flow-after-flow test.

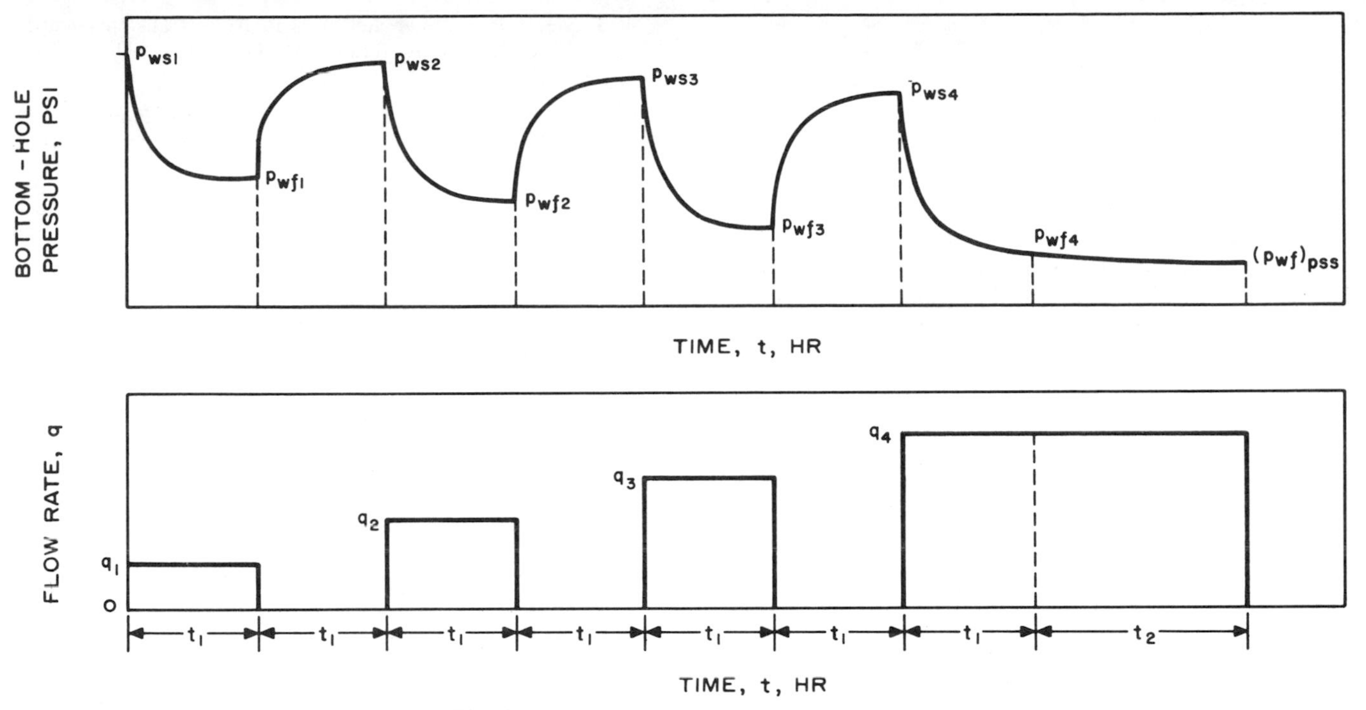

Fig. 4.17 Pressure-rate history for a modified-isochronal-flow test.

If deliverability test data are not available for a solution gas drive reservoir, it still may be possible to predict a well's deliverability by using the "inflow performance relationship" (IPR) proposed by Vogel[21] and the modification to the IPR proposed by Standing.[22] Vogel used computer simulation techniques to demonstrate that many solution gas drive reservoirs operating below the bubble point have an inflow performance relationship given by

$$q_o = \frac{J^* \bar{p}}{1.8}\left[1 - 0.2\left(\frac{p_{wf}}{\bar{p}}\right) - 0.8\left(\frac{p_{wf}}{\bar{p}}\right)^2\right], \quad \ldots \quad (4.50)$$

where q_o is the oil flow rate (STB/D) occurring at bottom-hole pressure p_{wf} and J^* is a productivity index. Given a stabilized q_o and the corresponding $\bar{p}$ and p_{wf}, it is possible to calculate J^* from Eq. 4.50. Then, to estimate q_o at another stabilized pressure, one uses Eq. 4.50 with the experimentally determined J^*. Standing[22] indicated that, as the reservoir is depleted, it is necessary to modify Eq. 4.50 because of changes in relative permeability and fluid properties. He suggested estimating a future value of the productivity index from the present value by using

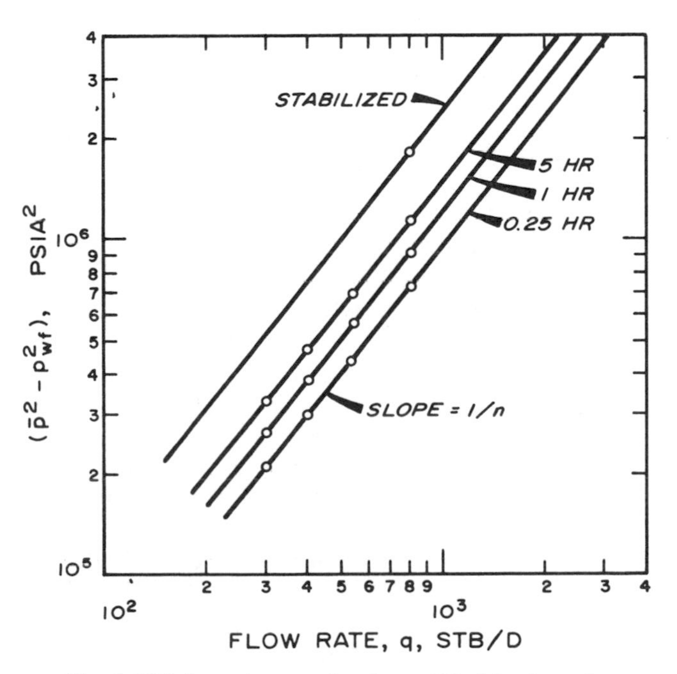

Fig. 4.18 Schematic example of a modified-isochronal-test data plot.

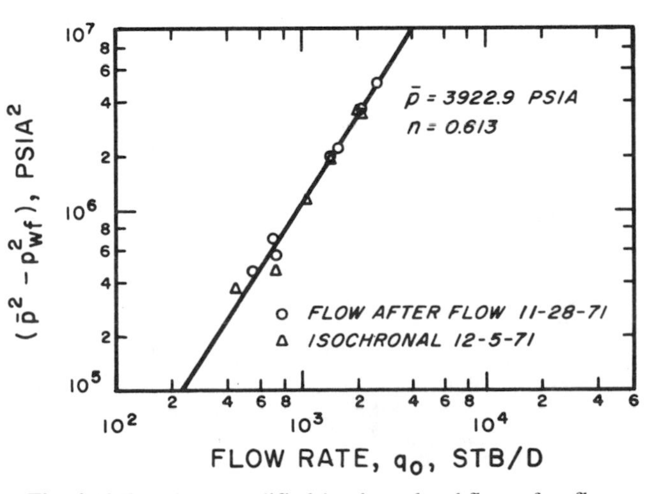

Fig. 4.19 Four-hour modified-isochronal and flow-after-flow deliverability curves. Field data from a saturated reservoir. After Fetkovich.[20]

$$J_F^* = J_p^* \frac{\left(\frac{k_{ro}}{\mu_o B_o}\right)_F}{\left(\frac{k_{ro}}{\mu_o B_o}\right)_p}, \quad \ldots \ldots \ldots \ldots \ldots \ldots \ldots \ldots (4.51)$$

where the subscript F refers to some time in the future and the subscript p refers to data at the present time. Values of k_{ro}, μ_o, and B_o in the future are estimated from material balance relationships. The procedure for developing a future IPR is (1) estimate J_p^* from current production data and Eq. 4.50; (2) estimate J_F^* from Eq. 4.51; and (3) estimate the future IPR (q_o) by using J_F^* in Eq. 4.50. Standing[22] gives an example calculation.

It is emphasized that deliverability tests are not transient well tests. They do not yield estimates of skin and formation permeability; rather, they provide empirical relationships between flow rate and drawdown for stabilized oil wells. They do include the nonideal conditions existing in the reservoir, particularly saturation distributions and variation of fluid properties with pressure. Deliverability testing thus may be valuable in helping predict future production rates as a function of available pressure differential.

4.9 Factors Complicating Multiple-Rate Testing

Multiple-rate tests exhibit their greatest advantage when changing wellbore storage makes normal transient test analysis difficult or impossible. That is because such tests can eliminate changes in the wellbore storage coefficient — even though the effects of wellbore storage still exist. Multiple-rate tests also reduce the loss of current production. However, such tests are difficult to control since they are flowing tests. Rate fluctuations are difficult to measure, especially on a continuous basis. The analysis techniques are much more bothersome and difficult than those for constant-rate tests; they frequently require the use of a computer.

To assure the best possible multiple-rate test, the engineer must have an idea of a well's flow characteristics. The rate change imposed must be large enough to give significant change in a pressure transient behavior of the well. The effect of rate change on pressure response must be estimated from Eq. 4.1, Eq. 4.6, or Eq. 4.13. Normally, rate is changed by a factor of two or more.

References

1. Matthews, C. S. and Russell, D. G.: *Pressure Buildup and Flow Tests in Wells,* Monograph Series, Society of Petroleum Engineers of AIME, Dallas (1967) **1,** Chap. 6.
2. Odeh, A. S. and Jones, L. G.: "Pressure Drawdown Analysis, Variable-Rate Case," *J. Pet. Tech.* (Aug. 1965) 960-964; *Trans.,* AIME, **234.** Also *Reprint Series, No. 9 — Pressure Analysis Methods,* Society of Petroleum Engineers of AIME, Dallas (1967) 161-165.
3. Russell, D. G.: "Determination of Formation Characteristics From Two-Rate Flow Tests," *J. Pet. Tech.* (Dec. 1963) 1347-1355; *Trans.,* AIME, **228.** Also *Reprint Series, No. 9 — Pressure Analysis Methods,* Society of Petroleum Engineers of AIME, Dallas (1967) 136-144.
4. Doyle, R. E. and Sayegh, E. F.: "Real Gas Transient Analysis of Three-Rate Flow Tests," *J. Pet. Tech.* (Nov. 1970) 1347-1356.
5. Jacob, C. E. and Lohman, S. W.: "Nonsteady Flow to a Well of Constant Drawdown in an Extensive Aquifer," *Trans.,* AGU (Aug. 1952) 559-569.
6. Gladfelter, R. E., Tracy, G. W., and Wilsey, L. W.: "Selecting Wells Which Will Respond to Production-Stimulation Treatment," *Drill. and Prod. Prac.,* API (1955) 117-129.
7. Winestock, A. G. and Colpitts, G. P.: "Advances in Estimating Gas Well Deliverability," *J. Cdn. Pet. Tech.* (July-Sept. 1965) 111-119.
8. Ramey, H. J., Jr.: "Verification of the Gladfelter-Tracy-Wilsey Concept for Wellbore Storage Dominated Transient Pressures During Production," *J. Cdn. Pet. Tech.,* (April-June 1976) 84-85.
9. Pinson, A. E., Jr.: "Conveniences in Analyzing Two-Rate Flow Tests," *J. Pet. Tech.* (Sept. 1972) 1139-1141.
10. Earlougher, R. C., Jr.: "Estimating Errors When Analyzing Two-Rate Flow Tests," *J. Pet. Tech.* (May 1973) 545-547.
11. Odeh, A. S. and Jones, L. G.: "Two-Rate Flow Test, Variable-Rate Case — Application to Gas-Lift and Pumping Wells," *J. Pet. Tech.* (Jan. 1974) 93-99; *Trans.,* AIME, **257.**
12. Slider, H. C.: "Application of Pseudo-Steady-State Flow to Pressure-Buildup Analysis," paper SPE 1403 presented at the SPE-AIME Regional Meeting, Amarillo, Tex., Oct. 27-28, 1966.
13. Slider, H. C.: "A Simplified Method of Pressure Buildup Analysis for a Stabilized Well," *J. Pet. Tech.* (Sept. 1971) 1155-1160; *Trans.,* AIME, **251.**
14. Earlougher, Robert C., Jr.: "Variable Flow Rate Reservoir Limit Testing," *J. Pet. Tech.* (Dec. 1972) 1423-1429.
15. Kazemi, Hossein: "Discussion of Variable Flow Rate Reservoir Limit Testing," *J. Pet. Tech.* (Dec. 1972) 1429-1430.
16. Rawlins, E. L. and Schellhardt, M. A.: "Back-Pressure Data on Natural-Gas Wells and Their Application to Production Practices," *Monograph 7,* USBM (1936).
17. *Theory and Practice of the Testing of Gas Wells,* 3rd ed., Pub. ERCB-75-34, Energy Resources and Conservation Board, Calgary, Alta., Canada (1975).
18. Cullender, M. H.: "The Isochronal Performance Method of Determining the Flow Characteristics of Gas Wells," *Trans.,* AIME (1955) **204,** 137-142. Also *Reprint Series, No. 9 — Pressure Analysis Methods,* Society of Petroleum Engineers of AIME, Dallas (1967) 203-208.
19. Katz, Donald L., Cornell, David, Kobayashi, Riki, Poettmann, Fred H., Vary, John A., Elenbaas, John R., and Weinaug, Charles F.: *Handbook of Natural Gas Engineering,* McGraw-Hill Book Co., Inc., New York (1959) Chap. 11.
20. Fetkovich, M. J.: "The Isochronal Testing of Oil Wells," paper SPE 4529 presented at the SPE-AIME 48th Annual Fall Meeting, Las Vegas, Nev., Sept. 30-Oct. 3, 1973.
21. Vogel, J. V.: "Inflow Performance Relationships for Solution-Gas Drive Wells," *J. Pet. Tech.* (Jan. 1968) 83-92; *Trans.,* AIME, **243.**
22. Standing, M. B.: "Concerning the Calculation of Inflow Performance of Wells Producing Solution Gas Drive Reservoirs," *J. Pet. Tech.* (Sept. 1971) 1141-1142.

Chapter 5

Pressure Buildup Testing

5.1 Introduction

Pressure buildup testing, probably the most familiar transient well testing technique, has been treated widely in the literature.[1-10] This type of testing was first introduced by the groundwater hydrologists,[2] but has been used extensively in the petroleum industry.

Pressure buildup testing requires shutting in a producing well. The most common and simplest analysis techniques require that the well produce at a constant rate, either from startup or long enough to establish a stabilized pressure distribution (t_{pss}), before shut-in. Fig. 5.1 schematically shows rate and pressure behavior for an ideal pressure buildup test. In that figure, and throughout this monograph, t_p is the production time and Δt is running shut-in time. The pressure is measured immediately before shut-in and is recorded as a function of time during the shut-in period. The resulting pressure buildup curve is analyzed for reservoir properties and wellbore condition; the methods used most are described in this chapter.

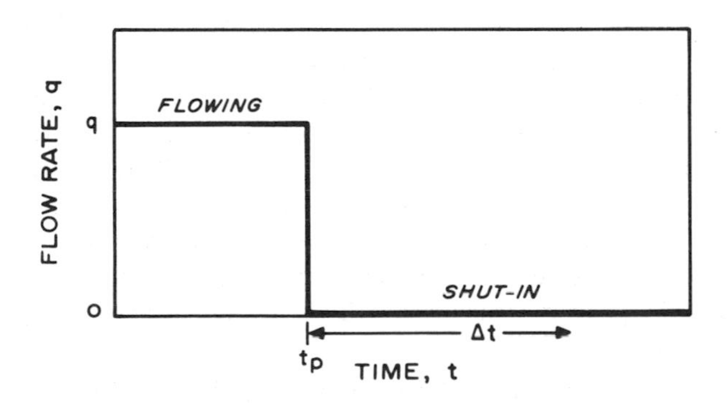

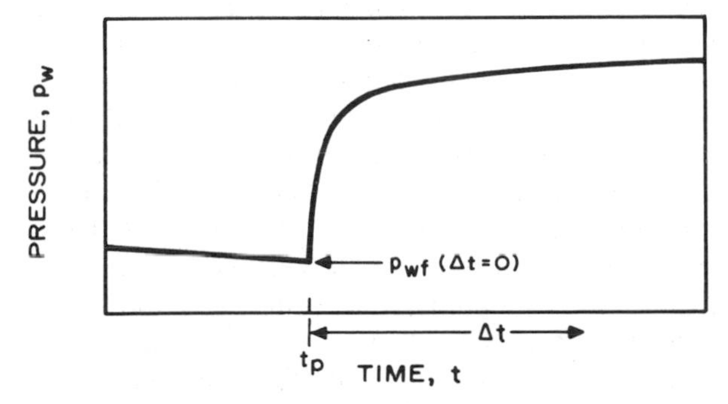

Fig. 5.1 Idealized rate and pressure history for a pressure buildup test.

As in all transient well tests, knowledge of surface and subsurface mechanical conditions is important in buildup-test data interpretation. Therefore, it is recommended that tubing and casing sizes, well depth, packer locations, etc., be determined before data interpretation starts. Short-time pressure observations usually are necessary for complete delineation of wellbore storage effects. Data may be needed at intervals as short as 15 seconds for the first few minutes of some buildup tests. As the test progresses, the data-collection interval can be expanded.

Stabilizing the well at a constant rate before testing is an important part of a pressure buildup test. If stabilization is overlooked or is impossible, standard data analysis techniques may provide erroneous information about the formation. Thus, it is important to determine the degree and adequacy of the stabilization; one way is to check the length of the pre-shut-in constant-rate period against the time required for stabilization, as given by Eqs. 2.40 and 2.42. For wells with significantly varying rates, buildup test analysis is still possible using the variable-rate methods of Chapter 4 or the modifications of those methods presented in Section 5.4.

5.2 Pressure Buildup Test Analysis During the Infinite-Acting Period

For any pressure-buildup testing situation, bottom-hole shut-in pressure in the test well may be expressed by using the principle of superposition for a well producing at rate q until time t_p, and at zero rate thereafter. At any time after shut-in,

$$p_{ws} = p_i - \frac{141.2\, qB\mu}{kh} \{p_D([t_p + \Delta t]_D) - p_D(\Delta t_D)\}\ , \qquad (5.1)$$

where p_D is the applicable dimensionless-pressure function and t_D is as defined by Eq. 2.3a:

$$t_D = \frac{0.0002637\, kt}{\phi \mu c_t r_w^2} \qquad (5.2)$$

During the infinite-acting time period, after wellbore storage effects have diminished and providing there are no major induced fractures, p_D in Eq. 5.1 may be replaced by the logarithmic approximation to the exponential integral, Eq. 2.5b:

$$p_D = \frac{1}{2}(\ln t_D + 0.80907) \quad \ldots \ldots \ldots \ldots \ldots (5.3)$$

Eq. 5.3 applies when $t_D > 100$, which occurs after a few minutes for most unfractured systems. By using Eqs. 5.2 and 5.3, Eq. 5.1 may be rewritten:

$$p_{ws} = p_i - m \log\left(\frac{t_p + \Delta t}{\Delta t}\right). \quad \ldots \ldots \ldots \ldots \ldots (5.4)$$

Eq. 5.4 describes a straight line with intercept p_i and slope $-m$, where

$$m = \frac{162.6 qB\mu}{kh} \quad \ldots \ldots \ldots \ldots \ldots (5.5)$$

Eq. 5.4 indicates that a plot of observed shut-in bottom-hole pressure, p_{ws}, vs $\log\left[(t_p + \Delta t)/\Delta t\right]$ should have a straight-line portion with slope $-m$ that can be used to estimate reservoir permeability,

$$k = \frac{162.6 qB\mu}{mh} \quad \ldots \ldots \ldots \ldots \ldots (5.6)$$

Both Theis[2] and Horner[6] proposed estimating permeability in this manner. The p_{ws} vs $\log\left[(t_p + \Delta t)/\Delta t\right]$ plot is commonly called the Horner plot (graph, method) in the petroleum industry; that terminology is used in this monograph.

Fig. 5.2 is a schematic Horner plot of pressure buildup data. The straight-line section is shown. As indicated by Eq. 5.4, this straight-line portion of the Horner plot may be extrapolated to $(t_p + \Delta t)/\Delta t = 1$, $\{\log\left[(t_p + \Delta t)/\Delta t\right] = 0\}$, the equivalent of infinite shut-in time, to obtain an estimate of p_i. That is an accurate estimate only for short production periods. However, the extrapolated pressure value is useful for estimating average reservoir pressure, as indicated in Chapter 6.

In Fig. 5.2, as in all other Horner plots in this monograph, the abscissa has been reversed so it increases from right to left in keeping with common practice. The reverse plotting, which is mathematically equivalent to plotting $\log\left[\Delta t/(t_p + \Delta t)\right]$, causes time to increase from left to right (see upper scale in Fig. 5.2) and gives the buildup curve the shape one would expect. However, it means that the slope, which normally would be thought of as positive, *is negative*. In Fig. 5.2, the slope is -42 psi/cycle, so $m = 42$ psi/cycle.

A result of using the superposition principle is that the skin factor, s, does not appear in the general pressure-buildup equation, Eq. 5.1. As a result, skin factor does not appear in the simplified equation for the Horner plot, Eq. 5.4. That means the Horner-plot slope is not affected by the skin factor; however, the skin factor still *does* affect the *shape* of the pressure buildup data. In fact, an early-time deviation from the straight line can be caused by skin factor as well as by wellbore storage, as indicated in Fig. 5.2 (also see Fig. E.1). The deviation can be significant for the large negative skins that occur in hydraulically fractured wells. In any case, the skin factor does affect flowing pressure before shut-in, so skin may be estimated from the buildup test data plus the flowing pressure immediately before the buildup test:[1,8,11]

$$s = 1.1513\left[\frac{p_{1hr} - p_{wf}(\Delta t = 0)}{m} - \log\left(\frac{k}{\phi\mu c_t r_w^2}\right) + 3.2275\right]. \quad \ldots \ldots \ldots \ldots \ldots (5.7)$$

In Eq. 5.7, $p_{wf}(\Delta t = 0)$ is the observed flowing bottom-hole pressure immediately before shut-in and $-m$ is the slope of the Horner plot. As a result of assumptions made in deriving Eq. 5.7,[1,8] the value of p_{1hr} must be taken from the Horner straight line. Frequently, pressure data do not fall on the straight line at 1 hour because of wellbore storage effects that allow afterflow into the well, or large negative skin factors resulting from induced fracturing, etc. In that case, the semilog line must be extrapolated to 1 hour and the pressure read. Fig. 5.2 shows the correct way to determine p_{1hr}.

In different types of transient test analysis, the slope is sometimes $+m$ and sometimes $-m$; additionally, m sometimes includes a minus sign (compare Eq. 3.6 for drawdown testing with Eq. 5.5 for buildup). This may cause some confusion in transient well test analysis. Confusion may be avoided by realizing (1) permeability must always be positive, so the sign of m may be determined from Eq. 5.5 (or its equivalent for other types of testing); (2) the first term inside the brackets of the skin equation, $\left[p_{1hr} - p_{wf}(\Delta t = 0)\right]/m$, is usually positive (the exception occurs in hydraulically fractured wells with $s << 0$); and (3) production rates are positive while injection rates are negative. There should not be a problem with analysis equations if the correct definition of m and its relation to the slope of the data

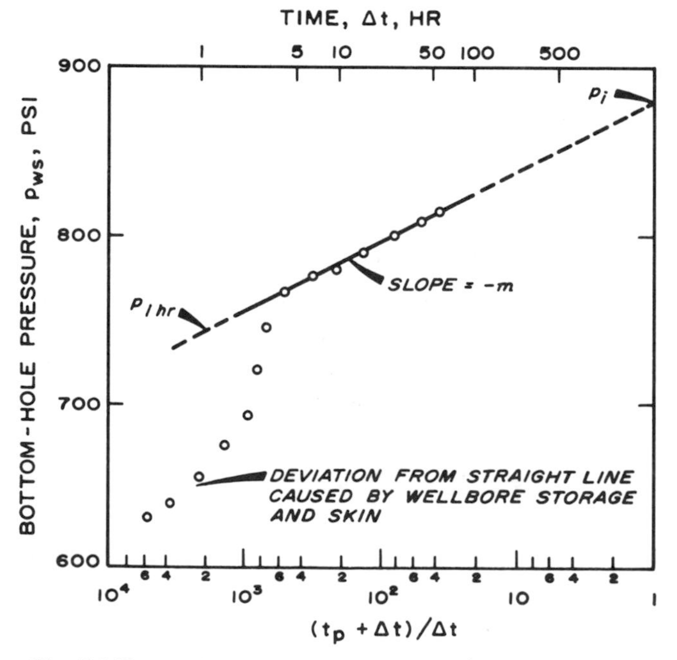

Fig. 5.2 Horner plot of pressure buildup data showing effects of wellbore storage and skin.

plot are used with the correct signs in the analysis equations. Appendix E gives appropriate signs for common analysis techniques for different tests.

Eq. 5.7 provides a good estimate of the skin factor as long as $t_p >> 1$ hour. But when t_p is on the order of 1 hour (for example, in drillstem testing), Eq. 5.7 should be replaced by

$$s = 1.1513 \left[\frac{p_{1\text{hr}} - p_{wf}(\Delta t{=}0)}{m} + \log\left(\frac{t_p + 1}{t_p}\right) - \log\left(\frac{k}{\phi \mu c_t r_w^2}\right) + 3.2275 \right] \quad (5.8)$$

The proper t_p to use for a given well in a multiple-well reservoir has been a matter of frequent concern to practicing engineers. Eqs. 5.1 and 5.4 assume a constant production rate from time 0 to time t_p, not often a realistic assumption. Horner[6] indicated that t_p often can be approximated as the cumulative production since completion divided by the rate immediately before shut-in (when rate varies). Except shortly after well completion, it appears desirable as a matter of general practice to approximate t_p using cumulative production since the last pressure equalization (or some other convenient, relatively short time in terms of reservoir depletion) rather than total cumulative production:

$$t_p = \frac{24 V_P}{q} \quad (5.9)$$

In Eq. 5.9, V_P is the cumulative volume produced since the *last pressure equalization* and q is the constant rate just before shut in. If t_p on this basis is significantly greater than t_{pss}[12,13] (for example, by a factor of 2), then replotting using t_{pss} (Eq. 2.24 for a closed boundary or Eq. 2.42 for a constant-pressure boundary) may be justified. For closed boundary conditions, a Horner plot *using* t_{pss} as opposed to a Miller-Dyes-Hutchinson (MDH) plot (Section 5.3) tends to prolong the straight-line portion of the buildup curve. However, the principal importance of using t_{pss} usually is in minimizing errors in estimating average reservoir pressure (Section 6.3).

When the time at constant rate immediately before shut-in is significantly less than t_{pss} and the rate variation is significant (for example, 20 to 50 percent), accurate values of permeability, skin, and static pressure generally can be obtained only by using the methods of superposition discussed in Section 5.4. When the time at a constant production rate is significantly less than t_{pss}, but is still large (more than about four times the buildup time of interest), reasonably accurate values of skin and permeability should still be obtainable with the normal Horner plot using Eq. 5.9, even though values of estimated static pressure could be poor. This comment applies to systems with negligible fracturing in which wellbore storage effects either have died out or have been properly adjusted for (Section 11.2).

Even though the well is shut in during pressure buildup testing, the afterflow caused by wellbore storage has a significant influence on pressure buildup data. Fig. 5.2 schematically shows that pressure points fall below the semilog straight line while wellbore storage is important. The duration of those effects may be estimated by making the log-log data plot described in Section 2.6. For pressure buildup testing, plot $\log[p_{ws} - p_{wf}(\Delta t{=}0)]$ vs $\log \Delta t$. When wellbore storage dominates, that plot will have a unit-slope straight line; as the semilog straight line is approached, the log-log plot bends over to a gently curving line with a low slope (see Fig. 2.10). In all pressure-buildup test analyses, the log-log data plot should be made before the straight line is chosen on the semilog data plot, since it is often possible to draw a semilog straight line through wellbore-storage-dominated data. This phenomenon occurs because wellhead shut-in does not correspond to sand-face shut-in. When the surface valve is closed, fluid continues to flow into the wellbore from the formation. Thus, pressure does not build up as fast as we might expect. As the flow rate drops off to zero, the pressure increases rapidly to approach the theoretically predicted level. The semilog data plot is steep and nearly linear during this time, and may be analyzed incorrectly. The analyzable data occur after the data-plot slope has become less steep, as indicated in Fig. 5.2.

When wellbore storage effects last so long that the semilog straight line does not develop, it may be possible to analyze the test data by using type-curve matching techniques in a manner similar to that described in Section 3.3, with $\Delta p = p_{ws} - p_{wf}(\Delta t = 0)$. The type curves of Figs. C.6,[14] C.8,[15] and C.9[16,17] are particularly useful for pressure buildup testing, providing a significant change in wellbore storage coefficient is not involved (see Section 11.2). It cannot be overemphasized that type-curve matching *should not be used for test analysis if semilog analysis techniques can be applied*. Type-curve matching generally gives only approximate results (within a factor of 2 or 3). Refs. 15 and 16 give examples of type-curve matching for pressure buildup analysis. The Gladfelter-Tracy-Wilsey[1,18] or Russell[1,9] approaches can also give good results for data nearing the semilog straight line (after q is less than 20 percent of the previous rate). However, curve-matching techniques, particularly Fig. C.8, can also give good quantitative results in this region.

Example 5.1 Pressure Buildup Test Analysis—Horner Method

Table 5.1 shows pressure buildup data from an oil well with an estimated drainage radius of 2,640 ft. Before shut-in the well had produced at a stabilized rate of 4,900 STB/D for 310 hours. Known reservoir data are

$$\begin{aligned}
\text{depth} &= 10{,}476 \text{ ft} \\
r_w &= (4.25/12) \text{ ft} \\
c_t &= 22.6 \times 10^{-6} \text{ psi}^{-1} \\
q_o &= 4{,}900 \text{ STB/D} \\
h &= 482 \text{ ft} \\
p_{wf}(\Delta t = 0) &= 2{,}761 \text{ psig} \\
\mu_o &= 0.20 \text{ cp} \\
\phi &= 0.09 \\
B_o &= 1.55 \text{ RB/STB} \\
\text{casing ID} &= (6.276/12) \text{ ft} \\
t_p &= 310 \text{ hours.}
\end{aligned}$$

Wellbore storage affects transient pressure behavior and,

therefore, should be considered in all transient test analyses. Failure to do so may result in analyzing the wrong portion of the data. Fig. 5.3, the log-log plot of the buildup data in Table 5.1, is used to check the significance of wellbore storage. Since there is no unit-slope line, we conclude that dominant wellbore storage has ended by 0.1 hour. However, the rapid pressure increase shown in Fig. 5.4 does indicate that wellbore storage or skin effects are significant until about 0.75 hour. The data obtained after 0.75 hour can be analyzed.

The Horner plot is shown as Fig. 5.4. Residual wellbore storage or skin effects at shut-in times of less than 0.75 hour are apparent. The straight line, drawn after $\Delta t = 0.75$ hour, has a slope of -40 psig/cycle, so $m = 40$ psig/cycle.

Eq. 5.6 is used to estimate permeability:

$$k = \frac{162.6(4{,}900)(1.55)(0.20)}{(40)(482)} = 12.8 \text{ md}.$$

Skin factor is estimated from Eq. 5.7 using $p_{1hr} = 3{,}266$ psig from Fig. 5.4:

$$s = 1.1513 \left[\frac{3{,}266 - 2{,}761}{40} - \log \left(\frac{(12.8)(12)^2}{(0.09)(0.20)(22.6 \times 10^{-6})(4.25)^2} \right) + 3.2275 \right] = 8.6.$$

We can estimate Δp across the skin from Eq. 2.9:

$$\Delta p_s = \frac{(141.2)(4{,}900)(1.55)(0.20)(8.6)}{(12.8)(482)} = 300 \text{ psi}.$$

TABLE 5.1—PRESSURE BUILDUP TEST DATA FOR EXAMPLE 5.1, $t_p = 310$ HOURS.

Δt (hours)	$t_p+\Delta t$ (hours)	$\frac{(t_p+\Delta t)}{\Delta t}$	p_{ws} (psig)	$p_{ws} - p_{wf}$ (psig)
0.0	—	—	2,761	—
0.10	310.10	3,101	3,057	296
0.21	310.21	1,477	3,153	392
0.31	310.31	1,001	3,234	473
0.52	310.52	597	3,249	488
0.63	310.63	493	3,256	495
0.73	310.73	426	3,260	499
0.84	310.84	370	3,263	502
0.94	310.94	331	3,266	505
1.05	311.05	296	3,267	506
1.15	311.15	271	3,268	507
1.36	311.36	229	3,271	510
1.68	311.68	186	3,274	513
1.99	311.99	157	3,276	515
2.51	312.51	125	3,280	519
3.04	313.04	103	3,283	522
3.46	313.46	90.6	3,286	525
4.08	314.08	77.0	3,289	528
5.03	315.03	62.6	3,293	532
5.97	315.97	52.9	3,297	536
6.07	316.07	52.1	3,297	536
7.01	317.01	45.2	3,300	539
8.06	318.06	39.5	3,303	542
9.00	319.00	35.4	3,305	544
10.05	320.05	31.8	3,306	545
13.09	323.09	24.7	3,310	549
16.02	326.02	20.4	3,313	552
20.00	330.00	16.5	3,317	556
26.07	336.07	12.9	3,320	559
31.03	341.03	11.0	3,322	561
34.98	344.98	9.9	3,323	562
37.54	347.54	9.3	3,323	562

Thus, pressure drop across the skin in this damaged well is about one-half the total pressure drop. The flow efficiency may be estimated from Eq. 2.12, using $\bar{p} = 3{,}342$ psig (as estimated in Example 6.1, Section 6.3). Flow efficiency is calculated as

$$\frac{3{,}342 - 2{,}761 - 300}{3{,}342 - 2{,}761} = 0.48.$$

This suggests that the production rate could be more than doubled by simply removing the damage, or possibly could be tripled with an acid or fracture treatment, depending on conditions around the well and on rock type.

5.3 Pressure Buildup Test Analysis in Finite and Developed Reservoirs

When wells being tested do not act like a single well in an infinite system, the equations in Section 5.2 require modification. In this section, we consider pressure buildup testing of a single well in a bounded reservoir and of a well in a developed reservoir. During the initial discussion, we will not consider the effects of changing offset-well drainage areas on the developed reservoir situation.

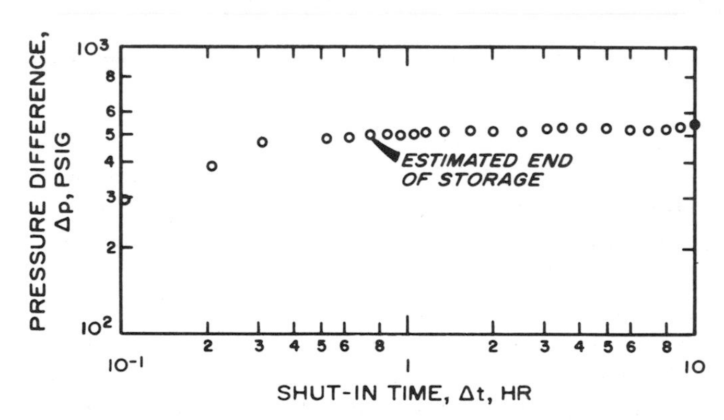

Fig. 5.3 Log-log data plot for the buildup test of Example 5.1.

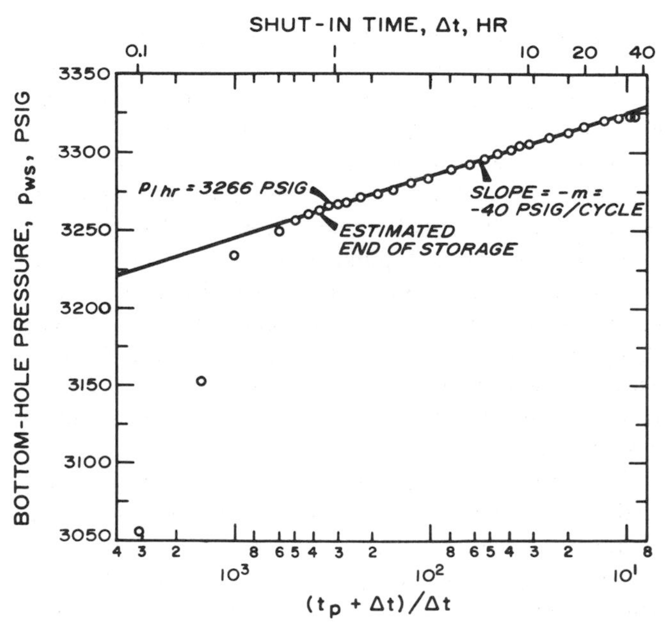

Fig. 5.4 Horner plot of pressure buildup data for Example 5.1.

Horner Plot

The Horner pressure-buildup test data analysis can be used to estimate permeability and skin in finite reservoirs just as in infinite-acting reservoirs, since boundary effects influence only late-time data. The data plot is as described in Section 5.2 and Fig. 5.2; Eqs. 5.6, 5.7, and 5.9 apply. Section 5.2 states that an estimate of p_i is obtained by extrapolating the straight-line section of the Horner plot to infinite shut-in time. For finite and developed reservoirs, the extrapolated pressure is not a good estimate of p_i and generally has been called the "false pressure," p^*.[1,6,9,10] Fig. 5.5 shows pressure buildup data for a well in a finite reservoir. The extrapolated false pressure, p^*, is higher than the average pressure at the instant of shut-in unless the drainage region is highly skewed.

Using the concept of the false pressure, we may rewrite Eq. 5.4:

$$p_{ws} = p^* - m \log \left(\frac{t_p + \Delta t}{\Delta t} \right) \quad . \quad . \quad . \quad (5.10)$$

Ramey and Cobb[10] show that p^* is related to p_i by

$$p^* = p_i - \frac{141.2\, qB\mu}{kh} \left[p_D(t_{pD}) - \frac{1}{2} (\ln t_{pD} + 0.80907) \right] \quad . \quad . \quad . \quad (5.11)$$

When the logarithmic approximation, Eq. 5.3, can be used for $p_D(t_{pD})$ in Eq. 5.11, p^* is identical to p_i.

Eq. 5.10 indicates that the normal Horner plot, p_{ws} vs log $[(t_p + \Delta t)/\Delta t]$, should have a straight-line section with slope $-m$, as schematically illustrated in Figs. 5.2 and 5.5. Although it is commonly believed that the Horner plot should be used only for new wells or when t_p is relatively small, Ramey and Cobb[10] and Cobb and Smith[19] indicate that the Horner plot may *always* be used for pressure-buildup data analysis. However, since it requires more work than the Miller-Dyes-Hutchinson method, the Horner plot is generally not used unless $t_p < t_{pss}$.

Miller-Dyes-Hutchinson Analysis

The Horner plot may be simplified if $\Delta t << t_p$. In that case, $t_p + \Delta t \simeq t_p$ and

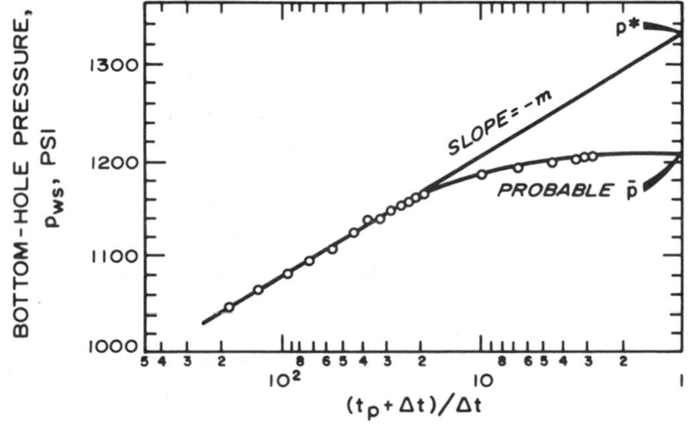

Fig. 5.5 Horner plot of typical pressure buildup data from a well in a finite reservoir. After Matthews and Russell.[1]

$$\log \left(\frac{t_p + \Delta t}{\Delta t} \right) \simeq \log t_p - \log \Delta t \quad . \quad . \quad . \quad (5.12)$$

If Eq. 5.12 is used in Eq. 5.10, then

$$p_{ws} = p_{1\mathrm{hr}} + m \log \Delta t \quad . \quad . \quad . \quad (5.13)$$

Eq. 5.13 indicates that a plot of p_{ws} vs log Δt should be a straight line with slope $+m$, where m is given by Eq. 5.5. Permeability may be estimated from Eq. 5.6, and the skin factor may be estimated from Eq. 5.7. The p_{ws} vs log Δt plot is commonly called the Miller-Dyes-Hutchinson (MDH) plot.[1,10] We use this terminology throughout this monograph. The false pressure may be estimated from the MDH plot by using

$$p^* = p_{1\mathrm{hr}} + m \log(t_p + 1)$$

$$\simeq p_{1\mathrm{hr}} + m \log(t_p) \quad . \quad . \quad . \quad (5.14)$$

Fig. 5.2 indicates that some minimum shut-in time is required before pressure-buildup data fall on the Horner straight line. The same is true for the MDH plot. The beginning of the MDH semilog straight line may be estimated by making the log-log data plot and observing when the data points reach the slowly curving low-slope line, about 1 to 1.5 cycles in time after the end of the unit-slope straight line. Alternatively, the time to the beginning of the semilog straight line for either the Horner or the MDH plot may be estimated from Eq. 2.22.[20]

$$\Delta t_D = 50\, C_D e^{0.14s}, \quad . \quad . \quad . \quad (5.15a)$$

or, in hours,

$$\Delta t = \frac{170{,}000\, C e^{0.14s}}{(kh/\mu)} \quad . \quad . \quad . \quad (5.15b)$$

For fractured wells, Δt estimated using a C based on wellbore storage volume rather than a C derived from a log-log plot unit slope (see Eq. 2.20) will tend to be a minimum value owing to neglect of any fracture storage volume.

Fig. 5.5 indicates that after some shut-in time, the pressure begins to fall below the semilog straight line. This is true for both Horner- and MDH-type plots. The end time of the semilog straight line may be estimated from

$$\Delta t = \frac{\phi \mu c_t A}{0.0002637\, k} (\Delta t_{DA})_{esl}, \quad . \quad . \quad . \quad (5.16)$$

where $(\Delta t_{DA})_{esl}$, the dimensionless shut-in time at the end of the semilog straight line, depends on reservoir shape and well location. Ramey and Cobb[10] and Cobb and Smith[19] present $(\Delta t_{DA})_{esl}$ data for a variety of shapes and well locations for both Horner- and MDH-type plots. Fig. 5.6 gives $(t_{DA})_{esl}$ data for a Horner plot for the shapes and well locations in Table 5.2. Fig. 5.7 gives the information for an MDH plot. Both figures identify the time when the data-plot slope deviates from the correct slope by about 5 percent. Cobb and Smith[19] (preprint version only) also give the time to the end of the semilog straight line for 2-, 10-, 15-, 20-, and 40-percent deviation. Similar data for a square with constant-pressure boundaries and the well at the center are available[21,22] and are included as Shape 7 in Figs. 5.6 and 5.7.

A comparison of Fig. 5.6 with Fig. 5.7 shows that, in closed systems, $(\Delta t_{DA})_{esl}$ is never longer for an MDH plot than for a Horner plot. For symmetric closed systems, the straight line will be prolonged by the Horner plot for producing times t_p up to $4t_{pss}$. For asymmetrical systems, the advantage is not so great. Practically speaking, the Horner plot is superior from the standpoint of straight-line duration for $t_p < t_{pss}$; otherwise, the MDH plot is equally good and is much easier to prepare. However, the figures do show that the MDH plot has a longer straight line than the Horner plot for a square with constant-pressure boundaries when $t_{pDA} > 0.15$.

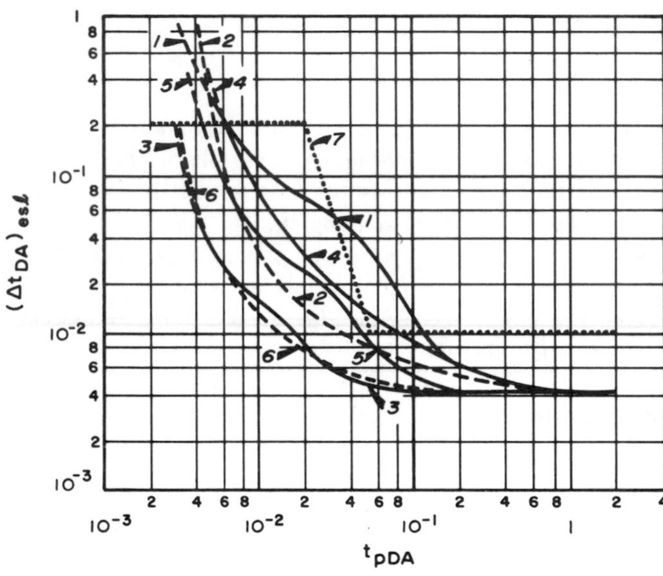

Fig. 5.6 Dimensionless time to end of Horner straight line for shapes identified in Table 5.2. Data of Cobb and Smith[19] and Kumar and Ramey.[22]

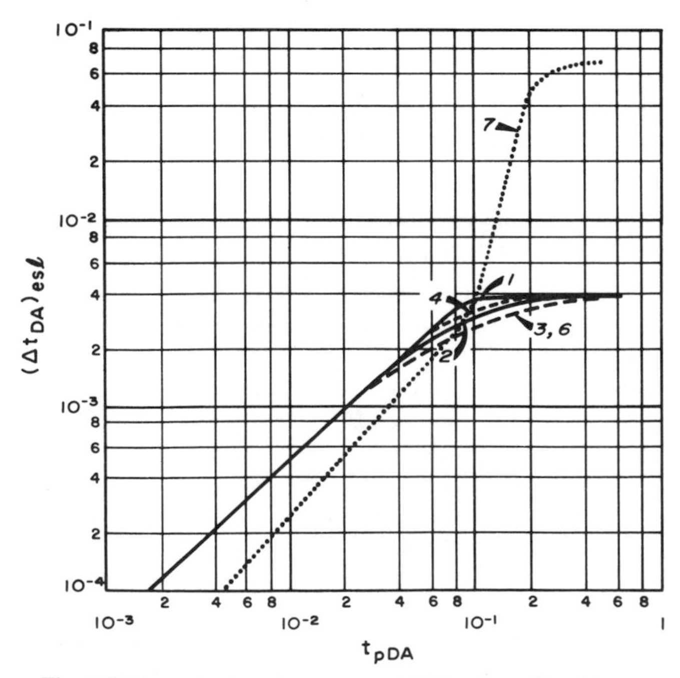

Fig. 5.7 Dimensionless time to end of Miller-Dyes-Hutchinson straight line for shapes identified in Table 5.2. Data of Cobb and Smith[19] and Kumar and Ramey.[22]

For the closed systems given in Table 5.2, Cobb and Smith[19] show that the average reservoir pressure is reached after a dimensionless shut-in time given by

$$\Delta t_{DA} = (t_{DA})_{pss}, \quad \text{(5.17)}$$

where $(t_{DA})_{pss}$ is given in Table C.1. In highly asymmetrical systems, or in systems with eccentric well locations, $\bar{p}$ is reached slightly before this time.

Example 5.2 Pressure Buildup Test Analysis—Miller-Dyes-Hutchinson Method

We use the buildup test data of Example 5.1 shown in Table 5.1. The log-log data plot, Fig. 5.3, shows that wellbore storage effects have died out after 0.75 hour. The MDH plot, p_{ws} vs log Δt, shown in Fig. 5.8 has a straight line with

$$m = 40 \text{ psig/cycle},$$

and,

$$p_{1hr} = 3{,}266 \text{ psig}.$$

Using Eq. 5.6,

$$k = \frac{162.6(4{,}900)(1.55)(0.20)}{(40)(482)} = 12.8 \text{ md},$$

the result obtained in Example 5.1. The skin factor may be estimated as in Example 5.1 to give the same result.

TABLE 5.2—SHAPES USED IN FIGS. 5.6 AND 5.7.

CURVE NUMBER	SHAPE
1	Square 1 × 1, well at center
2	Square 1 × 1, well off center
3	Square 1 × 1, well in quadrant
4	Rectangle 2 × 1, well at center
5	Rectangle 2 × 1, well off center
6	Rectangle 4 × 1, well at center
7	Square 1 × 1, well at center (constant-pressure boundaries)

——— NO FLOW

- - - - - - CONSTANT PRESSURE

The end of the MDH straight line may be estimated with Eq. 5.16 and Fig. 5.7. Using Eq. 2.3b and data from Example 5.1,

$$t_{pDA} = \frac{(0.0002637)(12.8)(310)}{(0.09)(0.20)(22.6 \times 10^{-6})(\pi)(2{,}640)^2}$$

$$= 0.117.$$

For Curve 1 in Fig. 5.7, $(\Delta t_{DA})_{esl} \simeq 0.004$. We approximate the well in the center of a circle by a well in the center of a square, since the two systems behave similarly (see Table C.1). Using Eq. 5.16,

$$\Delta t = \frac{(0.09)(0.20)(22.6 \times 10^{-6})(\pi)(2{,}640)^2(0.004)}{(0.0002637)(12.8)}$$

$$= 10.6 \text{ hours}.$$

The straight line chosen uses more than a full cycle of data before 10 hours, so the analysis should be correct. Actually, the straight line in Fig. 5.8 appears to last to about 20 hours. This may indicate that Fig. 5.7 provides conservative estimates, or that the value of A or c_t used in this example is too small. The end of the straight line for the Horner plot is estimated in a similar manner to be about 26 hours, which agrees well with Fig. 5.4.

Extended Muskat Analysis

In 1937, Muskat[3] proposed plotting pressure buildup data as $\log(\bar{p} - p_{ws})$ vs Δt. Subsequent theoretical studies[10,19,21] indicate that this graph should be used with caution and only as a *late-time* analysis method. Because of the long shut-in times usually required for pressure buildup data to reach the Muskat straight line, the method has limited value for pressure-buildup test analysis. However, it appears to be more practical for analyzing pressure buildup data in producing wells in water-drive reservoirs and filled-up waterfloods because of the longer duration of the Muskat straight line in those systems.[21,22]

The Muskat method uses a trial-and-error plot with several $\bar{p}$ estimates; a straight line is obtained for the correct $\bar{p}$. Fig. 5.9 is a schematic illustration of the extended Muskat data plot.[1,9,10] If the assumed $\bar{p}$ is too high, the plot will be concave upward; if $\bar{p}$ is too low, the plot will be concave downward.

The intercept at $\Delta t = 0$, $(\bar{p} - p_{ws})_{int}$, of the correct straight line of an extended Muskat plot may be used to estimate reservoir permeability from[10]

$$k = \frac{141.2\, qB\mu}{h(\bar{p} - p_{ws})_{int}}\, p_{DMint}\,(t_{pDA}). \quad \ldots \ldots \ldots \ldots \quad (5.18)$$

The dimensionless Muskat intercept, p_{DMint}, which is a function of dimensionless producing time, is given by Ramey and Cobb[10] for a single unfractured well in the center of a closed-square drainage system, and by Kumar and Ramey[22] for a square with a constant-pressure boundary. Fig. 5.10 shows the data for both systems. For the closed-square system,

$$p_{DMint}(t_{pDA} > 0.1) = 0.67, \quad \ldots \ldots \ldots \ldots \quad (5.19a)$$

if producing time exceeds the time to pseudosteady state. For the constant-pressure boundary system,

$$p_{DMint}(t_{pDA} > 0.25) = 1.34, \quad \ldots \ldots \ldots \ldots \quad (5.19b)$$

when producing time exceeds the time required to reach steady state. Matthews and Russell[1] (Page 31) and Russell[9] give data indicating that $p_{DMint}(t_{pDA} > 0.1) = 0.84$ for a closed circular system. The large difference between values of 0.67 and 0.84 for a square and a circular system operating at pseudosteady state is cause for concern. Under most circumstances, those two systems should behave almost identically. Yet, for the Muskat method of analysis, the appropriate factor to use in the analysis varies by about 25 percent. This is definitely an indication that the accuracy of the method is open to question.

The value of the Muskat intercept for the square with the constant-pressure boundary (Fig. 5.10) is significantly different from the value for the closed square. That results from the different behavior of the systems and because the Muskat plot for the constant-pressure boundary system uses p_e, the constant-boundary pressure, rather than $\bar{p}$, the average pressure at the time of shut-in.

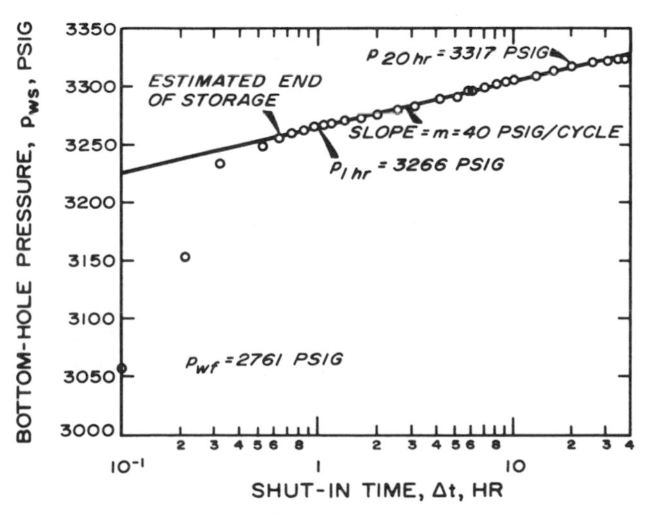

Fig. 5.8 Miller-Dyes-Hutchinson plot for the buildup test of Example 5.2.

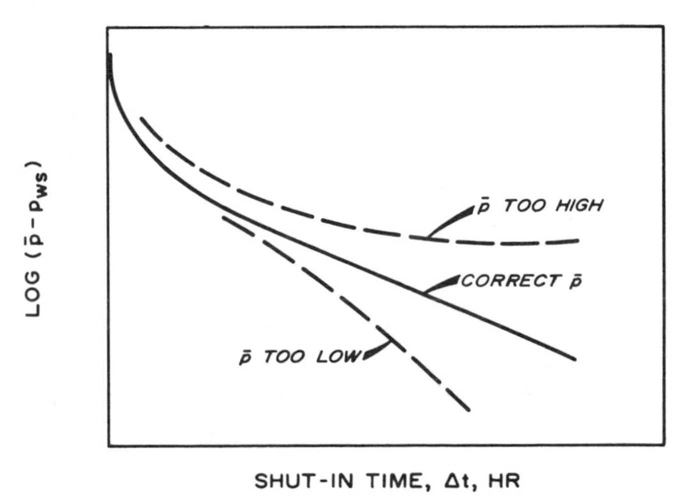

Fig. 5.9 Schematic extended Muskat data plot for pressure buildup test analysis.

The slope of the Muskat-plot straight line may be used to estimate drainage area. For a closed square,[10]

$$A = \frac{-0.00471\,k}{\phi\mu c_t m_M}, \qquad (5.20a)$$

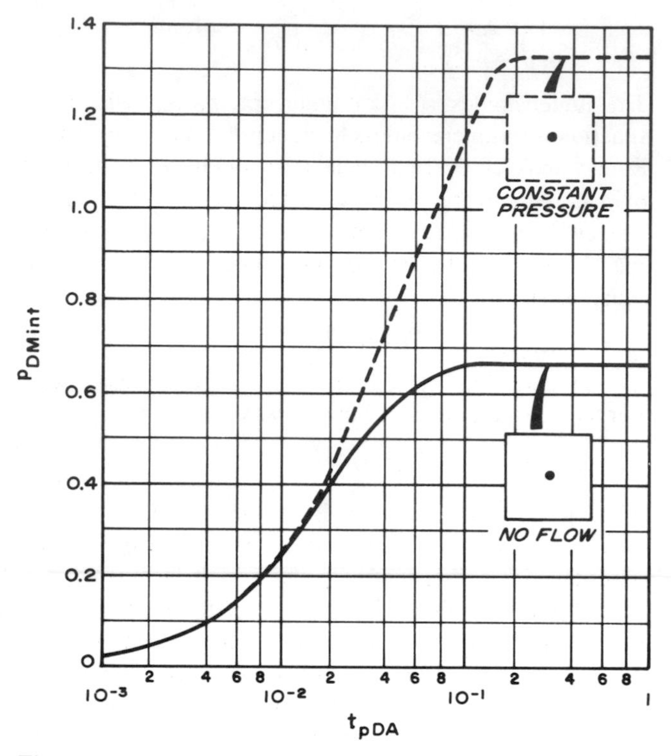

Fig. 5.10 Muskat dimensionless-intercept pressure for a well in the center of a closed or constant-pressure-boundary square. Data of Ramey and Cobb[10] and Kumar and Ramey.[22]

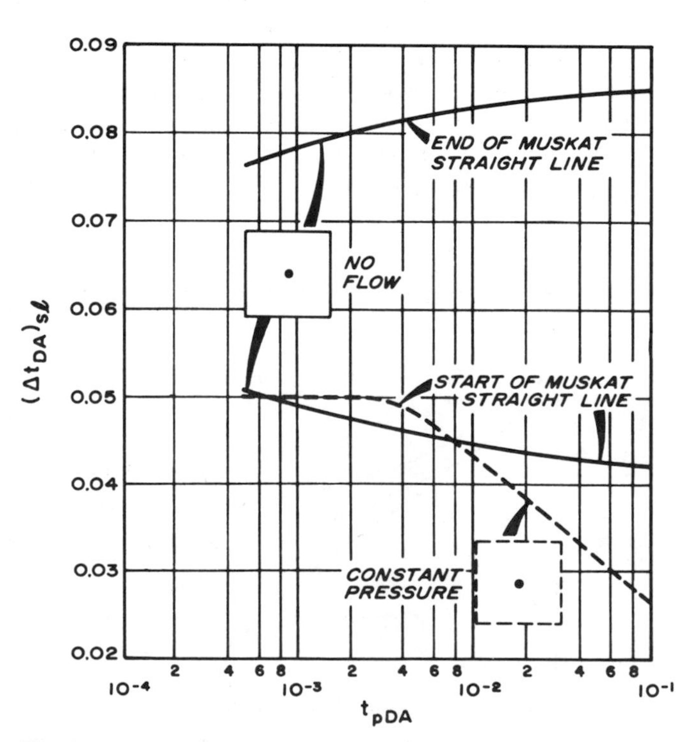

Fig. 5.11 Time to beginning and end of the Muskat straight line for a well in the center of a closed or constant-pressure boundary square. Data of Ramey and Cobb[10] and Kumar and Ramey.[22]

and for a square with constant pressure boundaries,

$$A = \frac{-0.00233\,k}{\phi\mu c_t m_M}. \qquad (5.20b)$$

In Eq. 5.20, m_M is the slope of the Muskat plot and is a negative number. Matthews and Russell[1] (Page 31) and Russell[9] indicate that the constant in Eq. 5.20a is -0.00528 for a closed circular system with the well at the center. Again, we see a large discrepancy in the values used for analyzing pressure buildup data in square and circular systems. That discrepancy in both Eqs. 5.19a and 5.20a may indicate that very accurate results should not be expected when using the Muskat analysis method.

The beginning and end of the Muskat straight line may be estimated from

$$\Delta t = \frac{\phi\mu c_t A}{0.0002637\,k}(\Delta t_{DA})_{sl}, \qquad (5.21)$$

where $(\Delta t_{DA})_{sl}$ is shown in Fig. 5.11. Data for both the start and the end of the straight line are given for the closed square. For the square with constant-pressure boundaries, Fig. 5.11 indicates only the start of the straight line; it can be expected to end at $\Delta t_{DA} = 0.25$.[21,22]

Example 5.3 Pressure Buildup Test Analysis—Extended Muskat Method

A Muskat plot of the pressure data of Example 5.1 is shown in Fig. 5.12. Only four points define the straight line in Fig. 5.12, so we should immediately be suspicious of analysis results. From Fig. 5.12,

$$(\bar{p} - p_{ws})_{int} = 31.4 \text{ psig},$$

and

$$m_M = -0.00586 \text{ cycle/hr}.$$

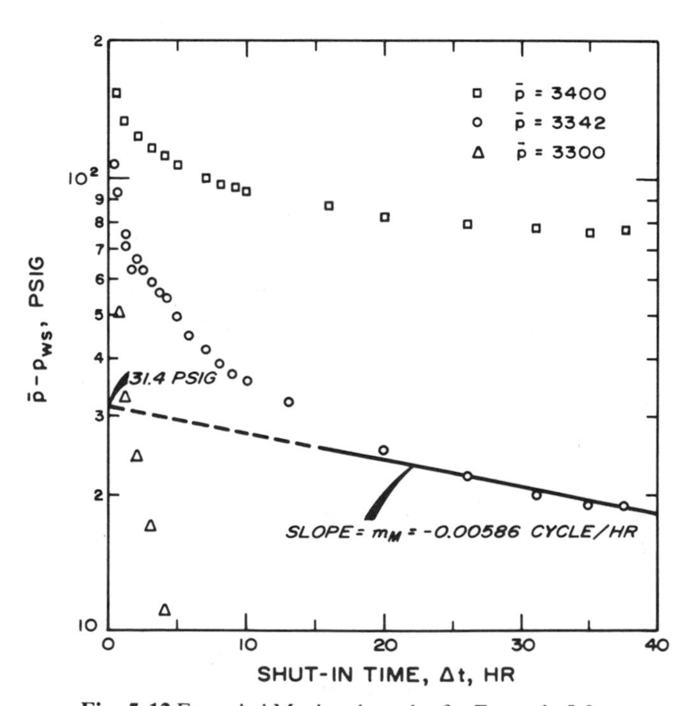

Fig. 5.12 Extended Muskat data plot for Example 5.3.

We will use Eq. 5.18 to estimate permeability from the intercept, but we must first obtain $p_{D\text{Mint}}$ (t_{pDA}) from Fig. 5.10. Thus, we assume a single well in the center of a closed square; normally, that would be a reasonable approximation to the circular system, but as shown with Eqs. 5.19 and 5.20, it may not be so reasonable for the extended Muskat analysis. Using Eq. 2.3b and an estimated permeability of 12 md,

$$t_{pDA} \simeq \frac{(0.0002637)(12)(310)}{(0.09)(0.20)(22.6 \times 10^{-6})(\pi)(2{,}640)^2}$$

$$\simeq 0.11.$$

Thus, for a first estimate, $p_{D\text{Mint}} = 0.67$ from Eq. 5.19a and Fig. 5.10. Using Eq. 5.18,

$$k = \frac{(141.2)(4{,}900)(1.55)(0.67)(0.20)}{(31.4)(482)} = 9.5 \text{ md}.$$

Since this is lower than the value used to estimate t_{pDA}, we need to iterate. Repeating the computations, $t_{pDA} \simeq 0.087$, $p_{D\text{Mint}} \simeq 0.665$, and $k = 9.4$, which is acceptable agreement.

This value of k is about 27 percent lower than estimated by the Horner and MDH methods. This is because the Muskat straight line is poorly defined in this case and apparently is not drawn through data from the correct time interval (estimated to be about 150 to 300 hours later in this example). Ramey and Cobb[10] indicate that only *very* late buildup data are straightened by the Muskat plot; we may not have enough of those data. Unfortunately, data at such times also may be influenced by interference from other wells or other boundary effects.

In spite of the problems with the data, we estimate A and the duration of the straight line to illustrate the method. Using Eq. 5.20 and the slope of the straight line in Fig. 5.12,

$$A = \frac{(-0.00471)(9.4)}{(-0.00586)(0.20)(0.09)(22.6 \times 10^{-6})}$$

$$= 18.6 \times 10^6 \text{ sq ft}.$$

From data provided for Example 5.1,

$$A = \pi\,(2{,}640)^2 = 21.9 \times 10^6 \text{ sq ft}.$$

So, the extended Muskat analysis is low by about 15 percent.

Fig. 5.11 indicates that for $t_{pDA} = 0.09$, the straight line should meet the requirement $0.042 < (\Delta t_{DA})_{sl} < 0.085$. Using Eq. 5.21,

$$\left[\frac{(0.09)(0.20)(22.6 \times 10^{-6})(\pi)(2{,}640)^2}{(0.0002637)(9.4)}\right](0.042)$$

$$< \Delta t < [3{,}590](0.085).$$

Thus,

$$151 < \Delta t < 305 \text{ hours}.$$

This clearly indicates that even though the extended Muskat plot *appears* to have a straight line, that straight line does not occur at the right shut-in time for analysis. In this case, careful analysis (using Eq. 5.21 and Fig. 5.11 to check the limits on the straight line) indicates that the results should be considered as incorrect — at least when analysis techniques for a closed square are used.

If, instead of using the Ramey and Cobb[10] data for a closed square system, we use the Matthews and Russell[1] data for a closed circular system, we would estimate

$$k = 11.9 \text{ md},$$

using a value of 0.84 in Eq. 5.19 rather than 0.67. Using a value of -0.00528 as the constant in Eq. 5.20a, we estimate

$$A = 20.8 \times 10^6 \text{ sq ft}.$$

These values are within 7 and 5 percent of the correct values, respectively. Thus, we see that we get much better results by using the analysis method for the closed circle than the closed square. This may or may not be coincidental. Unfortunately, little information is available for analyzing Muskat plots for closed, circular, or other systems.

Example 5.3 illustrates that results from the extended Muskat analysis method may be incorrect. Fig. 5.11 shows that it takes a long time for the extended Muskat straight line to develop; for a well in the center of a closed square, the well must be shut in almost half as long as it would take for that well to reach pseudosteady state under producing conditions. Detailed data are not available for shut-in requirements for other well locations and drainage shapes, but we can expect the requirements to be as severe as those for a well in a closed square. The relatively large difference in the analysis constants for the square and circular systems is an indication that systems with other shapes may have significantly different constants in the analysis equations. That can be expected because the Muskat analysis technique applies to data taken late in the buildup period, when the system boundaries have their largest effect on pressure behavior.

Fig. 5.12 also indicates the difficulty of identifying the Muskat straight line. The uppermost set of points, with $\bar{p} =$ 3,400 psi, appears to be straight for the last four points. Certainly it is as straight as the line for the assumed correct $\bar{p}$ in Example 5.3. The set of points for the assumed correct $\bar{p}$ appears to have a second straight-line section from about 8 to 20 hours. Only experience and the use of Eq. 5.21 and Fig. 5.11 would prevent drawing and analyzing that straight line. Eqs. 5.18 and 5.20 indicate that a straight line drawn from 8 to 20 hours would yield even lower permeability and drainage area.

Because of the long time required for pressure buildup data to reach the Muskat straight line, and because small geometry differences have a large effect on analysis results, the Muskat method has limited value for pressure-buildup test analysis.

Developed Reservoir Effects

When the test well is producing at pseudosteady state before a buildup test, or when it is experiencing a pressure decline because of production from other wells in the reservoir, the analysis methods discussed previously may give incorrect results. In such cases, it is best to use Eq. 5.1 in a more general form. Slider[23,24] has suggested analysis techniques that apply in these situations; Sections 3.4 and 4.5 give them for drawdown and multiple-rate testing. A similar approach is used for pressure buildup testing, as explained below.

Extrapolate the flowing pressure into the buildup period to estimate $p_{w\,ext}$. (Fig. 3.8 gives nomenclature for a flowing test.) Then estimate the difference between the observed shut-in pressure and the extrapolated flowing pressure, $\Delta p_{\Delta t}$, and plot $\Delta p_{\Delta t}$ vs log Δt. The data plotted should follow the relationship

$$\Delta p_{\Delta t} = p_{ws} - p_{w\,ext} = \Delta p_{1hr} + m \log \Delta t, \quad \ldots\ldots (5.22)$$

a straight line on a $\Delta p_{\Delta t}$ vs log Δt plot with slope m given by Eq. 5.5 and intercept given by

$$\Delta p_{1hr} = m\left[\log\left(\frac{k}{\phi\mu c_t r_w^2}\right) - 3.2275 + 0.86859\,s\right]. \quad \ldots\ldots (5.23)$$

Permeability is estimated from Eq. 5.6 and skin factor is estimated from

$$s = 1.1513\left[\frac{\Delta p_{1hr}}{m} - \log\left(\frac{k}{\phi\mu c_t r_w^2}\right) + 3.2275\right]. \quad \ldots\ldots (5.24)$$

If pressure decline at the test well is linear before the buildup test, then Eq. 5.22 becomes

$$p_{ws} - m^*\Delta t = \Delta p_{1hr}^* + m \log \Delta t, \quad \ldots\ldots (5.25)$$

where m^* is the linear rate of pressure change before the buildup test:

$$m^* = \frac{dp_{wf}}{dt} \quad \text{when } t < t_p\,. \quad \ldots\ldots (5.26)$$

Normally, m^* is a negative number. The value of Δp_{1hr}^* in Eq. 5.25 is derived from Eq. 5.23 and the equation for the extrapolated linear pressure behavior:

$$\Delta p_{1hr}^* = p_{wf}(\Delta t = 0) + m\left[\log\left(\frac{k}{\phi\mu c_t r_w^2}\right) - 3.2275 + 0.86859\,s\right]. \quad \ldots\ldots (5.27)$$

Thus, when pressure declines linearly before testing, a plot of $(p_{ws} - m^*\Delta t)$ vs log Δt should be a straight line. Skin factor may be estimated from Eq. 5.7 with Δp_{1hr}^* in place of p_{1hr}. Permeability is obtained from Eq. 5.6.

Example 5.4 Pressure Buildup Test Analysis—Developed System Analysis

Figs. 5.13 and 5.14 show simulated pressure buildup data for a well in a developed reservoir. All wells were produced at a constant rate for 15 days; then, one well was shut in for buildup while the others continued to produce. Known data are

$k = 15$ md	$q = 51$ STB/D
$h = 11$ ft	$B = 1.21$ RB/STB
$\mu = 1.37$ cp	$c_t = 26.7 \times 10^{-6}$ psi^{-1}
$\phi = 0.173$	$s = 0$
$r_w = 0.286$ ft	$p_{wf}(\Delta t{=}0) = 2{,}634$ psig.

Just before shut-in, bottom-hole flowing pressure was declining linearly at 7.836 psi/D, so $m^* = -0.3265$ psi/hr.

Fig. 5.13 is a plot of $\log[p_{ws} - p_{wf}(\Delta t = 0)]$ vs log Δt. It shows that wellbore storage effects are important initially, but die out after about 4 hours of shut-in time. The first two data points are on the unit-slope straight line; thus, Eq. 2.20 can be used to estimate the wellbore storage coefficient. From Fig. 5.13, $\Delta p = 20$ psig at $\Delta t = 0.1$ hour. Then,

$$C = \frac{(51)(1.21)}{24}\,\frac{0.1}{20} = 0.0129 \text{ RB/psig}.$$

For an oil gravity of 30 °API, $\rho = 54.7$ lb$_m$/cu ft and Eq. 2.16 may be used to estimate pipe size for a rising fluid level:

$$V_u = (0.0129)\,\frac{54.7}{144}\,\frac{32.17}{32.17} = 0.0049 \text{ bbl/ft}.$$

This corresponds roughly to 2-in. tubing.

Fig. 5.14 is a semilog plot of data from this test. Both the correct quantity accounting for the declining reservoir pres-

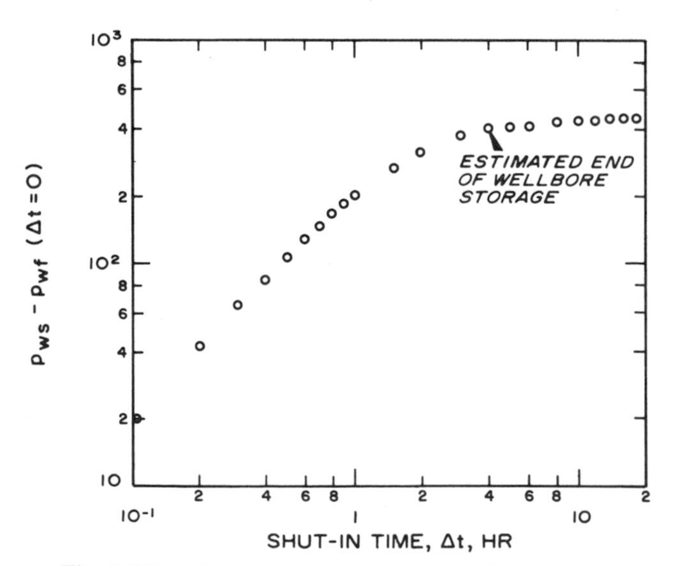

Fig. 5.13 Log-log data plot for a pressure buildup test in a developed reservoir, Example 5.4.

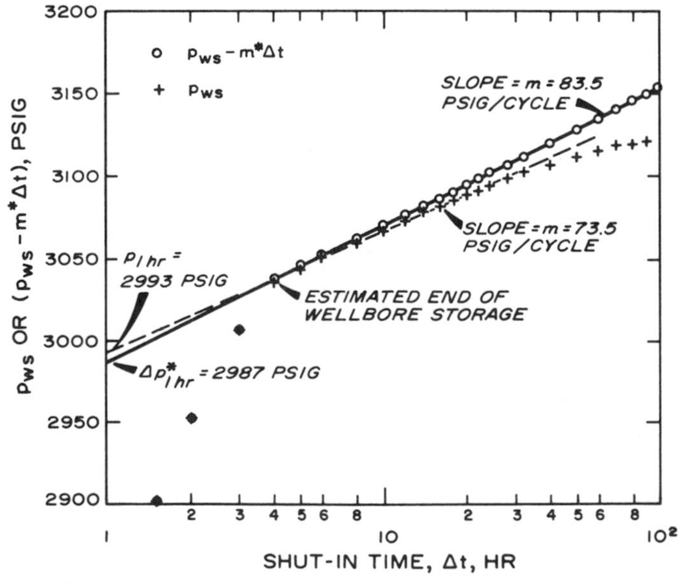

Fig. 5.14 Data plot for a pressure buildup test in a developed reservoir, Example 5.4.

sure, $p_{ws} - m^*\Delta t$, and the normal MDH plot are shown. For the correct data plot (circles in Fig. 5.14),

$m = 83.5$ psig/cycle,

and

$\Delta p_{1hr}{}^* = 2{,}987$ psig.

We estimate permeability from Eq. 5.6:

$$k = \frac{(162.6)(51)(1.21)(1.37)}{(83.5)(11)} = 14.97 \text{ md.}$$

This is within 0.2 percent of the input value, 15 md. The skin factor is estimated from Eq. 5.7 using $\Delta p_{1hr}{}^*$ in place of p_{1hr}:

$$s = 1.1513\left[\frac{2{,}987-2{,}634}{83.5} - \log\left(\frac{14.97}{(0.173)(1.37)(26.7\times10^{-6})(0.286)^2}\right) + 3.2275\right] = -0.007.$$

The input value is zero.

If the MDH plot (+'s in Fig. 5.14) is analyzed with Eqs. 5.6 and 5.7, we obtain k = 17.0 (a 13.3-percent error) and s = 0.7. In addition to providing more correct results, the $(p_{ws} - m^*\Delta t)$ plot has a semilog straight line of longer duration than the MDH plot. In this example, the line ends after about 24 hours for the MDH plot while data still fall on the semilog straight line after 100 hours for the correct data plot.

5.4 Buildup Test Analysis When Rate Varies Before Testing

Strictly speaking, the Horner and Miller-Dyes-Hutchinson plotting techniques apply only for a constant production rate preceding the buildup test. However, as indicated by Eq. 5.9, variable-rate conditions may be handled approximately in many circumstances. Nevertheless, in buildup tests with relatively short flow periods or with widely varying rate before shut-in, it is important to include the effects of rate variation on test analysis. For *infinite-acting systems* and unfractured wells, a variation of Eq. 4.1 may be used:

$$p_{ws} = p_i - m\sum_{j=1}^{N}\frac{q_j}{q_N}\log\left(\frac{t_N - t_{j-1} + \Delta t}{t_N - t_j + \Delta t}\right). \quad \text{(5.28)}$$

Fig. 5.15 identifies the nomenclature for the variable-rate period. Eq. 5.28 indicates that a plot of p_{ws} vs the summation term on the right-hand side should yield a straight-line portion with slope $-m$ given by Eq. 5.5 (with the final rate, q_N, used in place of q) and intercept p_i. Permeability is estimated from Eq. 5.6, and skin factor from Eq. 5.7 if $(t_N - t_{N-1}) >> 1$ hour or from Eq. 5.8 with t_p replaced by $(t_N - t_{N-1})$.

Odeh and Selig[25] propose a method similar to the Horner method for analyzing variable-rate buildup tests when the production period is shorter than the shut-in period. They suggest calculating a modified production time,

$$t_p{}^* = 2\left\{t_p - \frac{\sum_{j=1}^{N} q_j(t_j{}^2 - t_{j-1}{}^2)}{2\sum_{j=1}^{N} q_j(t_j - t_{j-1})}\right\}, \quad \text{(5.29)}$$

and a modified flow rate

$$q^* = \frac{1}{t_p{}^*}\sum_{j=1}^{N} q_j(t_j - t_{j-1}). \quad \text{(5.30)}$$

The summation term in Eq. 5.30 is the total stock-tank volume produced. The normal Horner analysis is used with $t_p{}^*$ and q^* in place of t_p and q.

Example 5.5 Pressure Buildup Test Analysis—Variable-Rate Analysis

Odeh and Selig[25] give the pressure buildup and pretest rate data in Table 5.3. They also indicate that

$B = 1.0$ RB/STB and $\mu = 0.6$ cp.

Fig. 5.16 is a plot of the data in Table 5.3, using Eq. 5.28. The summation term in that equation is written as follows for those data (N = 3):

$$\sum_{j=1}^{N}\frac{q_j}{q_N}\log\left(\frac{t_N - t_{j-1} + \Delta t}{t_N - t_j + \Delta t}\right)$$

$$= \frac{478.5}{159.5}\log\left(\frac{9-0+\Delta t}{9-3+\Delta t}\right) + \frac{319.0}{159.5}\log\left(\frac{9-3+\Delta t}{9-6+\Delta t}\right) + \frac{159.5}{159.5}\log\left(\frac{9-6+\Delta t}{9-9+\Delta t}\right)$$

$$= 3\log\left(\frac{9+\Delta t}{6+\Delta t}\right) + 2\log\left(\frac{6+\Delta t}{3+\Delta t}\right) + \log\left(\frac{3+\Delta t}{\Delta t}\right).$$

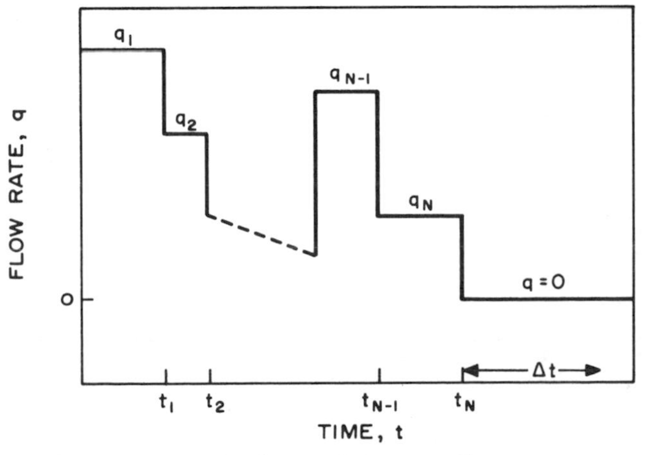

Fig. 5.15 Schematic of rate variation preceding a pressure buildup test.

For $\Delta t = 2$, this becomes

$$3 \log \left(\frac{9+2}{6+2}\right) + 2 \log \left(\frac{6+2}{3+2}\right) + \log \left(\frac{3+2}{2}\right)$$

$$= 0.4149 + 0.4082 + 0.3979$$

$$= 1.2210.$$

Table 5.3 summarizes the calculations. The slope in Fig. 5.16 gives $m = 153$ psi/cycle. Using Eq. 5.6 with the last rate,

$$kh = \frac{(162.6)(159.5)(1.0)(0.6)}{153} = 102 \text{ md ft.}$$

Odeh and Selig[25] indicate that the correct result for these simulated data is 106 md ft. They used the Horner plot to get 77 md ft and the modified analysis method of Eqs. 5.29 and 5.30 to get 97 md ft. It does appear that the varying rate before shut-in must be considered in analyzing this test.

Example 5.5 demonstrates that applying Eq. 5.28 can be tedious. Generally, a computer would be used to calculate the summation term in that equation.

5.5 Choice of Analysis Techniques

From a practical viewpoint, the Miller-Dyes-Hutchinson analysis technique is preferred because of its ease of use. When $t_p > t_{pss}$, the MDH method gives answers for permeability and skin that are just as good as those given by the Horner method. However, for short production times, the Horner method should be used since the semilog straight line with the correct slope is longer than the MDH plot. Recommended engineering practice is the following:

1. Use the MDH method as a first-pass method unless $t_p < t_{pss}$, or unless the system can be approximated by a well in the center of a square with constant-pressure boundaries (as in a five-spot, filled-up waterflood).

2. Use the Horner method for a second pass if circumstances dictate, or as a first pass if t_p is small.

When a test well in a developed reservoir has an extrapolatable flowing pressure decline, the developed-reservoir test analysis methods, Eqs. 5.22 and 5.25, should be used. Significant analysis errors can result if normal Horner or MDH plotting is used in such cases. When flow rate varies significantly before buildup testing, a variable flow-rate analysis technique, such as Eqs. 5.28 through 5.30, should be used.

The extended Muskat analysis method uses pressure buildup data occurring in the transition between the normal Horner or MDH straight line and the onset of average reservoir pressure.[10] This technique is not recommended except in unusual circumstances because of (1) the long testing time required for the extended Muskat method to exhibit the straight line; (2) the ease of observing an apparent straight line at incorrect average pressures or at times before the straight line actually starts; (3) frequent distortion caused by interference effects of other wells; and (4) the significant effect of system shape on the coefficients in the analysis equations. The extended Muskat method appears to be more applicable to wells in waterfloods and in water-drive reservoirs that more closely approach constant pressure than a no-flow boundary condition.[21,22]

5.6 Factors Complicating Pressure Buildup Testing

Frequently, pressure buildup tests are not as simple as discussed in the preceding examples. Many factors can influence the shape of a pressure buildup curve. An unusual shape may require explanation to complete a proper analysis, or it may prevent a proper analysis. In addition to wellbore storage effects, hydraulic fractures, particularly in low-permeability formations, can have a major effect on buildup-curve shape and analysis. Chapter 11 gives a more detailed discussion of both these factors.

One example of a pressure buildup curve that has an unusual shape when analyzed by the Horner or Miller-Dyes-Hutchinson methods is a test run with a nonstabilized rate before testing. It is important to recognize that condition

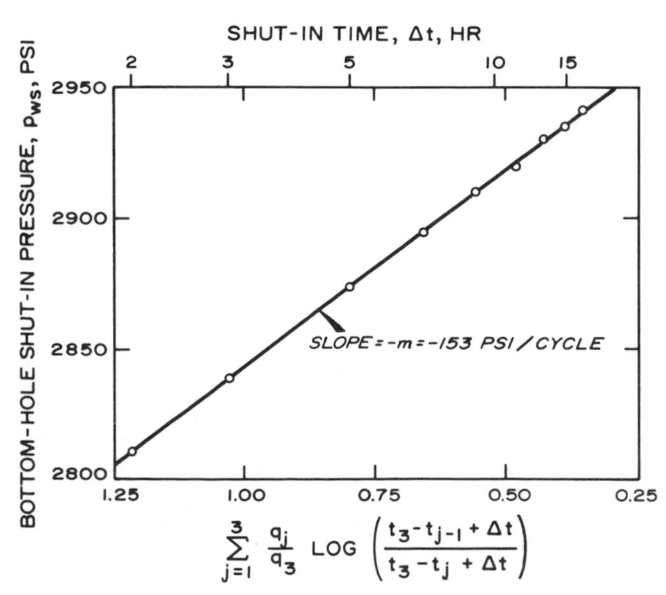

Fig. 5.16 Variable-rate pressure-buildup analysis plot for Example 5.5. Data of Odeh and Selig.[25]

TABLE 5.3—RATE AND PRESSURE DATA FOR EXAMPLE 5.5. Data of Odeh and Selig.[25]

Rate History

j	t_j (hours)	q_j (STB/D)
0	0	0
1	3	478.5
2	6	319.0
3	9	159.5

Buildup Data and Computations

Δt (hours)	$3 \log \left(\frac{9+\Delta t}{6+\Delta t}\right)$	$2 \log \left(\frac{6+\Delta t}{3+\Delta t}\right)$	$\log \left(\frac{3+\Delta t}{\Delta t}\right)$	Σ	p_{ws} (psig)
2	0.4149	0.4082	0.3979	1.2210	2,813
3	0.3748	0.3522	0.3010	1.0280	2,838
5	0.3142	0.2766	0.2041	0.7949	2,872
7	0.2705	0.2279	0.1549	0.6533	2,895
9	0.2375	0.1938	0.1249	0.5563	2,910
11	0.2117	0.1686	0.1047	0.4851	2,919
13	0.1910	0.1493	0.0902	0.4305	2,930
15	0.1740	0.1339	0.0792	0.3871	2,935
17	0.1597	0.1214	0.0706	0.3517	2,942

and consider it in the analysis, as discussed in Section 5.4. Other practical problems also can be troublesome. These include a bottom-hole pressure gauge in poor working condition, a leaking pump or lubricator, problems resulting from pump pulling before gauge placement, etc. Additionally, wells with a high gas-oil ratio can exhibit humping during pressure buildup[26] (see Fig. 2.13). In such cases, the bottom-hole pressure increases to a peak, decreases, and finally increases in a normal manner. In some situations, segregation of water and oil in the wellbore can produce a hump.

The shape of the pressure buildup curve also can be affected by fluid and rock interfaces, water-oil contacts, layering, and lateral fluid or rock heterogeneities. (See Chapter 10.) Wellbore storage, wellbore damage or improvement, and geometry of the drainage area can also affect the shape of a buildup curve. For examples, see Refs. 19, 20, and 27, as well as Chapter 11.

References

1. Matthews, C. S. and Russell, D. G.: *Pressure Buildup and Flow Tests in Wells,* Monograph Series, Society of Petroleum Engineers of AIME, Dallas (1967) **1,** Chap. 3.
2. Theis, Charles V.: "The Relation Between the Lowering of the Piezometric Surface and the Rate and Duration of Discharge of a Well Using Ground-Water Storage," *Trans.*, AGU (1935) 519-524.
3. Muskat, Morris: "Use of Data on the Build-Up of Bottom-Hole Pressures," *Trans.*, AIME (1937) **123,** 44-48. Also *Reprint Series, No. 9 — Pressure Analysis Methods,* Society of Petroleum Engineers of AIME, Dallas (1967) 5-9.
4. van Everdingen, A. F. and Hurst, W.: "The Application of the Laplace Transformation to Flow Problems in Reservoirs," *Trans.*, AIME (1949) **186,** 305-324.
5. Miller, C. C., Dyes, A. B., and Hutchinson, C. A., Jr.: "The Estimation of Permeability and Reservoir Pressure From Bottom-Hole Pressure Build-Up Characteristics," *Trans.*, AIME (1950) **189,** 91-104. Also *Reprint Series, No. 9 — Pressure Analysis Methods,* Society of Petroleum Engineers of AIME, Dallas (1967) 11-24.
6. Horner, D. R.: "Pressure Build-Up in Wells," *Proc.*, Third World Pet. Cong., The Hague (1951) Sec. II, 503-523. Also *Reprint Series, No. 9 — Pressure Analysis Methods,* Society of Petroleum Engineers of AIME, Dallas (1967) 25-43.
7. Perrine, R. L.: "Analysis of Pressure Buildup Curves," *Drill. and Prod. Prac.*, API (1956) 482-509.
8. Matthews, C. S.: "Analysis of Pressure Build-Up and Flow Test Data," *J. Pet. Tech.* (Sept. 1961) 862-870. Also *Reprint Series, No. 9 — Pressure Analysis Methods,* Society of Petroleum Engineers of AIME, Dallas (1967) 111-119.
9. Russell, D. G.: "Extensions of Pressure Build-Up Analysis Methods," *J. Pet. Tech.* (Dec. 1966) 1624-1636; *Trans.*, AIME, **237.** Also *Reprint Series, No. 9 — Pressure Analysis Methods,* Society of Petroleum Engineers of AIME, Dallas (1967) 175-187.
10. Ramey, H. J., Jr., and Cobb, William M.: "A General Buildup Theory for a Well in a Closed Drainage Area," *J. Pet. Tech.* (Dec. 1971) 1493-1505; *Trans.*, AIME, **251.**
11. van Everdingen, A. F.: "The Skin Effect and Its Influence on the Productive Capacity of a Well," *Trans.*, AIME (1953) **198,** 171-176. Also *Reprint Series, No. 9 — Pressure Analysis Methods,* Society of Petroleum Engineers of AIME, Dallas (1967) 45-50.
12. Pinson, A. E., Jr.: "Concerning the Value of Producing Time Used in Average Pressure Determinations From Pressure Buildup Analysis," *J. Pet. Tech.* (Nov. 1972) 1369-1370.
13. Kazemi, Hossein: "Determining Average Reservoir Pressure From Pressure Buildup Tests," *Soc. Pet. Eng. J.* (Feb. 1974) 55-62; *Trans.*, AIME, **257.**
14. Agarwal, Ram G., Al-Hussainy, Rafi, and Ramey, H. J., Jr.: "An Investigation of Wellbore Storage and Skin Effect in Unsteady Liquid Flow: I. Analytical Treatment," *Soc. Pet. Eng. J.* (Sept. 1970) 279-290; *Trans.*, AIME, **249.**
15. Earlougher, Robert C., Jr., and Kersch, Keith M.: "Analysis of Short-Time Transient Test Data by Type-Curve Matching," *J. Pet. Tech.* (July 1974) 793-800; *Trans.*, AIME, **257.**
16. McKinley, R. M.: "Wellbore Transmissibility From Afterflow-Dominated Pressure Buildup Data," *J. Pet. Tech.* (July 1971) 863-872; *Trans.*, AIME, **251.**
17. McKinley, R. M.: "Estimating Flow Efficiency From Afterflow-Distorted Pressure Buildup Data," *J. Pet. Tech.* (June 1974) 696-697.
18. Gladfelter, R. E., Tracy, G. W., and Wilsey, L. E.: "Selecting Wells Which Will Respond to Production-Stimulation Treatment," *Drill. and Prod. Prac.*, API (1955) 117-129.
19. Cobb, William M. and Smith, James T.: "An Investigation of Pressure Buildup Tests in Bounded Reservoirs," paper SPE 5133 presented at the SPE-AIME 49th Annual Fall Meeting, Houston, Oct. 6-9, 1974 (an abridged version appears in *J. Pet. Tech.*, Aug. 1975, 991-996; *Trans.*, AIME, **259**).
20. Chen, Hsiu-Kuo and Brigham, W. E.: "Pressure Buildup for a Well With Storage and Skin in a Closed Square," paper SPE 4890 presented at the SPE-AIME 44th Annual California Regional Meeting, San Francisco, April 4-5, 1974.
21. Ramey, Henry J., Jr., Kumar, Anil, and Gulati, Mohinder S.: *Gas Well Test Analysis Under Water-Drive Conditions,* AGA, Arlington, Va. (1973).
22. Kumar, Anil and Ramey, Henry J., Jr.: "Well-Test Analysis for a Well in a Constant-Pressure Square," paper SPE 4054 presented at the SPE-AIME 47th Annual Fall Meeting, San Antonio, Tex., Oct. 8-11, 1972 (an abridged version appears in *Soc. Pet. Eng. J.*, April 1974, 107-116).
23. Slider, H. C.: "Application of Pseudo-Steady-State Flow to Pressure-Buildup Analysis," paper SPE 1403 presented at SPE-AIME Regional Meeting, Amarillo, Tex., Oct. 27-28, 1966.
24. Slider, H. C.: "A Simplified Method of Pressure Buildup Analysis for a Stabilized Well," *J. Pet. Tech.* (Sept. 1971) 1155-1160; *Trans.*, AIME, **251.**
25. Odeh, A. S. and Selig, F.: "Pressure Build-Up Analysis, Variable-Rate Case," *J. Pet. Tech.* (July 1963) 790-794; *Trans.*, AIME, **228.** Also *Reprint Series, No. 9 — Pressure Analysis Methods,* Society of Petroleum Engineers of AIME, Dallas (1967) 131-135.
26. Stegemeier, G. L. and Matthews, C. S.: "A Study of Anomalous Pressure Build-Up Behavior," *Trans.*, AIME (1958) **213,** 44-50. Also *Reprint Series, No. 9 — Pressure Analysis Methods,* Society of Petroleum Engineers of AIME, Dallas (1967) 75-81.
27. Ramey, H. J., Jr., and Earlougher, R. C., Jr.: "A Note on Pressure Buildup Curves," *J. Pet. Tech.* (Feb. 1968) 119-120.

Chapter 6

Estimating Average Reservoir Pressure

6.1 Introduction

Average reservoir pressure, $\bar{p}$, is used for characterizing a reservoir, computing oil in place, and predicting future reservoir behavior. It is fundamental in understanding much reservoir behavior in primary recovery, secondary recovery, and pressure-maintenance projects.

Average pressure for a reservoir without water influx is sometimes defined as the pressure the reservoir would reach if all wells were shut in for infinite time. An equivalent definition, assuming uniform compressibility, is the average pressure obtained by planimetering an isobaric map of the reservoir. For some purposes, such a volumetrically weighted average pressure over the entire productive area is satisfactory. In some applications, separate volumetrically weighted average pressures are desired for gas cap and oil areas using appropriate volumes and compressibilities. At other times, the average pressure at a boundary is desired for computing influx. In yet other situations, other types of averages for various areas or fluid types may be desired for use in models that use different reservoir-performance prediction techniques. Thus, the method of averaging can depend on the use intended.

In high-permeability reservoirs, observed pressures after 24 to 72 hours of shut-in sometimes may be used as control points for isobaric maps with acceptable accuracy. However, most situations require some correction of observed pressures to estimate average pressure near the well. In this regard, one common estimate is the average pressure for a well's drainage region. With the increasing use of reservoir simulation, it is often necessary to estimate average pressure around a well in an area equivalent to the model's grid blocks. In general, only drainage-region average pressure can be estimated from well tests in developed reservoirs.

If a single well in a multiple-well reservoir is shut in, its pressure will eventually decline as a result of withdrawals from other wells, so its pressure will not level out at the true average pressure of the well's drainage region at the instant of shut-in. Nevertheless, each well's drainage region — or, for that matter, any other composite region — does have an average pressure at a given time.

Section 5.2 indicates that the average reservoir pressure in an infinite-acting reservoir ($p^* = p_i = \bar{p}$) may be estimated by extrapolating the straight-line portion of a Horner plot for a shut-in well to $\left[(t_p + \Delta t)/\Delta t\right] = 1$, as indicated in Fig. 6.1. Estimating $\bar{p}$ is more complex for bounded drainage regions since the pressure normally falls below the semilog straight line. This chapter presents methods for estimating average drainage-region pressure from individual-well pressure buildup data.

6.2 Estimating Drainage Volume

Most reservoirs are depleted by several producing wells. In closed, single-phase, constant-compressibility systems containing only production wells, each well drains only a portion of the reservoir.[1,2] At any given time the drainage limits of a well are, from a mathematical standpoint, equivalent to physical barriers to flow located around the well. The average pressure estimate for a well applies only to its drainage region as it existed at the instant of shut-in. Matthews and Russell[1] and Matthews and Lefkovits[2] treat the concept of drainage volume in detail.

In single-phase depletion systems, for wells put on production at the same time, each well initially drains an equal *volume* of the reservoir without regard to production rate. Matthews and Russell[1] present a series of figures showing drainage boundary movement from initial production to

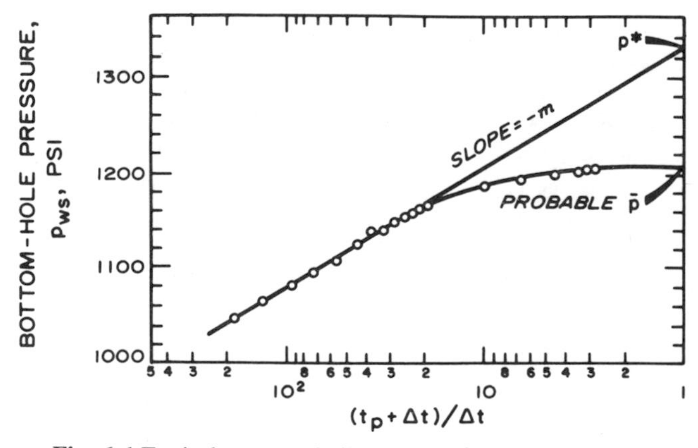

Fig. 6.1 Typical pressure-buildup curve for a well in a finite reservoir. After Matthews and Russell.[1]

pseudosteady-state conditions for unequal well production rates. Under pseudosteady-state conditions, the reservoir pore volume drained by a well is proportional to that well's production rate:[1,2]

$$V_{pi} = \frac{V_{pt} q_i}{q_t} \quad \text{. . . . . (6.1)}$$

This concept is based on pore volume and *not* on area; variations in thickness, porosity, and fluid saturation must be considered when changing drainage volumes estimated from Eq. 6.1 to drainage areas. *Eq. 6.1 applies to single-phase, constant-compressibility systems only.* When two or three fluid phases are present, the relationship between drainage volume and production rate may be much more complex because of (1) the complex nature of multiple-phase fluid flow through porous media; and (2) the possibility that some fluids may be produced from portions of the reservoir where other fluids are not mobile. For example, a well producing oil, water, and gas may produce oil (plus solution gas) from a portion of the pay section; water from below the water-oil contact, from the transition zone, or from coning; and gas from the gas cap, from the transition zone, or from coning. Under such conditions, it should be apparent that the simple approach indicated by Eq. 6.1 would not be necessarily correct.

Drainage regions in water-drive and pressure-maintained reservoirs are considerably more complex than those in depletion-type reservoirs. This subject is discussed in more detail in Section 6.4.

In spite of the difficulties of defining drainage volumes, there appear to be many situations where weighting estimated average drainage-region pressures proportional to production rates will provide workable estimates of volumetrically weighted average reservoir pressure. Using such a procedure, the drainage-region average pressure at the instant of shut-in would be estimated using methods presented in Sections 6.3 and 6.4. For that computation, individual-well drainage regions of an assumed nominal shape are required. The size of those regions would be estimated by using a fraction of the total reservoir hydrocarbon pore volume based on representative production rates before shut-in, as indicated by Eq. 6.1.[2] Then, the over-all average volumetrically weighted reservoir pressure would be estimated by weighting each drainage-region average pressure by the production rate in that drainage region.

Normally, pressure surveys include only a portion of the wells in a given reservoir. In such a case, it is often acceptable to compute average pressures based on nominal drainage areas related to the well pattern (for example, a 40-acre square). Such pressures then can be spotted on maps and contoured to obtain a volumetrically weighted average reservoir pressure. The desired pressure for contouring purposes is the average pressure in the vicinity of each well at some point in time. That pressure will differ from the pressure over the well's actual drainage region when the drainage region is highly skewed as the result of unequal production rates, irregular well locations, incomplete reservoir development, water influx, etc. Both approaches to estimating average reservoir pressure are pragmatic; in many cases, a result with a possible ±20-percent error is better than no result at all. It is hoped that the errors resulting from inaccuracies in evaluating drainage-region shapes will be in both directions and will tend to be self-cancelling on an over-all reservoir basis.

6.3 Estimating Drainage-Region Average Pressure

In his classic paper, Horner[3] presented a method for estimating average or initial reservoir pressure in an infinite system. That technique provides realistic estimates of average pressure for tests with short production periods, such as drillstem tests (see Chapter 8). Horner stated, however, that the extrapolation method does not apply to closed or multiple-well systems without correction. He said that, in closed, single-well reservoirs, the average pressure generally would be below the extrapolated false pressure, p^*, shown in Fig. 6.1. In fact, $\bar{p}$ is usually less than p^*, but it may be greater than p^* in highly skewed drainage shapes. This section presents several methods for using observed pressure-buildup data to estimate drainage-region average pressures. Those drainage-region average pressures then may be used to estimate a volumetrically weighted average reservoir pressure, as outlined in Section 6.2.

Matthews-Brons-Hazebroek Method

In 1954, Matthews, Brons, and Hazebroek[4] presented a technique for estimating average reservoir pressure from buildup tests in bounded drainage regions. The limitations of the method result from assumptions of no variation in drainage region, fluid mobility, or fluid compressibility. However, those limitations can be overcome effectively in multiple-well reservoirs (as discussed in Section 5.2 and later in this section) by using a production time, t_p, equal to t_{pss} rather than the longer actual production time. The Matthews-Brons-Hazebroek (MBH) technique does provide a way to estimate $\bar{p}$ for a well in almost any position within a variety of bounded drainage shapes. In using the method, the engineer ideally would divide the reservoir into drainage *areas*. Such division can be quite time-consuming, so nominal drainage regions based on well spacing or some other convenient parameter are often used in practice. When that is done, it should be recognized that significant differences in late-time buildup-curve appearance can result from unusually shaped drainage regions (normally, a result of high degrees of production imbalance or very irregular well spacing). Such circumstances could justify more precise evaluation of drainage-area shape as outlined in Section 6.2 and by Matthews, Brons, and Hazebroek,[4] depending on the intended use of the resulting average pressures.

To estimate drainage-volume average pressure by the MBH method, first extrapolate the Horner pressure-buildup plot (Section 5.2) to obtain the false pressure, p^* (Fig. 6.1). (Or p^* may be estimated from the Miller-Dyes-Hutchinson plot and Eq. 5.14). Then, average pressure is estimated from

$$\bar{p} = p^* - \frac{m}{2.3025} p_{D\text{MBH}}(t_{pDA}), \quad \text{. . . . . (6.2)}$$

where m is from the slope of the Horner (or Miller-Dyes-Hutchinson) straight line, Eq. 5.5:

$$m = \frac{162.6 qB\mu}{kh} \quad \ldots\ldots\ldots\ldots\ldots\ldots\ldots\ldots (6.3)$$

In Eq. 6.2, $p_{DMBH}(t_{pDA})$ is the MBH dimensionless pressure determined at the dimensionless time corresponding to t_p:

$$t_{pDA} = \frac{0.0002637\, kt_p}{\phi\mu c_t A} \quad \ldots\ldots\ldots\ldots\ldots\ldots\ldots (6.4)$$

Figs. 6.2 through 6.5 give MBH dimensionless pressures for several drainage-area shapes and well locations.

As in the normal Horner analysis technique, we have assumed that the well produces at constant rate q from time zero to t_p, when the well is shut in for buildup testing. If the rate is not constant, t_p is estimated as discussed in Section 5.2 (before Eq. 5.9). In such a case, one normally estimates t_p from Eq. 5.9:

$$t_p = \frac{24\, V_P}{q}, \quad \ldots\ldots\ldots\ldots\ldots\ldots\ldots\ldots\ldots (6.5)$$

where V_P is the cumulative volume produced since the *last pressure equalization* and q is the constant rate just before shut-in. Cumulative production from the last pressure survey is normally used because it is convenient. The important fact is that the Horner plot is based on superposition for an infinite-acting system, both *before* and *after* shut-in. If the system is not infinite-acting before shut-in, that must be considered in analysis, and either t_p or the analysis technique must be modified. The p_{DMBH} values from Figs. 6.2 through 6.5 include such a consideration.

Pinson[5] and Kazemi[6] indicate that t_p should be compared with the time required to reach pseudosteady state:

$$t_{pss} = \frac{\phi\mu c_t A}{0.0002637\, k}(t_{DA})_{pss}. \quad \ldots\ldots\ldots\ldots\ldots (6.6)$$

$(t_{DA})_{pss} \simeq 0.1$ for a symmetric closed square or circle with the well at the center and is given in the "Exact for $t_{DA} >$" column of Table C.1 for other shapes. If $t_p >> t_{pss}$, then t_{pss} should ideally replace t_p both for the Horner plot and in Eqs. 6.2 and 6.4 for use with the MBH dimensionless-pressure curves, Figs. 6.2 through 6.5.

Practically, substituting t_{pss} for a t_p usually will not significantly improve estimates of static pressure unless t_p is greater than about five to ten times t_{pss}, although results are relatively more sensitive with high rates. As discussed in Chapter 5, for a closed boundary condition, use of t_{pss} with the Horner method can increase the duration of the semilog straight line as opposed to an MDH plot, and may sometimes be a justification for a Horner plot using t_{pss} in place of t_p, where t_p is only 1.5 to 2 times t_{pss}.

Because of compensating factors (lower value of p^* and corresponding smaller correction), any value of t_p used with the Matthews-Brons-Hazebroek approach theoretically will give identical results for average pressure.[5] Practically, a relatively short t_p can eliminate serious numerical errors in calculated static pressures. That includes errors caused by long extrapolations and deviations from theoretical assumptions, such as (1) lack of rate stabilization before shut-in; (2) migration and changing drainage areas in multiple-well reservoirs; and (3) time variations in system compressibility and mobility.

Example 6.1 Average Drainage-Region Pressure—Matthews-Brons-Hazebroek Method

We use the pressure-buildup test data of Examples 5.1 through 5.3 (Table 5.1). Pressure buildup data are plotted in Figs. 5.3 and 5.4. Other data are

$\phi = 0.09$ | $m = 40$ psig/cycle (Fig. 5.4)
$c_t = 22.6 \times 10^{-6}$ psi^{-1} | $t_p = 310$ hours
$k = 12.8$ md (Example 5.1) | $A = \pi r_e^2 = \pi(2{,}640)^2$ sq ft.
$\mu = 0.20$ cp

To see if we should use $t_p = 310$ hours, we estimate t_{pss} from Eq. 6.6 using $(t_{DA})_{pss} = 0.1$ from Table C.1:

$$t_{pss} = \frac{(0.09)(0.2)(22.6 \times 10^{-6})(\pi)(2{,}640)^2(0.1)}{(0.0002637)(12.8)}$$

$$= 264 \text{ hours.}$$

Thus, we could replace t_p by 264 hours in the analysis. However, since t_p is only about 1.17 t_{pss}, we expect no difference in $\bar{p}$ from the two methods,[6] so we use $t_p = 310$ hours. As a result, Fig. 5.4 applies.

Fig. 5.4 does not show p^* since $(t_p + \Delta t)/\Delta t$ does not go to 1.0. However, we may compute p^* from p_{ws} at $(t_p + \Delta t)/\Delta t = 10$ by extrapolating one cycle:

$$p^* = 3{,}325 + (1 \text{ cycle})(40 \text{ psi/cycle})$$

$$= 3{,}365 \text{ psig.}$$

Using Eq. 6.4,

$$t_{pDA} = \frac{(0.0002637)(12.8)(310)}{(0.09)(0.20)(22.6 \times 10^{-6})(\pi)(2{,}640)^2}$$

$$= 0.117.$$

From the curve for the circle in Fig. 6.2, $p_{DMBH}(t_{pDA} = 0.117) = 1.34$. Eq. 6.2 gives the average pressure:

$$\bar{p} = 3{,}365 - \frac{40}{2.3025}(1.34)$$

$$= 3{,}342 \text{ psig.}$$

This is 19 psi higher than the maximum pressure recorded.

Dietz Method

Dietz[7] presents a slightly different approach for estimating $\bar{p}$. He suggests extrapolating the straight-line portion of an MDH plot (p_{ws} vs log Δt) directly to $\bar{p}$. The Dietz approach assumes that the well has been produced at a constant rate long enough to reach pseudosteady state before shut-in,[7] and that a semilog straight line of appropriate slope will develop (for wells that are not highly stimulated, $s > -3$). Dietz determined the time when $\bar{p}$ may be read *directly* from the extrapolated semilog straight line:

$$(\Delta t)_{\bar{p}} = \frac{t_p}{C_A t_{pDA}} = \frac{\phi\mu c_t A}{0.0002637\, C_A k}. \quad \ldots\ldots\ldots (6.7a)$$

For a well centrally located in a closed-square drainage area, $C_A = 30.8828$, so

$$(\Delta t)_{\bar{p}_{\text{square}}} = 122.8\, \frac{\phi\mu c_t A}{k}. \quad \ldots\ldots\ldots\ldots\ldots (6.7b)$$

For other shapes, the shape factor, C_A, is taken from Table

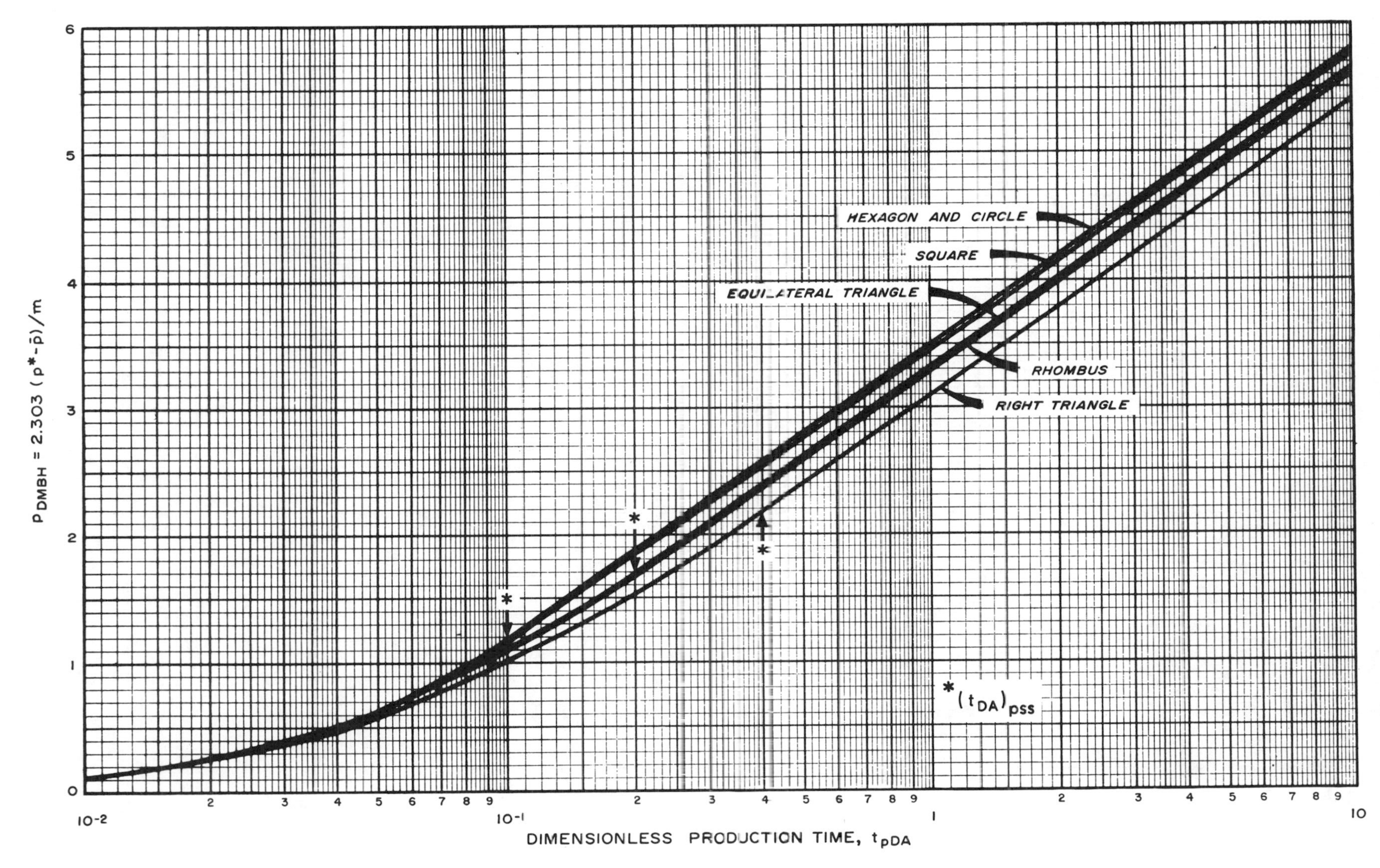

Fig. 6.2 Matthews-Brons-Hazebroek dimensionless pressure for a well in the center of equilateral drainage areas. After Matthews, Brons, and Hazebroek.[4]

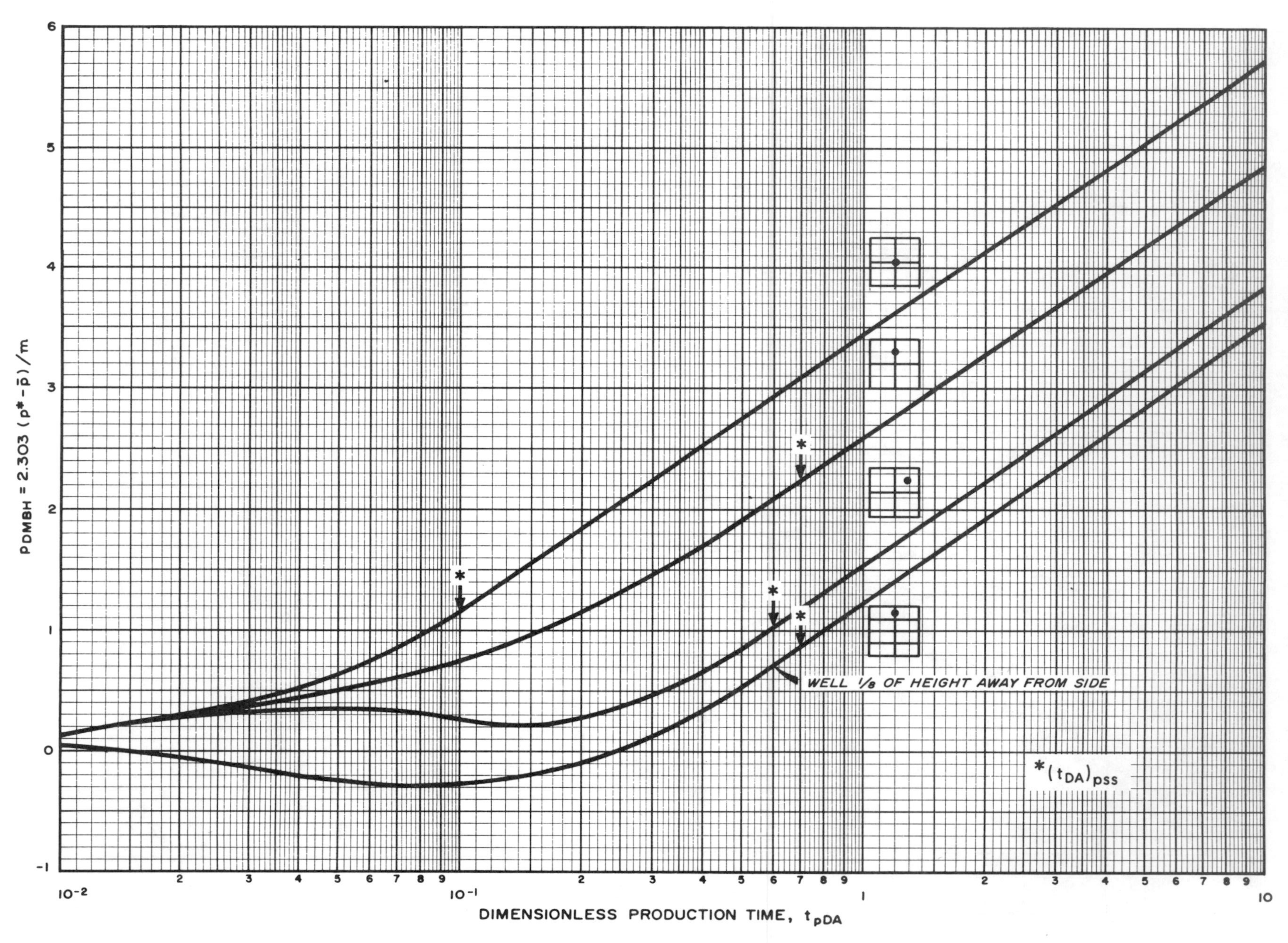

Fig. 6.3 Matthews-Brons-Hazebroek dimensionless pressure for different well locations in a square drainage area. After Matthews, Brons, and Hazebroek.[4]

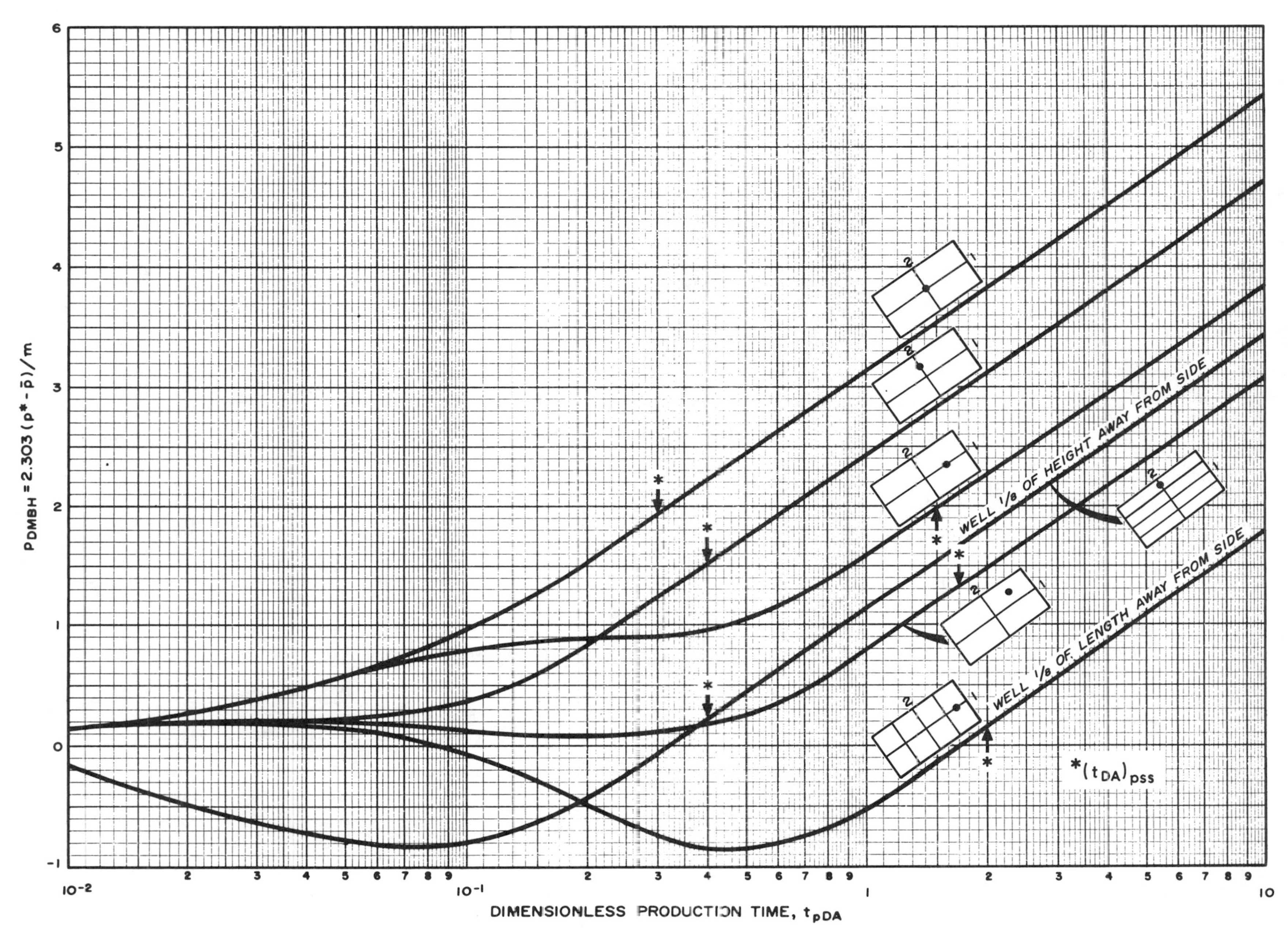

Fig. 6.4 Matthews-Brons-Hazebroek dimensionless pressure for different well locations in a 2:1 rectangular drainage area. After Matthews, Brons, and Hazebroek.[4]

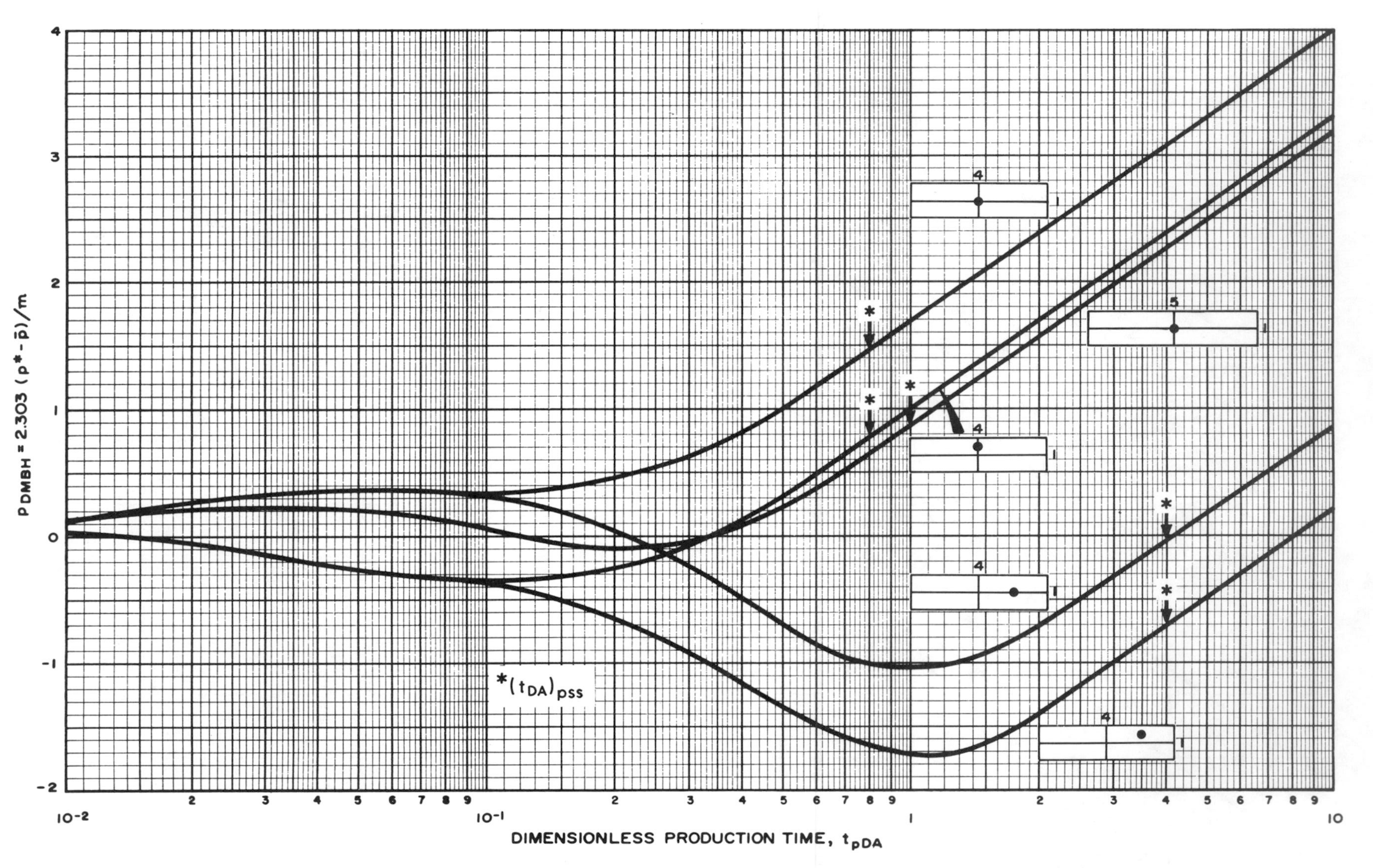

Fig. 6.5 Matthews-Brons-Hazebroek dimensionless pressure for different well locations in 4:1 and 5:1 rectangular drainage areas. After Matthews, Brons, and Hazebroek.[4]

C.1. The Dietz method has the advantage of being quick and simple, and usually is the preferred method for wells without a significant skin factor ($s > -3$ or $r_{wa} < 0.05\, r_e$) caused by acidizing or fracturing.

Ramey and Cobb[8] describe a method for extrapolating a *Horner plot* straight line to average reservoir pressure. For $t_p \geqslant t_{pss}$, they show that

$$\left(\frac{t_p + \Delta t}{\Delta t}\right)_{\bar{p}} = C_A t_{pDA} = \frac{0.0002637\, k t_p C_A}{\phi \mu c_t A} \quad \text{. . . (6.8a)}$$

For a centrally located well in a closed-square drainage area, $C_A = 30.8828$ and

$$\left(\frac{t_p + \Delta t}{\Delta t}\right)_{\bar{p}} = \frac{0.008144\, k t_p}{\phi \mu c_t A} \quad \text{. . . (6.8b)}$$

Eq. 6.8 reduces to Eq. 6.7 when $(t_p + \Delta t) \simeq t_p$. When $t_p < t_{eia}$, which may be estimated from Eq. 2.8, Ramey and Cobb show[8]

$$\left(\frac{t_p + \Delta t}{\Delta t}\right)_{\bar{p}} = e^{4\pi t_{pDA}}. \quad \text{. . . (6.9)}$$

The Matthews-Brons-Hazebroek,[4] Dietz,[7] and Ramey-Cobb[8] methods for estimating average reservoir pressure require a certain amount of knowledge about the drainage region. In particular, one must be able to approximate the boundary shape and well location and know that the boundary is effectively a no-flow boundary at the time the well is shut in. In many cases, one does not have such information with any degree of accuracy. Fortunately, under most circumstances it is permissible to assume a regular drainage shape based on the well pattern and use that when estimating average reservoir pressure with any of the three methods.

Example 6.2 Average Drainage-Region Pressure—Dietz Method

We again use the buildup test of Examples 5.1 through 5.3 and Example 6.1. Pertinent data are given in Example 6.1. The shape factor, C_A, for a closed circular reservoir is 31.62 (Table C.1). Using Eq. 6.7a,

$$(\Delta t)_{\bar{p}} = \frac{(0.09)(0.20)(22.6 \times 10^{-6})(\pi)(2{,}640)^2}{(0.0002637)(12.8)(31.62)}$$

$$= 83.5 \text{ hours.}$$

The MDH plot, Fig. 5.8, does not extend to 83.5 hours, but the straight line may be extrapolated to that time. From Fig. 5.8, $p_{ws} = 3{,}302$ psig at 8.35 hours, so extrapolating 1 cycle to 83.5 hours,

$$\bar{p} = 3{,}302 + (1 \text{ cycle})(40 \text{ psig/cycle})$$

$$= 3{,}342 \text{ psig,}$$

the result obtained in Example 6.1.

The equation proposed by Ramey and Cobb,[8] Eq. 6.8a, also may be used:

$$\left(\frac{t_p + \Delta t}{\Delta t}\right)_{\bar{p}} = \frac{(0.0002637)(12.8)(310)(31.62)}{(0.09)(0.20)(22.6 \times 10^{-6})(\pi)(2{,}640)^2}$$

$$= 3.71.$$

Extrapolating the Horner-plot straight line from 37.1 to 3.71 (Fig. 5.4) gives

$$\bar{p} = 3{,}302 + (1 \text{ cycle})(40 \text{ psig/cycle})$$

$$= 3{,}342 \text{ psig.}$$

Miller-Dyes-Hutchinson Method

Miller, Dyes, and Hutchinson[9] published a technique for estimating $\bar{p}$ for closed circular drainage regions from the MDH data plot (p_{ws} vs log Δt). The MDH average reservoir-pressure analysis method applies directly only to wells operating at pseudosteady state before the buildup test. (Ramey and Cobb[8] show how to use the MDH method for a closed square drainage region when $t_p < t_{pss}$.)

To use the MDH method to estimate average drainage-region pressure for a circular or square system producing at pseudosteady state before shut-in, choose *any convenient time* on the *semilog straight line,* Δt, and read the corresponding pressure, p_{ws}. Then calculate the dimensionless shut-in time based on the drainage area:

$$\Delta t_{DA} = \Delta t_D \left(\frac{r_w^2}{A}\right) = \frac{0.0002637\, k(\Delta t)}{\phi \mu c_t A} \quad \text{. . . (6.10)}$$

That dimensionless time is used with the upper curve of Fig. 6.6 to determine an MDH dimensionless pressure, $p_{D\,MDH}$. (Note that Fig. 6.6 is based on Δt_{DA} rather than on Δt_{De}, as is commonly done. That allows inclusion of the square systems.) The average reservoir pressure in the closed drainage region is estimated from

$$\bar{p} = p_{ws} + \frac{m\, p_{D\,MDH}(\Delta t_{DA})}{1.1513} \quad \text{. . . (6.11)}$$

In Eq. 6.11, p_{ws} is the pressure read from the MDH semilog straight line at any convenient shut-in time, Δt, and $p_{D\,MDH}$ is taken from the upper curve in Fig. 6.6 for the same Δt. The lower curves in Fig. 6.6 apply for estimating boundary pressures, p_e, in water-drive reservoirs and are discussed in Section 6.4.

Example 6.3 Average Drainage-Region Pressure—Miller-Dyes-Hutchinson Method

We consider the same buildup test as in Examples 6.1 and 6.2. The drainage area of the well is approximated by a circle with $r_e = 2{,}640$ ft. We choose $\Delta t = 20$ hours on the straight-line section of the MDH buildup curve in Fig. 5.8. From Eq. 6.10,

$$\Delta t_{DA} = \frac{(0.0002637)(12.8)(20)}{(0.09)(0.20)(22.6 \times 10^{-6})(\pi)(2{,}640)^2}$$

$$= 0.0076.$$

From the upper curve in Fig. 6.6, $p_{D\,MDH}$ at $\Delta t_{DA} = 0.0076$ is 0.78. From Fig. 5.8, p_{ws} at $\Delta t = 20$ hours is 3,317 psig and $m = 40$ psig/cycle. Then, using Eq. 6.11,

$$\bar{p} = 3{,}317 + (40)(0.78)/1.1513$$

$$= 3{,}344 \text{ psig.}$$

This compares with an average reservoir pressure of 3,342 psig estimated by the other methods.

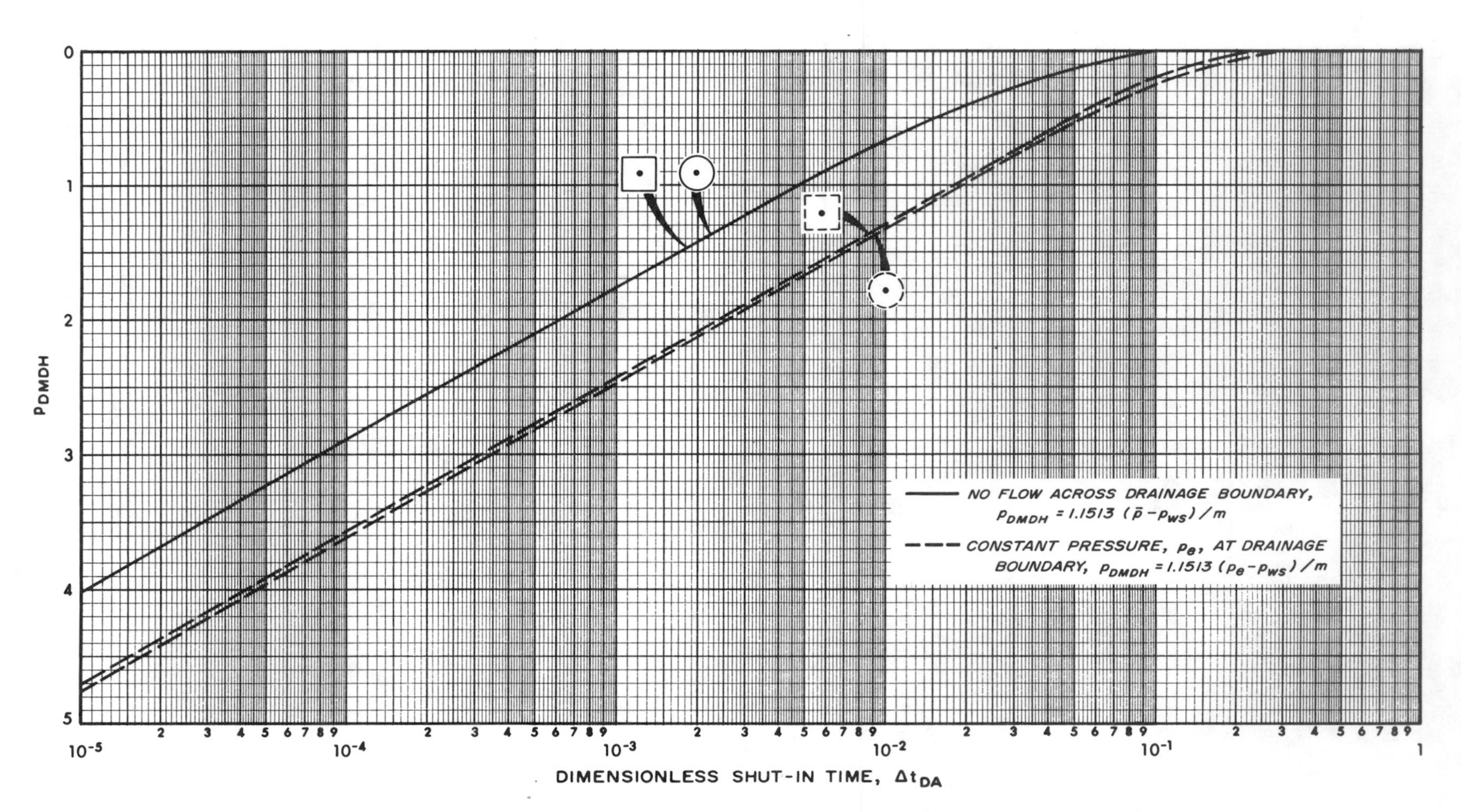

Fig. 6.6 Miller-Dyes-Hutchinson dimensionless pressure for circular and square drainage areas. Data of Miller, Dyes, and Hutchinson,[9] Perrine,[10] and Kumar and Ramey.[11]

Correcting Spot Pressure Readings to Average Drainage-Region Pressure

There are a surprising number of situations where a single pressure point, or "spot pressure," is the only pressure information available about a well. In older fields, the areal and chronological pressure coverage often consists only of spot readings. Brons and Miller[12] show two ways to estimate average pressure for single-well, closed drainage areas from such measurements. Their method is similar to the Dietz method in that it assumes the wells are producing at pseudosteady state before shut-in. It also assumes that the spot pressure measurement falls on the semilog straight-line portion of the buildup curve (that is, it is between the wellbore- and boundary-effects portions of the buildup curve). The technique requires knowledge or reasonable estimates of the shut-in time when pressure was measured, the flow rate, formation thickness, formation volume factor, fluid viscosity, and drainage-region shape and well location. Two approaches can be used. In the first, an estimate of formation permeability is required. When that approach is used, average drainage-region pressure is estimated from the spot pressure reading at shut-in time, Δt, using

$$\bar{p} = p_{ws}(\Delta t) + \frac{162.6\, qB\mu}{kh} \log\left(\frac{1}{\Delta t_{DA}\, C_A}\right), \qquad \text{(6.12a)}$$

or

$$\bar{p} = p_{ws}(\Delta t) + \frac{162.6\, qB\mu}{kh} \log\left(\frac{\phi\mu c_t A}{0.0002637\, k\, \Delta t\, C_A}\right). \qquad \text{(6.12b)}$$

For a closed square drainage region, $C_A = 30.8828$ and

$$\bar{p} = p_{ws}(\Delta t) + \frac{162.6\, qB\mu}{kh} \log\left(\frac{122.8\, \phi\mu c_t A}{k\, \Delta t}\right). \qquad \text{(6.12c)}$$

Note that Eq. 6.12 is independent of skin factor or apparent wellbore radius. However, it should not be expected to apply when $s < -3$.

If an estimate of permeability is not available, another approach may be used. That approach does have the drawback of requiring knowledge of both the skin factor (or apparent wellbore radius) and the flowing pressure before shut-in. Brons and Miller[12] showed that the k/μ term could be eliminated, so

$$\log\left(\frac{qB}{h}\right) = \log\left[(\bar{p} - p_{ws}) + (p_{ws} - p_{wf})\right] + \log\left(\frac{23.32\, \phi c_t A}{C_{Af}}\right) - \log(\Delta t) - \frac{f(\bar{p} - p_{ws})}{(\bar{p} - p_{ws}) + (p_{ws} - p_{wf})}, \qquad \text{(6.13a)}$$

where

$$f = \log\left(\frac{2.2458\, A}{r_w^2\, C_A}\right) + 0.86859\, s. \qquad \text{(6.13b)}$$

Unfortunately, solving Eq. 6.13 for $\bar{p}$ is difficult without a computer, or at least a programmable calculator. Brons and Miller suggested a graphical approach. Fig. 6.7 is a graphical representation of Eq. 6.13a (although different from the one proposed by Brons and Miller[12]). That figure applies only for the reservoir properties indicated on it, but the technique is general. To estimate $\bar{p}$, choose values for the reservoir properties indicated in Fig. 6.7 and calculate qB/h for a series of assumed values of $(p_{ws} - p_{wf})$ and $(\bar{p} - p_{ws})$. Once a figure like Fig. 6.7 has been constructed, qB/h and $(p_{ws} - p_{wf})$ are entered to determine $\bar{p} - p_{ws}$ (and, thus, $\bar{p}$). Fig. 6.7 is shown only for illustrative purposes; it is not meant to apply for general analyses.

Typical or average values for compressibility, porosity, and shape factor (commonly 30.8 to 31.6) often can be used with acceptable accuracy in the Brons-Miller technique in a given field situation. That is especially true since only average pressure in the vicinity of the well is usually desired for contouring. Engineering judgment is involved in any simplification and allowances should always be made for major changes in compressibility and mobility. Thus, λ_t and c_t should be used for multiple-phase reservoir flow, as indicated in Section 2.11. Normally, one or several simple graphs similar to Fig. 6.7 (for different skin factors) can be constructed for a given reservoir. Then, average pressure

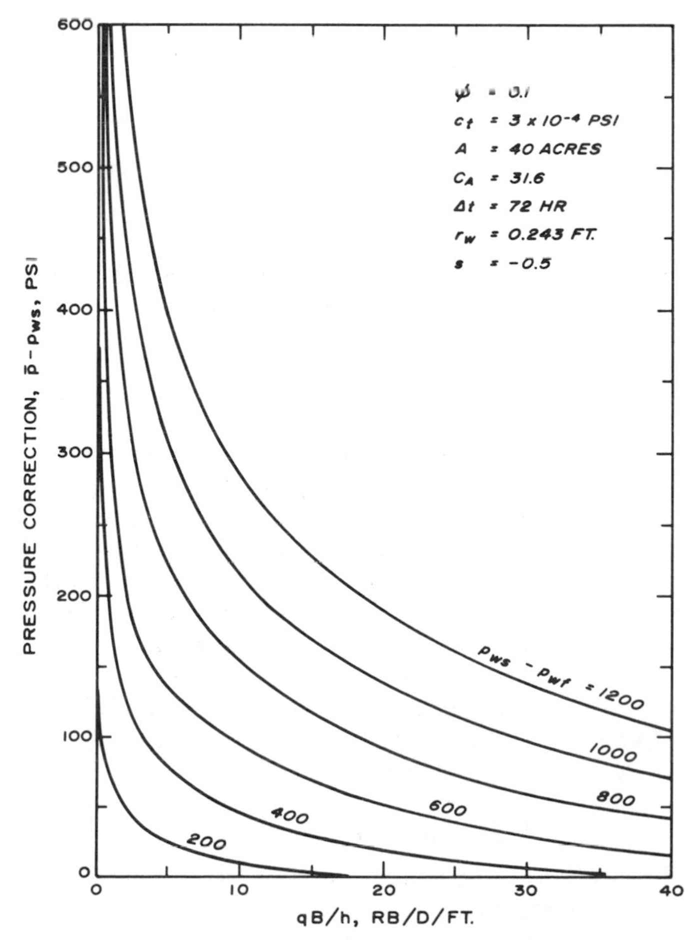

Fig. 6.7 Graph for estimating correction to a spot pressure reading to estimate average reservoir pressure. Brons-Miller[12] data and method.

may be estimated using the equivalent of Fig. 6.7, or Eq. 6.13a may be used with actual data. Of course, if an estimate of mobility is available, Eq. 6.12 may be used, thus greatly simplifying the analysis.

The method of Eq. 6.13a and Fig. 6.7 does not require knowledge of k/μ because that has been implicitly developed from previous flow rate and observed pressure data. The merit of such an approach is that it offers a way to correct spot pressure readings to average drainage-region pressures with limited data. The major disadvantages are the inherent inaccuracy in the implicitly estimated k/μ and that skin factor must be estimated.

An approach somewhat similar to the Brons-Miller approach may be devised that uses the MDH dimensionless pressure. To use that approach, we write the expression for k/μ used implicitly by Brons and Miller:[12]

$$\frac{k}{\mu}=\frac{162.6\,qB}{(p_{ws}-p_{wf})h}f, \quad \ldots \ldots \quad (6.14)$$

where f is given by Eq. 6.13b. In Eq. 6.14, p_{ws} is the observed spot-pressure reading and, *for purposes of this equation,* is taken to be equivalent to the average reservoir pressure. We assume that p_{ws} is on the semilog straight line and use the MDH method to estimate average reservoir pressure. Thus, Eq. 6.11,

$$\bar{p}=p_{ws}+\frac{m\,p_{D\mathrm{MDH}}(\Delta t_{DA})}{1.1513}, \quad \ldots \ldots \quad (6.11)$$

is used to estimate $\bar{p}$. To obtain $p_{D\mathrm{MDH}}$, use Fig. 6.6 where Δt_{DA} is estimated from

$$\Delta t_{DA}\simeq\frac{(0.0002637)(162.6\,qB)f}{(p_{ws}-p_{wf})\,h\phi c_t A}\,\Delta t, \quad \ldots \ldots \quad (6.15a)$$

$$\simeq\frac{0.04288\,qBf}{(p_{ws}-p_{wf})\,h\phi c_t A}\,\Delta t, \quad \ldots \ldots \quad (6.15b)$$

and p_{ws} is the shut-in pressure observed at time Δt; f is given by Eq. 6.13b. The slope used in Eq. 6.11 is estimated from

$$m=\frac{p_{ws}-p_{wf}}{f}. \quad \ldots \ldots \quad (6.16)$$

It should be clear that $\bar{p}$ estimated by this method is only an approximation, and applies only to a circular or square drainage region. If there is a significant difference between $\bar{p}$ and p_{ws}, a second iteration should be made. In that iteration, the value for $\bar{p}$ just estimated replaces p_{ws} in Eqs. 6.15 and 6.16. Then, a new value of $p_{D\mathrm{MDH}}$ can be obtained from Fig. 6.6 and a second estimate of $\bar{p}$ can be made using Eq. 6.11. Normally, a second iteration is not warranted, given the imprecision of other data used. Note that the f term in Eqs. 6.14 through 6.16 does depend on drainage area, wellbore radius, skin factor, and shape factor.

The Extended Muskat Method

As indicated in Section 5.3, Muskat[13] showed that a plot of log $(\bar{p}-p_{ws})$ vs Δt should give a straight line that can be used for analyzing pressure buildup data. That plot uses late-time pressure data corresponding to the curved tail-end seen on the Horner and MDH plots when the well is shut in long enough. The Muskat technique has been discussed and extended by others.[8,14,15] Section 5.3 explains the technique for estimating permeability and average reservoir pressure. The extended Muskat analysis applies for any length of producing time before shut-in, but requires very long shut-in times for the straight line to develop. Thus, the straight line is frequently not observed. Fig. 5.11 and Eq. 5.21 may be used to estimate the physical time range during which the Muskat straight line will occur. Example 5.3 illustrates the Muskat analysis technique for the data of Examples 6.1 through 6.3. For those examples, shut-in time was insufficient to develop the correct Muskat straight line, so an incorrect result was obtained.

Other Methods

Based on Muskat's work, Arps and Smith[16] suggest plotting dp_{ws}/dt vs p_{ws} during the late-transient buildup period to estimate average reservoir pressure. The plot should yield a straight line that, when extrapolated to zero, provides an estimate of $\bar{p}$. When applicable, the Arps-Smith technique should give reliable results. However, it suffers from the same problem as the Muskat technique: it requires a very long shut-in period.

Recommended Methods

In general, the Matthews-Brons-Hazebroek method for estimating average reservoir pressure can be expected to give the most reliable results for a variety of drainage shapes and production times, providing a t_p on the order of t_{pss} is used where production time is large. It is superior to the other methods discussed in its flexibility and is the only method that encompasses the entire range of t_p from infinite-acting through pseudosteady state. The Dietz-type methods of Eqs. 6.7 through 6.9 are also valid when production is at pseudosteady state before shut-in. The Dietz method appears to be the more practical for routine engineering application except for a short period immediately after well completion. For wells with a skin factor more negative than about -3, both methods require further modification since the correct semilog straight line may not develop for normal data plots (see Chapter 11).

Dynamic Pressure Used for Numerical Reservoir Simulation

One current common use of reservoir pressure data is for comparison with numerical simulator results during a history-matching process.[17] The pressure computed by a reservoir simulator to be compared with reservoir pressure is not usually a well's drainage-region average pressure. Rather, it is the average pressure of the *node* (grid block) that contains the well in the reservoir simulator.[17] The corresponding reservoir pressure may be estimated from pressure buildup tests by extrapolating the semilog straight line to the correct time and directly reading the "dynamic" pressure. The time is given by[17]

$$\Delta t_{\mathrm{dyn}}=\frac{200\,\phi c_t \Delta x^2}{(k/\mu)_t}, \quad \ldots \ldots \quad (6.17)$$

where Δx is the length of the side of a square node in the reservoir simulator. Although Eq. 6.17 assumes that the well is in the center of a square node that is relatively small with respect to the well's drainage region, the equation may be used with reasonable accuracy for wells slightly off center in rectangular nodes. In that case, Δx^2 is replaced by $\Delta x \Delta y$, the product of the length and width of the node.

If a complete pressure buildup curve is not available, Eq. 6.17 cannot be used directly. However, if a single (spot) shut-in pressure reading is available for the well and that pressure point is assumed to be on the semilog straight line, Eq. 6.17 still may be used in modified form, and the dynamic pressure may be estimated from[18]

$$(p_{ws})_{\text{dyn}} = (p_{ws})_{\text{OB}} + \frac{162.6\, qB}{(k/\mu)_t\, h} \log\left(\frac{200\, \phi c_t \Delta x^2}{(k/\mu)_t\, (\Delta t)_{\text{OB}}}\right) . \qquad (6.18)$$

Eq. 6.18 assumes that a reasonable value of total mobility is known. It should be used only when no better method is available for estimating the dynamic pressure for history-matching purposes. Clearly, the approach of Eq. 6.18 could be modified to be similar to the Brons-Miller method, thus eliminating the need for an estimate of k/μ.

6.4 Water-Drive Reservoirs

Water-drive reservoirs (that is, reservoirs in direct communication with an active aquifer) behave quite differently from closed reservoirs, particularly with respect to well drainage regions and flow patterns. Water-drive reservoirs range from those with complete pressure maintenance (equivalent to a constant pressure at the reservoir aquifer boundary with a nearby replenishable outcrop) to those that provide only a small amount of water influx to the producing reservoir. The analysis techniques in this section apply only to those edge water-drive reservoirs that provide complete pressure maintenance. In practice, pressure at the reservoir-aquifer boundary is generally neither constant nor uniform. It changes in time as a result of (1) movement of the reservoir transient into the aquifer, (2) finite aquifer extent, and (3) interference from withdrawals or injection in the aquifer. However, the concepts presented here are usually applied when the pressure at the water-oil contact has stabilized at nearly a constant value or when further declines in field boundary pressure and internal pressure are in step so that the internal field pressure distribution is constant. Material in this section is directly applicable only to constant-compressibility and unit-mobility-ratio, water-oil systems.

Estimating Drainage Volume

Steady-state drainage regions for wells in a water-drive reservoir do not resemble drainage areas during pseudosteady-state conditions in closed systems.[1,2,19] Ramey, Kumar, and Gulati[19] have studied steady-state drainage areas in water-drive reservoirs with from 1 to 35 production wells. Two general conclusions apply to all unit-mobility-ratio reservoir-aquifer systems with constant pressure at the oil-water contact: (1) flow near each well is radial—but the radial flow pattern may not extend for a very great distance; and (2) all producing wells have *direct* communication with the constant-pressure boundary in some way. Matthews and Russell[1] suggest techniques for estimating drainage areas under water-drive conditions; Ramey, Kumar, and Gulati[19] propose more accurate techniques for idealized cases and present results for many systems. Their results for a square with constant-pressure boundaries may be most directly applicable to injection and production wells in five-spot waterfloods with near unit mobility ratio, and for nominal well drainage areas in strong water-drive reservoirs.

Fig. 6.8 shows steady-state drainage regions for a 15-well, water-drive reservoir with wells producing at different rates. While that is an extreme case, each well's drainage area does extend to the constant-pressure boundary. Because of such behavior, individual-well drainage regions in water-drive reservoirs tend to be more irregular than those in reservoirs with no influx. Similar drainage-area distortions also may occur in reservoirs with incomplete development and highly imbalanced withdrawal rates with respect to total fluids in place.

Estimating Permeability, Skin Factor, and End of the Semilog Straight Line

Pressure-buildup analysis methods for permeability and skin factor are the same for water-drive reservoirs as for closed reservoirs. Either the Horner- or the MDH-type plot may be used. The slope and p_{1hr} values from the data plot are used to estimate permeability from Eq. 5.6 and skin factor from Eq. 5.7. Kumar and Ramey[11] show that, for a square with constant-pressure boundaries (equal influx on all sides), with the well in the center, the MDH straight line lasts longer than the Horner straight line if production time before shut-in is long enough ($t_{pDA} > 0.25$). This is in contrast to the behavior of closed systems. However, because of highly irregular drainage areas, which can change

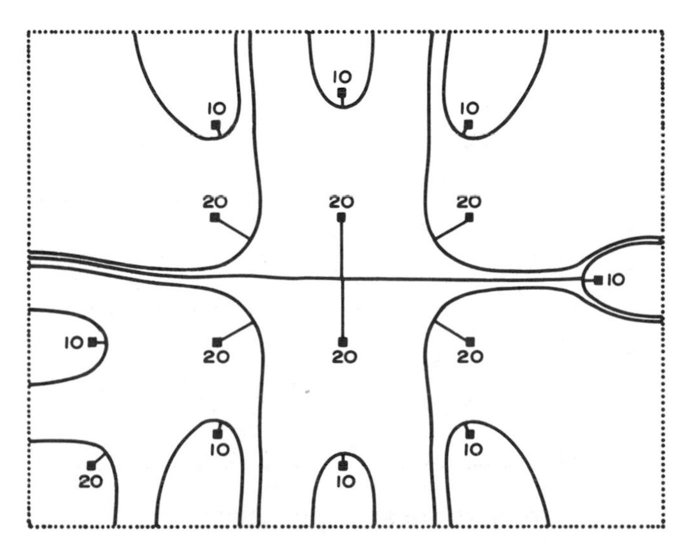

Fig. 6.8 Steady-state drainage regions for a 15-well, nonideal water-drive reservoir, wells producing at rates shown, dotted boundary at constant pressure. After Ramey, Kumar, and Gulati,[19] courtesy AGA.

fairly rapidly when a well is shut in, the semilog straight line of the infinite-acting period usually ends earlier for wells in water-drive reservoirs than for wells with no-flow boundaries. Eq. 5.16 and Figs. 5.6 and 5.7 provide a way to estimate the shut-in time when the semilog straight line ends for this system and for closed systems. The extended Muskat method may be used to estimate k by using Eq. 5.18 and Fig. 5.10. Refs. 11 and 19 give additional information, and Ref. 19 considers other fixed shapes and mixed boundary conditions.

Estimating Boundary and Average Pressure

Several pressures are important in water-drive systems, including the initial pressure and the average pressure for the hydrocarbon producing area as a function of time. In addition, the average pressure as a function of time at the original water-oil contact may be useful. In other instances, separate average pressures over some encroached area or produced inner area may be needed for past-performance matching and predictions depending on the technique used. Several techniques may be used to estimate average and boundary pressures for water-drive systems. For short production times, the Horner and Matthews-Brons-Hazebroek methods are probably the best. The Dietz and Miller-Dyes-Hutchinson methods both apply for $t_{pDA} > 0.25$ and $s > -3$. The Muskat method also applies and may have more utility for water-drive systems than for closed systems.

The *Matthews-Brons-Hazebroek* method is applied to a square with constant-pressure boundaries in much the same way as it is applied to a closed system. Pressure buildup data are plotted on semilog paper vs $\left[(t_p + \Delta t)/\Delta t\right]$. The resulting semilog straight line is extrapolated to $\left[(t_p + \Delta t)/\Delta t\right] = 1$ to obtain p^*. Then, the boundary pressure is computed from

$$p_e = p^* - \frac{m \, p_{DMBHe}(t_{pDA})}{2.3025}, \qquad (6.19a)$$

and the average pressure at the instant of shut-in is computed from

$$\bar{p} = p^* - \frac{m \, \bar{p}_{DMBH}(t_{pDA})}{2.3025}. \qquad (6.19b)$$

The MBH dimensionless pressures, p_{DMBHe} and $\bar{p}_{DMBH}$, are given in Fig. 6.9. Unfortunately, such data are not available for systems other than a well centered in a square with a constant-pressure boundary.

The *Dietz* method also may be applied to water-drive reservoirs. The semilog straight line is extrapolated to a time given by Eq. 6.7a, and $\bar{p}$ is read directly from the graph at that time. Table C.1 shows that the appropriate shape factor for a circular reservoir with constant-pressure boundaries is 19.1. Ref. 11 indicates that the shape factor for a square, constant-pressure-boundary reservoir is 19.5. Using the latter value in Eq. 6.7, the time to read $\bar{p}$ for a square system with constant-pressure boundaries from the extrapolated semilog straight line is given by

$$(\Delta t)_{\bar{p}} = 195 \frac{\phi \mu c_t A}{k}. \qquad (6.20)$$

Shape factors are not available for other shapes with constant-pressure boundaries. The Dietz method applies only when $t_{pDA} > 0.25$ and $s > -3$.

The *Miller-Dyes-Hutchinson* approach also may be used for water-drive systems. A normal MDH plot of buildup pressure vs log Δt is made and the system boundary pressure is then estimated from

$$p_e = p_{ws}(\Delta t) + \frac{m \, p_{DMDH}(\Delta t_{DA})}{1.1513}. \qquad (6.21)$$

In Eq. 6.21, $p_{ws}(\Delta t)$ is the shut-in pressure at any time, Δt, on the semilog straight line; m is from the slope of the semilog straight line; Δt_{DA} is from Eq. 6.10; and p_{DMDH} is from the dashed curve on Fig. 6.6 for either the circular or square system. The calculation technique is the same as the MDH method, Section 6.3.

The *extended Muskat method* also may be used to estimate p_e for closed systems. Kumar and Ramey[11] show empirically that this method does give valid results for p_e when late-time data are analyzed. They also state that the method seems to be more applicable in water-drive reservoirs than in closed reservoirs. The method is basically the same as described in Section 6.3. Kumar and Ramey give techniques for estimating formation permeability using the Muskat plot.

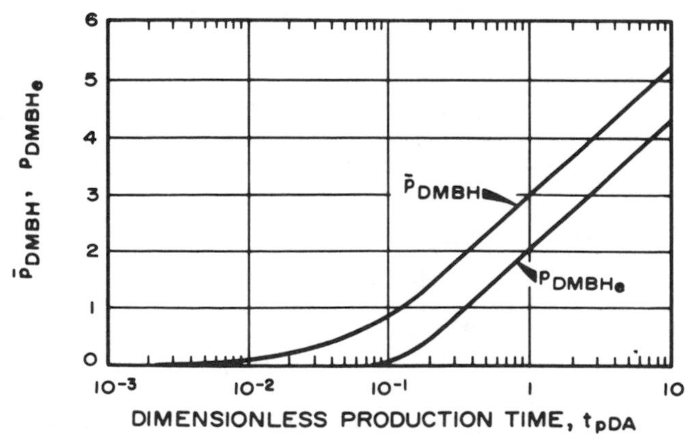

Fig. 6.9 Matthews-Brons-Hazebroek dimensionless pressures for a well in the center of a constant-pressure square. Upper curve is based on $\bar{p}$ at time of shut-in; lower curve is based on p_e, the pressure at the boundary. After Kumar and Ramey.[11]

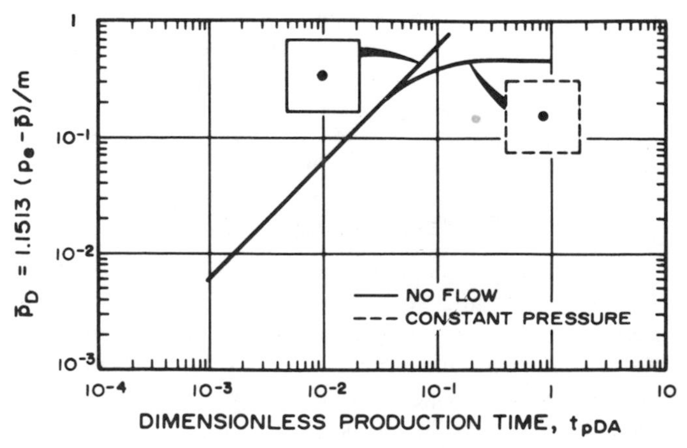

Fig. 6.10 Dimensionless average pressures for square systems. After Kumar and Ramey.[11]

Several other techniques also may be used to estimate average system pressures and boundary pressures. Given a boundary pressure, the average system pressure at the instant of shut-in may be estimated from[11]

$$\bar{p} = p_e - \frac{m\,\bar{p}_D(t_{pDA})}{1.1513}, \quad \ldots \ldots \ldots \ldots \ldots \ldots \ldots \ldots \ldots \ldots \quad (6.22)$$

where the average dimensionless pressure, $\bar{p}_D$, is given in Fig. 6.10. Another approach uses the well's flowing pressure, the buildup slope, and other data in

$$\bar{p} = p_{wf}(\Delta t = 0) + m\left[\log\left(\frac{A}{r_w^2}\right) + \log\left(\frac{2.2458}{C_A'}\right) + 0.86859\,s\right] \quad \ldots \ldots \ldots \ldots \quad (6.23)$$

For this equation to give acceptable results, it is important that the skin factor be included. The shape factor, C_A', is 19.1 or 19.5 for a circular or square system.[11] Eq. 6.23 applies only when t_{pDA} (Eq. 6.4) is greater than 0.25.

Kumar and Ramey indicate that the drainage-region average pressure at the time of shut-in may be read directly from the semilog straight line at the value of Δt or $(t_p + \Delta t)/\Delta t$ obtained from Fig. 6.11. t_{pDA} is calculated from Eq. 6.4. For the MDH plot, it is necessary to estimate Δt from $(\Delta t_{DA})_{\bar{p}}$ by using Eq. 2.3b.

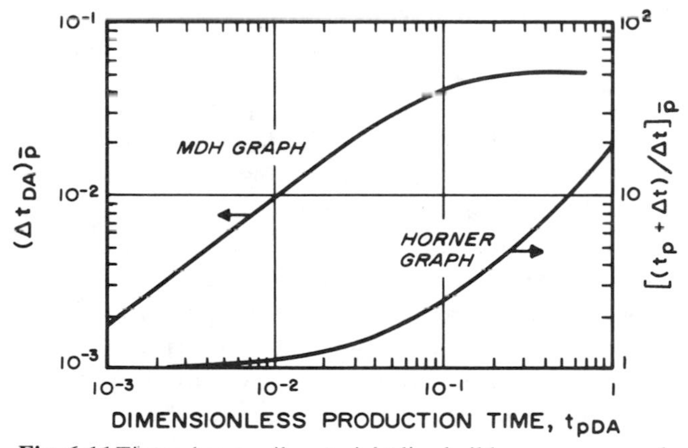

Fig. 6.11 Time when semilog straight-line buildup pressure equals average pressure at instant of shut-in. Constant-pressure square system. After Kumar and Ramey.[11]

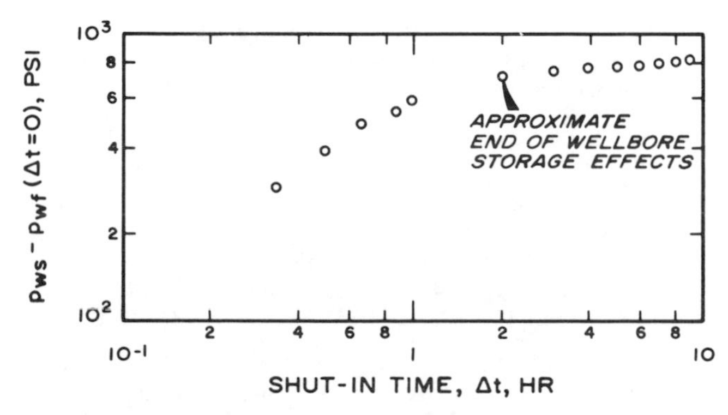

Fig. 6.12 Log-log data plot for a buildup test on a well in a constant-pressure square. Example 6.4. Data from Kumar and Ramey.[11]

TABLE 6.1—PRESSURE BUILDUP DATA FOR A WELL IN THE CENTER OF A SQUARE WITH CONSTANT-PRESSURE BOUNDARIES.[11] EXAMPLE 6.4.

Shut-in Time, Δt (hours)	Pressure, p_{ws} (psi)
0	3,561
0.333	3,851
0.500	3,960
0.667	4,045
0.883	4,104
1.0	4,155
2.0	4,271
3.0	4,306
4.0	4,324
5.0	4,340
6.0	4,352
7.0	4,363
8.0	4,371
9.0	4,380
10.0	4,387
20.0	4,432

Example 6.4 Average Drainage-Region Pressure and Pressure Buildup Analysis for a Water-Drive Reservoir

Kumar and Ramey[11] provide the simulated pressure buildup data in Table 6.1 for a well in the center of a square drainage area with constant-pressure boundaries. Other pertinent data are

$t_p = 4{,}320$ hours $= 180$ days $B = 1.136$ RB/STB
$q = 350$ STB/D $h = 49$ ft
$\mu = 0.80$ cp $r_w = 0.29$ ft
$c_t = 17 \times 10^{-6}$ psi^{-1} $\phi = 0.23$
$A = 7.72$ acres, well in center of a square with constant-pressure boundaries, 580×580 ft
$= 336{,}400$ sq ft.

Fig. 6.12 is the log-log plot of data from Table 6.1. Wellbore storage effects appear to diminish after about 2 hours of shut-in. Because of the long production time, the MDH plot shown in Fig. 6.13 is adequate for analysis. Fig. 6.13 indicates

$$m = 152 \text{ psi/cycle}, \quad \text{and} \quad p_{1hr} = 4{,}236 \text{ psi}.$$

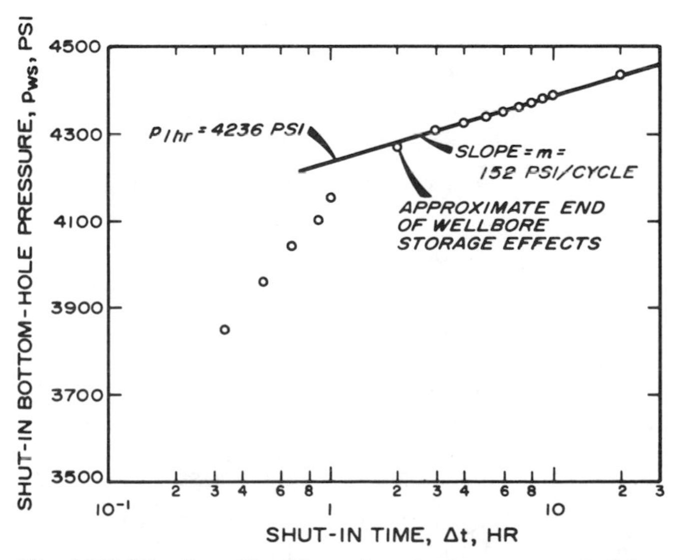

Fig. 6.13 Miller-Dyes-Hutchinson data plot for a pressure buildup test on a well in a constant-pressure square. Example 6.4. Data from Kumar and Ramey.[11]

We estimate permeability from Eq. 5.6:

$$k = \frac{(162.6)(350)(1.136)(0.80)}{(152)(49)}$$

$$= 6.9 \text{ md}.$$

Skin factor is estimated from Eq. 5.7:

$$s = 1.1513 \left[\frac{4,236 - 3,561}{152} - \log\left(\frac{6.9}{(0.23)(0.80)(17 \times 10^{-6})(0.29)^2}\right) + 3.2275\right]$$

$$= 0.3.$$

We may estimate the system boundary pressure, p_e, by the MDH method, Eq. 6.21. We use the square-drainage-region, constant-pressure-boundary curve of Fig. 6.6 to get $p_{D\text{MDH}}$. In this case, we use Δt = 10 hours and Eq. 6.4 to calculate

$$\Delta t_{DA} = \frac{(0.0002637)(6.9)(10)}{(0.23)(0.80)(17 \times 10^{-6})(336;400)}$$

$$= 0.0173.$$

Fig. 6.6 indicates that $p_{D\text{MDH}}$ = 1.01 at Δt_{DA} = 0.0173. Using Eq. 6.21 and p_{ws} = 4,388 psi (Fig. 6.13) at Δt = 10 hours,

$$p_e = 4,388 + \frac{(152)(1.01)}{1.1513}$$

$$= 4,521 \text{ psi},$$

which is 89 psi higher than the maximum recorded pressure.

The average pressure at the time of shut-in is estimated from Eq. 6.22 and Fig. 6.10. To use Fig. 6.10, we estimate t_{pDA} from Eq. 6.4.

$$t_{pDA} = \frac{(0.0002637)(6.9)(4,320)}{(0.23)(0.80)(17 \times 10^{-6})(336,400)}$$

$$= 7.47.$$

From Fig. 6.10, $\bar{p}_D = 0.478$ for $t_{pDA} > 1$, so using Eq. 6.22,

$$\bar{p} = 4,521 - \frac{(152)(0.478)}{1.1513}$$

$$= 4,458 \text{ psi}.$$

Alternatively, we could use Fig. 6.11 to get $(\Delta t_{DA})_{\bar{p}}$ = 0.0513. This implies that

$$(\Delta t)_{\bar{p}} = \frac{(0.23)(0.80)(17 \times 10^{-6})(336,400)(0.0513)}{(0.0002637)(6.9)}$$

$$= 29.7 \text{ hours}.$$

Extrapolating the straight line in Fig. 6.13, $\bar{p}$ = 4,458 psi.

Kumar and Ramey[11] illustrate Horner plotting and the Matthews-Brons-Hazebroek method (Fig. 6.9) for this example.

6.5 Factors Complicating Average-Pressure Estimation

Since most reservoirs have many wells, the general pressure level in the reservoir usually continues to decrease during buildup tests on a single well. That is not important in the early stages of a buildup test; however, if the well is shut in for several days, readjustment of the drainage areas of offsetting wells and the general pressure decline may begin to affect the buildup response at the shut-in well. Methods for analysis under such conditions are given in Section 5.3. The fact that this happens indicates that an average pressure for a well's drainage region or an entire reservoir applies only at a point in time. In particular, as a result of changing drainage regions and continued reservoir depletion, we cannot expect a single well in a developed reservoir to reach the average pressure applicable to its drainage area at the instant of shut-in. The alternative is estimating drainage-region average pressure by using the Matthews-Brons-Hazebroek approach and then *volumetrically* averaging the pressures for each drainage region to obtain the over-all reservoir average pressure. Such an approach may not be viable as normal engineering practice because of practical difficulties in estimating drainage areas, since they vary in response to changing operating practices. Fortunately, pressure buildup is usually rapid in high-permeability reservoirs, so observed pressures may be representative of drainage-region average pressure. In lower-permeability reservoirs, wells tend to have slower and more incomplete buildups during reasonable shut-in periods. Therefore, a different approach is indicated in such reservoirs. The most general approach is either full or key-well buildup surveys to estimate average reservoir pressures for each well's nominal drainage region (40-acre square, for example) using techniques given in this chapter. The resulting pressures may be contoured, assuming them to be at the well points, and then volumetrically weighed by any of several acceptable techniques.

A common mistake made in pressure buildup analysis is to begin thinking that the mathematical conveniences used in analysis exist in reality. In fact, in any developed reservoir, the only true no-flow boundaries are the physical boundaries. While there may be effective no-flow boundaries between producing wells when all wells are producing, those boundaries move (drainage shapes change) as flow rates vary and as wells are shut in and put back on production. It is a common error in pressure analysis methods to assume that when one well is shut in, it will always act like a well in a closed drainage system; in fact, that only happens for single wells in very small reservoirs.

References

1. Matthews, C. S. and Russell, D. G.: *Pressure Buildup and Flow Tests in Wells,* Monograph Series, Society of Petroleum Engineers of AIME, Dallas (1967) **1,** Chap. 4.
2. Matthews, C. S. and Lefkovits, H. C.: "Studies on Pressure Distribution in Bounded Reservoirs at Steady State," *Trans.,* AIME (1955) **204,** 182-189.
3. Horner, D. R.: "Pressure Build-Up in Wells," *Proc.,* Third World Pet. Cong., The Hague (1951) Sec. II, 503-523. Also *Reprint Series, No. 9 — Pressure Analysis Methods,* Society of Petroleum Engineers of AIME, Dallas (1967) 25-43.
4. Matthews, C. S., Brons, F., and Hazebroek, P.: "A Method for Determination of Average Pressure in a Bounded Reservoir," *Trans.,* AIME (1954) **201,** 182-191. Also *Reprint Series No. 9 — Pressure Analysis Methods,* Society of Petroleum Engineers of AIME, Dallas (1967) 51-60.

5. Pinson, A. E., Jr.: "Concerning the Value of Producing Time Used in Average Pressure Determinations From Pressure Buildup Analysis," *J. Pet. Tech.* (Nov. 1972) 1369-1370.
6. Kazemi, Hossein: "Determining Average Reservoir Pressure From Pressure Buildup Tests," *Soc. Pet. Eng. J.* (Feb. 1974) 55-62; *Trans.*, AIME, **257.**
7. Dietz, D. N.: "Determination of Average Reservoir Pressure From Build-Up Surveys," *J. Pet. Tech.* (Aug. 1965) 955-959; *Trans.*, AIME, **234.**
8. Ramey, H. J., Jr., and Cobb, William M.: "A General Buildup Theory for a Well in a Closed Drainage Area," *J. Pet. Tech.* (Dec. 1971) 1493-1505; *Trans.*, AIME, **251.**
9. Miller, C. C., Dyes, A. B., and Hutchinson, C. A., Jr.: "The Estimation of Permeability and Reservoir Pressure From Bottom-Hole Pressure Build-Up Characteristics," *Trans.*, AIME (1950) **189,** 91-104. Also *Reprint Series, No. 9 — Pressure Analysis Methods,* Society of Petroleum Engineers of AIME, Dallas (1967) 11-24.
10. Perrine, R. L.: "Analysis of Pressure Buildup Curves," *Drill. and Prod. Prac.*, API (1956) 482-509.
11. Kumar, Anil and Ramey, Henry J., Jr.: "Well-Test Analysis for a Well in a Constant-Pressure Square," paper SPE 4054 presented at the SPE-AIME 47th Annual Fall Meeting, San Antonio, Tex., Oct. 8-11, 1972 (an abridged version appears in *Soc. Pet. Eng. J.*, April 1974, 107-116).
12. Brons, F. and Miller, W. C.: "A Simple Method for Correcting Spot Pressure Readings," *J. Pet. Tech.* (Aug. 1961) 803-805; *Trans.*, AIME, **222.**
13. Muskat, Morris: "Use of Data on the Build-Up of Bottom-Hole Pressures," *Trans.*, AIME (1937) **123,** 44-48. Also *Reprint Series, No. 9 — Pressure Analysis Methods,* Society of Petroleum Engineers of AIME, Dallas (1967) 5-9.
14. Matthews, C. S.: "Analysis of Pressure Build-Up and Flow Test Data," *J. Pet. Tech.* (Sept. 1961) 862-870. Also *Reprint Series, No. 9 — Pressure Analysis Methods,* Society of Petroleum Engineers of AIME, Dallas (1967) 111-119.
15. Russell, D. G.: "Extensions of Pressure Build-Up Analysis Methods," *J. Pet. Tech.* (Dec. 1966) 1624-1636; *Trans.*, AIME, **237.** Also *Reprint Series, No. 9 — Pressure Analysis Methods,* Society of Petroleum Engineers of AIME, Dallas (1967) 175-187.
16. Arps, J. J. and Smith, A. E.: "Practical Use of Bottom-Hole Pressure Buildup Curves," *Drill. and Prod. Prac.*, API (1949) 155-165.
17. van Poollen, H. K., Breitenbach, E. A., and Thurnau, D. H.: "Treatment of Individual Wells and Grids in Reservoir Modeling," *Soc. Pet. Eng. J.* (Dec. 1968) 341-346.
18. Earlougher, Robert C., Jr.: "Comparing Single-Point Pressure Buildup Data With Reservoir Simulator Results," *J. Pet. Tech.* (June 1972) 711-712.
19. Ramey, Henry J., Jr., Kumar, Anil, and Gulati, Mohinder S.: *Gas Well Test Analysis Under Water-Drive Conditions,* AGA, Arlington, Va. (1973) Chaps. 4-7.

Chapter 7

Injection Well Testing

7.1 Introduction

In many reservoirs, the number of injection wells approaches the number of producing wells, so the topic of testing those wells is important. That is particularly true when tertiary recovery projects are being considered or are in progress. When an input well receives an expensive fluid, its ability to accept that fluid uniformly for a long time is important to the economics of the tertiary recovery project. In particular, increasing wellbore damage must be detected and corrected promptly.

The information available about injection well testing is much less abundant than information about production well testing. Matthews and Russell[1] summarize injection well testing, but emphasize falloff testing. Injectivity testing is rarely discussed in the literature, but it can be important.[2] Falloff testing is treated[3-7] rather thoroughly, particularly for systems with unit mobility ratio. Gas-well falloff testing, especially in association with in-situ combustion, also has been discussed.[8,9]

Injection-well transient testing and analysis are basically simple — as long as the *mobility ratio* between the injected and the in-situ fluids is about unity. Fortunately, that is a reasonable approximation for many waterfloods. It also is a reasonable approximation in watered-out waterfloods that initially had mobility ratios significantly different from unity, and early in the life of tertiary recovery projects when so little fluid has been injected that it appears only as a skin effect. When the unit-mobility-ratio condition is satisfied, injection well testing for liquid-filled systems is analogous to production well testing. Injection is analogous to production (but the rate, q, used in equations is negative for injection while it is positive for production), so an injectivity test (Section 7.2) parallels a drawdown test (Chapter 3). Shutting in an injection well results in a pressure falloff (Section 7.3) that is analogous to a pressure buildup (Chapter 5). The equations for production well testing in Chapters 3 through 5 apply to injection well testing as long as sign conventions are observed. The analogy should become clear in the next two sections.

When the unit-mobility-ratio assumption is not satisfied, the analogy between production well testing and injection well testing is not so complete. In that situation, analysis depends on the relative sizes of the water bank and the oil bank; generally, analysis is possible only when $r_{ob} > 10r_{wb}$ (see Section 7.5). Fracturing effects, which can have a significant effect on analysis, are discussed in Section 11.3.

Reservoirs with injection wells can reach true steady-state conditions when total injection rate equals total production rate. In that situation, or when the situation is approached, the steady-state analysis techniques of Section 7.7 may be useful.

7.2 Injectivity Test Analysis in Liquid-Filled, Unit-Mobility-Ratio Reservoirs

Injectivity testing is pressure transient testing during injection into a well. It is analogous to drawdown testing, for both constant and variable injection rates. Although sometimes called "injection pressure buildup" or simply "pressure buildup," we prefer to use the term "injectivity testing" to avoid confusion with production-well pressure buildup testing. This section applies to liquid-filled reservoirs with mobility of the injected fluid essentially equal to the mobility of the in-situ fluid. If the unit-mobility-ratio condition is not satisfied, results of analysis by techniques in this section may not be valid. Even in that situation, if the radius of investigation is not beyond the water (injected-fluid) bank, valid analysis can be made for permeability and skin, but not necessarily for static reservoir pressure.

Fig. 7.1 shows an ideal rate schedule and pressure response for injectivity testing. The well is initially shut in and pressure is stabilized at the initial reservoir pressure, p_i. At time zero, injection starts at constant rate, q. Fig. 7.1 illustrates the convention that $q < 0$ for injection. It is advisable to monitor the injection rate carefully so the methods of Chapter 4 (variable-rate analysis) may be applied if the rate varies significantly.

Since unit-mobility-ratio injection well testing is analogous to production well testing, the analysis methods in Chapters 3 and 4 for drawdown and multiple-rate testing may be applied directly to injection well testing. Of course, while pressure at a production well declines during drawdown, pressure at an injection well increases during injection. That difference is accounted for in the analysis methods by using $q < 0$ for injection and $q > 0$ for production.

For the constant-rate injectivity test illustrated in Fig. 7.1, the bottom-hole injection pressure is given by Eq. 3.5:

$$p_{wf} = p_{1hr} + m \log t . \qquad (7.1)$$

Eq. 7.1 indicates that a plot of bottom-hole injection pressure vs the logarithm of injection time should have a straight-line section, as shown in Fig. 7.2. The intercept, p_{1hr}, is given by Eq. 3.7; the slope is m and is given by Eq. 3.6:

$$m = \frac{-162.6\, qB\mu}{kh} . \qquad (7.2)$$

As in drawdown testing, wellbore storage may be an important factor in injection well testing. Often, reservoir pressure is low enough so that there is a free liquid surface in the shut-in well. In that case, the wellbore storage coefficient is given by Eq. 2.16 and can be expected to be relatively large. Therefore, we recommend that all injectivity test analyses start with the $\log(p_{wf} - p_i)$ vs $\log t$ plot so the duration of wellbore storage effects may be estimated as explained in Sections 2.6 and 3.2. As indicated in Fig. 7.2, wellbore effects may appear as a semilog straight line on the p_{wf} vs $\log t$ plot; if such a line is analyzed, low values of permeability will be obtained and calculated skin factor will be shifted in the negative direction. Eq. 3.8 may be used to estimate the beginning of the semilog straight line shown in Fig. 7.2:

$$t > \frac{(200{,}000 + 12{,}000\, s)\, C}{(kh/\mu)} . \qquad (7.3)$$

Once the semilog straight line is determined, reservoir permeability is estimated from Eq. 3.9:

$$k = \frac{-162.6\, qB\mu}{mh} . \qquad (7.4)$$

Skin factor is estimated with Eq. 3.10:

$$s = 1.1513 \left[\frac{p_{1hr} - p_i}{m} - \log\left(\frac{k}{\phi \mu c_t r_w^2}\right) + 3.2275 \right] . \qquad (7.5)$$

Example 7.1 Injectivity Test Analysis in an Infinite-Acting Reservoir

Figs. 7.3 and 7.4 show pressure response data for an injectivity test in a waterflooded reservoir. Before the test, all wells in the reservoir had been shut in for several weeks and pressure had stabilized. Known reservoir data are

depth = 1,002 ft	h = 16 ft
$c_t = 6.67 \times 10^{-6}$ psi^{-1}	μ = 1.0 cp
ϕ = 0.15	B = 1.0 RB/STB
ρ_w = 62.4 lb$_m$/cu ft	q = −100 STB/D
p_i = 194 psig	r_w = 0.25 ft.

The well is completed with 2-in. tubing set on a packer. The reservoir had been under waterflood for several years. We can safely assume that the unit-mobility-ratio assumption is satisfied, since the test radius of investigation is less than the distance to the water bank, as shown by calculations later in this example.

The log-log data plot, Fig. 7.3, indicates that wellbore storage is important for about 2 to 3 hours. The deviation of the data above the unit-slope line suggests that the wellbore storage coefficient decreased at about 0.55 hour. Sections 2.6 and 11.2 and Figs. 2.12 and 11.5 through 11.7 discuss such changing wellbore storage conditions. The data in Fig. 7.3 start deviating upward from the unit-slope straight line when Δp = 230 psi and p_{wf} = 424 psig. Since the column of water in the well is equivalent to about 434 psi, it appears that the apparent decrease in storage coefficient corresponds to fillup of the tubing.

From the unit-slope portion of Fig. 7.3, Δp = 408 psig when Δt = 1 hour. Using Eq. 2.20, we estimate the apparent wellbore storage coefficient:

$$C = \frac{(100)(1.0)}{24} \frac{(1.0)}{(408)} = 0.0102 \text{ bbl/psi.}$$

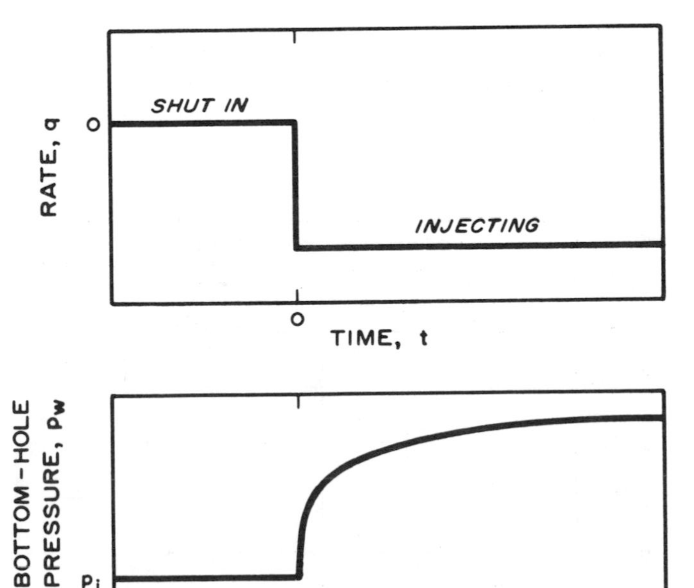

Fig. 7.1 Idealized rate schedule and pressure response for injectivity testing.

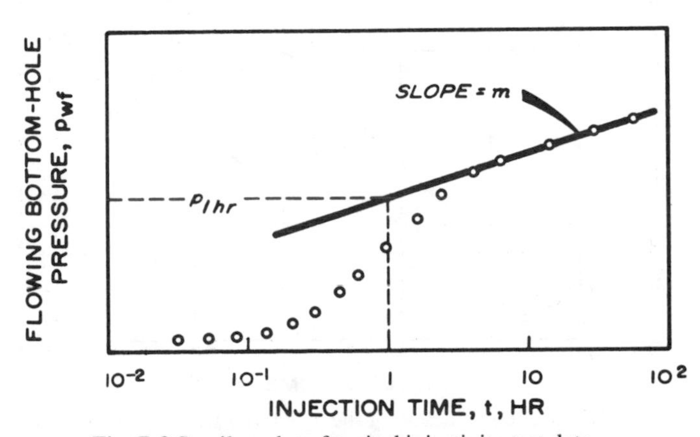

Fig. 7.2 Semilog plot of typical injectivity test data.

(C is always positive.) Wellbore capacity for a rising fluid level can be estimated (from Eq. 2.16) to get $V_u = 0.0044$ bbl/ft. Two-inch tubing has a capacity of about 0.004 bbl/ft, so the unit-slope straight line does correspond to a rising fluid level in the tubing. If we use $C = 0.0102$ in Eq. 7.3, or if we go 1 to 1.5 cycles in Δt after the data start deviating from the unit-slope line (Section 2.6), we would decide that the semilog straight line should not start for 5 to 10 hours of testing. Those rules indicate too long a time for a *decreasing* wellbore storage condition. Figs. 7.3 and 7.4 clearly show that wellbore storage effects have died out after about 2 to 3 hours.

Fig. 7.4 shows a semilog straight line through the data after 3 hours of injection. From this line, $m = 80$ psig/cycle and $p_{1\text{hr}} = 770$ psig. Permeability is estimated using Eq. 7.4:

$$k = \frac{-(162.6)(-100)(1.0)(1.0)}{(80)(16)} = 12.7 \text{ md}.$$

We may now determine if the unit-mobility-ratio analysis applies. The estimated permeability is used to estimate a radius of investigation from Eq. 2.41:

$$r_d \simeq 0.029 \sqrt{\frac{kt}{\phi \mu c_t}}$$

$$\simeq 0.029 \sqrt{\frac{(12.7)(7)}{(0.15)(1.0)(6.67 \times 10^{-6})}}$$

$$\simeq 273 \text{ ft}.$$

A volumetric balance provides an estimate of the distance to the water bank. The volume injected is

$$W_i = \frac{\pi r_{wb}^2 h \phi \Delta S_w}{5.6146},$$

so

$$r_{wb} = \sqrt{\frac{5.6146\, W_i}{\pi h \phi \Delta S_w}}.$$

Assuming that $\Delta S_w = 0.4$ and that injection has been under way for at least 2 years,

$$W_i \simeq (100 \text{ STB/D})(1.0 \text{ RB/STB})(2 \text{ years})(365 \text{ D/year})$$

$$\simeq 73{,}000 \text{ res bbl}$$

and

$$r_{wb} = \sqrt{\frac{(5.6146)(73{,}000)}{\pi(16)(0.15)(0.4)}} \simeq 369 \text{ ft}.$$

Since $r_d < r_{wb}$, we are justified in using the unit-mobility-ratio analysis.

Eq. 7.5 provides an estimate of the skin factor:

$$s = 1.1513 \left\{ \frac{770 - 194}{80} - \log \left[\frac{12.7}{(0.15)(1.0)(6.67 \times 10^{-6})(0.25)^2} \right] + 3.2275 \right\}$$

$$= 2.4.$$

The well is damaged; the pressure drop across the skin may be estimated from Eq. 2.9:

$$\Delta p_s = \frac{(141.2)(-100)(1.0)(1.0)(2.4)}{(12.7)(16)}$$

$$= -167 \text{ psi}.$$

The negative sign here indicates damage since the pressure decreases away from the well (in the positive r direction) for injection while it increases for production. This is seen by computing the flow efficiency from Eq. 2.12. Assume $\bar{p} = p_i = 194$ psi, since the reservoir is stabilized before injection. Using $p_{wf} = 835$ psig from the last available data point, the *flow efficiency* is

$$\frac{194 - 835 - (-167)}{194 - 835} = 0.74.$$

If we had ignored the sign on q when estimating Δp_s, we would have incorrectly computed a flow efficiency of 1.26, indicating improvement instead of damage.

Multiple-rate injection testing, constant-pressure injection testing, injectivity testing after falloff testing, etc., are all performed and analyzed as explained for production well

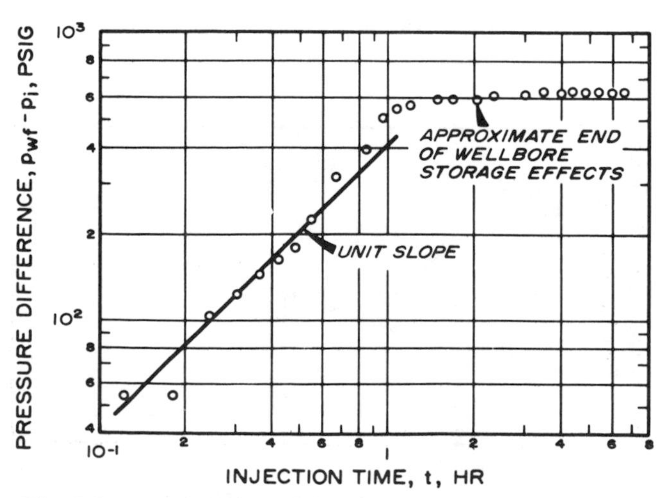

Fig. 7.3 Log-log data plot for the injectivity test of Example 7.1. Water injection into a reservoir at static conditions.

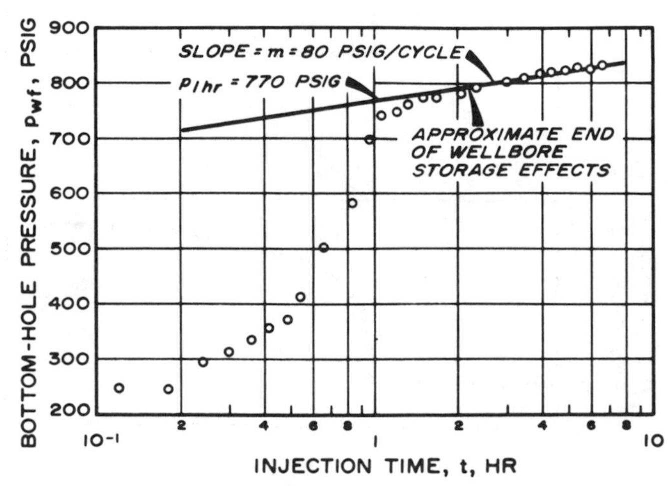

Fig. 7.4 Semilog plot for the injectivity test of Example 7.1. Water injection into a reservoir at static conditions.

testing in Chapters 3 and 4. Type-curve matching for injection well testing is done just as it is for production well testing (Section 3.3); the Δp used must be positive for plotting the log scale, although it is actually a negative number. The signs must be considered in analysis.

Eqs. 7.1 through 7.5 apply to injectivity testing in infinite-acting reservoirs, just as do Eqs. 3.5 through 3.10 for drawdown testing. When an injection well in a developed reservoir shows the effects of interference from other wells, the infinite-acting analysis may not be strictly applicable. In that case, the techniques presented in Section 3.4 should be used.

7.3 Falloff Test Analysis for Liquid-Filled, Unit-Mobility-Ratio Reservoirs

Falloff testing, illustrated schematically in Fig. 7.5, is analogous to pressure buildup testing in a production well. Injection is at a constant rate, q, until the well is shut in at time t_p. Pressure data taken immediately before and during the shut-in period are analyzed as pressure buildup data are analyzed. The pressure falloff behavior can be expressed by Eq. 5.10 for both infinite-acting and developed reservoirs:

$$p_{ws} = p^* - m \log\left(\frac{t_p + \Delta t}{\Delta t}\right) . \quad (7.6)$$

The false pressure, p^*, is equivalent to the initial pressure, p_i, for an infinite-acting system. As illustrated in Fig. 7.6, Eq. 7.6 indicates that a plot of p_{ws} vs $\log[(t_p + \Delta t)/\Delta t]$ should have a straight-line portion with intercept p^* at infinite shut in time $[(t_p + \Delta t)/\Delta t = 1]$ and with slope $-m$, where m is given by Eq. 5.5:

$$m = \frac{162.6\, qB\mu}{kh} . \quad (7.7)$$

As in buildup testing, the Horner graph is plotted with the horizontal scale increasing from right to left (Fig. 7.6). Thus, although the slope appears to be negative, it is actually positive because of the reverse plotting; m is negative since $m = -\text{slope}$.

As for other transient well tests, the log-log data plot should be made so the end of wellbore storage effects may be estimated and the proper semilog straight line (Fig. 7.6) can be chosen. Eq. 5.15b may be used to estimate the beginning of the semilog straight line for falloff testing:

$$t = \frac{170{,}000\, Ce^{0.14s}}{(kh/\mu)} , \quad (7.8)$$

but the log-log plot is preferred.

Once the correct semilog straight line has been determined, reservoir permeability and skin factor are estimated from Eqs. 5.6 and 5.7:

$$k = \frac{162.6\, qB\mu}{mh} , \quad (7.9)$$

and

$$s = 1.1513\left[\frac{p_{1hr} - p_{wf}(\Delta t = 0)}{m} - \log\left(\frac{k}{\phi\mu c_t r_w^2}\right) + 3.2275\right] . \quad (7.10)$$

As is the case in pressure buildup testing, if the injection rate varies before the falloff test, the equivalent injection time may be approximated from Eq. 5.9:

$$t_p = \frac{24\, V_P}{q} , \quad (7.11)$$

where V_P is the cumulative volume injected since the *last pressure equalization* and q is the constant rate just before shut-in. Comments made in Sections 5.2 and 6.3 about the proper t_p to use for a Horner-type analysis also apply here. In Eq. 7.11, the numerator is usually the cumulative injection since the last pressure equalization rather than the cumula-

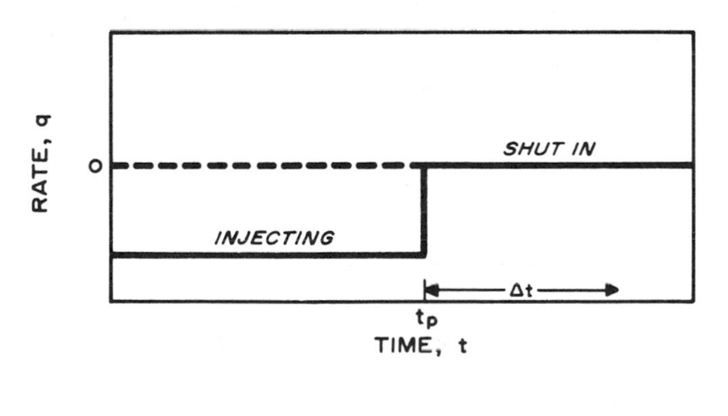

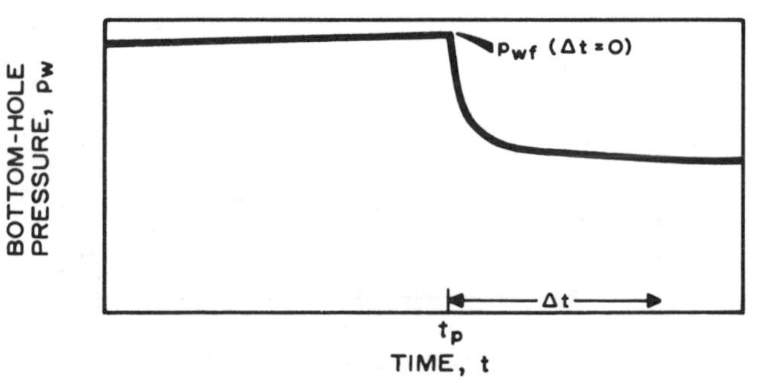

Fig. 7.5 Idealized rate schedule and pressure response for falloff testing.

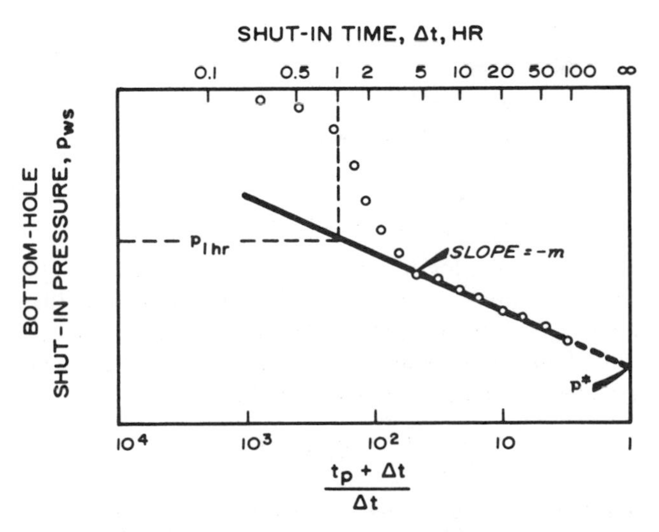

Fig. 7.6 Horner plot of a typical falloff test.

tive injection since the well was put on injection. If $t_p > 2t_{pss}$, then, for reasons discussed in Sections 5.2 and 6.3, the time to reach pseudosteady state (or steady state, which for a five-spot system[10] occurs at $t_{DA} = 0.25$ with A = the area per well, not per pattern) should be used[11,12] in place of t_p.

Miller-Dyes-Hutchinson-type plotting of falloff data, as suggested by Eq. 5.13,

$$p_{ws} = p_{1hr} + m \log \Delta t \; , \quad \ldots\ldots\ldots\ldots\ldots\ldots \quad (7.12)$$

also applies to falloff testing. The analysis method of Section 5.3 applies: m in Eq. 7.12 is the slope of the p_{ws} vs log Δt straight line and is defined by Eq. 7.7; k is estimated from Eq. 7.9; skin factor is estimated from Eq. 7.10; and the false pressure, p^*, may be estimated from Eq. 5.14. The end of the semilog straight line (either Horner or MDH) may be estimated by using Eq. 5.16 and Figs. 5.6 and 5.7. Because it is less work, the MDH plot is more practical unless t_p is less than about twice the maximum shut-in time. If necessary, the Horner plot may be used for a second pass to estimate average pressure.

Muskat-type plotting may be used to analyze pressure falloff tests, but this is generally not recommended since the boundary conditions in injection well testing are more complicated than the simple single-well closed systems assumed in the analysis technique described in Section 5.3. The information in Section 6.4 indicates that a Muskat plot may provide good results if there is essentially a constant-pressure boundary between production and injection wells.

Example 7.2 Pressure Falloff in a Liquid-Filled, Infinite-Acting Reservoir

During a stimulation treatment, brine was injected into a well and the falloff data shown in Figs. 7.7 through 7.9 were taken.[6] Other data include

$t_p = 6.82$ hours

total falloff time = 0.67 hour

$p_{tf}(\Delta t = 0) = 1{,}310$ psig
$q_w = -807$ STB/D
$B_w = 1.0$ RB/STB
$\mu_w = 1.0$ cp
$c_t = 1.0 \times 10^{-5}$ psi^{-1}
$c_w = 3.0 \times 10^{-6}$ psi^{-1}
$\phi = 0.25$
$r_w = 0.4$ ft
$\rho_w = 67.46$ lb$_m$/cu ft
$h = 28$ ft
depth = 4,819 ft
$A = 20$ acres = 871,200 sq ft.

Fig. 7.7 is the log-log plot for the test data. From the shape of the curve, it appears that the semilog straight line should begin by 0.1 to 0.2 hour. Using $\Delta p = 238$ psi and $\Delta t = 0.01$ hour from the unit-slope straight line, we estimate the wellbore storage coefficient from Eq. 2.20:

$$C = \frac{(807)(1.0)(0.01)}{(24)(238)} = 0.0014 \text{ RB/psi}.$$

C must be positive, so we disregard the sign convention here. Since wellhead pressure was always above atmospheric, the wellbore remained full during the test. Thus, Eq. 2.17 and a wellbore compressibility of $c_w = 3.0 \times 10^{-6}$ psi^{-1} can be used to estimate the wellbore volume corresponding to $C = 0.0014$ bbl/psi: $V_w = 467$ bbl. Using the depth of 4,819 ft, we compute a casing radius of 0.42 ft, which is too large for a hole of radius 0.4 ft. Nevertheless,

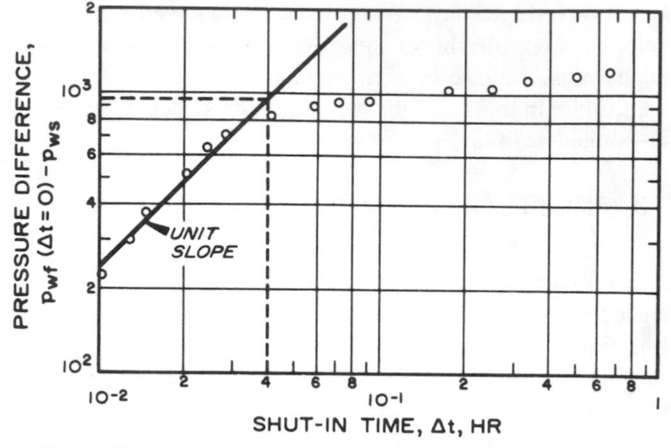

Fig. 7.7 Log-log data plot for a falloff test after brine injection, Example 7.2.

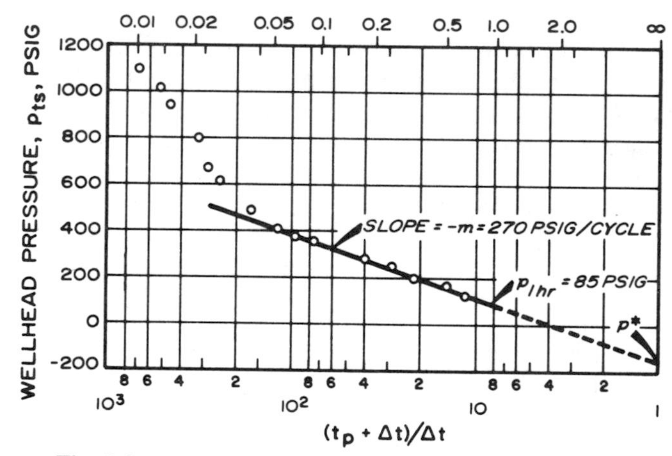

Fig. 7.8 Horner plot of pressure falloff after brine injection, Example 7.2.

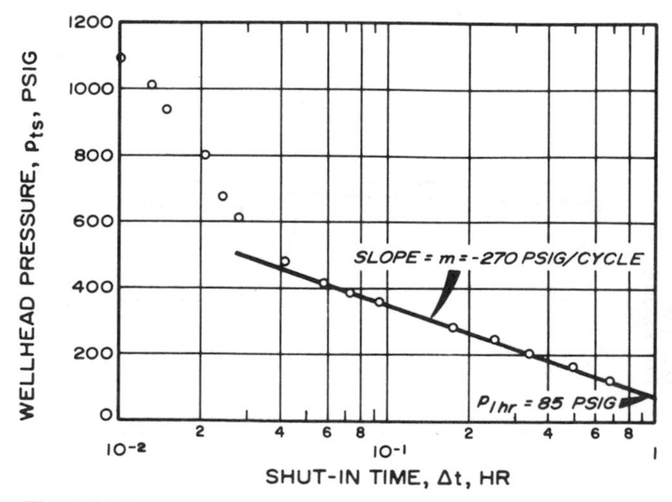

Fig. 7.9 Miller-Dyes-Hutchinson plot of pressure falloff after brine injection, Example 7.2.

the agreement is within reason. If the well was shut in at the injection pump rather than at the wellhead, the connecting lines would cause the storage coefficient to be larger than that resulting from the wellbore only. Unfortunately, we do not have all the information necessary to know if such speculation is correct. This clearly indicates the *need* for a diagram or a sketch of the well completion equipment and surface connecting lines.

Wellhead pressures are plotted vs $\log[(t_p + \Delta t)/\Delta t]$ in Fig. 7.8. That Horner plot can be used to estimate k, s, and p^*. Since the falloff time (0.67 hour) is much smaller than the flow time (6.82 hours), the log Δt (MDH) plot shown in Fig. 7.9 also may be used. The correct straight line in Figs. 7.8 and 7.9 indicates $m = -270$ psig/cycle and $p_{1hr} = 85$ psig. Thus, using Eq. 7.9,

$$k = \frac{(162.6)(-807)(1.0)(1.0)}{(-270)(28)} = 17.4 \text{ md}.$$

The skin factor is estimated from Eq. 7.10:

$$s = 1.1513 \left\{ \frac{85 - 1{,}310}{-270} - \log \left[\frac{17.4}{(0.25)(1.0)(1.0 \times 10^{-5})(0.4)^2} \right] + 3.2275 \right\}$$

$$= 0.15.$$

From Fig. 7.8, $p_{ts}^* = -151$ psig. This is the false pressure at the *surface*. Using the hydrostatic gradient of 0.4685 psi/ft and the depth of 4,819 ft, the initial bottom-hole pressure is estimated:

$$p^* = (4{,}819)(0.4685) - 151 = 2{,}107 \text{ psig}.$$

Since injection time t_p is short, we can safely assume that $p^* = \bar{p}$, so $\bar{p} = 2{,}107$ psig.

When the test well is operating at true steady state, falloff test analysis by the MDH technique *should be sufficient* and, thus, is the preferred method from the practical standpoint of less work. Shutting in the well will disturb steady-state conditions in the reservoir, and adjacent producing wells will eventually cause pressure decline in the test well. The pressure does not level off in a falloff test as it does in a buildup test. Pressure falloff will continue for a time, then the pressure will deviate *below* the semilog straight line rather than above it, as might be expected by analogy to pressure buildup testing. This is not a violation of the analogy between the tests — but is caused by interference from adjacent withdrawal wells; indeed, data in a pressure buildup test in an injection project will deviate *above* the semilog straight line when injection at adjacent wells continues.

If there is a general pressure increase or decrease at the injection well before falloff testing, the techniques described in the "Developed Reservoir Effects" portion of Section 5.3 may be applied. It is necessary, however, to recognize the basic differences in behavior between a reservoir with only producing wells and a reservoir with both injection and producing wells.

As indicated in Section 7.2, multiple-rate testing of injection wells is analogous to multiple-rate testing in production wells. Two-rate testing (two-rate falloff testing) is appropriate to eliminate changing wellbore storage during a falloff test (Section 11.2). A two-rate falloff test is run by injecting at a relatively high rate (but below parting pressure) and then decreasing the injection rate while observing the pressure decrease as a result of the rate decrease. If rates are chosen correctly, surface pressure is maintained and changing wellbore storage effects are eliminated. That can be important because injection wells frequently go on vacuum during a falloff test, resulting in an increasing wellbore storage condition (see Sections 2.6 and 11.2 and Figs. 2.12, 11.2, and 11.3, and Ref. 13) and essentially unanalyzable test data. A two-rate falloff test is analyzed like the two-rate production-well test described in Section 4.3. The data plot is made as suggested by Eq. 4.6 and the analysis is based on Eqs. 4.10 and 4.11. The simplified analysis technique in Section 4.3 also may be used when the conditions it assumes are satisfied.

Example 7.3 Two-Rate Falloff Test

An injectivity test was started on an injection well in a waterflooded reservoir before a tertiary recovery test. After a few hours of injection, it was evident that the −100 STB/D injection rate could not be maintained without exceeding fracture pressure, so the rate was reduced. Since an injectivity test had been planned, a bottom-hole pressure gauge was operating and the pressure data in Table 7.1 were obtained. Other data are

$t_1 = 371$ minutes $= 6.183$ hours
$q_1 = -100$ STB/D
$q_2 = -48.5$ STB/D
$B_w = 1.0$ RB/STB
$\mu_w = 1.0$ cp
$h = 20$ ft
$r_w = 0.39$ ft
$c_t = 7.0 \times 10^{-6}$ psi^{-1}
$\phi = 0.20$
$p_{wf}(\Delta t = 0) = 832$ psi.

The log-log data plot in Fig. 7.10 indicates that wellbore storage effects are insignificant after the first data point.

We analyze test data for formation properties by using equations for multiple-rate production-well testing. Eq. 4.6 applies for a two-rate test and the pressure data should be plotted vs $\{\log[(t_1 + \Delta t)/\Delta t] + (q_2/q_1)\log \Delta t\}$. Fig. 7.11 is

TABLE 7.1—TWO-RATE FALLOFF TEST DATA FOR EXAMPLE 7.3.
$q_1 = -100$ STB/D, $q_2 = -48.5$ STB/D, $t_1 = 6.183$ hours.

Δt (hours)	p_{w} (psi)	$\log\left(\frac{t_1+\Delta t}{\Delta t}\right)$	$\log \Delta t$	$\log\left(\frac{t_1+\Delta t}{\Delta t}\right) + \frac{q_2}{q_1}\log \Delta t$
0	831.8	—	—	—
0.167	661.3	1.580	−0.777	1.203
0.333	640.6	1.292	−0.478	1.060
0.500	631.3	1.126	−0.301	0.980
0.667	630.3	1.012	−0.176	0.927
0.833	625.1	0.925	−0.079	0.887
1.000	623.1	0.856	0.000	0.856
1.333	621.0	0.751	0.125	0.812
1.667	620.0	0.673	0.222	0.781
2.000	620.0	0.612	0.301	0.758
3.000	611.7	0.486	0.477	0.717
4.000	611.7	0.406	0.602	0.698
5.000	611.7	0.350	0.699	0.689

such a plot; it has a slope of $m_1' = 81$ psi/cycle and $p_{1hr} = 624$ psi. Using Eq. 4.10,

$$k = \frac{(-162.6)(-100)(1.0)(1.0)}{(81)(20)} = 10.0 \text{ md}.$$

The skin factor is estimated using Eq. 4.11:

$$s = 1.1513 \left\{ \left[\frac{-100}{-100-(-48.5)}\right] \left[\frac{832-624}{81}\right] - \log\left[\frac{10.0}{(0.2)(1.0)(7.0\times10^{-6})(0.39)^2}\right] + 3.2275 \right\}$$

$$= 0.6.$$

The two-rate falloff test in Example 7.3 eliminated high wellbore storage that had been previously observed in the well. Wellbore storage effects in that two-rate test were insignificant after about 15 minutes. That was accomplished with only a 6-hour duration for the initial rate; the entire test lasted only 12 hours.

7.4 Average and Interwell Reservoir Pressure

In finite, liquid-filled reservoirs of uniform mobility and $\phi c_t h$, the false pressure is obtained by extrapolating the straight-line portion of the Horner plot to $(t_p + \Delta t)/\Delta t = 1$. In new wells or wells with short injection times, $p^* \simeq p_i$. However, as is the case in pressure buildup analysis, p^* must be corrected to average reservoir pressure for finite reservoirs. A Matthews-Brons-Hazebroek-type[14] dimensionless pressure may be used to correct the false pressure to average pressure, as given by Eq. 6.2:

$$\bar{p} = p^* - \frac{m\, p_{D\text{MBH}}(t_{pDA})}{2.3025} \quad \text{. . . . . (7.13)}$$

Fig. 7.12 is the MBH dimensionless pressure correlation for a five-spot waterflood.[1] The area per well (one-half the five-spot area) is used in t_{pDA}. Similar correlations are not available for other waterflooding patterns.

Caution should be exercised in applying Eq. 7.13 and Fig. 7.12, and in attempting to apply the methods of Chapter 6 to falloff test analysis. Those methods apply only if (1) the reservoir behaves as if it is lithologically homogeneous and there is a single homogeneous fluid present (that is, the unit mobility ratio assumption must be satisfied); (2) no marked contrast in $\phi c_t h$ exists between the injected and original fluid; (3) noncommunicating layers of markedly different properties are not present; and (4) limited hydraulically induced fractures are present.

In reservoirs with composite fluid banks (Section 7.5), Eq. 7.13 and the methods of Chapter 6 yield incorrect results, especially before fillup. To remedy this situation, Hazebroek, Rainbow, and Matthews[4] proposed a procedure for estimating $\bar{p}$ from falloff tests run before fillup. Their procedure is essentially the same as the extended Muskat method in that $\log(p_{ws} - p_e)$ is plotted vs shut-in time, Δt. At late times, a straight line results when the correct value of p_e has been assumed. The approach must be applied to late-time data; in general, the time restrictions for the beginning of the Muskat straight line given by Eq. 5.21 and Fig. 5.11 for the square with constant-pressure boundaries must be satisfied. The actual numerical values may differ somewhat because of the circular nature of an expanding oil bank, but quantitative methods for estimating the beginning of the Muskat straight line for that situation are not available.

The interwell reservoir pressure sometimes may be used as an approximation of average reservoir pressure. In a five-spot pattern with unit mobility ratio, the pressure halfway between the injector and the producer is

$$\bar{p} = p_{wf}(\Delta t = 0) + \frac{162.6\, qB\mu}{kh}\left[\log\left(\frac{A}{r_w^2}\right) - 0.83867 + 0.86859\, s\right]$$

$$= p_{wf}(\Delta t = 0) + m\left[\log\left(\frac{A}{r_w^2}\right) - 0.83867 + 0.86859\, s\right], \quad \text{. . . . . (7.14)}$$

where A is the area within the five-spot pattern. If the skin

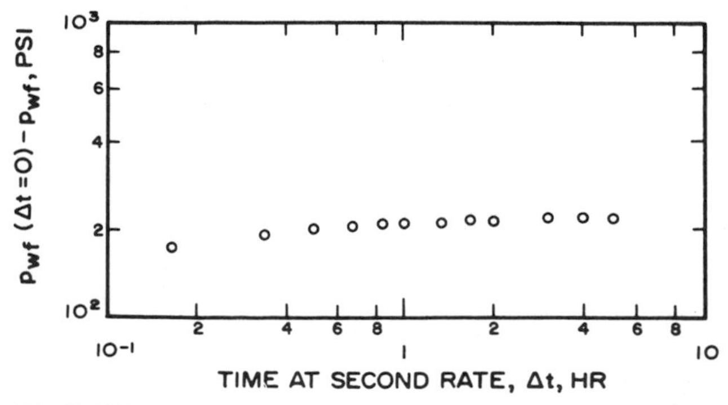

Fig. 7.10 Log-log data plot for the two-rate falloff test of Example 7.3. Water injection in a waterflooded reservoir.

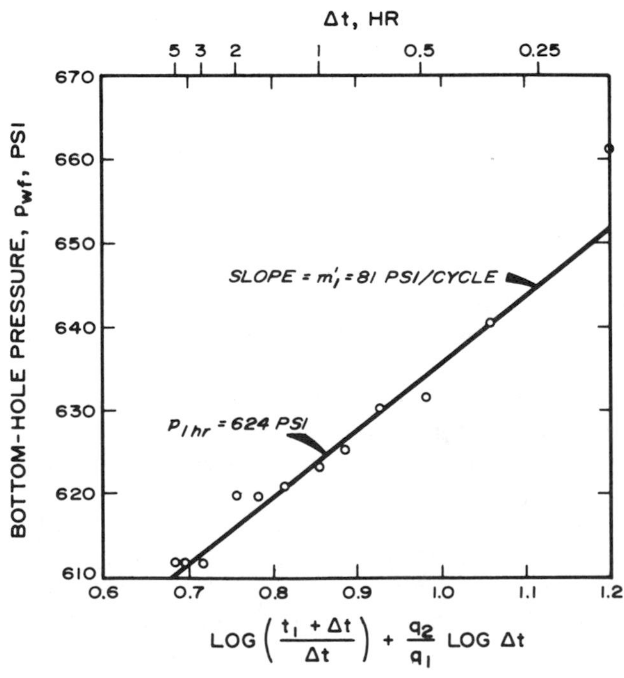

Fig. 7.11 Two-rate falloff-test data plot for Example 7.3. Water injection into a waterflooded reservoir.

factor is not the same in the production and injection wells, replace s in Eq. 7.14 by the average of the two skin factors for theoretically correct results. If the wellbore radii differ, replace r_w^2 by the product of the two r_w values. The coefficient in front of the brackets in Eq. 7.14 is m, obtained from the falloff curve, Eq. 7.7.

Another approximation to the interwell average pressure after fillup is just the arithmetic average of the pressures outside the skin zones at the stabilized injection well and the adjacent stabilized production wells. This pressure may be estimated from

$$\bar{p} \simeq \frac{1}{2}\left[(p_{wf} - \Delta p_s)_{\text{inj}} + \frac{1}{n}\sum_{i=1}^{n}(p_{wf} + \Delta p_s)_{\text{prod},i}\right], \quad (7.15)$$

where n is the number of producers surrounding the injector. Note that the pressure drop across the skin must be removed before the average reservoir (interwell) pressure is estimated. This is done because injectors and producers can be expected to have different skins and rates because of different wellbore conditions, net sand variations, or different operating practices. As used in Eq. 7.15, Δp_s is positive for damage and negative for improvement. This is a minor deviation from the strict sign interpretation used in Example 7.1.

Example 7.4 Estimating Average Pressure From a Falloff Test—Unit-Mobility-Ratio, Liquid-Filled System

The falloff test of Example 7.2 can be used to illustrate estimating $\bar{p}$ even though the injection time is very short. From Example 7.2, $p^* = -151$ psig at the surface or 2,107 psig at reservoir datum. Using Eq. 2.3b,

$$t_{pDA} = \frac{(0.0002637)(17.4)(6.82)}{(0.25)(1.0)(10^{-5})(871{,}200)} = 0.0144.$$

From Fig. 7.12, $p_{DMBH}(t_{pDA} = 0.0144) \simeq 0.001$ (read non-zero for illustrative purposes only). Then, applying Eq. 7.13,

$$p = 2{,}107 - \frac{(-270)(0.001)}{2.3025}$$

$$= 2{,}107 + 0.12$$

$$= 2{,}107 \text{ psi}.$$

Thus, the p^* value for this short injection time is a usable estimate of $\bar{p}$. Note that the correction to p^* is positive,

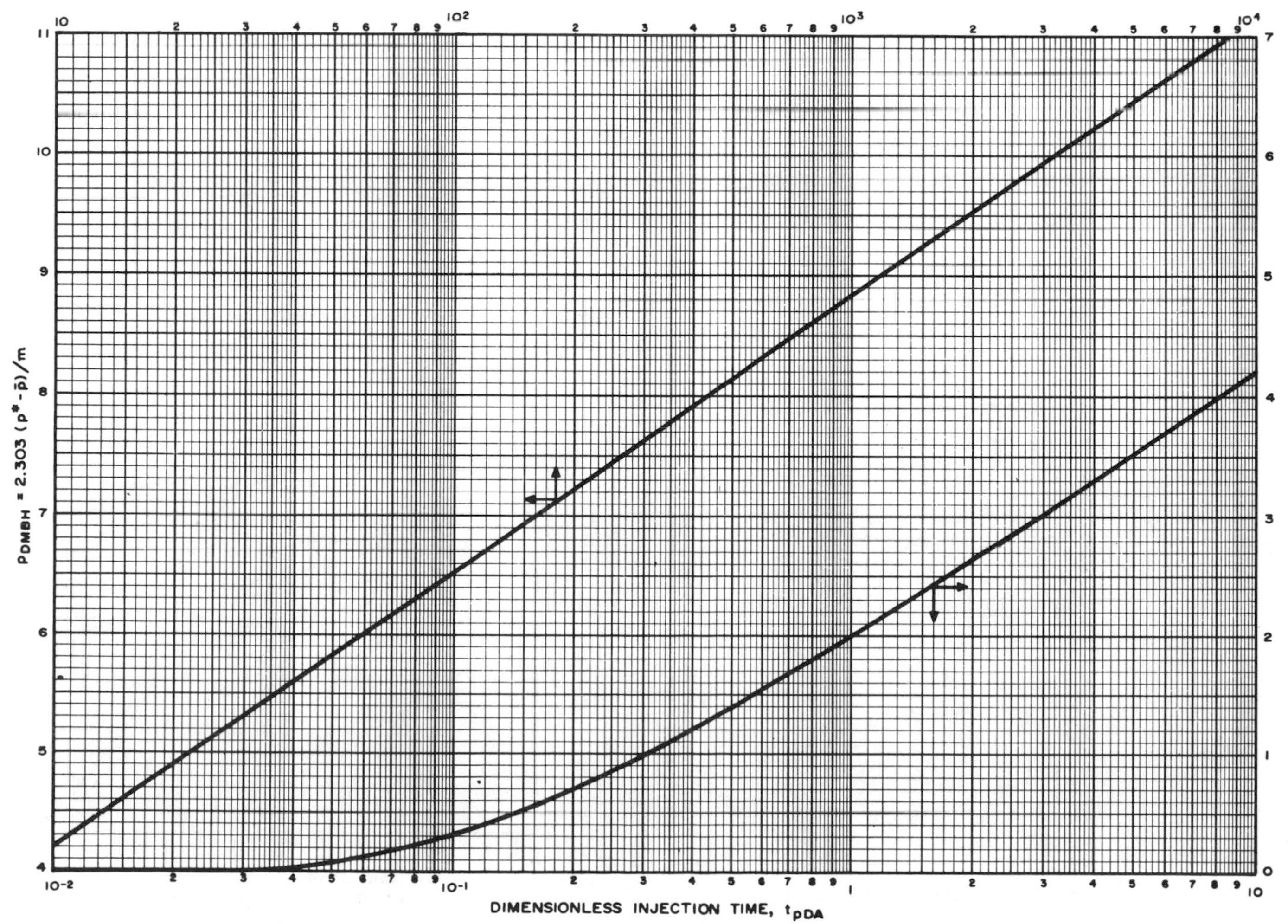

Fig. 7.12 Matthews-Brons-Hazebroek dimensionless pressure for the injection well of a five-spot waterflood, unit mobility ratio. After Matthews and Russell.[1]

indicating that $\bar{p} > p^*$ for falloff.

Eq. 7.14 cannot be used for this example since it assumes a steady-state pressure profile in the formation, which as previously noted, would not occur until a t_{pDA} of about 0.25 based on the area per well (not per pattern).

7.5 Composite System Testing—Non-Unit Mobility Ratio

This section considers transient test analysis for an injection well with the fluid distribution shown in Fig. 7.13. For injection wells, the locations of the banks shown in Fig. 7.13 move; we refer to such a moving bank system as a "composite system." Odeh[15] and Bixel and van Poollen[16] have studied production-well transient test behavior in reservoirs with physical radial discontinuities similar to those indicated in Fig. 7.13. They were concerned with physical discontinuities in the rock system rather than with moving fluid banks. Nevertheless, the analysis methods they present may be useful for injection well testing. Their results can be used to examine the effects of a wide range of porosity-compressibility (ϕc_t) products and mobility (k/μ) ratios or permeability changes when the second bank radius is large compared with the first bank.

Hazebroek, Rainbow, and Matthews[4] have presented material specifically for injection wells before fillup. They assume a significant gas saturation ahead of the oil bank and that the pressure at the leading edge of the oil bank is constant and dominated by the pressure in the gas phase. Although that assumption is not strictly correct, their method may be applied to estimate permeability, skin factor, and average drainage-area pressure from falloff test analysis. The analysis uses an extended Muskat-type plot of $\log(p_{ws} - p_e)$ vs Δt, with p_e being varied until a straight line is obtained with late-time falloff data. The slope and intercept of that plot may be used with correlations to estimate permeability and skin factor. If an independent estimate of the mobility ratio and the ϕc_t ratio between the two banks can be made, the permeability in each bank may be estimated. The Hazebroek-Rainbow-Matthews method is covered in detail in Chapter 8 of Ref. 1, and, therefore, is not repeated here. Hazebroek, Rainbow, and Matthews[4] conclude that permeability and skin factor for the zone near the well may be estimated equally well by their method or by the MDH and Horner techniques. It must be realized, however, that those two methods give the properties of the fluid bank within the radius of investigation given by Eq. 2.41. If the boundary between the inner and outer banks does not exceed that radius of investigation for the portion of the data analyzed, incorrect results will be obtained. When the distance to the boundary between the two regions is small compared with the radius of investigation, the permeability of the outer region will be obtained by transient data analysis; the skin factor will reflect the presence of the inner bank.

For very small injected volumes, the Hazebroek-Rainbow-Matthews technique is probably superior to normal techniques for estimating permeability of the inner zone and actual damage skin factor — when the important assumption of constant pressure at the outer edge of the oil bank is satisfied. For systems with large injection volumes, normal analysis methods should provide equally reliable data, with the possible exception of estimates of average reservoir pressure. In that case, the Hazebroek-Rainbow-Matthews method (or equivalently, the extended Muskat method) should be used for estimating average pressure if the methods of Section 7.4 cannot be applied. The following analysis approach[17,18] is preferred for fluid-filled systems, when it is applicable, to that of Hazebroek, Rainbow, and Matthews.

Fig. 7.14 shows typical expected injection well falloff behavior in a two-bank system, as presented by Merrill,

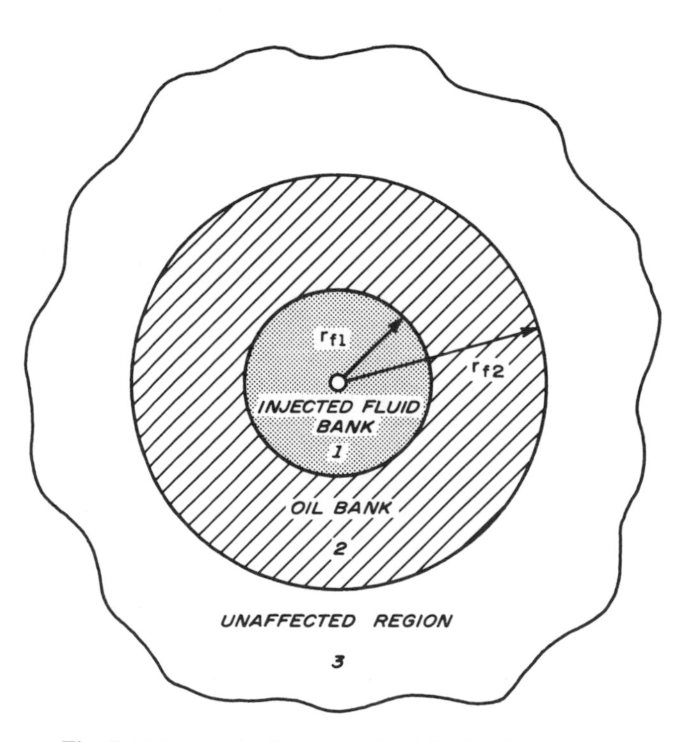

Fig. 7.13 Schematic diagram of fluid distribution around an injection well (composite reservoir).

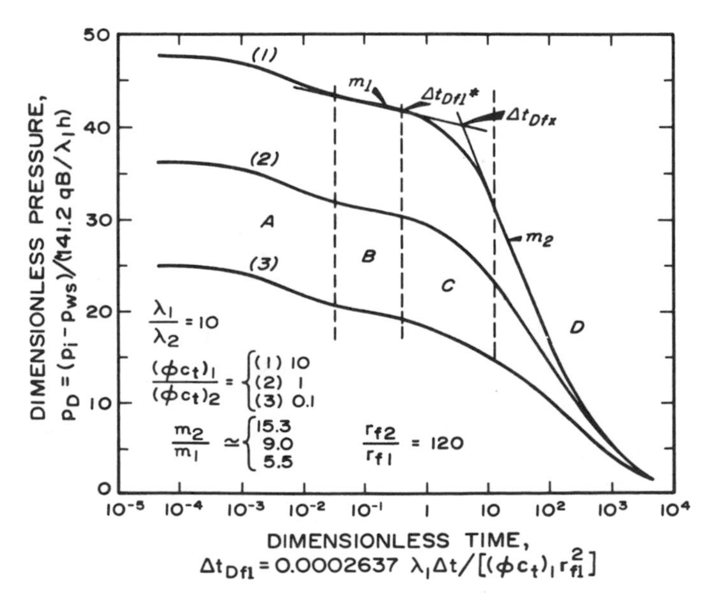

Fig. 7.14 Simulated pressure falloff data for a two-zone system. After Merrill, Kazemi, and Gogarty.[18]

Kazemi, and Gogarty[18] for a liquid-filled system. The three falloff curves apply for a mobility ratio $[M = \lambda_1/\lambda_2 = (k/\mu)_1/(k/\mu)_2]$ of 10 between the first and second bank; there is no third bank. The three curves apply for different ratios of porosity-compressibility product (ϕc_t) between the two zones. Wellbore storage effects are included. The "A" portion of Fig. 7.14 is dominated by wellbore storage effects; the "B" portion is a semilog straight line that provides information about the injected fluid bank, Region 1; the "C" portion is a transition as the second fluid bank begins to exert its influence on the falloff behavior; and the "D" portion of the curve includes a second semilog straight line whose slope is determined by properties of Regions 1 and 2.

Merrill, Kazemi, and Gogarty[18] propose methods for estimating both the location of the front of Region 1 in Fig. 7.13 and the permeability of the two fluid banks in a two-zone system. Their approach does not require previous knowledge of the mobility ratio, although an estimate of the ϕc_t ratio must be available. The data presented in Ref. 18 and here are based on computer simulations for which $r_{f2}/r_{f1} > 50$. Practically speaking, if $r_{f2}/r_{f1} > 10$, the techniques probably still apply. However, for lower values of r_{f2}/r_{f1}, chances of successful analysis are poor. The Merrill-Kazemi-Gogarty approach differs from the Hazebroek-Rainbow-Matthews approach in that it requires knowledge of neither the mobility ratio nor the location of the interior fluid front.

Merrill, Kazemi, and Gogarty[18] proposed two ways for estimating the distance to the front of Region 1 from a p_{ws} vs log Δt plot of falloff data. One approach is to use the extrapolated intersection time of the two semilog straight lines on the MDH plot, Δt_{fx}, with

$$r_{f1} = \sqrt{\frac{0.0002637\,(k/\mu)_1}{(\phi c_t)_1}\,\frac{\Delta t_{fx}}{\Delta t_{Dfx}}} \quad \text{. . . . . (7.16)}$$

Fig. 7.15 correlates Δt_{Dfx} with the ratio of the two semilog slopes from the falloff curve and the ϕc_t ratio in the two fluid banks. The second method uses the point of deviation of the observed pressure data from the first semilog straight line, $\Delta t_{f1}{}^*$, with

$$r_{f1} = \sqrt{\frac{0.0002637\,(k/\mu)_1}{(\phi c_t)_1}\,\frac{\Delta t_{f1}{}^*}{\Delta t_{Df1}{}^*}} \quad \text{. . . . . . . (7.17)}$$

Merrill, Kazemi, and Gogarty[18] show that $0.13 < \Delta t_{Df1}{}^* < 1.39$, with an average value of 0.389. This agrees quite well with an interpretive rule of thumb that the water-bank (first bank) slope normally will be valid to a time equivalent to Δt_{Df1} (based on r_{f1}, see Fig. 7.13) of about 0.25. $\Delta t_{Df1}{}^*$ does not correlate well with slope and specific storage ratios, so we do not recommend using Eq. 7.17 unless insufficient data are available to estimate Δt_{fx} for use in Eq. 7.16.

The permeability in the injected fluid bank may be estimated from the slope $(\pm m_1)$ of the first semilog straight line and Eq. 7.9. Skin factor is estimated from m_1, p_{1hr}, and Eq. 7.10. If $r_{f2} > 10 r_{f1}$, the mobility in the second zone may be estimated from

$$\left(\frac{k}{\mu}\right)_2 = \frac{(k/\mu)_1}{(\lambda_1/\lambda_2)}, \quad \text{. . . . . (7.18)}$$

where the mobility ratio, (λ_1/λ_2), is from either Fig. 7.16 or Fig. 7.17.[18] If both semilog straight lines appear and if it is possible to estimate the ratio of specific storage capacities, it is possible to estimate mobility or permeability in each zone. A common error in transient test analysis is to assume that each slope indicates the mobility of a particular fluid zone. Figs. 7.14, 7.16, and 7.17, and Eq. 7.18 clearly show this is not the case for the second zone. Further modifications are needed when $r_{f2} < 10 r_{f1}$, even for a liquid-filled, two-bank system. Unfortunately, it seems that reservoir simulation

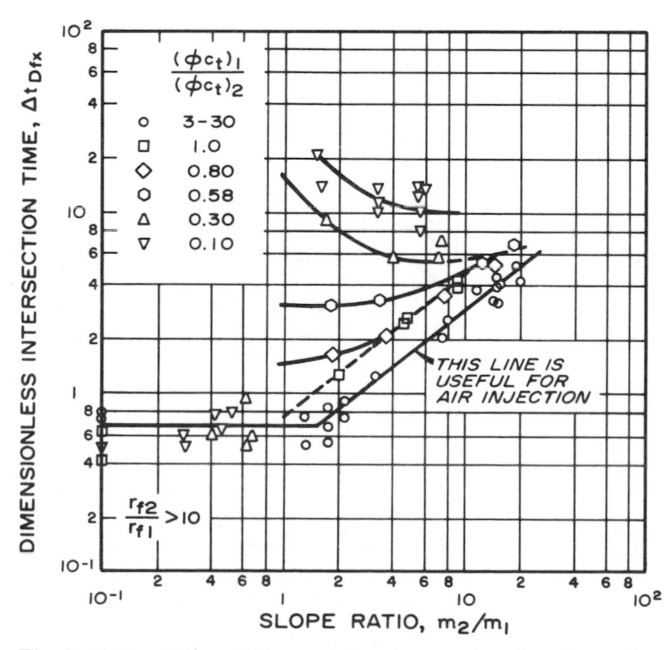

Fig. 7.15 Correlation of dimensionless intersection time, Δt_{Dfx}, for falloff data from a two-zone reservoir. After Merrill, Kazemi, and Gogarty.[18]

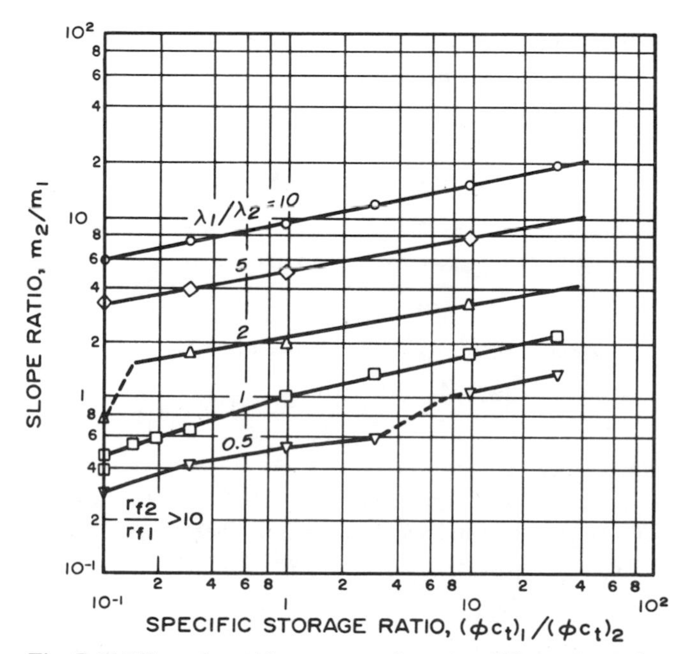

Fig. 7.16 Effect of specific storage ratio and mobility ratio on the slope ratio for falloff testing in a two-zone reservoir. After Merrill, Kazemi, and Gogarty.[18]

approaches are required for analysis of that common situation.

Merrill, Kazemi, and Gogarty[18] suggest a way to estimate the maximum wellbore storage coefficient that still allows the first semilog straight line to be observed. By using their approach, but substituting Eq. 2.22b as the criterion for the end of wellbore storage effects for pressure falloff testing, we see that

$$C \leq 5.9 \times 10^{-7} \left(\frac{k}{\mu}\right)_1 h\Delta t_{f1}^* \, e^{-0.14s}, \quad \ldots\ldots\ldots (7.19)$$

for the first semilog straight line to be detected. Eq. 7.19 allows for about 1 cycle of semilog straight line between die-out of afterflow and initial deviation caused by second-bank effects. That is a difficult criterion to achieve, especially if the boundary between the first and second banks is relatively close to the injection well.

Example 7.5 Pressure Falloff Analysis in a Two-Zone System

Fig. 7.18 is a semilog plot of simulated falloff data for a two-zone waterflood from Merrill, Kazemi, and Gogarty.[18] Data used in the simulation were

$r_w = 0.25$ ft
$r_{f1} = 30$ ft
$r_{f2} = r_e = 3{,}600$ ft, so $r_{f2}/r_{f1} = 120$
$(k/\mu)_1 = \lambda_1 = 100$ md/cp
$(k/\mu)_2 = \lambda_2 = 50$ md/cp
$(\phi c_t)_1 = 8.95 \times 10^{-7}$ psi^{-1}
$(\phi c_t)_2 = 1.54 \times 10^{-6}$ psi^{-1}
$q = -400$ STB/D
$B_w = 1.0$ RB/STB
$h = 20$ ft
$s = 0$
$C = 0.$

Since the data were simulated with no wellbore storage effect, we need not make the log-log data plot. Fig. 7.18, the MDH plot of the data, shows that $m_1 = -32.5$ psi/cycle and $m_2 = -60.1$ psi/cycle. Also,

$$m_2/m_1 = -60.1/(-32.5) = 1.85,$$

and

$$(\phi c_t)_1/(\phi c_t)_2 = 8.95 \times 10^{-7}/1.54 \times 10^{-6} = 0.581.$$

To estimate k/μ for Region 1 we use Eq. 7.9:

$$\left(\frac{k}{\mu}\right)_1 = \frac{162.6(-400)(1.0)}{(-32.5)(20)} = 100 \text{ md/cp},$$

the correct result.

To estimate $(k/\mu)_2$ we enter Fig. 7.17 with the slope and ϕc_t ratios above and read $\lambda_1/\lambda_2 = 2.0$. Then, from Eq. 7.18,

$$\left(\frac{k}{\mu}\right)_2 = \frac{100}{2.0} = 50 \text{ md/cp}.$$

We may use either Eq. 7.16 or Eq. 7.17 to estimate the location of the front of the water bank. From Fig. 7.18, the falloff plot, $\Delta t_{Dfx} = 0.095$ hour and $\Delta t_{f1}^* = 0.013$ hour. Using Fig. 7.15 with $m_2/m_1 = 1.85$ and $(\phi c_t)_2/(\phi c_t)_1 = 0.58$, we get $\Delta t_{Dfx} = 3.05$. Using Eq. 7.16,

$$r_{f1} = \sqrt{\frac{(0.0002637)(100)(0.095)}{(8.95 \times 10^{-7})(3.05)}} = 30 \text{ ft},$$

the value set in the simulation. To use Eq. 7.17 we must assume a value or a range for t_{Df1}^*. Using $0.13 \leq t_{Df1}^* \leq 1.39$ and $\overline{t_{Df1}^*} = 0.389$, we get

$$54 > r_{f1} > 17,$$

and

$$\bar{r}_{f1} = 31 \text{ ft}.$$

In this case the average value of t_{Df1}^* gave quite acceptable results, but that may be a coincidence.[18]

Although there is no wellbore storage in this example, we can estimate the maximum wellbore storage coefficient that

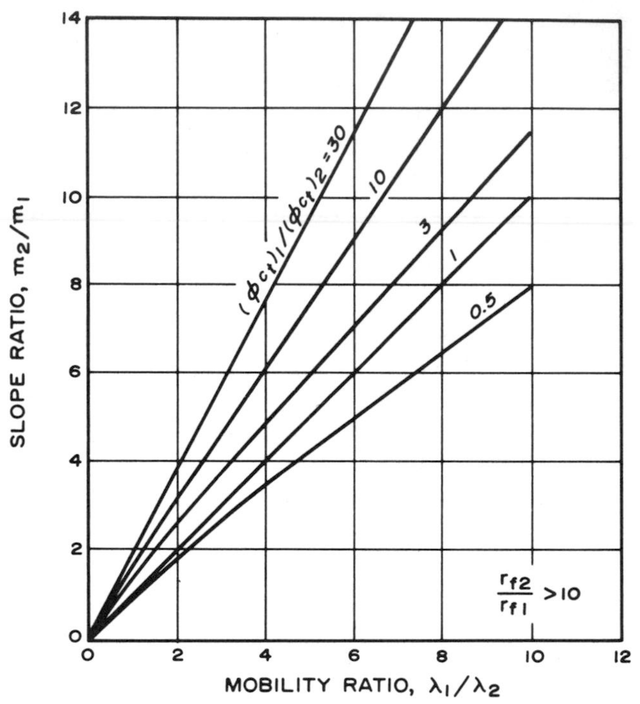

Fig. 7.17 Crossplot of data in Fig. 7.16. After Merrill, Kazemi, and Gogarty.[18]

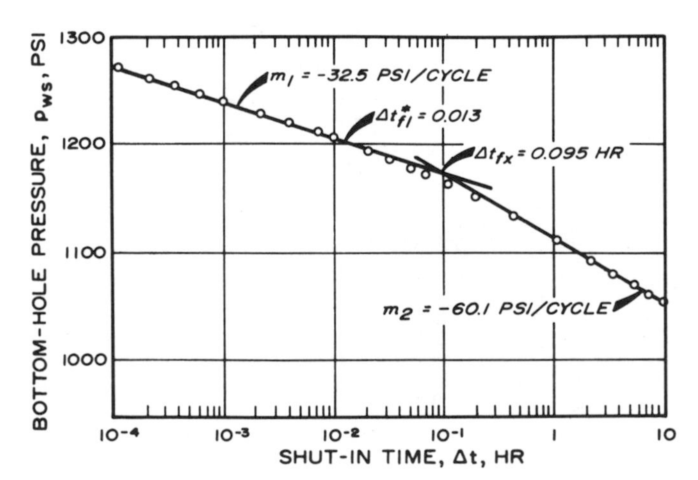

Fig. 7.18 Falloff test data for Example 7.5. After Merrill, Kazemi, and Gogarty.[18]

would not have obscured the first straight line on Fig. 7.18. Using Eq. 7.19 with $s = 0$,

$$C \leq 5.9 \times 10^{-7} (100)(20)(0.013)e^{(-0.14)(0)}$$
$$\leq 1.53 \times 10^{-5} \text{ RB/psi.}$$

If we assume that the wellbore is full of water of compressibility, $c_w = 3.0 \times 10^{-6}$ psi^{-1}, then from Eq. 2.17,

$$V_w \leq \frac{1.53 \times 10^{-5}}{3.0 \times 10^{-6}} = 5.1 \text{ bbl.}$$

Thus, if the well is completed with 2-in. tubing on a packer ($V_u \simeq 0.004$ bbl/ft), the maximum depth to meet the restriction on the wellbore storage coefficient would be about 1,300 ft — and this is assuming zero skin. For 3-in. tubing the depth must not exceed about 570 ft. If the skin factor had been 5.0, the depths for 2- and 3-in. tubing would be 635 and 282 ft, respectively.

This material applies to composite systems with two radial fluid zones, with the second zone large compared with the first. Based on a study of three-zone systems, Merrill, Kazemi, and Gogarty[18] conclude that the *only useful information* obtainable in such reservoirs is the mobility of the first zone and a rough estimate of its extent if there is a distinct contrast of mobility ratios. Reliable estimates of the mobilities and the locations of the second and third zones cannot be made with currently available technology. It is likely that the only way such estimates could be made would be by a matching process using a reservoir simulator such as that discussed in Ref. 17.

Merrill, Kazemi, and Gogarty[18] and Dowdle[19] propose methods for estimating the water saturation in the injected-fluid zone by combining Eq. 7.16, Fig. 7.15, and the material-balance equation. These methods apply at a fairly early stage of water injection into a previously liquid-filled reservoir as a result of the restriction $r_{f2} > 10r_{f1}$.

Type-curve matching may be applied to composite systems under certain circumstances. Bixel and van Poollen[16] propose such a method for analyzing pressure buildup tests with widely varying ϕc_t and k/μ ratios where the second zone is large.

One characteristic of water injection with a non-unit mobility ratio is that the injectivity tends to change as water enters the formation.[20] During the early stages of injection, this will appear as a changing skin factor. When enough fluid is injected to form a significantly large fluid bank around the injection well, the mobility of that bank will be detected by transient tests, and skin factor computed from the first-bank slope should not change unless the injected fluid is actually damaging or stimulating the wellbore.

7.6 A Pragmatic Approach to Falloff Test Analysis

From a practical point of view, a stepwise approach to pressure falloff analysis usually can be applied with acceptable results. The procedure is as follows:

1. Plot log Δp vs log Δt to determine when wellbore storage effects cease to be important. Use that plot to select the semilog straight line for the following step.

2. Regardless of the mobility ratio and whether the reservoir is filled up or not, make the MDH plot. Choose what appears to be the correct semilog straight line and estimate permeability and skin factor.

3. Calculate the expected end of the semilog straight line, assuming that it corresponds to

$$\Delta t_{Df1}{}^* \simeq 0.25 \quad \text{(7.20)}$$

Thus, the approximate end time of the semilog straight line may be estimated from

$$\Delta t^* \simeq \frac{950\,(\phi c_t)_1\, r_{f1}{}^2}{(k/\mu)_1}, \quad \text{(7.21)}$$

where $(k/\mu)_1$ is estimated from the MDH slope and r_{f1} is estimated independently such as by material balance. Eq. 7.21 is a reasonable rule-of-thumb estimate for both unfilled and filled systems operating at steady or pseudosteady conditions before shut-in.

4. If the apparent end of the MDH straight line does not correspond approximately to the time estimated in Step 3, additional steps can be taken to complete the process. This might include using the Horner method with t_p computed by normal methods, and t_{pss} computed from $(t_{DA})_{pss}$ using the area to the front of the oil bank. Also, the Hazebroek-Rainbow-Matthews method could be applied at this point, if necessary.

5. The average pressure, $\bar{p}$, may be estimated using the Matthews-Brons-Hazebroek, Fig. 7.12, or Hazebroek-Rainbow-Matthews (extended Muskat plot) methods. When the oil bank is relatively thin, the mobility ratio is near unity, and wellbore effects have died out, a simple Dietz-type extrapolation of the MDH straight line equivalent to a dimensionless time of 0.445 (based on radius of the oil bank) may be made to estimate $\bar{p}$:

$$(\Delta t)_{\bar{p}} = \frac{1{,}690\,(\phi c_t)_1\, r_{f2}{}^2}{(k/\mu)_1} \quad \text{(7.22)}$$

Eq. 7.22 can give a reasonable estimate of the reservoir pressure at the leading edge of the oil bank, assuming a constant pressure beyond that point.

6. When applicable, the Merrill-Kazemi-Gogarty[18] method can be used to estimate the second-bank mobility.

There can be errors in all the methods because of imprecise boundary conditions and assumptions used in deriving those techniques. Generally, the MDH method does give quite good values for mobility, unless the mobility ratio between the banks varies significantly from unity and the inner and outer banks are about the same size. The Hazebroek-Rainbow-Matthews approach is a late-time method based on a constant-pressure outer boundary condition. The worst errors in application of that method can be expected to occur when $t_{DAf2} < 0.44$. The calculation of the dimensionless time can prevent or reveal such application.

7.7 Series-of-Steady-State Analysis

Hall[20] proposed a technique for analyzing injection wells that basically assumes a series of steady-state injection conditions. Required data are cumulative volume injected and a good record of injection pressure. The technique presented here is a modified version of Hall's technique.[20] Eq. 2.2 is written for steady-state flow conditions and p_D is assumed to be independent of time. That assumption is not correct for long periods of time, but it is a workable approximation over reasonable time periods and does provide a simple method for monitoring injection-well performance. Using the constant p_D assumption, both sides of Eq. 2.2 can be multiplied by dt and integrated from time 0 to time t. The result is

$$\int_0^t p_{wf}dt - p_e t = \frac{141.2\mu(p_D+s)}{kh} W_i , \quad \ldots\ldots (7.23)$$

where W_i is the cumulative fluid injected at time t, a positive number. Usually, the integral on the left side of Eq. 7.23 can be approximated by a summation using wellhead pressure, p_{tf}, plus a constant fluid head term, Δp_{tw}, to approximate bottom-hole pressure, p_{wf}. Alternatively, it can be evaluated by planimetering a graph of injection pressure vs time. If p_{wf} on the left side of Eq. 7.23 is approximated by the surface injection pressure plus a constant fluid head term, the equation may be written as

$$\int_0^t p_{tf}dt - (p_e - \Delta p_{tw})t = \frac{141.2\mu(p_D+s)}{kh} W_i , \quad \ldots\ldots (7.24)$$

where Δp_{tw} is the constant fluid head between surface and bottom hole. If $(p_e - \Delta p_{tw})t$ is small compared with the integral in Eq. 7.24, as it often is for pumped-off waterfloods, a plot of the integral (or its approximation) vs cumulative water injection (called a "Hall plot") should give a straight line with slope

$$m_H = \frac{141.2\mu(p_D+s)}{kh} \text{ psi/(B/D)} , \quad \ldots\ldots (7.25)$$

as illustrated in Fig. 7.19. Eq. 7.25 assumes the integral term has units psi × days (not hours). If p_D and s are known, we should be able to estimate k/μ from Eq. 7.25. Or, if p_D and k/μ are known, we should be able to estimate s. However, we must obtain at least k/μ or s from a transient test to use Eq. 7.25 and we must be able to estimate p_D. If $(p_e - \Delta p_{tw})t$ is about 15 percent or more of the integral, its effect should be included in the data plot, or serious quantitative errors can result. In most cases, little error is caused by neglecting $(p_e - \Delta p_{tw})$. The error may be estimated by using two or three values in making the plot and observing the effect.

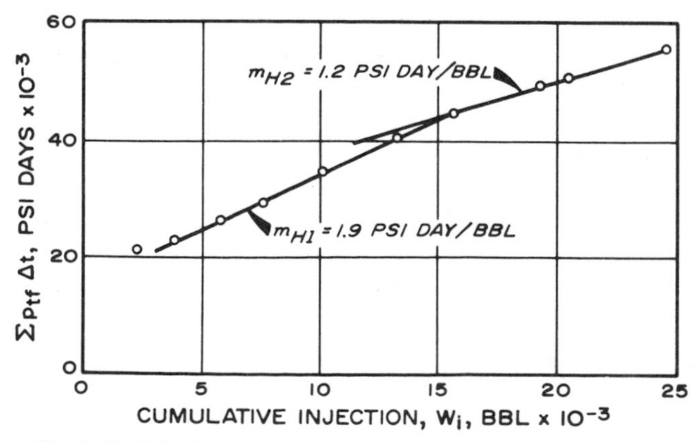

Fig. 7.19 Hall plot for a water injection well showing the effects of stimulation, Example 7.6.

The major benefit of the Hall plot is not from the single straight line, but from changes in the slope of the line. Changes in the slope of the Hall plot can be caused by changes in k/μ, s, or p_D. In any fluid injection operation, we expect k/μ to change in the vicinity of the well as fluid is injected and as the gas volume in the reservoir is filled up. As that happens, both k/μ and p_D change. In addition, p_D may change as a result of changes in operating practices or the addition of new offset production wells. Actual changes in the skin factor will also affect the slope of the Hall plot. Since Eq. 2.10 indicates that any change in permeability in the vicinity of the wellbore can be expressed as skin, we choose to show how to use the Hall plot to estimate changes in skin factor. Nevertheless, the plot can be used to estimate changes in any of the quantities of the right side of Eq. 7.25. The change in skin factor is estimated from the change in slope of the Hall plot:

$$s_2 = s_1 + \frac{kh}{141.2\mu}(m_{H2} - m_{H1}) , \quad \ldots\ldots (7.26)$$

where k/μ is supplied from transient test data. Another approach is to use the two slopes on a Hall plot, such as in Fig. 7.19, to estimate the ratio of the new flow efficiency to the old flow efficiency:

$$\frac{E_{f2}}{E_{f1}} = \frac{m_{H1}}{m_{H2}} . \quad \ldots\ldots (7.27)$$

While always in the right direction, the amount of change in the flow efficiency can be somewhat distorted if p_{tf} used in the Hall plot is significantly different from the true differential $(p_{wf} - p_e)$; that is, if $(p_e - \Delta p_{tw}) > 0.15\, p_{tf}$.

Example 7.6 Hall Method Steady-State Analysis

Fig. 7.19 is a Hall plot for a water injection well in a 1,000-ft deep, filled-up Illinois reservoir. In that reservoir $(p_e - \Delta p_{tw})$ is very small compared with p_{tf}, so the data plot shown is adequate. The injection well was shot with nitroglycerin on completion. It was stimulated with micellar solution[21] after a cumulative water injection of about 15,000 bbl. From transient testing before stimulation,

$$\left(\frac{kh}{\mu}\right)_1 = 280 \text{ md ft/cp},$$

and

$$s_1 = -1.12.$$

From transient testing several weeks after stimulation,

$$\left(\frac{kh}{\mu}\right)_2 = 290 \text{ md ft/cp},$$

and

$$s_2 = -2.3.$$

From Fig. 7.19,

$$m_{H1} = 1.9 \text{ psi/(B/D)},$$

and

$$m_{H2} = 1.2 \text{ psi/(B/D)}.$$

Applying Eq. 7.26,

$$s_2 = -1.12 + \frac{280}{141.2}(1.2 - 1.9)$$

$$= -2.5.$$

This compares favorably with $s = -2.3$ determined by transient testing several weeks after stimulation.

Muskat[22] has devised a method for analyzing water injection well data when the injection pressure, p_{wf}, is constant. He suggests that a plot of $1/q$, where q is the time varying injection rate, vs log W_i, where W_i is the cumulative volume injected, should be a straight line. The slope of the line is related to the permeability of the injected fluid, while the intercept is related to mobility ratio and saturations in the various bank areas.

7.8 Step-Rate Testing

A step-rate injectivity test is normally used to estimate fracture pressure in an injection well.[23] Such information is useful in waterfloods and is critically important in tertiary floods where it is important to avoid injecting expensive fluids through uncontrolled, artificially induced fractures.

A step-rate injectivity test is simple, inexpensive, and fast. Fluid is injected at a series of increasing rates, with each rate preferably lasting the same length of time. In relatively low-permeability formations ($k < 5$ md), each injection rate should last about 1 hour; 30-minute injection times are adequate for formations with permeability exceeding 10 md.[23] As few as four rates may be used, but normally six, seven, or eight rates are preferred. The analysis consists of plotting injection pressure at the end of each rate vs injection rate. It is preferable to plot bottom-hole pressure, but surface pressure may be used if it is positive throughout the test and friction effects are not significant. The plot should have two straight-line segments, as illustrated in Fig. 7.20. The break in the line indicates formation fracture pressure. (Unfortunately, it can also indicate the breakdown pressure of the cement bond. When the cement bond fails, the slope of the second straight line in Fig. 7.20 usually continues below the fracture pressure as the rate is decreased.) The fracture pressure may vary depending on fluid saturation conditions in the formation and long-term variations in reservoir pressure level with time.[23]

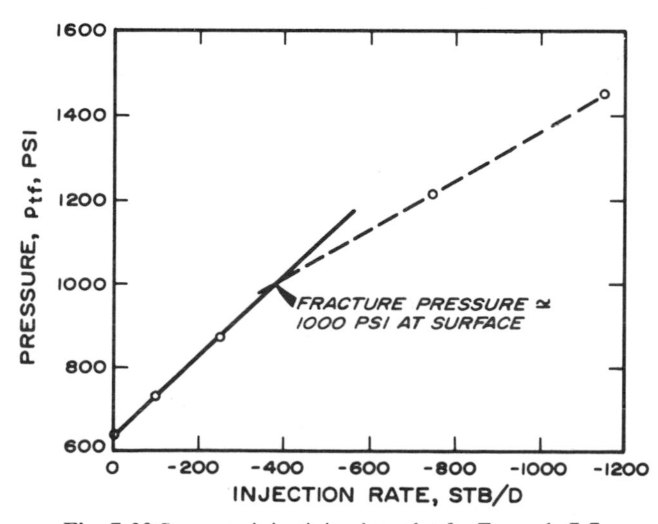

Fig. 7.20 Step-rate injectivity data plot for Example 7.7. Data of Felsenthal.[23]

TABLE 7.2—PRESSURE AND RATE DATA FOR A STEP-RATE INJECTIVITY TEST,[23] EXAMPLE 7.6.

t (hours)	q (STB/D)	p_{tf} (psi)	$\sum$ term, Eq. 4.1	$(p_i - p_{tf})/q$ (psi/STB/D)
0	0	642	—	—
0.5	−100	720	−0.301	0.780
1.0	−100	730	0.000	0.880
1.5	−250	856	−0.110	0.856
2.0	−250	874	0.120	0.928
2.25	−750	1,143	−0.335	0.668
2.50	−750	1,182	−0.112	0.720
3.00	−750	1,216	0.124	0.765
4.00	−1,150	1,450	0.246	0.703

Pressure data taken during each rate may be analyzed with a multiple-rate transient technique (Section 4.2) to estimate formation permeability and skin factor. Eqs. 4.1, 4.4, and 4.5 can be used, providing the effective wellbore radius was not already large because of previous fracture stimulation, thus making the line-source log approximation an inappropriate solution.

Example 7.7 Step-Rate Analysis

Felsenthal[23] provides the data in Table 7.2 for a step-rate test in a reservoir with the following properties:

$B_w = 1.0$ RB/STB
$\mu_w = 0.45$ cp
$h = 270$ ft
$\phi = 0.186$
$c_t = 1.5 \times 10^{-5}$ psi^{-1}
$r_w = 0.25$ ft
Depth = 7,260 ft
Injected-fluid pressure gradient = 0.433 psi/ft.

Fig. 7.20 shows the normal step-rate data plot, p_{tf} vs q. The break in the data indicates a surface fracture pressure of about 1,000 psi. The fracture gradient is estimated by dividing the bottom-hole fracture pressure by the depth. The fracture gradient is

$$[(0.433)(7{,}260) + 1{,}000]/7{,}260 = 0.57 \text{ psi/ft}.$$

The data in Table 7.2 also may be analyzed for formation properties by using the methods described in Section 4.2. The two right-hand columns in Table 7.2 contain the data to be plotted according to Eq. 4.1. Fig. 7.21 shows the data plot. The first four points, for the rates before the fracture was induced, fall on the expected straight line. That line has the properties

$$m' = 0.357\ [\text{psi/(STB/D)}]/\text{cycle},$$

and

$$b' = 0.885\ \text{psi/(STB/D)}.$$

We estimate formation permeability from Eq. 4.4:

$$k = \frac{(162.6)(1.0)(0.45)}{(0.357)(270)} = 0.76\ \text{md}.$$

The skin factor is estimated from Eq. 4.5:

$$s = 1.1513\left\{\frac{0.885}{0.357} - \log\left[\frac{0.76}{(0.186)(0.45)(1.5\times10^{-5})(0.25)^2}\right] + 3.2275\right\}$$

$$= -1.5.$$

In Fig. 7.21, the data points for $q = -750$ and $q = -1{,}150$ STB/D do not fall on the straight line. Those points correspond to data taken after the formation fractured (see Fig. 7.20). They do not fall on the initial straight line in Eq. 7.25 because the assumptions of radial, infinite-acting flow (used in Eq. 4.1) are not satisfied after the formation is fractured.

In this multiple-rate analysis, we have assumed a unit mobility ratio; there are no data to indicate the accuracy of that assumption.

References

1. Matthews, C. S. and Russell, D. G.: *Pressure Buildup and Flow Tests in Wells,* Monograph Series, Society of Petroleum Engineers of AIME, Dallas (1967) **1,** Chap. 8.

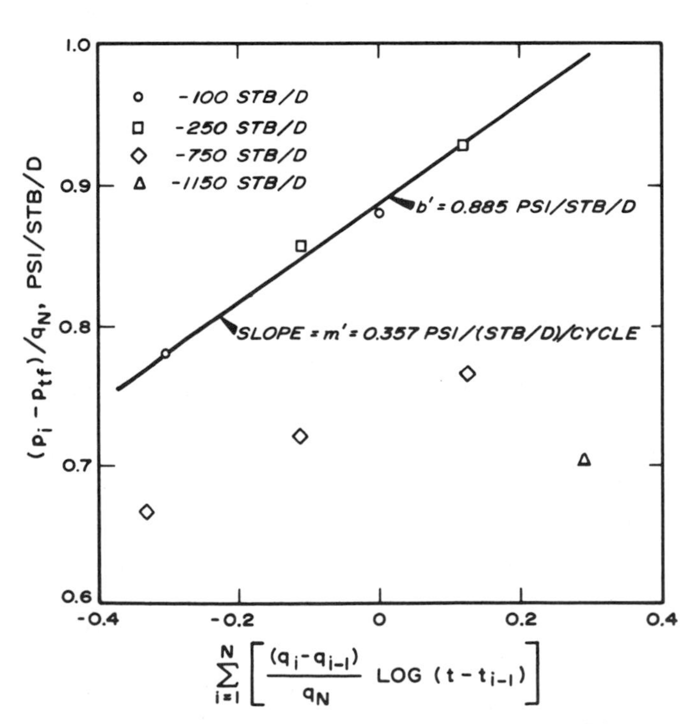

Fig. 7.21 Multiple-rate-type data plot for Example 7.7. Data of Felsenthal.[23]

2. Morse, J. V. and Ott, Frank III: "Field Application of Unsteady-State Pressure Analysis in Reservoir Diagnosis," *J. Pet. Tech.* (July 1967) 869-876.
3. Nowak, T. J. and Lester, G. W.: "Analysis of Pressure Fall-Off Curves Obtained in Water Injection Wells to Determine Injective Capacity and Formation Damage," *Trans.,* AIME (1955) **204,** 96-102. Also *Reprint Series, No. 9 — Pressure Analysis Methods,* Society of Petroleum Engineers of AIME, Dallas (1967) 61-67.
4. Hazebroek, P., Rainbow, H., and Matthews, C. S.: "Pressure Fall-Off in Water Injection Wells," *Trans.,* AIME (1958) **213,** 250-260.
5. Clark, K. K.: "Transient Pressure Testing of Fractured Water Injection Wells," *J. Pet. Tech.* (June 1968) 639-643; *Trans.,* AIME, **243.**
6. McLeod, H. O., Jr., and Coulter, A. W., Jr.: "The Stimulation Treatment Pressure Record — An Overlooked Formation Evaluation Tool," *J. Pet. Tech.* (Aug. 1969) 951-960.
7. Robertson, D. C. and Kelm, C. H.: "Injection-Well Testing To Optimize Waterflood Performance," *J. Pet. Tech.* (Nov. 1975) 1337-1342.
8. van Poollen, H. K.: "Transient Tests Find Fire Front in an In-Situ Combustion Project," *Oil and Gas J.* (Feb. 1, 1965) 78-80.
9. Kazemi, Hossein: "Locating a Burning Front by Pressure Transient Measurements," *J. Pet. Tech.* (Feb. 1966) 227-232; *Trans.,* AIME, **237.**
10. Ramey, Henry J., Jr., Kumar, Anil, and Gulati, Mohinder S.: *Gas Well Test Analysis Under Water-Drive Conditions,* AGA, Arlington, Va. (1973).
11. Pinson, A. E., Jr.: "Concerning the Value of Producing Time Used in Average Pressure Determinations From Pressure Buildup Analysis," *J. Pet. Tech.* (Nov. 1972) 1369-1370.
12. Kazemi, Hossein: "Determining Average Reservoir Pressure From Pressure Buildup Tests," *Soc. Pet. Eng. J.* (Feb. 1974) 55-62; *Trans.,* AIME, **257.**
13. Earlougher, Robert C., Jr., Kersch, K. M., and Ramey, H. J., Jr.: "Wellbore Effects in Injection Well Testing," *J. Pet. Tech.* (Nov. 1973) 1244-1250.
14. Matthews, C. S., Brons, F., and Hazebroek, P.: "A Method for Determination of Average Pressure in a Bounded Reservoir," *Trans.,* AIME (1954) **201,** 182-191. Also *Reprint Series, No. 9 — Pressure Analysis Methods,* Society of Petroleum Engineers of AIME, Dallas (1967) 51-60.
15. Odeh, A. S.: "Flow Test Analysis for a Well With Radial Discontinuity," *J. Pet. Tech.* (Feb. 1969) 207-210; *Trans.,* AIME, **246.**
16. Bixel, H. C. and van Poollen, H. K.: "Pressure Drawdown and Buildup in the Presence of Radial Discontinuities," *Soc. Pet. Eng. J.* (Sept. 1967) 301-309; *Trans.,* AIME, **240.** Also *Reprint Series, No. 9 — Pressure Analysis Methods,* Society of Petroleum Engineers of AIME, Dallas (1967) 188-196.
17. Kazemi, Hossein, Merrill, L. S., and Jargon, J. R.: "Problems in Interpretation of Pressure Fall-Off Tests in Reservoirs With and Without Fluid Banks," *J. Pet. Tech.* (Sept. 1972) 1147-1156.
18. Merrill, L. S., Jr., Kazemi, Hossein, and Gogarty, W. Barney: "Pressure Falloff Analysis in Reservoirs With Fluid Banks," *J. Pet. Tech.* (July 1974) 809-818; *Trans.,* AIME, **257.**
19. Dowdle, Walter L.: "Discussion of Pressure Falloff Analysis in Reservoirs With Fluid Banks," *J. Pet. Tech.* (July 1974) 818.

20. Hall, H. N.: "How to Analyze Waterflood Injection Well Performance," *World Oil* (Oct. 1963) 128-130.

21. Gogarty, W. B., Kinney, W. L., and Kirk, W. B.: "Injection Well Stimulation With Micellar Solutions," *J. Pet. Tech.* (Dec. 1970) 1577-1584.

22. Muskat, Morris: *Physical Principles of Oil Production,* McGraw-Hill Book Co., Inc., New York (1949) 682-686.

23. Felsenthal, Martin: "Step-Rate Tests Determine Safe Injection Pressures in Floods," *Oil and Gas J.* (Oct. 28, 1974) 49-54.

Chapter 8

Drillstem Testing

8.1 Introduction

A *drillstem test* (DST) is normally run in a zone of undetermined potential in a well being drilled, although DST's are sometimes run in known productive zones in development wells. A DST provides a temporary completion of the test interval; the drillstring serves as the flowstring. A good DST yields a sample of the type of reservoir fluid present, an indication of flow rates, a measurement of static and flowing bottom-hole pressure, and a short-term pressure transient test. The DST helps determine the possibility of commercial production by virtue of the types of fluids recovered and the flow rates observed. Analysis of the DST transient pressure data can provide an estimate of formation properties and wellbore damage. Those data may be used to estimate the well's flow potential with a regular completion that uses stimulation techniques to remove damage and increase effective wellbore size.

To run a drillstem test, a special DST tool is attached to the drillstring and lowered to the zone to be tested.[1-6] The tool isolates the formation from the mud column in the annulus, allows formation fluid to flow into the drillpipe, and continuously records the pressure during the test. Most DST's include a short production period (the initial flow period), a short shut-in period (the initial buildup), a longer flow period (the second flow period), and a longer shut-in period (the final buildup).[1,2] Fig. 8.1 is a schematic DST pressure chart for a two-cycle test (note that pressure increases downward in most DST charts shown in this chapter). The first cycle in Fig. 8.1 includes the initial flow and buildup periods, while the second cycle includes the second flow and final buildup periods. Early drillstem testing techniques used only one cycle with a longer flow duration. Fig. 8.2 shows tests with more than two cycles are possible.[1,3] (Note that pressure increases upward in that figure.)

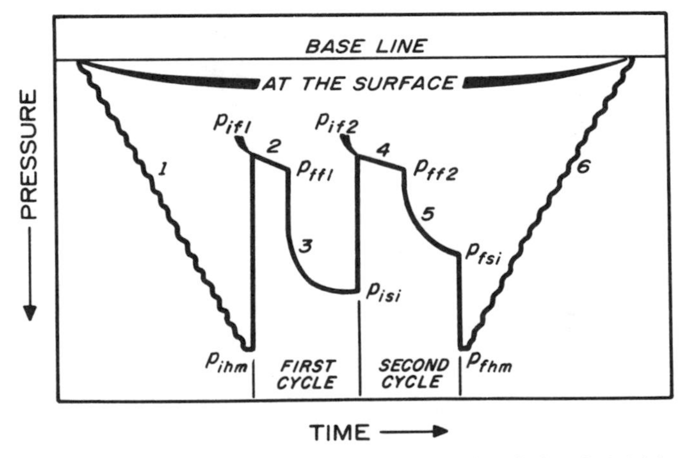

Fig. 8.1 Schematic of a DST chart: (1) going into hole; (2) initial flow period; (3) initial shut-in period; (4) final flow period; (5) final shut-in period; and (6) coming out of hole. p_{ihm} = initial hydrostatic mud pressure; p_{if1} = initial flowing pressure in first flow period; p_{ff1} = final flowing pressure in first flow period; p_{isi} = initial shut-in pressure; p_{if2} = initial flowing pressure in second flow period; p_{ff2} = final flowing pressure in second flow period; p_{fsi} = final shut-in pressure; and p_{fhm} = final hydrostatic mud pressure.

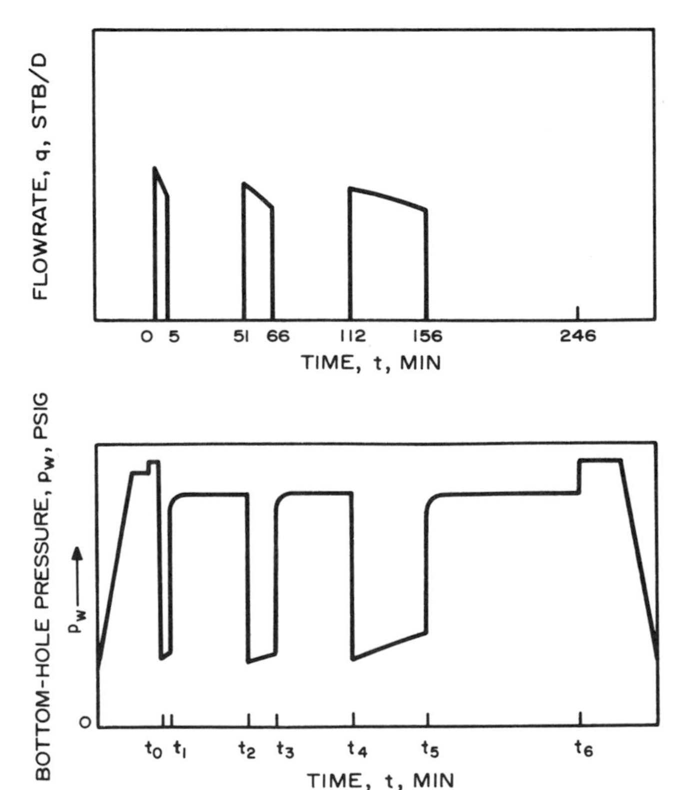

Fig. 8.2 Example of a three-cycle drillstem test. After McAlister, Nutter, and Lebourg.[3]

This chapter discusses the DST testing technique and presents methods for analyzing pressure data from both the flow and shut-in portions of a DST. A series of example DST pressure charts illustrates various DST operational conditions and test malfunctions. A short section on the wireline formation tester is also included.

8.2 Drillstem Testing Tools and Technique

Most drillstem testing tools include two or more clock-driven, Bourdon-tube recording pressure gauges, one or two packers, and a set of flow valves. The tool is attached to the drillstem and is lowered to the test interval where the packer is set, then the valves in the tool are opened and closed by manipulation of the drillpipe. The DST is run while the flow valves are being manipulated.

Fig. 8.3 shows typical drillstem testing tools used by the Halliburton Co. for the three basic types of tests: the single-packer test, the straddle-packer test, and the hook-wall packer test. Fig. 8.4 shows the operating states of a Halliburton DST tool during a test. Johnston Schlumberger, Lynes, Inc., Arrow Testers, and others also offer DST service and tools.

As shown in Fig. 8.3, the upper section of the DST tool is the same for all three test types. The uppermost part of the tool is an impact-reversing sub that allows produced fluids to be reverse-circulated out of the drillpipe (Fig. 8.4e). The closed in pressure (CIP) valve is the main flow-control valve in the DST tool string. In conjunction with the hydrospring tester valve, it allows the two flow periods, the two closed-in periods, and reverse circulation.

As shown in Fig. 8.4a, the CIP valve is open and the hydrospring tester valve is closed as the DST tool is run into the hole. The bypass ports are open while the tool is being run into and out of the hole to allow fluid to flow through the tool to help minimize pressure surges caused by running the relatively large-diameter packer.

The hydrospring tester valve is a hydraulic time-delay master valve that opens slowly and closes quickly. When the packer is set, weight is applied to cock the hydrospring and activate the hydraulic time delay. A few minutes later, the hydraulic time delay closes the bypass ports and then opens the hydrospring tester valve to start the DST. During the test, the CIP valve is closed and opened to cause the shut-in and flow periods (Figs. 8.4b and 8.4c). The hydrospring tester valve closes immediately when the weight of the pipe is picked up. Then the bypass ports open. With both the hydrospring tester and CIP valves closed, a fluid sample is isolated between those valves when the tool is removed from the hole (Fig. 8.4f).

The optional handling sub and choke assembly aids in making up the tool and also provides a receptacle for a down-hole choke, if such a device is desired. The DST tool contains two Bourdon-tube pressure recording elements. The upper element, in the flowstring, senses the pressure as fluid flows into the drillstring during the test. The lower pressure recorder, near the bottom of the tool (Figs. 8.3a through 8.3d), is "blanked off" from the flow portion of the system. It records the annulus pressure below the packer rather than the pressure of the fluid inside the drillstem. In a good test, the pressures from the two recorders will differ only by the hydrostatic head between them. In poor tests, it is often possible to determine the kind of malfunction by comparing the two pressure charts.

Many DST tools include hydraulic jars and safety joints to aid in removing a stuck tool. If the tool cannot be unstuck by jarring, the drillstring may be backed off at a safety joint, allowing recovery of the pipe and a portion of the tool.

Single-packer tests use either a nonrotating, expanding packer with a tail pipe extending to the bottom of the hole or a hook-wall packer. Both assemblies include a perforated

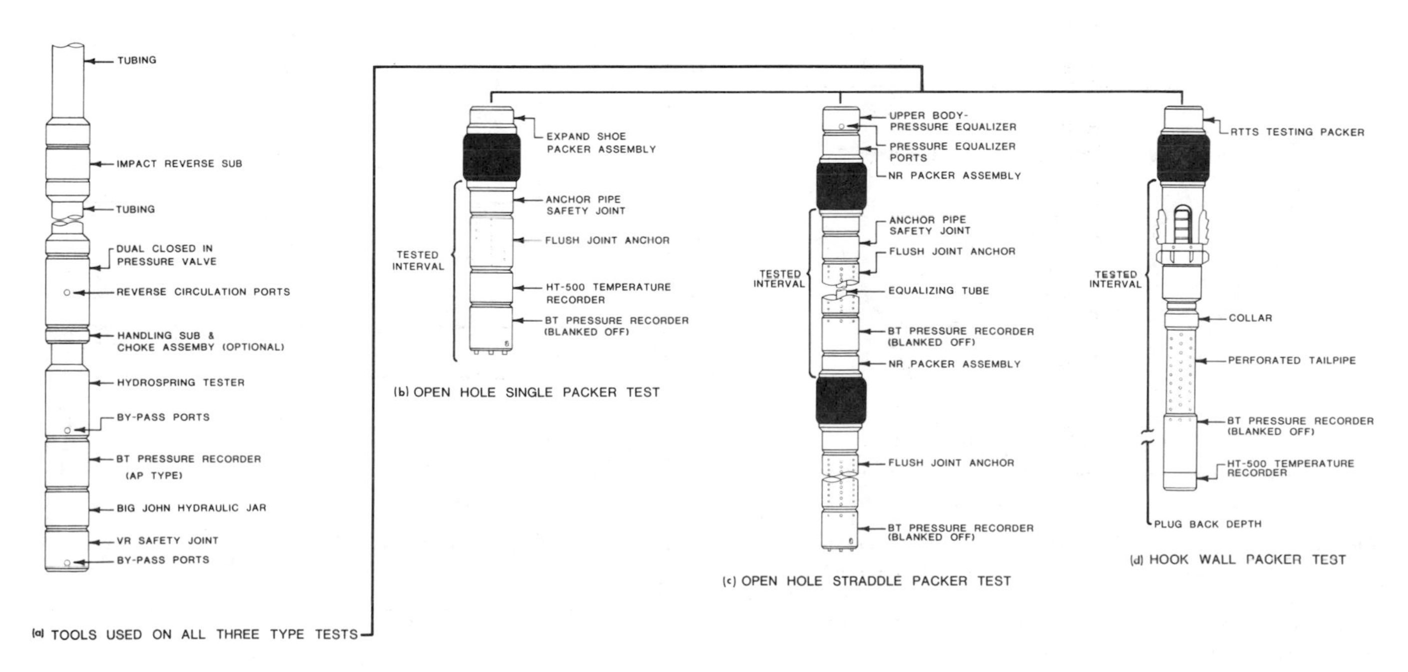

Fig. 8.3 Typical DST tools used for three types of tests. Upper assembly (left) is similar on all three test types. After Edwards and Shryock.[6] Courtesy *Petroleum Engineer*.

anchor pipe and a blanked-off pressure recorder. During the test, fluid flows from the formation through the perforated anchor pipe and into the drillstring. A temperature recorder may or may not be included in the tool string.

The straddle-packer test uses two packers, a perforated anchor pipe, and a blanked-off pressure recorder between the packers. An equalizing tube connects the annulus above the top packer to the hole below the bottom packer. The equalizing tube aids in bypassing wellbore fluid around the packers while running in and out of the hole and balances the load created on the drillstring by the annulus hydrostatic pressure during the test. A third pressure recorder may be included below the bottom packer to indicate whether that packer remains sealed throughout the test.

As indicated in Fig. 8.4a, the CIP valve is open and the hydrospring tester valve is closed while the tool is run into the hole. The bypass ports are open, so mud may flow both around the outside of the tool and through the packer while the tool is in motion. Both pressure recorders are in communication with the mud column and should record hydrostatic pressure as they are lowered into the hole (Fig. 8.1). When the packer is set, the bypass ports close and the hydrospring tester valve opens, resulting in the configuration shown in Fig. 8.4b. Both pressure recorders should show the same pressure response. To shut in the tool for a buildup, the CIP valve is closed (Fig. 8.4c). The CIP valve is opened for the second flow and closed for the second buildup. After the final buildup, the CIP valve remains closed and the hydrospring tester valve is closed, trapping a fluid sample under pressure. Then the bypass ports are opened, and pressure is equalized across the packer (Fig. 8.4d). The packer is unseated, the reverse circulating valve is opened, and mud is pumped down the annulus to displace the produced fluids up the drillstring for measurement at the surface (Fig. 8.4e). As the pipe and tool are removed from the hole (Fig. 8.4f), the mud in the drillstring is allowed to bleed into the annulus through the open reverse-circulating valve.

Hole condition may dictate the total time that the tool can remain in the hole, since a primary consideration is complete removal of the tools at the end of the test. Thus, conditions existing in the well may dictate relatively short testing times. Experience in the area is the best way to determine total allowable testing time. When allowed testing time is short, the division of the test between the various test periods is important. Pages 22 through 24 of Ref. 5 provide guidelines for choosing the length of the flow and shut-in periods in a DST — whether total test time is limited or not. Table 8.1 summarizes that material.

In a standard DST, the initial flow period is usually short, 5 to 10 minutes; the idea is simply to release the high hydrostatic mud pressure. The initial shut-in period should be sufficiently long to allow the measured pressure to approach *stabilized formation pressure*. Experience indicates that 1 hour is usually required for the initial shut-in period.[5,7] The second flow period should be long enough to allow flow

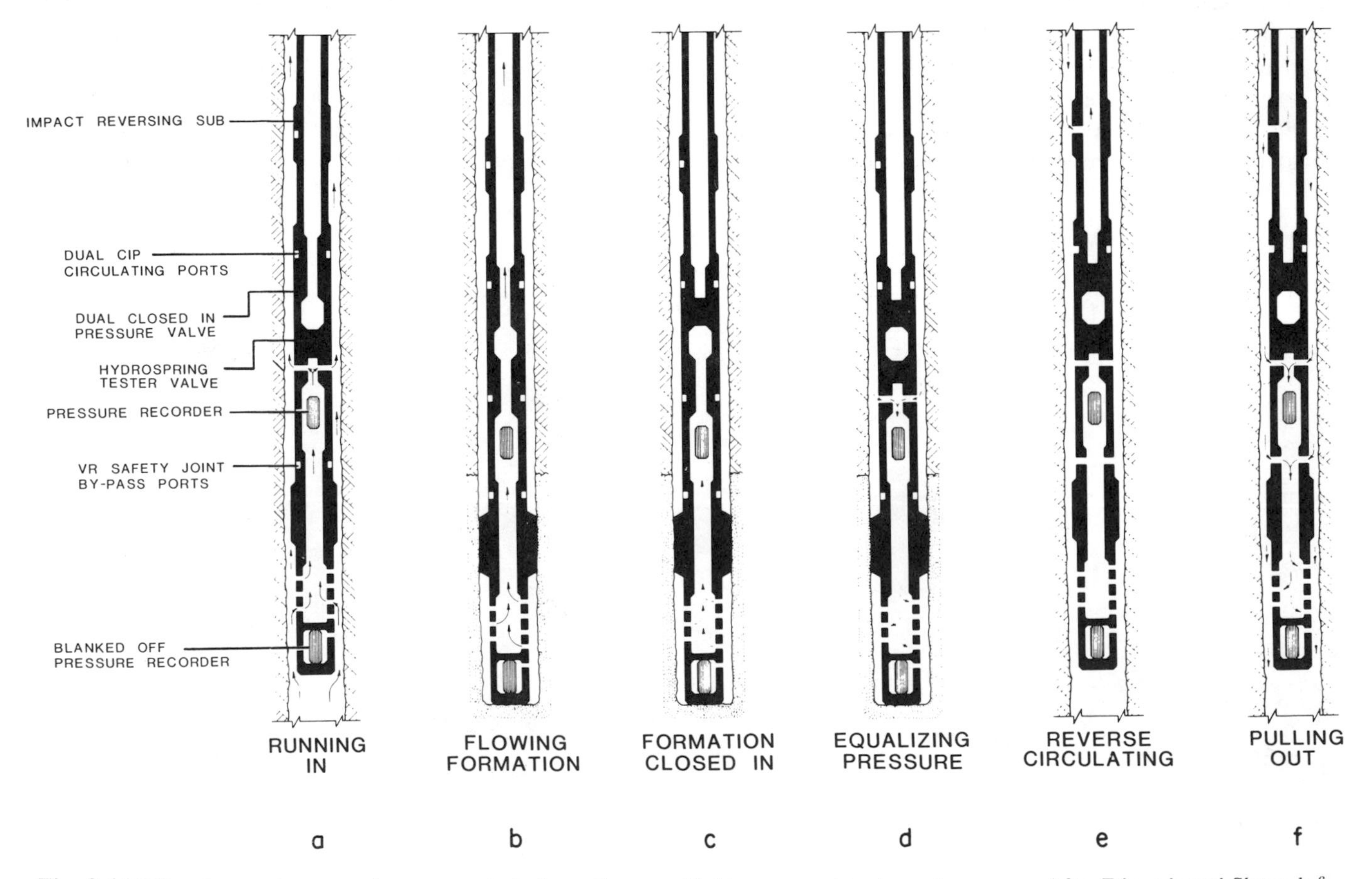

Fig. 8.4 DST-tool operating states for an open-hole formation test. Fluid movement is shown by arrows. After Edwards and Shryock.[6] Courtesy *Petroleum Engineer*.

stabilization; Table 8.1 provides guidelines. The length of the final shut-in period depends on test behavior during the final flow period. Recommendations are given in Table 8.1.

The multiflow evaluator, a tool that allows unlimited sequences of production and shut-in, has been available for drillstem testing since 1965.[3] The tool includes a fluid chamber to recover an uncontaminated formation-fluid sample under pressure at the end of the flow period.

8.3 Analyzing Drillstem-Test Pressure Data

Normal Drillstem-Test Pressure Buildup Analysis

Drillstem-test pressure buildup data are analyzed much like any other pressure buildup data; the techniques of Section 5.2 apply. In a DST, the flow period is about the same duration as the shut-in period, so pressure buildup data *must* be analyzed with the Horner plot, p_{ws} vs $\log[(t_p + \Delta t)/\Delta t]$. The value used for t_p is usually the length of the preceding flow period. However, if the initial flow period is very long, it is more accurate to use the *sum* of the flow-period lengths[3] for t_p for the final buildup.

In liquid-producing wells, the flow rate during a drillstem test decreases with time since the backpressure exerted on the formation face increases as the produced fluid moves up the drillstring. Flow rate may *stabilize if formation fluids flow to the surface*. The increasing flowing pressure is evident in Figs. 8.1 and 8.2. Normally, the decreasing flow rate over the flow period is neglected in analyzing DST pressure buildup data and the average flow rate over the flow period is used. Neglecting the flow-rate decrease is reasonable for a moderate bottom-hole pressure increase compared with total pressure drawdown, but it can lead to significant errors in the buildup analysis for high-productivity wells unless the well flows at the surface for a substantial portion of the flow period.[8]

If the pressure in the flowstring recorder increases linearly with time, liquid flow rate into the drillstring is constant (for a constant-inner-diameter drillstring) until liquid reaches the surface. Such a constant flow rate implies that flow rate is independent of drawdown, since bottom-hole flowing pressure is increasing. Eq. 2.2 indicates that flow rate from a porous medium to a wellbore must decrease with decreasing drawdown (increasing flowing bottom-hole pressure), so something other than the formation must be controlling the flow rate under such circumstances. The controlling factor is critical flow[9] (flow rate independent of pressure drop, see Section 13.6) through the perforations in the anchor pipe. In such an instance, the flowing pressure data from the flowstring recorder are useless, although shut-in data are analyzable. Fortunately, all data from the blanked-off recorder can be analyzed in the normal fashion.

Wellbore storage is not often significant in the buildup portion of a DST since the well is closed in near the formation face. However, if analysis results appear suspicious, the log-log data plot (Section 5.2) should be made to determine what part of the data should be analyzed. If thick sections are being tested in low-permeability or gas reservoirs, wellbore storage can be significant in a DST. Although production during the DST flow period appears to be *wellbore-storage dominated* until flow starts at the surface, the flow rate may be estimated, so normal analysis methods should apply if the varying rate is considered or if the rate variation is less than 5 to 10 percent.

If the shut-in period is long enough, and if wellbore storage is not dominant, a Horner plot of the buildup data should have a straight line section with slope $-m$, as indicated in Section 5.2. The value of m may be used to estimate permeability from Eq. 5.6:

$$k = \frac{162.6\, qB\mu}{mh} \quad \text{. . . . . (8.1)}$$

If μ and h are not known, kh/μ may be estimated by rearranging Eq. 8.1. The flow rate normally used is the average over t_p. The skin factor is estimated from Eq. 5.8:

$$s = 1.1513 \left[\frac{p_{1hr} - p_{wf}(\Delta t = 0)}{m} + \log\left(\frac{t_p + 1}{t_p}\right) - \log\left(\frac{k}{\phi\mu c_t r_w^2}\right) + 3.2275\right] \quad \text{. . . . . . (8.2)}$$

The term $\log[(t_p + 1)/t_p]$ is included since it may be important in drillstem testing. The term is normally neglected when $t_p >> 1$ or when the skin factor is high.

DST analyses commonly report *damage ratio*:

$$\frac{J_{ideal}}{J_{actual}} = \frac{\bar{p} - p_{wf}}{\bar{p} - p_{wf} - \Delta p_s}, \quad \text{. . . . . (8.3)}$$

where the pressure drop across the skin is computed from Eq. 2.9:

TABLE 8.1—RECOMMENDED FLOW AND SHUT-IN TIMES FOR DRILLSTEM TESTING WHEN EXPERIENCE IN THE AREA IS NOT AVAILABLE.
(Information from Pages 22-24 of Ref. 5.)

Test Period	Situation During Test	Recommended Time	Minimum Time (minutes)
Initial flow	All	Short — release hydrostatic mud pressure	3 to 5
Initial shut-in	All	60 minutes unless total test time is too short — 45 minutes then	30
Final flow	Strong, continuing blow	60 minutes	60
	Blow dies	Shut in when blow dies	
	Reservoir fluid produced at surface	60 minutes — longer to gauge flow rates if time is available	60
Final shut-in	Strong continuous blow during flow period	Shut-in time equal to flow time	45
	Blow dies during flow period	Minimum shut-in time of twice flow time*	Two times flow time
	Reservoir fluid produced at surface during flow period	Shut-in time equal to one-half flow time	30

*In this case there will be no buildup, so no buildup analysis. Flow-period data may be analyzed by the type-curve method described following Example 8.1.

$$\Delta p_s = \frac{141.2\,qB\mu}{kh}\,s \quad \text{(8.4)}$$

Initial, or average, reservoir pressure is estimated by extrapolating the Horner straight line to infinite shut-in time, $(t_p + \Delta t)/\Delta t = 1$. Since a DST is a short-duration test, there is generally no need to correct the extrapolated pressure for drainage shape as done in Chapter 6. The extrapolated p_i should be about the same for both the initial and final shut-in periods. If it is significantly different, a *very small reservoir* or a *bad test* is indicated. The definition of a significant difference depends on the reliability of the data and the buildup extrapolation, but a typical value might be 5 percent. When such a difference occurs, the test should be repeated with a longer final flow period, if possible.

If rate varies significantly during the flow period, then the multiple-rate analysis techniques in Chapter 4 should be used. Odeh and Selig[10] propose a simplified analysis technique that is useful for large rate variations when t_p is less than shut-in time. They suggest modifying t_p as given by Eq. 5.29:

$$t_p^* = 2\left[t_p - \frac{\sum_{j=1}^{N} q_j(t_j^2 - t_{j-1}^2)}{2\sum_{j=1}^{N} q_j(t_j - t_{j-1})}\right] \quad \text{(8.5)}$$

Similarly, q is modified as indicated by Eq. 5.30:

$$q^* = \frac{1}{t_p^*}\sum_{j=1}^{N} q_j(t_j - t_{j-1}) \quad \text{(8.6)}$$

The modified values, t_p^* and q^*, are used in the Horner plot and normal analysis given by Eqs. 8.1 through 8.4.

For all practical purposes, the radius of investigation during a DST is equivalent to the radius of drainage given in Eq. 2.41:

$$r_d = 0.029\sqrt{\frac{kt}{\phi\mu c_t}} \quad \text{(8.7)}$$

If a barrier to flow exists within the radius of investigation it might affect the semilog plot. In that case, the distance to the barrier may be estimated from material in Chapter 10 or Ref. 11. Generally, DST's are much too short to see the influence of a boundary. If changes in the slope of the semilog plot of DST data are interpreted as reservoir discontinuities, the results should be viewed with a great deal of skepticism.

Drillstem-Test Buildup Analysis With Limited Data

The analysis procedure explained previously cannot be used if the pressure data available are incomplete. That is usually the case immediately after the DST is completed, since the *full* pressure record is read in the service company offices, not at the wellsite. However, a few key data points are read at the wellsite and given to the engineer just after the test. These include the initial hydrostatic mud pressure, p_{ihm}; the initial shut-in pressure, p_{isi}; the pressure at the end of each flow period, p_{ff1} and p_{ff2}; the final shut-in pressure, p_{fsi}; and the final hydrostatic mud pressure, p_{fhm}. The flow and shut-in period durations are usually also reported. Such limited data may be used to estimate reservoir properties. The initial reservoir pressure is taken as

$$p_i \simeq \bar{p} \simeq p_{isi}. \quad \text{(8.8)}$$

The value of m for the semilog straight line is approximated by

$$m \simeq \frac{p_{isi} - p_{fsi}}{\log\left[(t_p + \Delta t)/\Delta t\right]}, \quad \text{(8.9)}$$

where Δt is the total final shut-in time (time when p_{fsi} was read). Permeability may be estimated from Eq. 8.1. (If the initial and final shut-in pressures are the same, m estimated from Eq. 8.9 will be zero and the approximate method will not be usable.) The *damage ratio* is estimated from[4]

$$\frac{J_{\text{ideal}}}{J_{\text{actual}}} \simeq \frac{0.183\,(p_{isi} - p_{ff2})}{m}, \quad \text{(8.10)}$$

or from[5]

$$\frac{J_{\text{ideal}}}{J_{\text{actual}}} \simeq \frac{p_{isi} - p_{ff2}}{m(4.43 + \log t_p)}, \quad \text{(8.11)}$$

where t_p is in hours. Eqs. 8.9 through 8.11 should be used only when more complete data are not available since they can be significantly in error.

Type-curve matching may be used to analyze pressure buildup data from drillstem tests. When wellbore storage is significant, the type curves in Appendix C, Figs. C.8 (Ref. 12) or C.9 (Ref. 13), may be useful. Type-curve methods are more useful for analyzing flow-period data, as discussed following Example 8.1.

Example 8.1 Drillstem Test Analysis by the Horner Method

Figs. 8.5 and 8.6 show DST data given by Ammann[7] for an open-hole test in the Arbuckle formation.

We first check the recorded hydrostatic mud pressure against the value calculated from gauge depth and mud density. From Fig. 8.6, the depth to the gauge is 4,174 ft and mud density is 10.1 lb_m/gal. Therefore, the hydrostatic mud pressure is

$$p_{hm} \simeq (4{,}174\text{ ft})\left(10.1\,\frac{\text{lb}_m}{\text{gal}}\right)\left(7.4805\,\frac{\text{lb}_m/\text{cu ft}}{\text{lb}_m/\text{gal}}\right)\left(\frac{0.43310\text{ psi/ft}}{62.3664\text{ lb}_m/\text{cu ft}}\right)$$

$$\simeq (4{,}174\text{ ft})(0.5247\text{ psig/ft})$$

$$\simeq 2{,}190\text{ psig}.$$

From Fig. 8.6,

$p_{ihm} = 2{,}314$ psig, so the deviation is 5.66 percent, and
$p_{fhm} = 2{,}290$ psig, so the deviation is 4.57 percent.

These deviations are primarily a result of errors and variation in the mud density. The difference of 1.04 percent between p_{ihm} and p_{fhm} may be a result of mud loss. Such differences in the range of 0.5 to 1 percent are indicative of the accuracy of p_i estimated from a DST.

Gauge No. 241				Depth 4174				Clock No. 1547				12 hour		Ticket No. 166710		
	First Flow Period		First Closed In Pressure			Second Flow Period		Second Closed In Pressure			Third Flow Period		Third Closed In Pressure			
	Time Defl. .000"	PSIG Temp. Corr.	Time Defl. .000"	$\text{Log}\frac{t+\theta}{\theta}$	PSIG Temp. Corr.	Time Defl. .000"	PSIG Temp. Corr.	Time Defl. .000"	$\text{Log}\frac{t+\theta}{\theta}$	PSIG Temp. Corr.	Time Defl. .000"	PSIG Temp. Corr.	Time Defl. .000"	$\text{Log}\frac{t+\theta}{\theta}$	PSIG Temp. Corr.	
0	.000	57	.000		35	.000	32	.000		145						
1	.008	35	.042		1664	.108	37	.0825		1669						
2	.016	32	.084		1701	.216	50	.165		1699						
3	.024	32	.126		1708	.324	65	.2475		1706						
4	.032	32	.168		1711	.432	80	.330		1711						
5	.040	35	.210		1713	.540	97	.4125		1713						
6			.252		1716	.648	112	.495		1713						
7			.294		1716	.756	130	.5775		1713						
8			.336		1716	.864	145	.660		1716						
9			.378		1716			.7425		1716						
10			.420		1718			.825		1718						
11																
12																
13																
14																
15																
Reading Interval	1		6			15		12								Minutes

REMARKS:

FORM 183-R1—PRINTED IN U.S.A. SPECIAL PRESSURE DATA LITTLE'S 96673 75C 8/74

Fig. 8.5 Pressure vs time measurements for a DST from the Arbuckle formation; Example 8.1. After Ammann.[7]

Following an initial flow period of 5 minutes, the shut-in period is 60 minutes (Fig. 8.5). Thus, we make a Horner plot of the final flow-period data with $t_p = 2$ hours, the length of the second flow period. Times in minutes are obtained by using the time-recording interval shown below each set of readings in Fig. 8.5. Fig. 8.7 is the Horner plot for data from both shut-in periods. Each shut-in has a straight line that extrapolates to $p_i = 1{,}722$ psig. Although not shown, the log-log data plot indicates no significant wellbore storage effects during the buildup periods.

FLUID SAMPLE DATA

Date 2-16-59 | Ticket Number 166710

Sampler Pressure ____ P.S.I.G. at Surface | Kind of Job Open Hole | Halliburton District Perry

Recovery: Cu. Ft. Gas; cc. Oil; cc. Water; cc. Mud; Tot. Liquid cc.

Tester C. E. Sims | Witness Mr. Johnson

Drilling Contractor

Gravity 44 ° API @ ___ °F.

Gas/Oil Ratio ____ cu. ft./bbl.

EQUIPMENT & HOLE DATA

Formation Tested Arbuckle

Elevation 1123 Ft.

Net Productive Interval 4182-4198 Ft.

RESISTIVITY | CHLORIDE CONTENT

Recovery Water ___ @ ___ °F. ___ ppm

Recovery Mud ___ @ ___ °F. ___ ppm

Recovery Mud Filtrate ___ @ ___ °F. ___ ppm

Mud Pit Sample ___ @ ___ °F. ___ ppm

Mud Pit Sample Filtrate ___ @ ___ °F. ___ ppm

All Depths Measured From Ground level

Total Depth 4198 Ft.

Main Hole/Casing Size 8-3/4"

Drill Collar Length 240' I.D. 2-1/2"

Drill Pipe Length ___ I.D. 4-1/2 API FH

Packer Depth(s) 4182 Ft.

Mud Weight 10.1 lb/gal vis 48 cp | Depth Tester Valve ___ Ft.

Cushion TYPE None AMOUNT ___ Ft. | Depth Back Pres. Valve | Surface Choke 5/8" | Bottom Choke 5/8"

Recovered 300 Feet of oil

Recovered 75 Feet of oil and gas cut mud

Recovered ___ Feet of

Recovered ___ Feet of

Recovered ___ Feet of

Remarks Set packer, opened tool and took a 5 minute first flow pressure. Rotated tool for a closed in pressure of 60 minutes. Tool opened with a good blow. Gas to surface in 4 minutes. Flowed test in 5 minutes on first flow pressure.

5-3/4" OD - 4-3/4" ID x 16' Perforated Anchor.

Legal Location Sec.-Twp.-Rng.: Sec. 13-22N-4W; Lease Name: C. K. Walker, North; Well No.: 76; Test No.: 1; Tested Interval: Arbuckle; Field Area: North Covington; Meas. From Tester Valve; County: Garfield; State: Oklahoma; Lease Owner/Company Name: Sinclair Oil and Gas Company

TEMPERATURE	Gauge No. 241, Depth: 4174 Ft.		Gauge No. 3142, Depth: 4193 Ft.		Gauge No. ___, Depth: ___ Ft.		TIME	
	12 Hour Clock		– Hour Clock		Hour Clock		Tool Opened	A.M. P.M.
Est. 110 °F.	Blanked Off –		Blanked Off no		Blanked Off		Opened Bypass	A.M. P.M.
Actual °F.	Pressures		Pressures		Pressures			
	Field	Office	Field	Office	Field	Office *	Reported Minutes	Computed Minutes
Initial Hydrostatic	2280	2314						
First Period Flow Initial	0	32					—	—
First Period Flow Final							5	
First Period Closed in	1720	1718					60	
Second Period Flow Initial		35					—	—
Second Period Flow Final	150	145					120	
Second Period Closed in	1695	1718					120	
Third Period Flow Initial							—	—
Third Period Flow Final								
Third Period Closed in								
Final Hydrostatic	2205	2290					—	—

FORM 181-R1—PRINTED IN U.S.A. FORMATION TEST DATA LITTLE'S 96671 10M 8/74

Fig. 8.6 Drillstem-test data sheet for a DST from the Arbuckle formation; Example 8.1. After Ammann.[7]

To analyze the pressure buildup data we must first estimate the average flow rate during each flow period. The drillpipe was initially empty for this test, so the pressure existing before opening the tool for the first flow was atmospheric. At the end of the first flow period, the pressure was 35 psi (Fig. 8.5). Assuming that all fluid flowing during that time was drilling mud, we may estimate the height of the mud column. From the calculation of hydrostatic mud pressure, the mud exerts a pressure of 0.5247 psi/ft of column height, so 35 psi is equivalent to 35/0.5247 = 67 ft of mud. Fig. 8.6 reports that 75 ft of oil- and gas-cut mud were recovered. This agrees well enough with the estimated 67-ft value to use that value for production during the first flow period. Fig. 8.6 reports that 240 ft of 2.5-in.-ID drill collar were used in the tool string. The capacity of the drill collar is 0.00607 bbl/ft, so 67 ft is equivalent to (67)(0.00607) = 0.407 bbl. Assuming that 0.407 bbl of fluid was produced

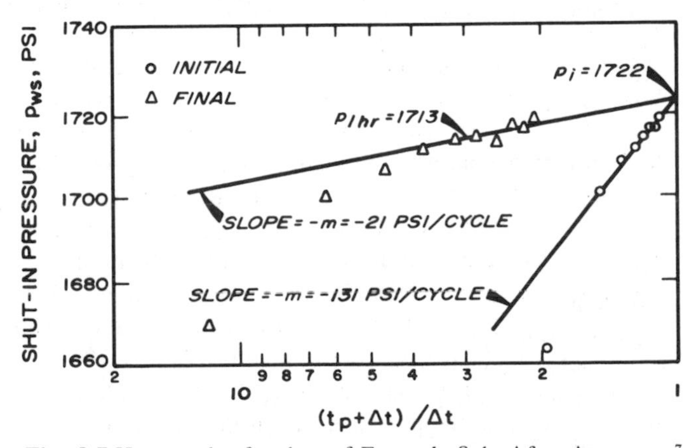

Fig. 8.7 Horner plot for data of Example 8.1. After Ammann.[7]

from the formation in the first 5-minute flow period, we estimate an initial rate of

$$q \simeq (0.407 \text{ bbl/5 min})(1{,}440 \text{ min/D}) = 117 \text{ STB/D}.$$

Eq. 8.1 now can be used to estimate kh/μ. Assuming B = 1.0 RB/STB and using m = 131 psi/cycle from Fig. 8.7,

$$\frac{kh}{\mu} \simeq \frac{(162.6)(117)(1)}{131} = 145 \text{ md ft/cp}.$$

If we take h = 16 ft, the tested interval,

$$\frac{k}{\mu} \simeq 145/16 \simeq 9.1 \text{ md/cp}.$$

The pressure increases from 35 psi (at the end of the first flow period) to 145 psi at the end of the second flow period. Measured oil gravity was 44 °API[7] so oil specific gravity is 0.806, corresponding to a gradient of about 0.349 psi/ft. Assuming that all the fluid produced was oil, the pressure increase of 145 − 35 = 110 psi corresponds to 315 ft of oil. Note there are only 240 ft of drill collar (Fig. 8.6); above that there is drillpipe with a capacity of 0.01422 bbl/ft.[7] Assuming that the oil flows through the dense mud, we can estimate the *volume of oil* in the pipe:

$$\begin{aligned} V_o &\simeq (240 \text{ ft drill collar} - 67 \text{ ft mud})(0.00607 \text{ bbl/ft}) \\ &\quad + \left[315 \text{ ft oil} - (240 - 67) \text{ ft oil in drill collar}\right] \\ &\quad \times (0.01422 \text{ bbl/ft}) \\ &\simeq 1.05 + 2.02 \\ &\simeq 3.07 \text{ bbl oil recovered}. \end{aligned}$$

Thus,

$$q_o \simeq \frac{3.07 \text{ bbl} \times 1{,}440 \text{ min/D}}{120 \text{ min}} \simeq 36.8 \text{ STB/D}.$$

Using Eq. 8.1 and m = 21 psi/cycle from Fig. 8.7, and assuming B = 1.0 RB/STB,

$$\frac{kh}{\mu} \simeq \frac{(162.6)(36.8)(1)}{21} \simeq 285 \text{ md ft/cp},$$

or

$$\frac{k}{\mu} = 17.8 \text{ md/cp}.$$

This is almost twice the value estimated from the first shut-in period — not unusual in drillstem-test analysis. Part of the discrepancy may be a result of an error in measurement of a flow-period length or in reported pipe sizes. In working this example, Ammann[7] states that there was no drill collar while his data indicate that there was. Most likely, this is where most of the discrepancy arises. Errors are also undoubtedly introduced by the assumption of the type of fluid entering the drillstring (all mud in the first flow period, all oil in the second flow period). Another possible source of some of the discrepancy may be that part of the production during the first flow period is a result of decompression of the wellbore fluid from hydrostatic mud pressure, about 2,300 psig, to the formation pressure of about 1,700 psig. The over-pressure in and near the wellbore can affect both the flow rate and the pressure during the first flow period. Generally, the results from the second flow period and associated buildup are more reliable. In any case, the material above illustrates the approach. Ammann[7] gives a more complete analysis.

To estimate skin factor we assume $\phi = 0.15$ and $c_t = 25 \times 10^{-6}$, and use Eq. 8.2 for the second flow period:

$$\begin{aligned} s &\simeq 1.1513 \left\{ \frac{1{,}713 - 145}{21} + \log\left[\frac{2+1}{1}\right] \right. \\ &\quad \left. - \log\left[\frac{17.8}{(0.15)(25 \times 10^{-6})(8.75/24)^2}\right] + 3.2275 \right\} \\ &\simeq 81.5. \end{aligned}$$

The well is severely damaged. We estimate pressure drop across the skin from Eq. 8.4:

$$\begin{aligned} \Delta p_s &\simeq \frac{(141.2)(36.8)(1)}{285}(81.5) \\ &= 1{,}486 \text{ psig}. \end{aligned}$$

From Eq. 8.3, the damage ratio is

$$\frac{1{,}722 - 145}{1{,}722 - 145 - 1{,}486} = 17.3.$$

This indicates the well is producing at only about 6 percent of its ideal capacity. Stimulation would be required for a successful completion.

Although we have sufficient data for a Horner-type analysis, we may apply Eqs. 8.10 and 8.11 to compare methods for estimating the damage ratio. From Eq. 8.10, the damage ratio is

$$\frac{0.183\,(1{,}718 - 145)}{21} = 13.7,$$

and from Eq. 8.11, it is

$$\frac{1{,}718 - 145}{21(4.43 + \log 2)} = 15.8.$$

These values agree reasonably well with the result estimated from the skin factor.

Example 8.1 illustrates some of the problems that can occur in DST pressure analysis. The problem of estimating flow rate is a real one, and must be dealt with by using pressures, densities, volumes of the various fluids produced, and pipe capacities — if flow does not occur at the surface. Inconsistent or inaccurate fluid-volume and pipe-size data, as occur in Example 8.1, make analysis difficult and should be avoided if possible. Normally, one does not analyze pressure data from the first flow and shut-in periods. Results from analyzing those data tend to be less accurate than results from analyzing the second flow and shut-in periods because of longer flow duration and likely absence of mud production during the second flow period.

Analyzing Flow-Period Data

If rate variation can be estimated during the flow period, it is possible to analyze pressure data from the flow period with methods given in Section 4.2. Such multiple-rate analyses can be particularly useful for wells with substantial flowing bottom-hole pressure increase that either do not flow to the surface or have insufficient surface flow time at a stable rate to provide reliable analysis results from the shut-in pressure data.

Occasionally, the pressure exerted by the produced fluid column can reach the reservoir pressure, causing production to stop during the flow period — the well kills itself. In such cases, data from the shut-in period cannot be analyzed. However, flow-period data can be analyzed by multiple-rate techniques (Section 4.2) or by type-curve matching techniques presented in Refs. 9 and 14 through 17. The type curves in Refs. 14 through 16 do not consider skin factor, so they are not recommended. Ramey, Agarwal, and Martin[9] provide type curves that include skin effect that may be used to analyze DST flow-period data *as long as flow does not reach the surface* and there is no significant change in the wellbore storage coefficient (pipe inner diamater). Figs. 8.8A through 8.8C* are the Ramey-Agarwal-Martin type curves. In those figures, the dimensionless pressure ratio is defined as

$$p_{DR} = \frac{p_D}{p_{Do}} = \frac{p_i - p_{wf}(t)}{p_i - p_o}, \qquad (8.12)$$

where p_o is the pressure existing in the drillstring im-

*See footnote on Page 24.

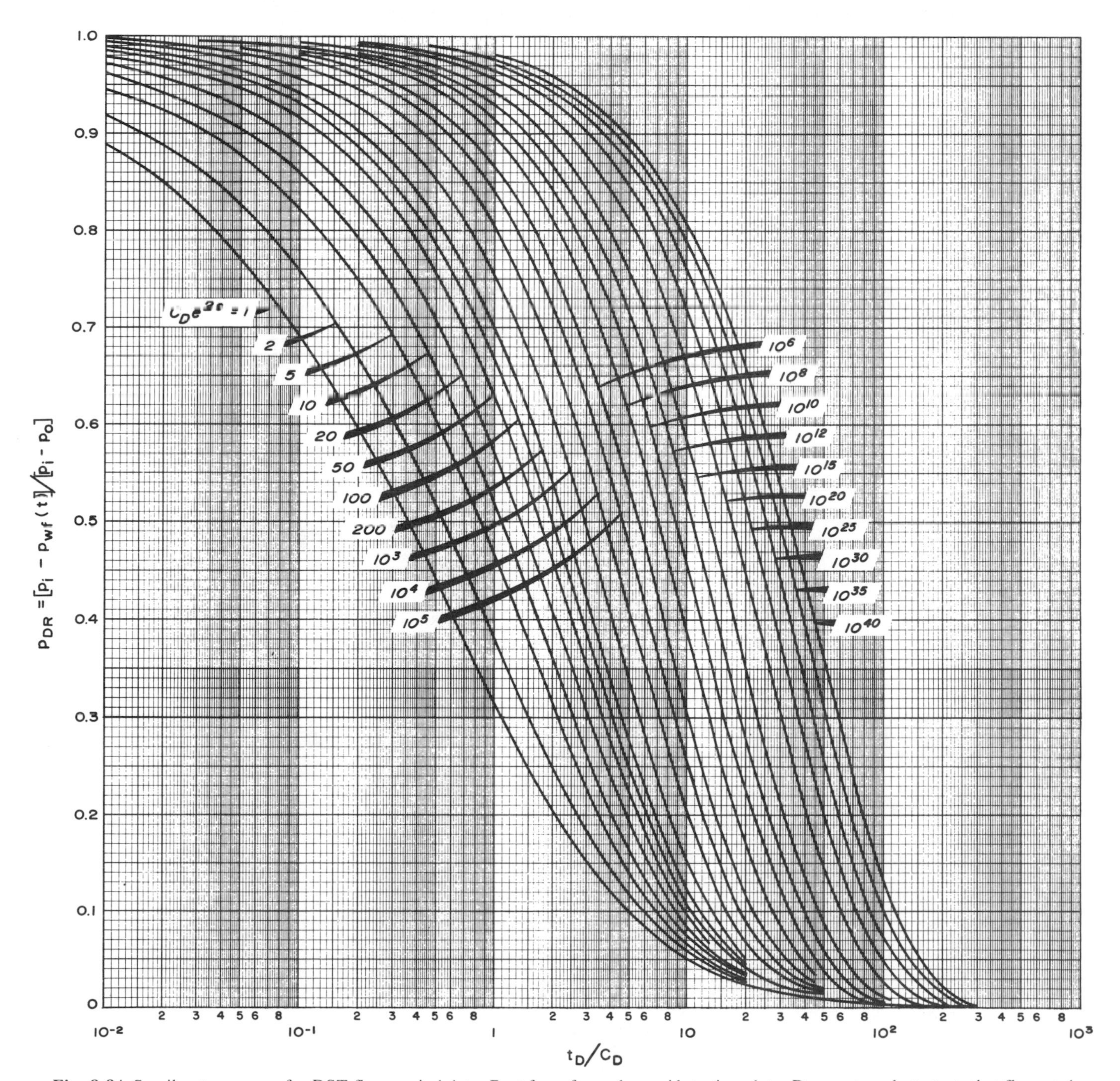

Fig. 8.8A Semilog type curve for DST flow-period data. Best form for early- *and* late-time data. Does not apply to tests that flow at the surface. After Ramey, Agarwal, and Martin.[9] Courtesy CIM.

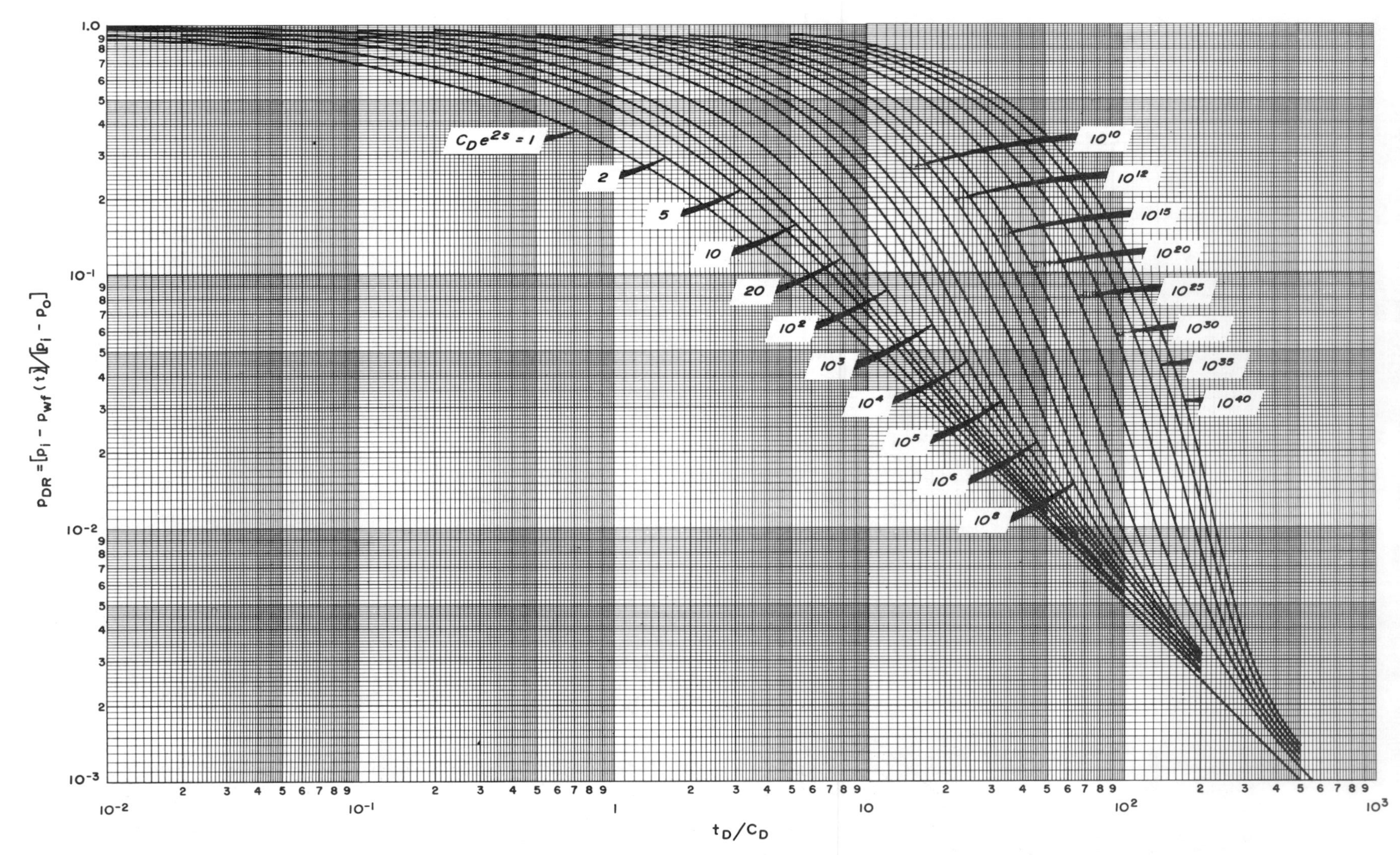

Fig. 8.8B Log-log type curve for DST flow-period data. Best form for late-time data. Does not apply to tests that flow at the surface. After Ramey, Agarwal, and Martin.[9] Courtesy CIM.

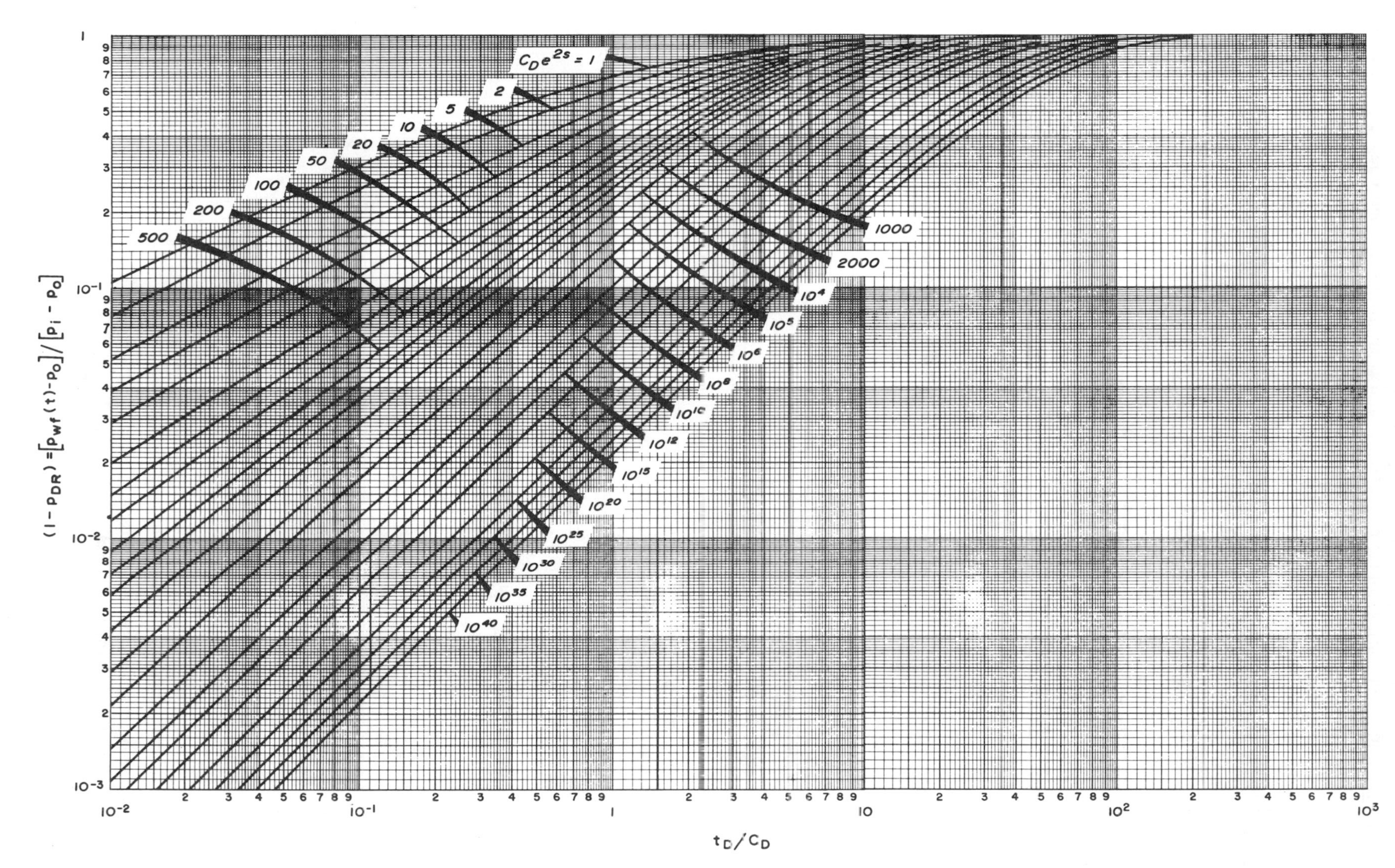

Fig. 8.8C Log-log type curve for DST flow-period data. Best form for early-time data. Does not apply to tests that flow at the surface. After Ramey, Agarwal and Martin.[9] Courtesy CIM.

mediately before the flow period begins. For the initial flow period, p_o would be atmospheric pressure or the pressure exerted by any fluid cushion in the drillstring; for the final flow period, p_o would be the pressure at the end of the first flow period. In Fig. 8.8, the dimensionless time is defined by Eq. 2.3a:

$$t_D = \frac{0.0002637\,kt}{\phi\mu c_t r_w^2}, \qquad (8.13)$$

and the dimensionless wellbore-storage coefficient is defined by Eq. 2.18:

$$C_D = \frac{5.6146\,C}{2\pi\phi c_t h r_w^2}. \qquad (8.14)$$

For a DST flow period, the wellbore storage coefficient usually results from a rising liquid level in the drillpipe. Thus, Eq. 2.16 applies:

$$C = \frac{V_u}{\left(\dfrac{\rho}{144}\dfrac{g}{g_c}\right)}, \qquad (8.15)$$

where V_u is the volume per unit length of the drillpipe in barrels per foot. Note that this monograph uses different units for C than do Ramey, Agarwal, and Martin.[9]

The type-curve matching technique is similar to the method described in Section 3.3, with one important simplification: the pressure ratio in Fig. 8.8 always goes from zero to one and is independent of flow rate and formation properties. Thus, when plotting data on the tracing paper laid over the grid of Fig. 8.8A, 8.8B, or 8.8C, the pressure scale is fixed. When the tracing-paper data plot is slid to match one of the type curves, only horizontal motion is used. That simplifies the matching technique. Once the experimental data have been matched to one of the type curves, data from both the overlay and the underlying type curve are read at a convenient match point. Three data items are required: the parameter on the curve matched, $(C_De^{2s})_M$; the time-scale match point, t_M, from the data plot; and the corresponding point from the type curve, $(t_D/C_D)_M$. Permeability may be estimated from the time-scale match point by using

$$k = 3{,}389\,\frac{\mu}{h}\,\frac{C}{t_M}\left(\frac{t_D}{C_D}\right)_M. \qquad (8.16)$$

It is not necessary to know the flow rate to estimate permeability by this method. It is necessary to estimate the wellbore storage coefficient from Eq. 8.15, so the fluid density must be known. Skin factor is estimated from the parameter on the curve matched:

$$s = \frac{1}{2}\ln\left[\frac{\phi c_t h r_w^2\,(C_De^{2s})_M}{0.89359\,C}\right]. \qquad (8.17)$$

As usual, it is necessary to have values for porosity, total system compressibility, formation thickness, and wellbore radius to estimate the skin factor. Damage ratio then may be estimated from Eq. 8.3.

Ramey, Agarwal, and Martin[9] suggest that all three type curves be used to analyze DST flow-period data. That requires plotting the data three times and making three curve matches. Eqs. 8.16 and 8.17 apply to all three curve matches. The semilog type curve, Fig. 8.8A, usually should be best when both early and relatively late-time data are available. Fig. 8.8B provides poor resolution of early-time data, while Fig. 8.8C is useful for early-time data.

Type curves in Figs. 8.8A through 8.8C can be used to conveniently estimate permeability and skin factor from DST flow-period data. However, they are not applicable when fluid influx to the drillstem is at essentially constant rate; that is, when flow occurs at the surface. They are also not applicable when the wellbore storage coefficient changes (because of pipe size or compressibility changes). Such changes are illustrated by Figs. 8.20 and 8.21.

Example 8.2 Analysis of Drillstem-Test Flow Data by Type-Curve Matching

Ramey, Agarwal, and Martin[9] give the pressure data in Table 8.2 for the second flow period of a DST. Other data are

$p_i = 3{,}475$ psig (initial shut-in pressure)
$p_o = 643$ psig
$r_w = 3.94$ in.
$V_u = 0.0197$ bbl/ft
$\phi = 0.16$
$c_t = 8.0 \times 10^{-6}$ psi^{-1}
$\mu = 1.0$ cp
$h = 17$ ft
$\rho = 52.78$ lb$_m$/cu ft.

TABLE 8.2—DST DATA FOR FLOW-PERIOD ANALYSIS OF EXAMPLE 8.2.
(From Ramey, Agarwal, and Miller.[9])

Time (minutes)	p_{wf} (psig)	$\dfrac{p_i - p_{wf}(t)}{p_i - p_o}$
0	643	1.0000
3	665	0.9922
6	672	0.9898
9	692	0.9827
12	737	0.9668
15	786	0.9495
18	832	0.9333
21	874	0.9184
24	919	0.9025
27	962	0.8874
30	1,005	0.8722
33	1,046	0.8577
36	1,085	0.8439
39	1,128	0.8287
42	1,170	0.8139
45	1,208	0.8005
48	1,248	0.7864
51	1,289	0.7719
54	1,318	0.7617
57	1,361	0.7465
60	1,395	0.7345
63	1,430	0.7221
66	1,467	0.7090
69	1,499	0.6977
72	1,536	0.6847
75	1,570	0.6727
78	1,602	0.6614
81	1,628	0.6522
84	1,655	0.6427
87	1,683	0.6328
90	1,713	0.6222
93	1,737	0.6137
96	1,767	0.6031
99	1,794	0.5936
102	1,819	0.5847
105	1,845	0.5756
108	1,869	0.5671
111	1,894	0.5583
114	1,917	0.5501
117	1,948	0.5392
120	1,969	0.5318

Fig. 8.9 shows the data of Table 8.2 matched to Fig. 8.8A. The matchpoint data are

$$(C_D e^{2s})_M = 10^{10},$$

$$(t_D/C_D)_M = 0.65,$$

and

$$t_M = 10 \text{ minutes} = 0.1667 \text{ hour}.$$

We estimate permeability from Eq. 8.16, but to do this the wellbore storage coefficient must first be estimated from Eq. 8.15:

$$C = \frac{0.0197}{\left(\frac{52.78}{144}\right)\left(\frac{32.17}{32.17}\right)} = 0.0537 \text{ bbl/psi}.$$

Then, using Eq. 8.16,

$$k = \frac{(3{,}389)(1.0)(0.0537)(0.65)}{(17)(0.1667)}$$

$$= 41.7 \text{ md}.$$

Using the parameter on the matched curve and Eq. 8.17,

$$s = \frac{1}{2} \ln \left[\frac{(0.16)(8.0 \times 10^{-6})(17)(3.94/12)^2(10^{10})}{(0.89359)(0.0537)}\right]$$

$$= 6.5.$$

Ramey, Agarwal, and Martin[9] indicate that the flow rate was constant owing to critical flow at early flow time. This curve-matching technique does not apply under constant-rate conditions, so the early data should not be considered in the curve match. Clearly, they do not match the entire $C_D e^{2s} = 10^{10}$ curve of Fig. 8.8A, as shown in Fig. 8.9. In fact, they do not completely match other curves in Fig. 8.8A. Unfortunately, several curves can be matched with the late-time data. The match shown in Fig. 8.9 is the lowest value of $C_D e^{2s}$ for which the most points matched the curve. At lower $C_D e^{2s}$ values fewer points match; at higher values no more points match.

Ramey, Agarwal, and Martin[9] report that (1) core analysis showed an average permeability of 35.4 md for the zone tested; and (2) a Horner-type analysis of the second shut-in period (with average flow rate) indicated $k = 22.2$ md, and a damage ratio of 1.16.

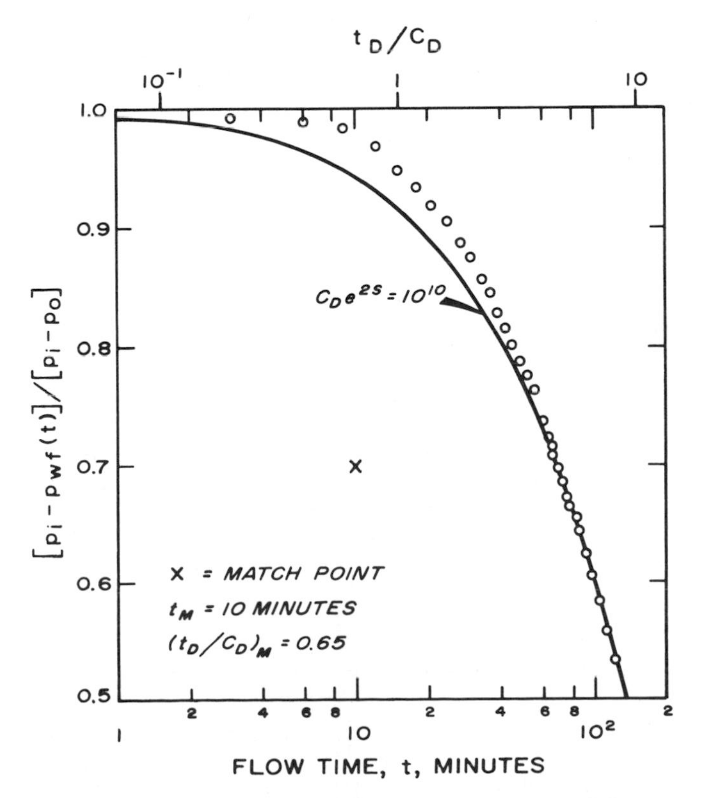

Fig. 8.9 Type-curve match for DST flow data of Example 8.2; type curve of Fig. 8.8A. After Ramey, Agarwal and Martin.[9]

Example 8.2 illustrates the mechanics of using Figs. 8.8A through 8.8C for analysis of DST flow-period data. It also illustrates that the technique should be used with caution. Periods of constant rate flow during a DST (manifested by a linear p vs t trace on the DST chart) are not unusual. When they occur, they can rule out analysis by curve matching with Figs. 8.8A through 8.8C — or at best make the results have doubtful validity.

Computer Matching Drillstem-Test Data

It does appear feasible to use all the data obtained during a DST for test analysis.[18] Such analysis requires a numerical reservoir simulator and uses a history-matching approach to vary formation properties until the DST pressure and rate behavior are matched by the simulator. Since the technique uses all the data, it should be particularly useful when conventional interpretation techniques cannot be applied with confidence.

8.4 Trouble Shooting Drillstem-Test Pressure Charts

Because of the complexity of the DST tool operation, there are many opportunities for test failure. Therefore, it is important to carefully examine the DST charts and decide if the test was mechanically and operationally successful. That should be done at the wellsite so that the option of rerunning the test may be exercised if necessary.

To recognize a poor DST, one must be familiar with DST chart characteristics. Murphy[19] and Timmerman and van Poollen[20] provide such information. A good DST chart has the following characteristics.

1. The pressure base line is straight and clear.
2. Recorded initial and final hydrostatic mud pressures are the same and are consistent with depth and mud weight.
3. Flow and buildup pressures are recorded as smooth curves.

Frequently, bad hole conditions, tool malfunctions, and other difficulties can be identified from the DST charts. Figs. 8.10 through 8.23 illustrate many situations. The captions explain the characteristic indicated by each figure.

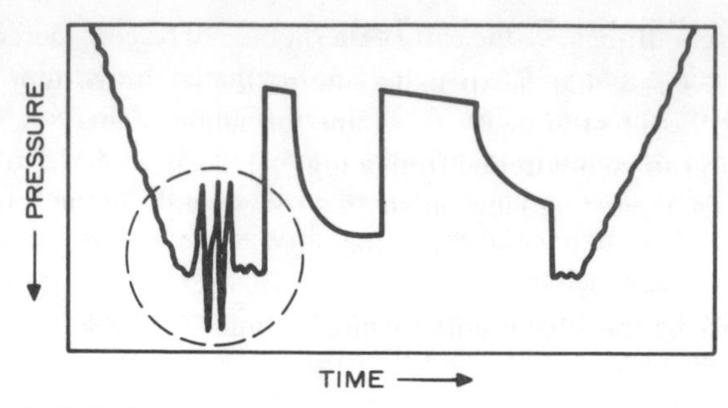

Fig. 8.10 Tight hole condition. This may cause pressure surging or tool sticking.

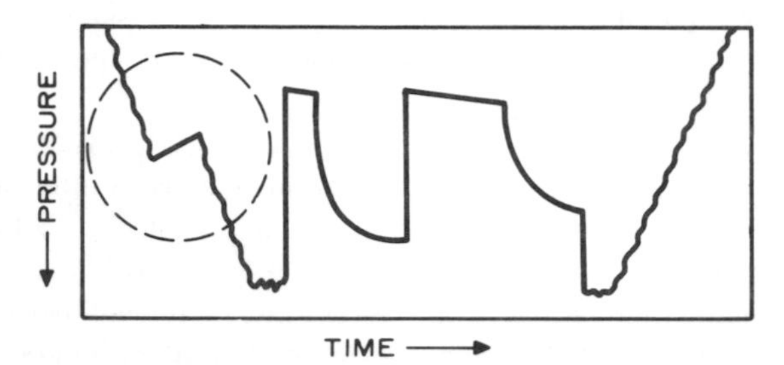

Fig. 8.11 Leaking drillpipe or mud loss to some formation, or both, are indicated by the hydrostatic mud-pressure decrease shown here. A leaking drillpipe may be confirmed if an abnormally large amount of mud is recovered with the produced fluids. In this case test data must be disregarded.

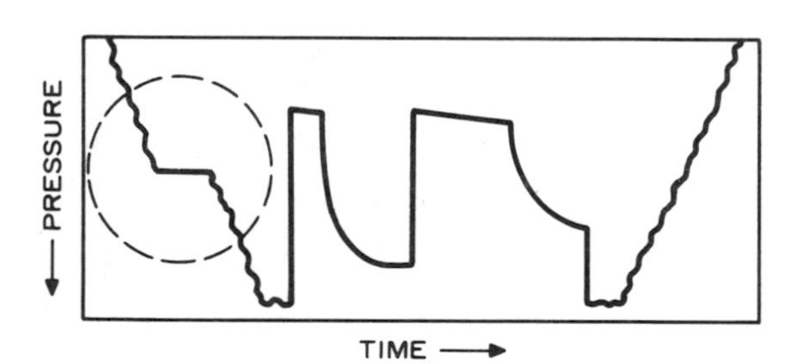

Fig. 8.12 Delay while going into the hole without mud loss.

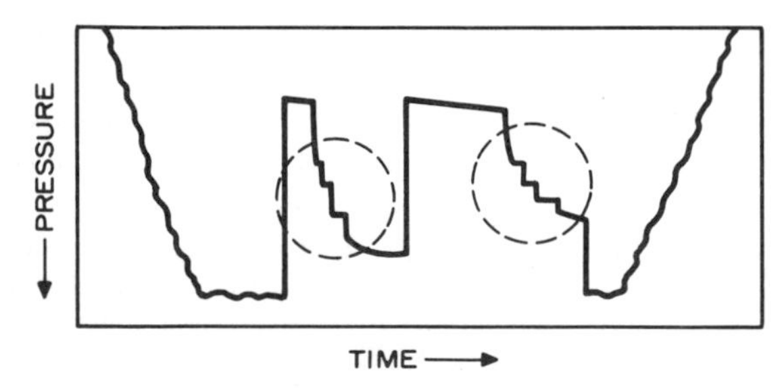

Fig. 8.13 The stair-stepping pattern in the buildup curves indicates a malfunctioning pressure gauge or recorder. Such test data cannot be analyzed.

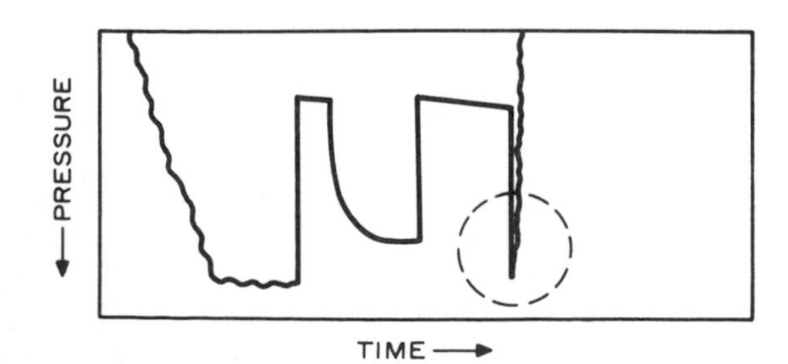

Fig. 8.14 Clock stopped.

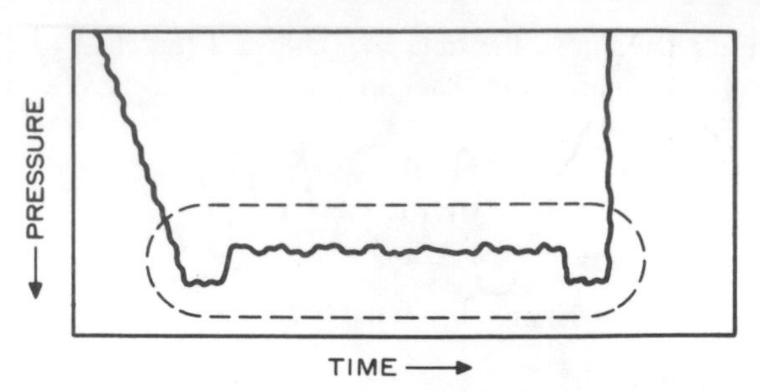

Fig. 8.15 Clock ran away.

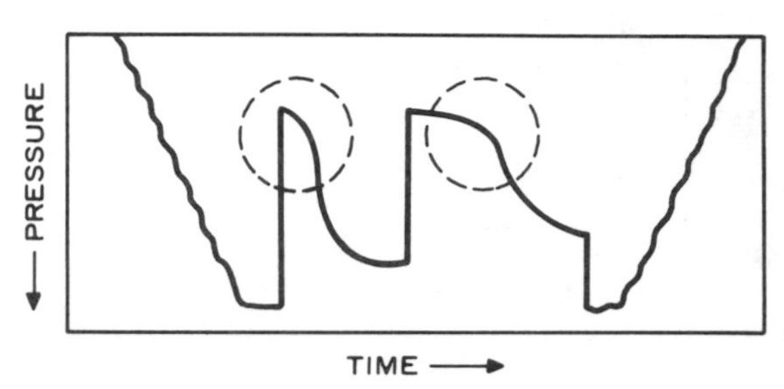

Fig. 8.16 The S shape of the latter part of the flow curve and the early part of the buildup curve indicates fluid communication around the packer. This may be caused by a fracture or a poorly seated packer.

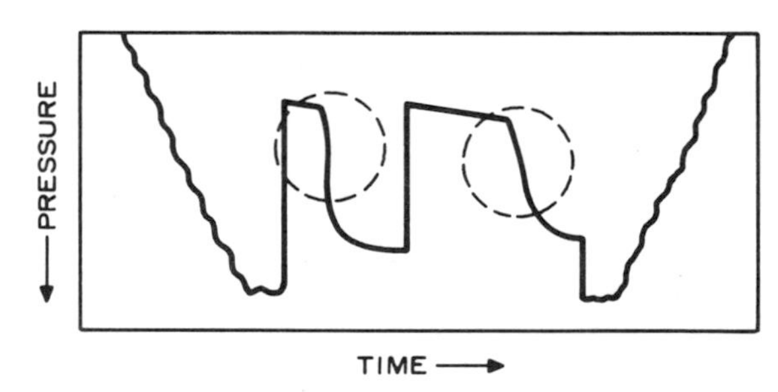

Fig. 8.17 An S shape occurring only in the buildup portion of the curve indicates gas is going into solution in the wellbore. This mechanism is characterized by a *sharp transition* between the flow and buildup curves.

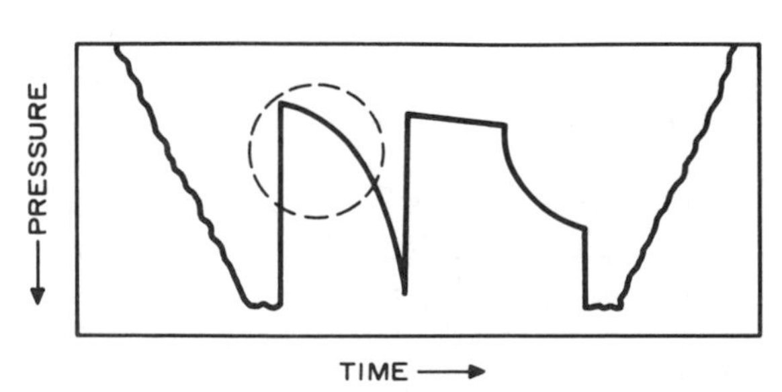

Fig. 8.18 An S-shaped curve occurs only in the first flow period when the volume below the closed-in pressure valve is large compared with the volume of the fluid flowed during the flow period. A wellbore storage effect caused by the relatively large volume between the hydrospring tester and CIP valves.

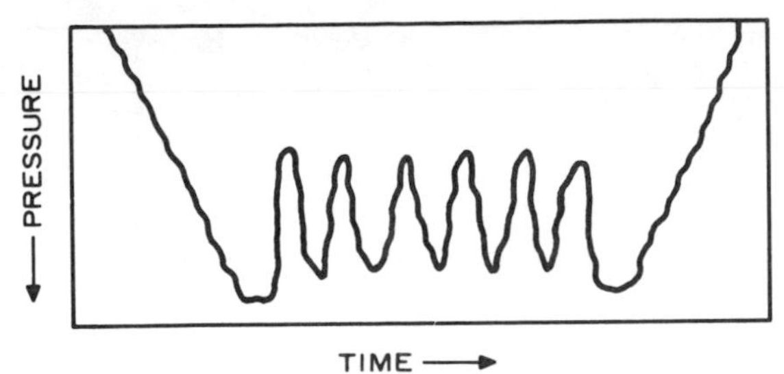

Fig. 8.19 This behavior indicates a plugged bottom-hole choke or perforated anchor. The up-and-down nature of the pressure curve is caused by momentary breakthrough and release of the pressure.

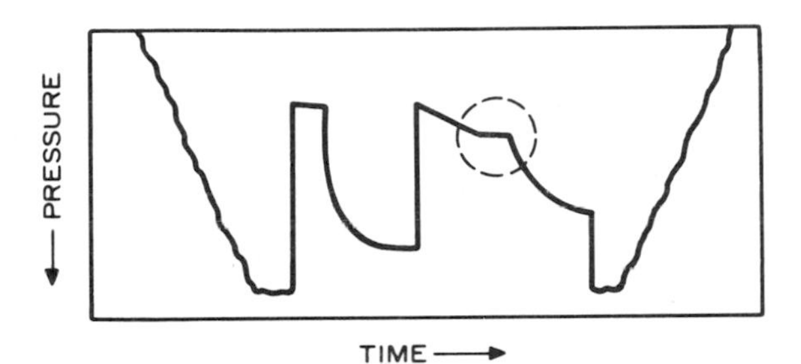

Fig. 8.20 The flat portion in the second flow period indicates the well is flowing at the surface.

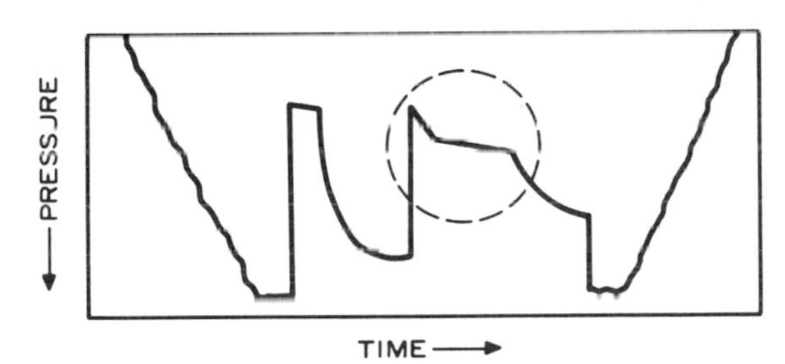

Fig. 8.21 A decrease in slope in either flow period indicates fillup of the drill collar and transition to a drillpipe of a larger internal diameter.

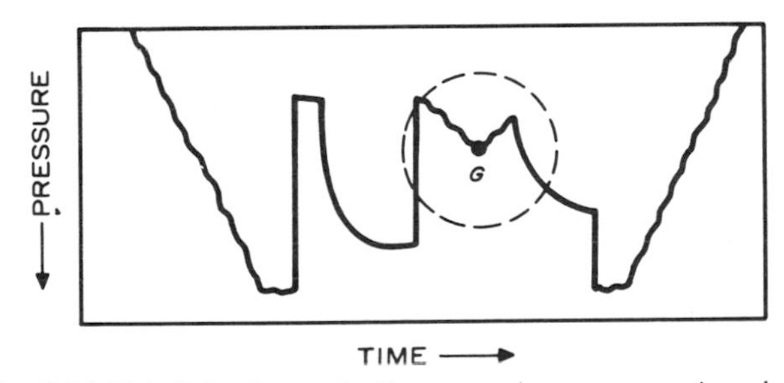

Fig. 8.22 This behavior typically occurs in gas reservoirs when flow occurs at the surface. The pressure decrease at Point G is caused by the water cushion flowing at the surface, which decreases average density of the flowing column.

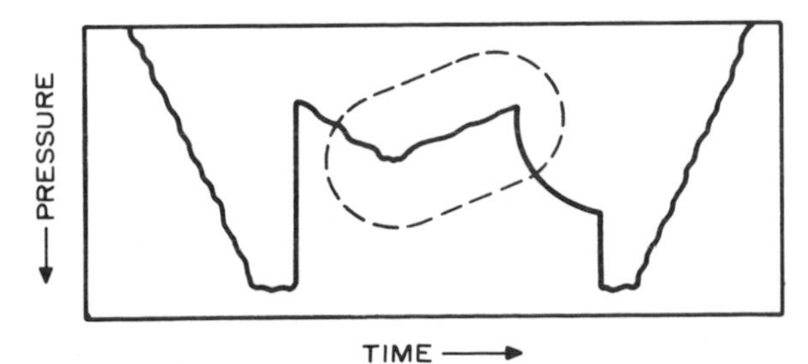

Fig. 8.23 The rippled appearance in the flow curve indicates that gas has broken through the liquid in the drillstring and the well is flowing by heads.

8.5 Wireline Formation Tests

A quick and inexpensive alternative to a drillstem test may be a test run with a wireline formation tester.[21-24] That tool, run on a wireline from a logging truck, includes a pad on an expanding mechanism that presses against the formation face, a means for establishing fluid communication between the formation and the tool, a sample chamber, and a pressure transducer with surface recorder.

The tester is lowered into the well on a logging cable while the mechanism is collapsed. The tool is located opposite the formation to be tested, the mechanism is expanded, and fluid communication is established. Formation fluid flows into the chamber and the pressure response is recorded. A new version of the tool may be set and used at several locations during a single run.[24]

Interpretation of wireline-formation-tester pressure data is semiqualitative, so the information obtained is inferior to that of a normal DST. More recent tools with larger fluid chambers ($\sim$12 gal vs 2.75 to 5 gal) tend to give reasonable fluid-recovery and p_i results. Although permeability may be estimated from the wireline formation tester, the degree of uncertainty is high. The skin factor cannot be estimated. A general rule for interpretation based on experience in Canada and in the Rocky Mountain region is presented in Refs. 22 and 23.

References

1. Matthews, C. S. and Russell, D. G.: *Pressure Buildup and Flow Tests in Wells,* Monograph Series, Society of Petroleum Engineers of AIME, Dallas (1967) **1**, Chap. 9.

2. van Poollen, H. K.: "Status of Drill-Stem Testing Techniques and Analysis," *J. Pet. Tech.* (April 1961) 333-339. Also *Reprint Series, No. 9 — Pressure Analysis Methods,* Society of Petroleum Engineers of AIME, Dallas (1967) 104-110.

3. McAlister, J. A., Nutter, B. P., and Lebourg, M.: "A New System of Tools for Better Control and Interpretation of Drill-Stem Tests," *J. Pet. Tech.* (Feb. 1965) 207-214; *Trans.,* AIME, **234.**

4. Edwards, A. G. and Winn, R. H.: "A Summary of Modern Tools and Techniques Used in Drill Stem Testing," Publication T-4069, Halliburton Co., Duncan, Okla. (Sept. 1973).

5. "Review of Basic Formation Evaluation," Form J-328, Johnston Schlumberger, Houston (1974).

6. Edwards, A. G. and Shryock, S. H.: "New Generation Drill Stem Testing Tools/Technology," *Pet. Eng.* (July 1974) 46, 51, 56, 58, 61.

7. Ammann, Charles B.: "Case Histories of Analyses of Characteristics of Reservoir Rock From Drill-Stem Tests," *J. Pet. Tech.* (May 1960) 27-36.

8. Kazemi, Hossein: "Damage Ratio From Drill-Stem Tests With Variable Back Pressure," paper SPE 1458 presented at the SPE-AIME California Regional Meeting, Santa Barbara, Nov. 17-18, 1966.

9. Ramey, Henry J., Jr., Agarwal, Ram G., and Martin, Ian: "Analysis of 'Slug Test' or DST Flow Period Data," *J. Cdn. Pet. Tech.* (July-Sept. 1975) 37-42.

10. Odeh, A. S. and Selig, F.: "Pressure Build-Up Analysis, Variable-Rate Case," *J. Pet. Tech.* (July 1963) 790-794; *Trans.*, AIME, **228.** Also *Reprint Series, No. 9 — Pressure Analysis Methods*, Society of Petroleum Engineers of AIME, Dallas (1967) 131-135.

11. Gibson, J. A. and Campbell, A. T., Jr.: "Calculating the Distance to a Discontinuity From D.S.T. Data," paper SPE 3016 presented at the SPE-AIME 45th Annual Fall Meeting, Houston, Oct. 4-7, 1970.

12. Earlougher, Robert C., Jr., and Kersch, Keith M.: "Analysis of Short-Time Transient Test Data by Type-Curve Matching," *J. Pet. Tech.* (July 1974) 793-800; *Trans.*, AIME, **257.**

13. McKinley, R. M.: "Wellbore Transmissibility From Afterflow-Dominated Pressure Buildup Data," *J. Pet. Tech.* (July 1971) 863-872; *Trans.*, AIME, **251.**

14. Papadopulos, Istavros S. and Cooper, Hilton H., Jr.: "Drawdown in a Well of Large Diameter," *Water Resources Res.* (1967) **3,** No. 1, 241-244.

15. Cooper, Hilton H., Jr., Bredehoeft, John D., and Papadopulos, Istavros S.: "Response of a Finite-Diameter Well to an Instantaneous Charge of Water," *Water Resources Res.* (1967) **3,** No. 1, 263-269.

16. Kohlhaas, Charles A.: "A Method for Analyzing Pressures Measured During Drillstem-Test Flow Periods," *J. Pet. Tech.* (Oct. 1972) 1278-1282; *Trans.*, AIME, **253.**

17. Ramey, Henry J., Jr., and Agarwal, Ram G.: "Annulus Unloading Rates as Influenced by Wellbore Storage and Skin Effect," *Soc. Pet. Eng. J.* (Oct. 1972) 453-462; *Trans.*, AIME, **253.**

18. Brill, J. P., Bourgoyne, A. T., and Dixon, T. N.: "Numerical Simulation of Drillstem Tests as an Interpretation Technique," *J. Pet. Tech.* (Nov. 1969) 1413-1420.

19. Murphy, W. C.: "The Interpretation and Calculation of Formation Characteristics From Formation Test Data," Pamphlet T-101, Halliburton Co., Duncan, Okla. (1970).

20. Timmerman, E. H. and van Poollen, H. K.: "Practical Use of Drill-Stem Tests," *J. Cdn. Pet. Tech.* (April-June 1972) 31-41.

21. Moran, J. H. and Finklea, E. E.: "Theoretical Analysis of Pressure Phenomena Associated With the Wireline Formation Tester," *J. Pet. Tech.* (Aug. 1962) 899-908; *Trans.*, AIME, **225.**

22. Burnett, O. W. and Mixa, E.: "Application of the Formation Interval Tester in the Rocky Mountain Area," *Drill. and Prod. Prac.*, API (1964) 131-140.

23. Banks, K. M.: "Recent Achievements With the Formation Tester in Canada," *J. Cdn. Pet. Tech.* (July-Sept. 1963) 84-94.

24. Schultz, A. L., Bell, W. T., and Urbanosky, H. J.: "Advances in Uncased-Hole, Wireline Formation-Tester Techniques," *J. Pet. Tech.* (Nov. 1975) 1331-1336.

Chapter 9

Multiple-Well Testing

9.1 Introduction

A multiple-well transient test, including both interference (Section 9.2) and pulse (Section 9.3) tests, directly involves more than one well. In an interference test, a long-duration rate modification in one well creates a pressure interference in an observation well that can be analyzed for reservoir properties. A pulse test provides equivalent data by using shorter-rate pulses (with smaller observed pressure changes), but the analysis technique is more complicated. While numerous variations are possible, this chapter presents only basic techniques for analyzing simple interference and pulse tests. For more complex tests, material in Chapter 2 may be used to devise an appropriate analysis technique. Computer assistance may be helpful in analyzing multiple-well tests, but usable answers can be easily computed in many cases using techniques presented in this chapter.

The multiple-well test requires at least one active (producing or injecting) well and at least one pressure-observation well. Fig. 9.1 schematically illustrates two wells being used in an interference or pulse test in a large reservoir. The observation well is shut in for pressure measurement. (Theory does not preclude an active observation well, but practical considerations rule that out.) In multiple-well testing, the flow rate at the active well is varied while bottom-hole pressure response is measured at the observation wells.

Fig. 9.1 Active and observation wells in an interference or pulse test.

Fig. 9.2 is a schematic illustration of rate history at the active well and pressure response at both the active and observation wells.

Multiple-well testing has the advantage of generally investigating more reservoir than a single-well test.[1-4] Although it is a common belief that interference testing provides information about only the region between the wells, test results are actually influenced by a much larger region. Vela and McKinley[5] show that the influence region for pulse

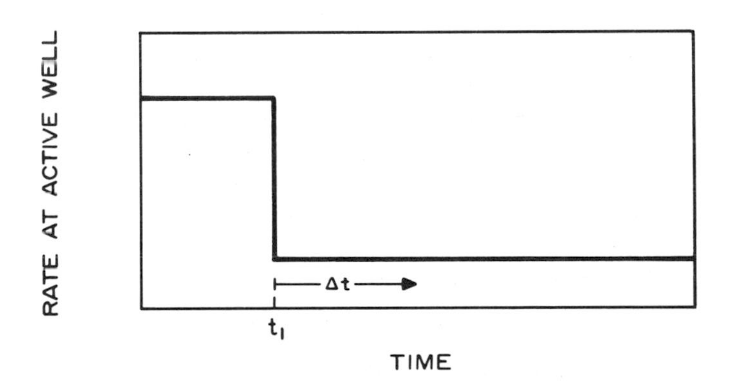

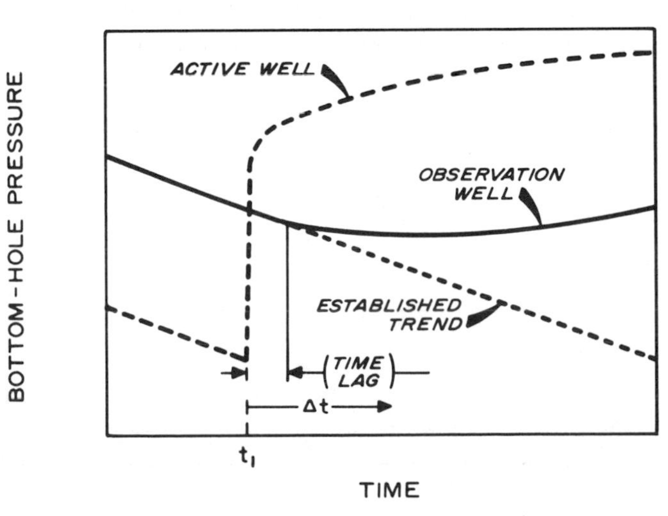

Fig. 9.2 Schematic illustration of rate history and pressure response for an interference test.

testing is approximately as indicated in Fig. 9.3. The radius of influence is given by

$$r_{inf} = 0.029 \sqrt{\frac{kt}{\phi \mu c_t}} \quad \text{. (9.1)}$$

We can expect a similar influence region in normal interference testing. The major difference is that the testing time, t, is much larger in interference testing than in pulse testing. Thus, r_{inf} and the total influence region are substantially larger in interference testing than in pulse testing. In general, we cannot estimate quantitative areal variations in permeability and porosity-compressibility product without using some type of reservoir simulator.[6] That is partially because of inhomogeneities and anisotropy and partly because of the nonuniqueness of analysis techniques.[6-8] If the reservoir can be assumed to be homogeneous, then it may be possible to estimate anisotropic reservoir properties[9-11] by using multiple observation wells.

The skin effect does not influence a multiple-well test since the skin affects only the active well — as long as the skin is concentrated directly around the well. However, a large negative skin or a fracture can affect observation-well response.[12] Wellbore storage effects are minimized by interference and pulse testing, but are not completely eliminated. More research is required to understand the effect of wellbore storage on multiple-well testing.

Usually, both the mobility-thickness product, kh/μ, and the porosity-compressibility-thickness product, $\phi c_t h$, may be estimated from a multiple-well test. In some cases, reservoir extent (Section 9.2) or anisotropic permeability values and orientation (Section 9.4) may be estimated. In reservoirs with fluid-fluid contacts (gas cap-oil zone, for example) in the influence region, interference and pulse tests may yield unreliable or meaningless results because of the different properties in the different fluid regions.

9.2 Interference-Test Analysis

Type-Curve Matching

Type-curve matching is applied to interference-test analysis in basically the same manner as it is applied to drawdown testing (Section 3.3). Fortunately, type-curve matching is simpler for interference testing than for single-well testing because there is usually only one type curve (Fig. C.2) to consider for infinite-acting systems.

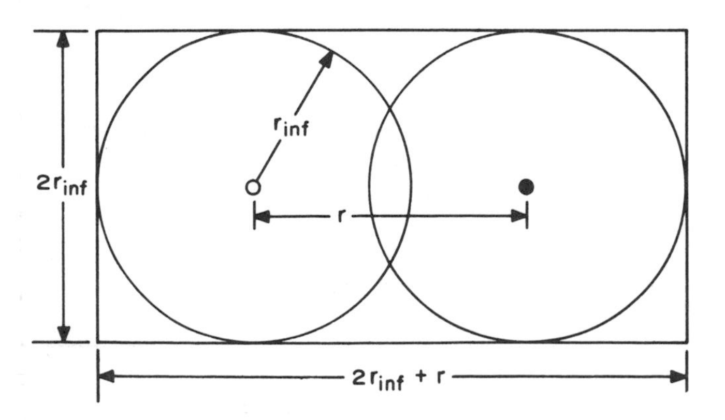

Fig. 9.3 Approximate influence region in interference and pulse testing. After Vela and McKinley.[5]

To analyze an interference test by type-curve matching, one plots the observation-well pressure data as Δp vs t on tracing paper laid over Fig. C.2, using the technique described in Section 3.3 and Fig. 3.5. The major grid lines are traced and the grid of Fig. C.2 is used to locate the data points, without regard for the curve on Fig. C.2. Then the tracing paper is slid horizontally and vertically until the data points match the exponential-integral curve on Fig. C.2, as illustrated in Fig. 9.4. When the data are matched to the curve, a convenient match point such as that shown in Fig. 9.4 is chosen, and match-point values from the tracing paper and the underlying type-curve grid are read. Permeability is estimated from

$$k = \frac{141.2\, qB\mu}{h} \frac{(p_D)_M}{\Delta p_M} \text{ , } \quad \text{. . . . . (9.2)}$$

and the porosity-compressibility product is estimated from

$$\phi c_t = \frac{0.0002637}{r^2} \frac{k}{\mu} \frac{t_M}{(t_D/r_D{}^2)_M} \text{ . } \quad \text{. . . . . (9.3)}$$

The type-curve analysis method is simple, fast, and accurate when the exponential integral p_D applies; that is, when $r_D = r/r_w > 20$ and $t_D/r_D{}^2 > 0.5$.

If the active well is shut in after time t_1, the resulting change in pressure at the observation well may be further analyzed, thus improving the accuracy of the analysis. (Here we assume that the active well is producing or injecting during the interference test. Of course, it could have been shut in to create the interference test. In that case, the analogy to shutting in a producing or injecting well is resumption of production or injection after the shut-in period.) Fig. 9.5 schematically illustrates how the data are used when the rate condition preceding the test is resumed. The data during the first portion of the test are matched to the type curve of Fig. C.2, as indicated by the data points that fall on the solid line in Fig. 9.5. After the change in rate, the difference between the extrapolated, matched type curve and the actual data points, $\Delta p_{\Delta t}$, is determined from the data plot; Δt is the time from the change in rate at the active well to the time the data are taken. It can be shown by superposition that

$$\Delta p_{\Delta t} = \Delta p_{w\,ext} - \Delta p_{ws} \text{ , } \quad \text{. . . . . (9.4a)}$$

$$= \frac{141.2\, qB\mu}{kh} p_D(\Delta t_D, r_D) \text{ , } \quad \text{. . . . . (9.4b)}$$

where p_D is simply the exponential integral (Fig. C.2). Thus, when $\Delta p_{\Delta t}$ is plotted vs Δt on the same data plot, the points should fall on the curve matched by the original data. If they do not, then either (1) the original data were not matched correctly and the match should be repeated until the two portions of the data fall on the same curve; or (2) something else is influencing the response at the interference well that precludes correct type-curve matching. The following example illustrates interference-test analysis by type-curve matching for a 48-hour injection period followed by a long falloff.

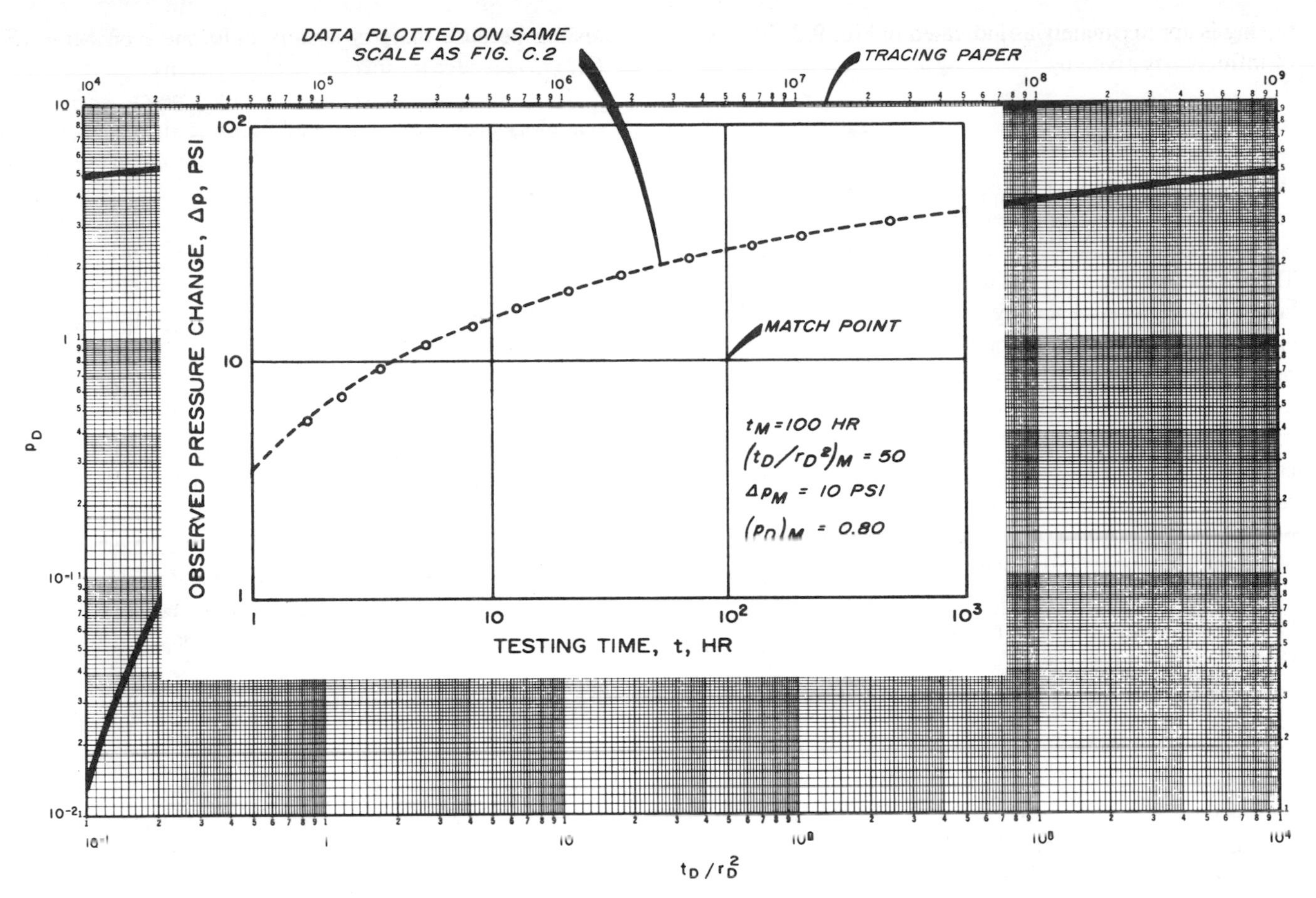

Fig. 9.4 Illustration of type-curve matching for an interference test using the type curve of Fig. C.2.

Example 9.1 Interference-Test Analysis by Type-Curve Matching

During an interference test, water was injected into Well A for 48 hours. The pressure response in Well B, 119 ft away, was observed for 148 hours. Known reservoir properties are

depth = 2,000 ft	$B_w = 1.0$ RB/STB
$q = -170$ B/D	$\mu_w = 1.0$ cp
$h = 45$ ft	$r = 119$ ft
$p_i = 0$ psig	$c_t = 9.0 \times 10^{-6}$ psi^{-1}.
$t_1 = 48$ hours	

Observed pressure data are given in Table 9.1.

The open circles in Fig. 9.6 show the data match to the type curve of Fig. C.2. We have plotted Δp as positive, even though it is actually negative, since it is not possible to take logarithms of negative numbers. We could remember that Δp is negative to prevent confusion in signs in the analysis; however, that is not a major problem because we know that both k and ϕc_t must be positive. The line in Fig. 9.6 was traced from the type curve of Fig. C.2, using the overlay technique for data points before 48 hours. Although we would prefer to have more than four data points, the match is acceptable. At the *match point* marked,

TABLE 9.1—INTERFERENCE DATA FOR EXAMPLES 9.1 AND 9.3, INJECTION FOR 48 HOURS.

t (hours)	p_w (psig)	$\Delta p = p_i - p_w$ (psi)
0.0	0	—
4.3	22	−22
21.6	82	−82
28.2	95	−95
45.0	119	−119
48.0	Injection Ends	
51.0	109	−109
69.0	55	−55
73.0	47	−47
93.0	32	−32
142.0	16	−16
148.0	15	−15

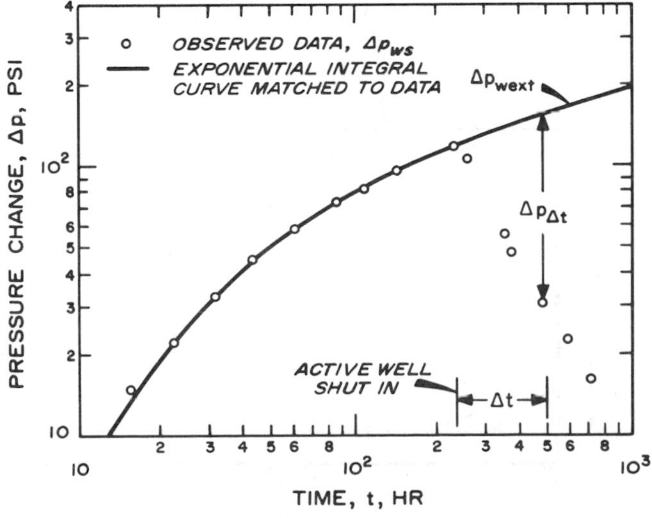

Fig. 9.5 Schematic observation-well pressure response for an interference test with the active well shut in after 240 hours.

$$\Delta p_M = -100 \text{ psig at } (p_D)_M = 0.96,$$

and

$$t_M = 10 \text{ hours at } (t_D/r_D{}^2)_M = 0.94.$$

Using Eq. 9.2,

$$k = \frac{141.2(-170)(1.0)(1.0)}{(45)} \frac{(0.96)}{(-100)}$$

$$= 5.1 \text{ md}.$$

From Eq. 9.3,

$$\phi c_t = \frac{(0.0002637)(5.1)}{(119)^2(1.0)} \frac{10}{0.94}$$

$$= 1.01 \times 10^{-6} \text{ psi}^{-1},$$

and

$$\phi = \frac{1.01 \times 10^{-6}}{9.0 \times 10^{-6}} = 0.11.$$

We can estimate the accuracy of the above analysis by using the data from the declining-pressure part of the test ($t > 48$ hours). We extrapolate the solid line in Fig. 9.6 by tracing the curve from Fig. C.2, and estimate the difference between Δp_{ext} and the observed Δp, $\Delta p_{\Delta t}$. Table 9.2 shows the computations.

The plus symbols in Fig. 9.6 are a plot of $\Delta p_{\Delta t}$ vs Δt. The points fall on just about the same line as the circles, so we can be confident of the analysis results.

Example 12.1 treats the same interference-test data using computer analysis techniques.

Earlougher and Ramey[13] provide dimensionless pressure data that are useful for interference-test analysis in bounded systems. They give p_D at several observation-well locations for a variety of closed rectangular systems with one active well. Fig. 9.7 shows p_D values for a variety of observation-well positions in a 2:1 rectangle. Fig. 9.8 shows the location of the active well in Fig. 9.7 and defines the dimensionless distance coordinates. Note that the dimensionless pressure response at each observation well deviates upward from the exponential-integral solution, as we would expect in a closed system. The parameters on the curves in Fig. 9.7 are the dimensionless locations of the observation wells and the system area divided by the square of the distance between the active and observation wells, A/r^2.

If the system shape and approximate active-well location can be estimated, figures such as Fig. 9.7 may be used for type-curve matching interference data in bounded systems. Such matching may require much patience, since data for more than 150 type curves are given in Ref. 13.

The reciprocity principle[7,13] can be helpful for interference-test analysis in bounded systems when using type

TABLE 9.2—CALCULATION OF $\Delta p_{\Delta t}$ FOR SHUT-IN PERIOD DATA OF EXAMPLE 9.1.

t (hours)	$\Delta t = t-48$ (hours)	$\Delta p = p_i - p_{ws}$ (psi)	Δp_{wext}, Fig. 9.6 (psi)	$\Delta p_{\Delta t}$, Eq. 9.4 (psi)
51	3	−109	−124	−15
69	21	−55	−140	−85
73	25	−47	−142	−95
93	45	−32	−155	−123
142	94	−16	−177	−161
148	100	−15	−179	−164

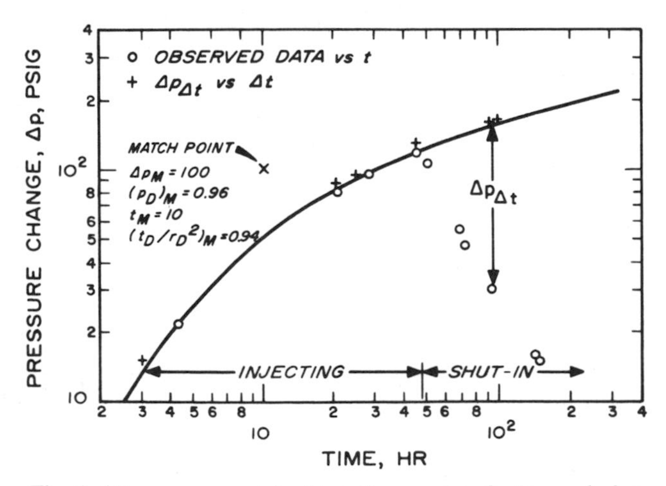

Fig. 9.6 Type-curve match of interference data for Example 9.1.

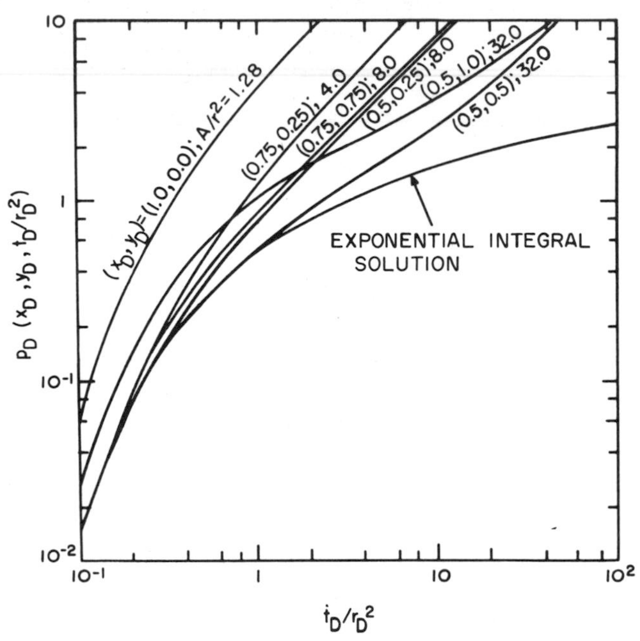

Fig. 9.7 Type-curve plot of p_D vs $t_D/r_D{}^2$ for various observation-well locations in a closed 2:1 rectangle with the active well at $x_{Dw}=0.5$, $y_{Dw}=0.75$. After Earlougher and Ramey.[13] Courtesy CIM.

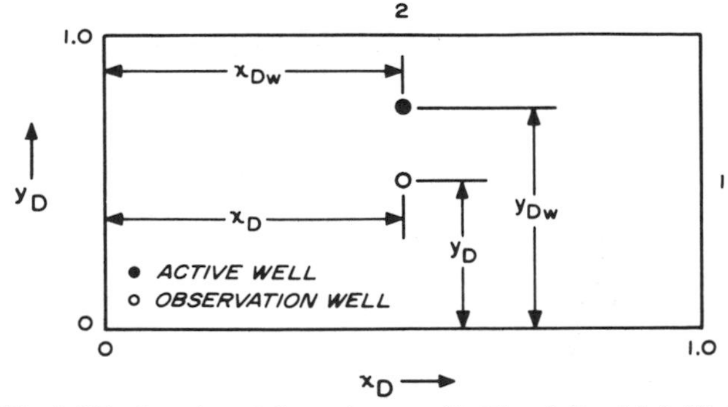

Fig. 9.8 Explanation of dimensions used in Figs. 9.7 and 9.9. Well locations for data match of Example 9.2.

curves such as those in Fig. 9.7. Stated simply, the reciprocity principle says that the pressure response at Observation Well A caused by production at rate q from Well B is equal to the pressure response that would be caused at Observation Well B by production at rate q from Well A. The principle is valid if pressure behavior in the system satisfies the normal diffusivity equation (Eq. 2.1) and if compressibility, permeability, viscosity, and porosity are not pressure-sensitive. The reciprocity principle tells us that it makes no difference which well is the active well and which well is the observation well; the same reservoir properties should be determined for testing either way. McKinley, Vela, and Carlton[7] have demonstrated that principle in the field for pulse testing.

If interference data are matched to a type curve such as one of those in Fig. 9.7, it may be possible to estimate reservoir area from the A/r^2 parameter for the matched curve, as illustrated in the following example.

Example 9.2 Interference-Test Analysis in a Bounded System

Simulated pressure data for an interference test in a closed reservoir[13] are given in Table 9.3. One well produces at 427 STB/D while the second well, 340 ft distant, remains shut in as an observation well. The wells lie in a structure such that a good approximation is a 2:1 closed rectangle with the producing well located at $x_D = 0.5$, $y_D = 0.75$. The size of the reservoir is not known. Other data are

$h = 23$ ft $\qquad r_w = 0.27$ ft
$\mu_o = 0.8$ cp $\qquad \phi = 0.12$
$B_o = 1.12$ RB/STB $\qquad c_t = 8.3 \times 10^{-6}$ psi^{-1}
$q_o = 427$ STB/D $\qquad r = 340$ ft.

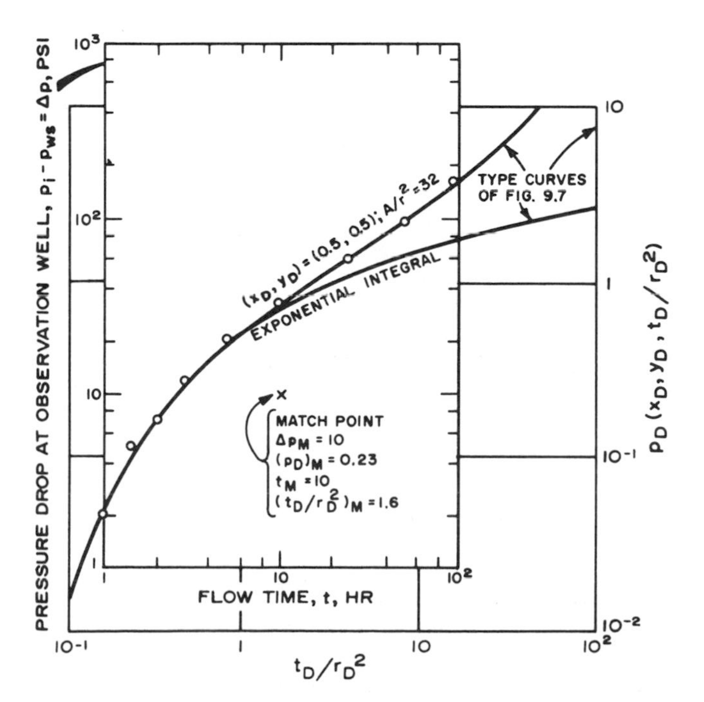

Fig. 9.9 Type-curve match for interference analysis of Example 9.2. After Earlougher and Ramey.[13] Courtesy CIM.

Fig. 9.9 shows the pressure differences of Table 9.3 matched with one of the type curves of Fig. 9.7. The data match the curve for $x_D = 0.5$, $y_D = 0.5$ reasonably well. At the pressure match point indicated in Fig. 9.9,

$$\Delta p_M = 10 \text{ psi at } (p_D)_M = 0.23,$$

and

$$t_M = 10 \text{ hours at } (t_D/r_D{}^2)_M = 1.6.$$

The permeability is estimated from Eq. 9.2:

$$k = \frac{(141.2)(427)(1.12)(0.8)(0.23)}{(23)(10)}$$

$$= 54.0 \text{ md.}$$

Using Eq. 9.3,

$$\phi c_t = \frac{(0.0002637)(54.0)}{(340)^2(0.8)} \; \frac{10}{1.6}$$

$$= 9.62 \times 10^{-7} \text{ psi}^{-1},$$

and

$$\phi = 0.116.$$

From the matched curve, $A/r^2 = 32$. Thus, we can estimate the reservoir size:

$$A = 32(340)^2 = 3.70 \times 10^6 \text{ sq ft}$$

$$= 84.9 \text{ acres.}$$

The exponential-integral type curve may not apply for interference-test analysis in some situations. Fig. 9.10 schematically illustrates how the dimensionless pressure deviates from the exponential-integral solution when the active well has a vertical fracture or has a high wellbore storage coefficient. The severity of the fracturing effect[12] depends on both the length of the fracture and the distance between the active well and the observation well. Similar deviations from the exponential-integral solution may be caused by wells with a large negative skin effect ($s << 0$) distributed some distance into the formation. Wellbore storage effects cause the pressure response at the observation well to fall below the exponential-integral solutions at early times because the change in the sand-face rate is less than the change in the surface rate. The degree of deviation from the exponential-integral solution depends on both the wellbore storage coefficient and the distance between the wells. Un-

TABLE 9.3—PRESSURE INTERFERENCE DATA AT OBSERVATION WELL OF EXAMPLE 9.2.
Data of Earlougher and Ramey.[13]

t (hours)	$p_i - p_{ws}$ (psi)
0	0
1	2
1.5	5
2	7
3	12
5	21
10	33
24	62
48	100
96	170

fortunately, very little information is currently available about these phenomena.*

Semilog Analysis Methods

If the two wells in an interference test are much closer together than the distance to the nearest boundary or to another active well (by a factor of about 10) in the system, the pressure response at the observation well eventually will be described by the logarithmic approximation to the exponential integral (Eq. 2.5b). Then the observation-well pressure would be approximated by

$$p_{ws}(t,r) = p_{1\text{hr}} + m \log t \ . \quad \ldots\ldots\ldots\ldots\ldots (9.5)$$

Eq. 9.5 is strictly valid for $t_D/r_D^2 > 100$, where

$$\frac{t_D}{r_D^2} = \frac{0.0002637\, kt}{\phi\mu c_t r^2} \ . \quad \ldots\ldots\ldots\ldots\ldots (9.6)$$

The restriction may be reduced to $t_D/r_D^2 > 10$ with only 1-percent deviation. If t_D/r_D^2 exceeds 2 or 3, Eq. 9.5 is often adequate for data analysis. When Eq. 9.5 can be used for interference-test analysis, observed pressures are plotted as p_{ws} vs $\log t$ during the initial phase of an interference test. Such a plot should have a semilog straight-line portion with slope m given by

$$m = \frac{-162.6\, qB\mu}{kh} \ , \quad \ldots\ldots\ldots\ldots\ldots (9.7)$$

and intercept $p_{1\text{hr}}$ given by

$$p_{1\text{hr}} = p_i + m\left[\log\left(\frac{k}{\phi\mu c_t r^2}\right) - 3.2275\right] \ . \quad \ldots\ldots (9.8)$$

The skin factor does not appear in Eq. 9.8 since fluid is flowing only at the active well and not into or out of the observation well. The semilog straight-line slope may be used to estimate the system permeability,

$$k = \frac{-162.6\, qB\mu}{mh} \ , \quad \ldots\ldots\ldots\ldots\ldots (9.9)$$

and the reservoir porosity-compressibility product,

$$\phi c_t = \frac{k}{r^2\mu}\ \text{antilog}\left(\frac{p_i - p_{1\text{hr}}}{m} - 3.2275\right) . \quad \ldots (9.10)$$

It is usually necessary to extrapolate the pressure response to estimate $p_{1\text{hr}}$ (which must be on the semilog straight line), since it usually takes many hours to reach a semilog straight-line response at an observation well.

If the observation well is shut in after time t_1, then normal superposition may be used for analysis of observed pressures. After a long shut-in time, the pressure behavior at the observation well will be given by

$$p_{ws}(t_1 + \Delta t, r) = p_i + m \log\left(\frac{t_1 + \Delta t}{\Delta t}\right) . \quad \ldots (9.11)$$

Eq. 9.11 indicates that a plot of observed pressure after the active well is shut in vs $\log[(t_1 + \Delta t)/\Delta t]$ should have a semilog straight-line portion with slope = m. Permeability is estimated from Eq. 9.9 and the porosity-compressibility product is estimated from

$$\phi c_t = \frac{k}{r^2\mu}\ \text{antilog}\left[\frac{p_{1\text{hr}} - p_{ws}(\Delta t = 0)}{m} - \log\left(\frac{t_1 + 1}{t_1}\right) - 3.2275\right] . \quad \ldots\ldots\ldots\ldots (9.12)$$

As before, the semilog straight line must be extrapolated to $p_{1\text{hr}}$. The term $\log[(t_1 + 1)/t_1]$ is normally small and often can be neglected.

As with single-well testing, the signs of Δp and the rate must be considered when analyzing interference tests, since tests are commonly run with either production or injection at the active well. Eqs. 9.10 and 9.12 will give incorrect results if the signs are not treated properly. No simple rule of thumb can always be applied to prevent errors, but we can state that (1) k must *always* be positive, so the sign of m must be opposite the sign of q; and (2) except when r is small or k is very large, the $(p_i - p_{1\text{hr}})/m$ term in Eq. 9.10 (or its equivalent in Eq. 9.12) will be positive. Normally, ϕc_t is of the order of 10^{-7} for liquid-filled systems; it may exceed 10^{-4} for systems with free-gas saturation.

Example 9.3 Interference-Test Analysis Using Long-Time Equations

The interference test of Example 9.1 is analyzed using Eqs. 9.5 through 9.12. The observed pressure data are given

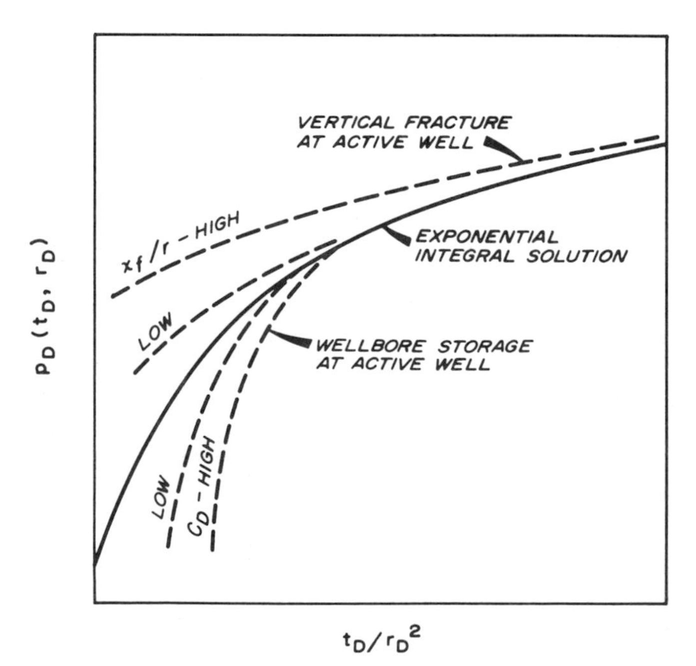

Fig. 9.10 Schematic illustration of the effect of vertical fractures and wellbore storage on observation-well response during an interference test.

*A paper that treats this subject in more detail appeared just before publication of this monograph — Jargon, J. R.: "Effect of Wellbore Storage and Wellbore Damage at the Active Well on Interference Test Analysis," *J. Pet. Tech.* (Aug. 1976) 851-858. Jargon presents several type curves that show the influence of storage and damage at the active well on observation-well response. He indicates that active-well wellbore storage and wellbore damage effects are not important at the observation well when

$$\frac{t_D}{r_D^2} > (230 + 15\,s)\left(\frac{C_D}{r_D^2}\right)^{0.86} .$$

Before that time, kh/μ estimated from the exponential-integral type curve is low and $\phi c_t h$ estimated is high.

in Table 9.1; other data are given in Example 9.1.

Fig. 9.11 is a semilog plot of the observation-well pressure data. The last three points in the injection part of the test appear to form a straight line, so we attempt analysis. Since there are only three data points on the straight line, and since t_D/r_D^2 (calculated from results below) is only about 5, the analysis is tentative. Type-curve matching is preferred in such a situation. From Fig. 9.11,

$$m = 120 \text{ psig/cycle},$$

and

$$p_{ws}(t = 10 \text{ hours}) = 41 \text{ psig},$$

so extrapolating one cycle,

$$p_{1hr} = 41 - 120 = -79 \text{ psig}.$$

Using Eq. 9.9,

$$k = \frac{(-162.6)(-170)(1.0)(1.0)}{(120)(45)}$$

$$= 5.1 \text{ md}.$$

We estimate ϕc_t from Eq. 9.10:

$$\phi c_t = \frac{5.1}{(119)^2(1.0)} \text{antilog}\left[\frac{0-(-79)}{120} - 3.2275\right]$$

$$= 9.71 \times 10^{-7} \text{ psi}^{-1},$$

and

$$\phi \simeq 0.11.$$

We can also analyze the decreasing-pressure portion of the test. Fig. 9.12 is a plot of the type suggested by Eq. 9.11. The slope taken from the last four points, 111 psi/cycle, may be used in Eq. 9.9 to estimate

$$k = \frac{(-162.6)(-170)(1.0)(1.0)}{(111)(45)} = 5.5 \text{ md}.$$

We estimate p_{1hr} by extrapolating the line in Fig. 9.12. At 1 hour, $(t_1 + \Delta t)/\Delta t = 49$. We read $p = 72$ psig from Fig. 9.12 at $(t_1 + \Delta t)/\Delta t = 4.9$. Then, extrapolating one cycle,

$$p_{1hr} = 72 + 111 = 183 \text{ psi}.$$

Fig. 9.11 indicates that $p_{ws}(\Delta t = 0) = 123$ psi. Then, from Eq. 9.12,

$$\phi c_t = \frac{5.5}{(119)^2(1.0)} \text{antilog}\left[\frac{183-123}{111} - \log\left(\frac{48+1}{48}\right) - 3.2275\right]$$

$$= 7.82 \times 10^{-7} \text{ psi}^{-1}.$$

Both k and ϕc_t are close to the values computed from the increasing-pressure portion of the analysis. The injection-period semilog analysis agrees well with the type-curve analysis in spite of the fact that (Eq. 9.6)

$$\frac{t_D}{r_D^2} = \frac{(0.0002637)(5.1)(48)}{(9.71 \times 10^{-7})(1.0)(119)^2} = 4.7,$$

for the injection period.

When there is significant rate variation at the active well during an interference test, analysis becomes much more difficult. In this case, complete superposition must be used; but because the logarithmic approximation to the exponential-integral solution generally will not apply, analysis techniques are not so simple. Under these circumstances, results generally must be analyzed by using computer methods such as those discussed by Jahns[6] and Earlougher and Kersch.[10]

9.3 Pulse Testing

Pulse testing is a special form of multiple-well testing first described by Johnson, Greenkorn, and Woods.[14-16] The technique uses a series of short-rate pulses at the active well. Pulses generally are alternating periods of production (or

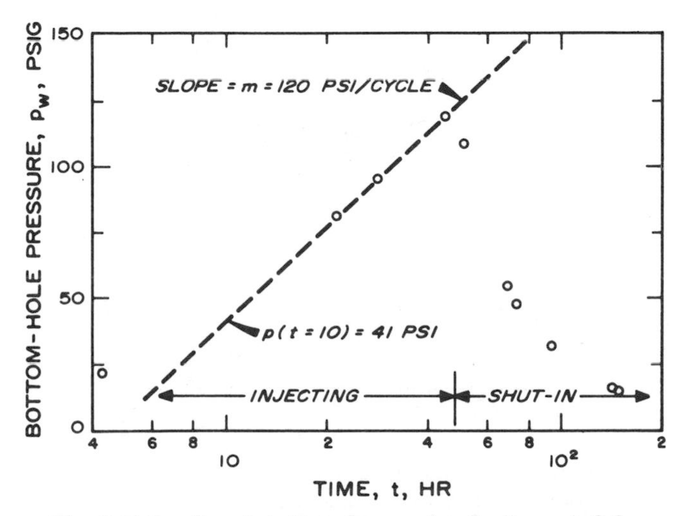

Fig. 9.11 Semilog plot of interference data for Example 9.3.

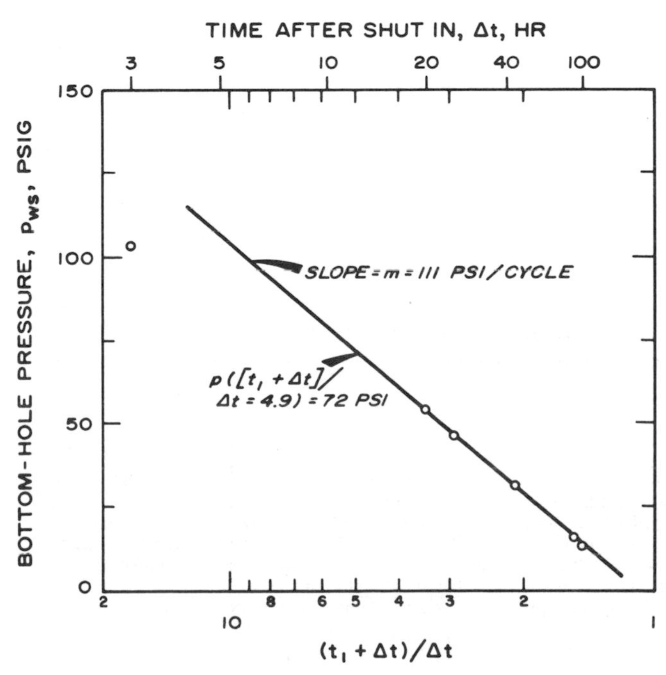

Fig. 9.12 Decreasing pressure portion of observation-well data from the interference test of Example 9.3.

injection) and shut-in, with the same rate during each production (injection) period. The pressure response to the pulses is measured at the observation well. Because the pulses are of short duration, the pressure responses are small, sometimes less than 0.01 psi. Therefore, special pressure-measuring equipment is usually required.[17,18]

The main advantages of pulse testing result from the short pulse length. Infinite-system equations (Eq. 2.5) usually apply at such short times, regardless of the system size. A pulse test may last from a few hours to a few days, so it should disrupt normal operations only slightly compared with interference testing. Reservoir pressure trends and noise are automatically removed by the analysis technique given in this section.

Fig. 9.13 schematically illustrates pulse testing for a two-well system. The figure is for a producing well that is pulsed by shutting in, continuing production, shutting in, continuing production, etc. The upper portion of the figure shows the constant production rate before the test and the rate pulses. The lower portion of the figure illustrates the pressure behavior at the observation well and correlates the pressure pulses with the rate pulses. Although the flow time and shut-in time are equal in Fig. 9.13, pulse testing can be done with unequal flow and shut-in times. However, all flow times must be the same and all shut-in times must be the same.

Two characteristics of the pressure response at the observation well are used for pulse-test analysis. One is the time lag, the time between the *end* of a pulse and the pressure peak caused by the pulse (Fig. 9.14). A time lag is associated with each pulse. (The pulse responses shown in Fig. 9.14 are not necessarily representative of an actual pulse-test response; the relationship of the pulses has been drawn to simplify defining the terms used in pulse-test analysis.) Time lags for the first and fourth pulses are shown in Fig. 9.14; similar definitions apply for the second, third, fifth, sixth, seventh, and eighth pulses in that figure.

The second variable used in pulse-test analysis is the amplitude of the pressure response, Δp_1, Δp_2, etc., shown in Fig. 9.14. We determine pulse-response amplitude by first constructing the tangent between the two peaks (or valleys) on either side of the pulse to be measured. Then we draw a line parallel to that tangent at the peak of the subject response. The pressure amplitude is the *vertical* distance between the two lines. The same approach applies to both peaks and valleys. Strictly speaking, the Δp value for the first peak in Fig. 9.14 is negative, as it is for all other odd peaks; Δp is positive for even responses. In the analysis methods presented in this section, the *sign convention is eliminated,* since it tends to be confusing. Rather, we designate that the sign of Δp is the same as the sign of the rate in the flowing part of the test, so $\Delta p/q > 0$.

Pulse-test analysis techniques have been outlined by Johnson, Greenkorn, and Woods,[14] Culham,[19] Startzman,[20] and Brigham and Kamal.[21,22] The Kamal-Brigham technique[22] has the advantage of being flexible and convenient to use for hand analysis, so it is presented here. Many analysis techniques require use of a digital computer.

The Kamal-Brigham analysis method uses several definitions. The ratio of pulse length to the total cycle length is defined as

$$F' = \frac{\Delta t_P}{\Delta t_C}, \qquad (9.13)$$

where Δt_P and Δt_C are indicated in Fig. 9.14. (This nomenclature deviates from that used by Johnson, Greenkorn, and Woods.[14]) The dimensionless time lag is defined analogously to any dimensionless time,

$$(t_L)_D = \frac{0.0002637\,kt_L}{\phi\mu c_t r_w^2}, \qquad (9.14)$$

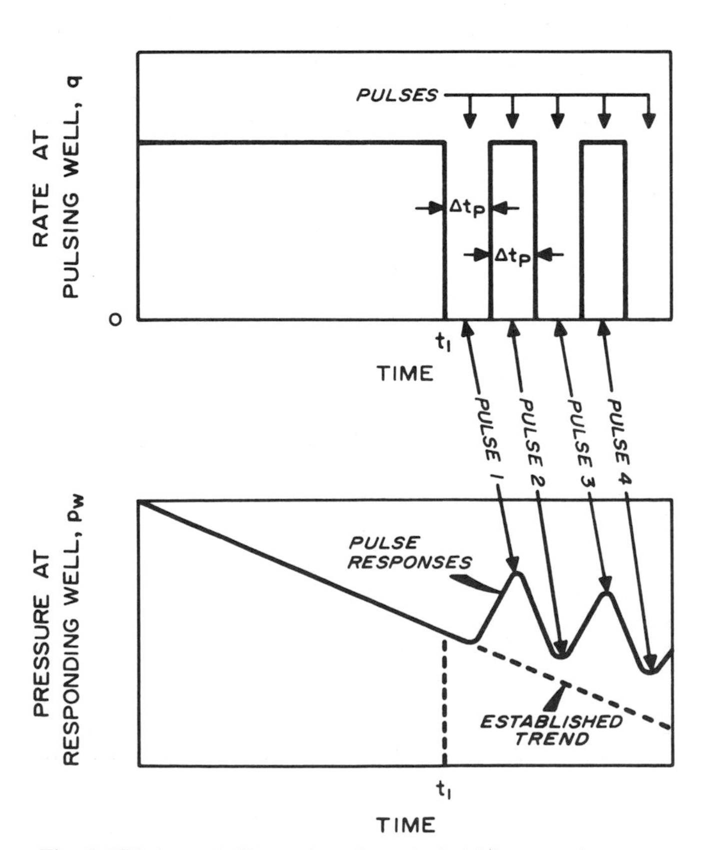

Fig. 9.13 Schematic illustration of rate (pulse) history and pressure response for a pulse test.

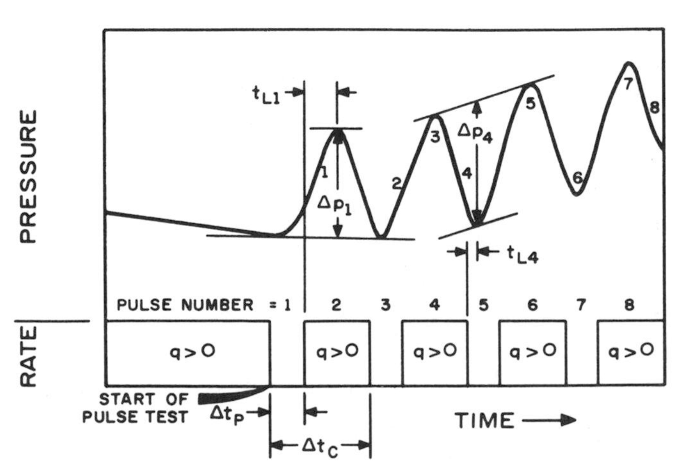

Fig. 9.14 Schematic pulse-test rate and pressure history showing definition of time lag (t_L) and pulse-response amplitude (Δp).

using the wellbore radius of the active well. The dimensionless distance between the active and observation wells is

$$r_D = r/r_w, \quad \text{(9.15)}$$

where r is indicated in Fig. 9.1. The dimensionless-pressure response amplitude is

$$\Delta p_D = \frac{kh\,\Delta p}{141.2\,qB\mu}, \quad \text{(9.16)}$$

where q is the rate at the active well while it is active. We use the sign convention that $\Delta p/q$ is always positive.

The time lag and pressure response amplitude from one or more pulse responses are used to estimate reservoir properties from a pulse test. The permeability is estimated from

$$k = \frac{141.2\,qB\mu\,\{\Delta p_D[t_L/\Delta t_C]^2\}_{\text{Fig.}}}{h\Delta p[t_L/\Delta t_C]^2}. \quad \text{(9.17)}$$

In Eq. 9.17, Δp and t_L are from the observation-well response for the pulse being analyzed; Δt_C is the cycle length at the pulsing well; and $\{\Delta p_D[t_L/\Delta t_C]^2\}_{\text{Fig.}}$ is from Fig. 9.15, 9.16, 9.17, or 9.18 for the appropriate values of $t_L/\Delta t_C$ and F'. It is necessary to use the figure specifically applying to the pulse being analyzed. Fig. 9.15 is for analysis of the *first odd pulse* (first pulse); Fig. 9.16 is for the *first even pulse* (second pulse); Fig. 9.17 is for all *odd pulses after the first* (3, 5, 7, ...); and Fig. 9.18 is for all *even pulses after the first* (4, 6, 8, ...).

Once the permeability is estimated from Eq. 9.17, we estimate the porosity-compressibility product from

$$\phi c_t = \frac{0.0002637\,k\,t_L}{\mu r^2\{(t_L)_D/r_D^2\}_{\text{Fig.}}} \quad \text{(9.18)}$$

In Eq. 9.18 the value of $\{(t_L)_D/r_D^2\}_{\text{Fig.}}$ is from Fig. 9.19, 9.20, 9.21, or 9.22. As before, the figure used depends on whether the first odd or even pulse or one of the remaining odd or even pulses is being analyzed.

Once pulse-test data are available and plotted and time lags and pressure responses are measured, pulse-test analysis by the Kamal-Brigham technique is rapid. It is good practice to analyze several pulses to get an idea of the reliability of the results.

Prats and Scott[23] have presented preliminary data showing the effect of wellbore storage at the observation well on pulse-test response. Wellbore storage effects at the observation well increase the time lag (t_L) and reduce the response amplitude (Δp) of the first pulse. Data are not presented for later pulses, but they can be expected to be affected, also. Using the Prats and Scott[23] data, we can approximate that the effect of wellbore storage at the responding well will result in less than a 5-percent increase in time lag and a virtually unaffected response amplitude when the distance between pulsed and responding wells satisfies

$$r_D > 34\,C_D^{0.54}. \quad \text{(9.19a)}$$

Eq. 9.15 gives r_D using r_w of the *observation* well and C_D is from Eq. 2.18. In terms of physical quantities, Eq. 9.19a becomes

$$r > 32\,r_w \left(\frac{C}{\phi c_t h r_w^2}\right)^{0.54}, \quad \text{(9.19b)}$$

or approximately,

$$r > 32 \left(\frac{C}{\phi c_t h}\right)^{0.54}. \quad \text{(9.19c)}$$

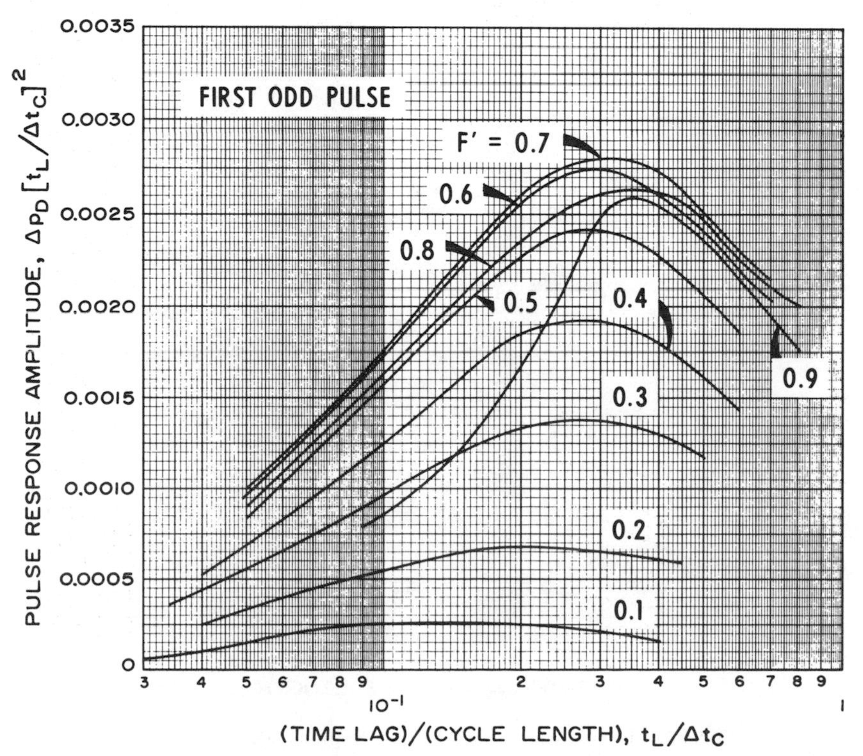

Fig. 9.15 Pulse testing: relation between time lag and response amplitude for first odd pulse. After Kamal and Brigham.[22]

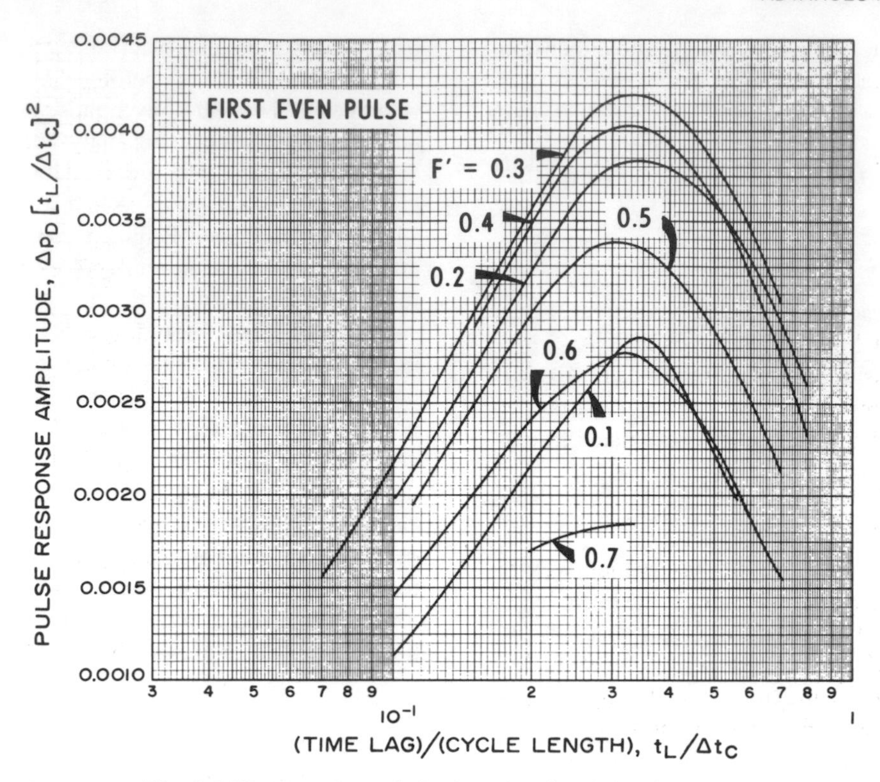

Fig. 9.16 Pulse testing: relation between time lag and response amplitude for first even pulse. After Kamal and Brigham.[22]

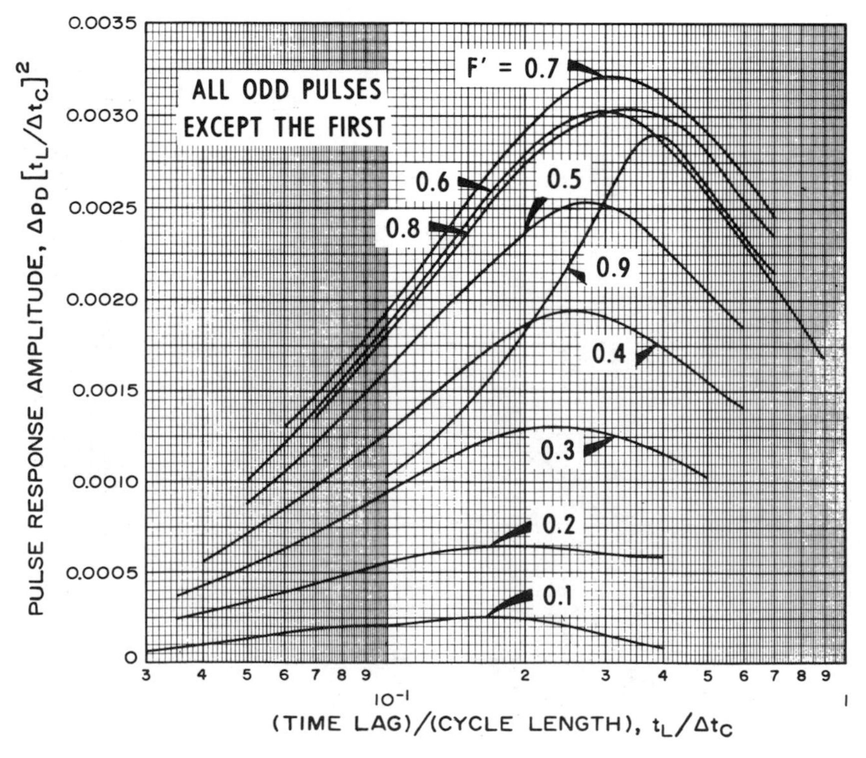

Fig. 9.17 Pulse testing: relation between time lag and response amplitude for all odd pulses after the first. After Kamal and Brigham.[22]

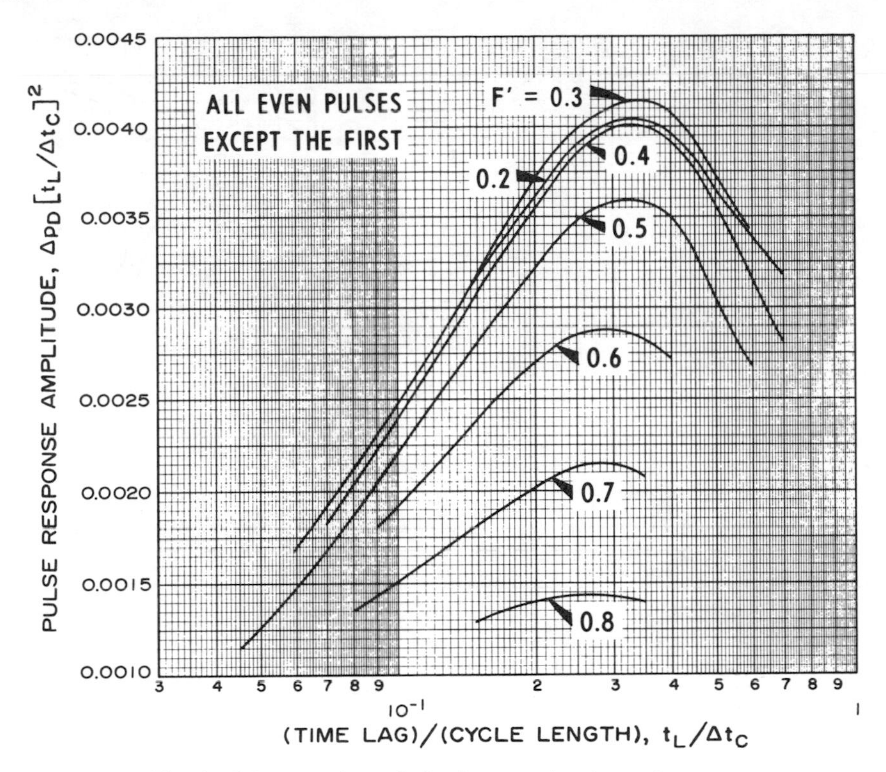

Fig. 9.18 Pulse testing: relation between time lag and response amplitude for all even pulses after the first. After Kamal and Brigham.[22]

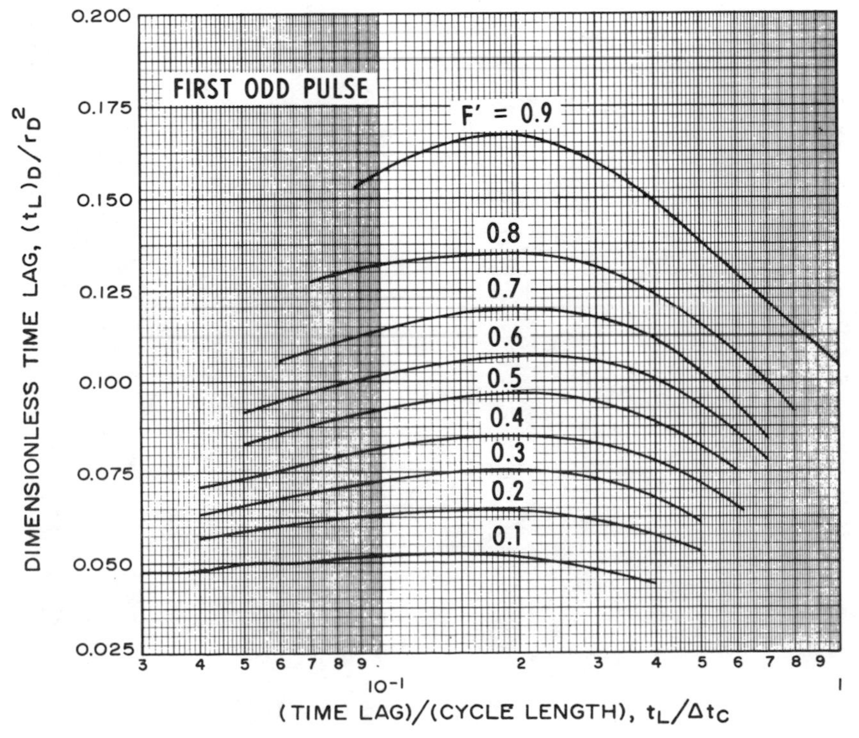

Fig. 9.19 Pulse testing: relation between time lag and cycle length for first odd pulse. After Kamal and Brigham.[22]

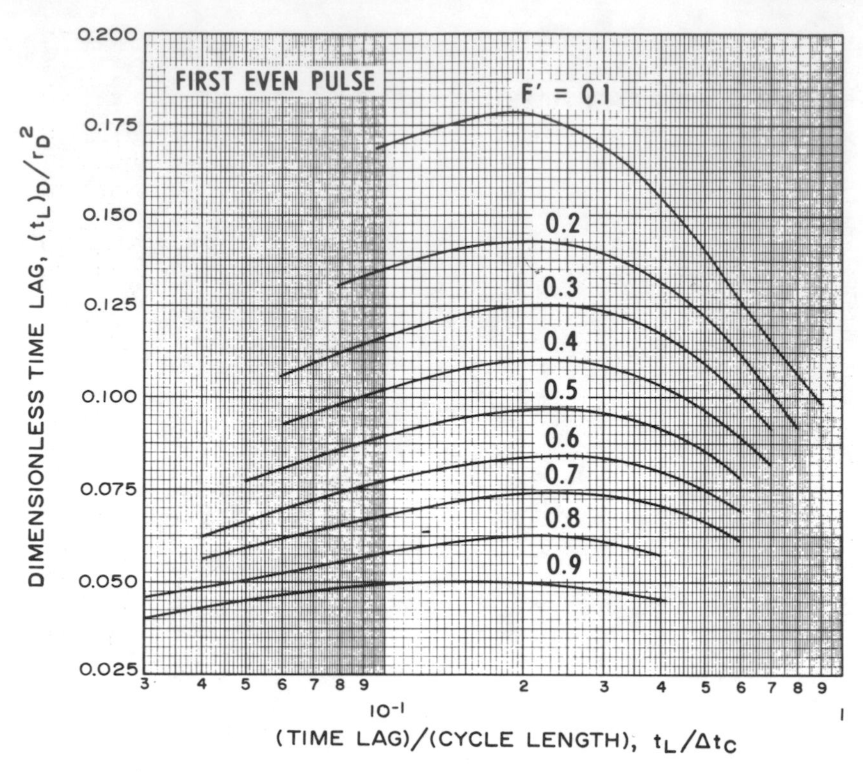

Fig. 9.20 Pulse testing: relation between time lag and cycle length for first even pulse. After Kamal and Brigham.[22]

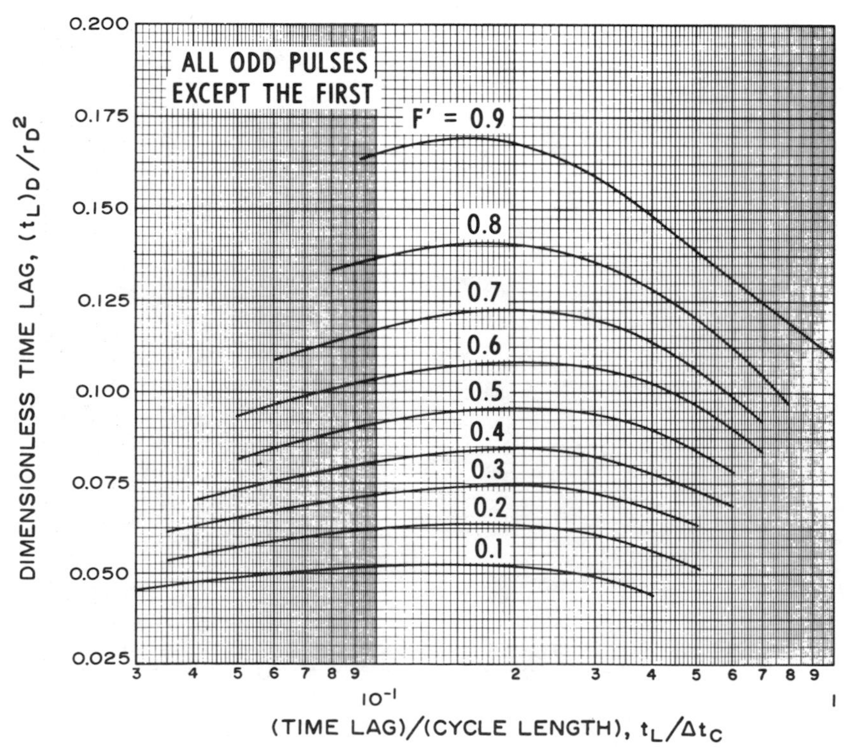

Fig. 9.21 Pulse testing: relation between time lag and cycle length for all odd pulses after the first. After Kamal and Brigham.[22]

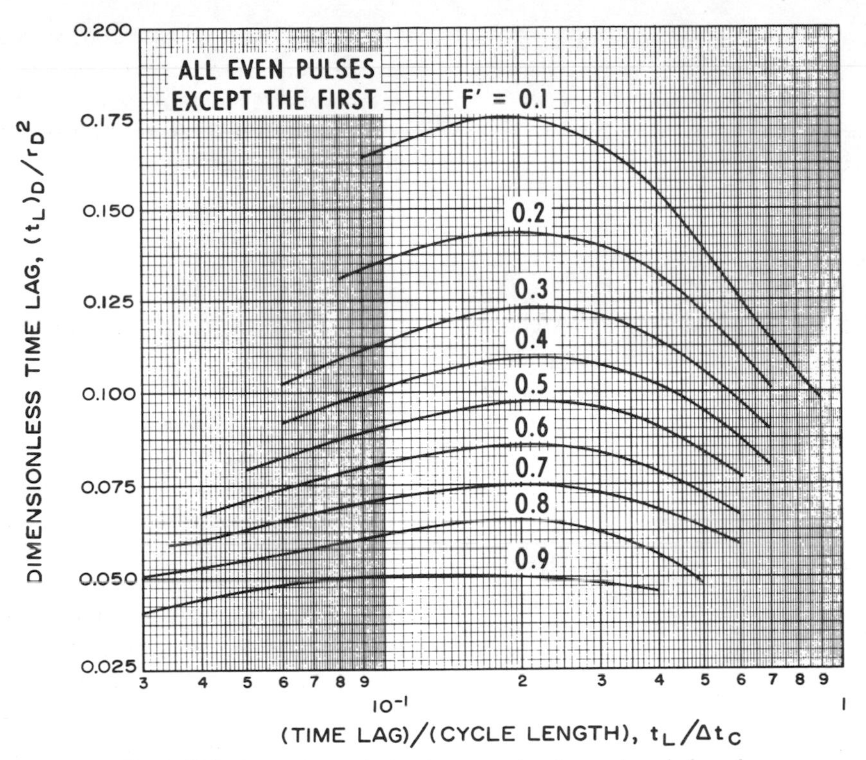

Fig. 9.22 Pulse testing: relation between time lag and cycle length for all even pulses after the first. After Kamal and Brigham.[22]

Wellbore storage at the active well also can be expected to influence pulse-test response; details are not currently available.*

Example 9.4 Pulse-Test Analysis

Fig. 9.23 shows pulse-test data from McKinley, Vela, and Carlton.[7] Producing Wells A8 and A9 were tested by pulsing Well A8 (Δt_P = 1 hour, q = 550 STB/D) and observing the response at Well A9 (plus symbols in Fig. 9.23). At some other time, Well A9 was pulsed (Δt_P = 1 hour, q = 450 STB/D), and the response at Well A8 was observed (open circles in Fig. 9.23). Reservoir data include B = 1.0 RB/STB and r = 1,320 ft (40-acre spacing).

We analyze the second peak (*third pulse response*) of Fig. 9.23 for illustration. From each set of data on the figure,

$$\frac{\Delta p}{q} = 7.6 \times 10^{-4} \text{ psi/(STB/D)},$$

and

$$t_L = 0.15 \text{ hours}.$$

For these tests, Δt_C = 2 hours and Δt_P = 1 hour. Thus,

$$t_L/\Delta t_C = \frac{0.15}{2.00} = 0.075,$$

and

$$F' = 1.0/2.0 = 0.5.$$

There is not enough information to estimate permeability from Eq. 9.17, but the equation can be rearranged to compute kh/μ. Use Fig. 9.17 for all odd pulses after the first with F' = 0.5 and $t_L/\Delta t_C$ = 0.075 and read $\{\Delta p_D[t_L/\Delta t_C]^2\}_{\text{Fig.}}$ = 0.00128. Then,

$$\frac{kh}{\mu} = \frac{(141.2)(1.0)\{0.00128\}}{(7.6 \times 10^{-4})[0.075]^2} = 42{,}300 \text{ md ft/cp}.$$

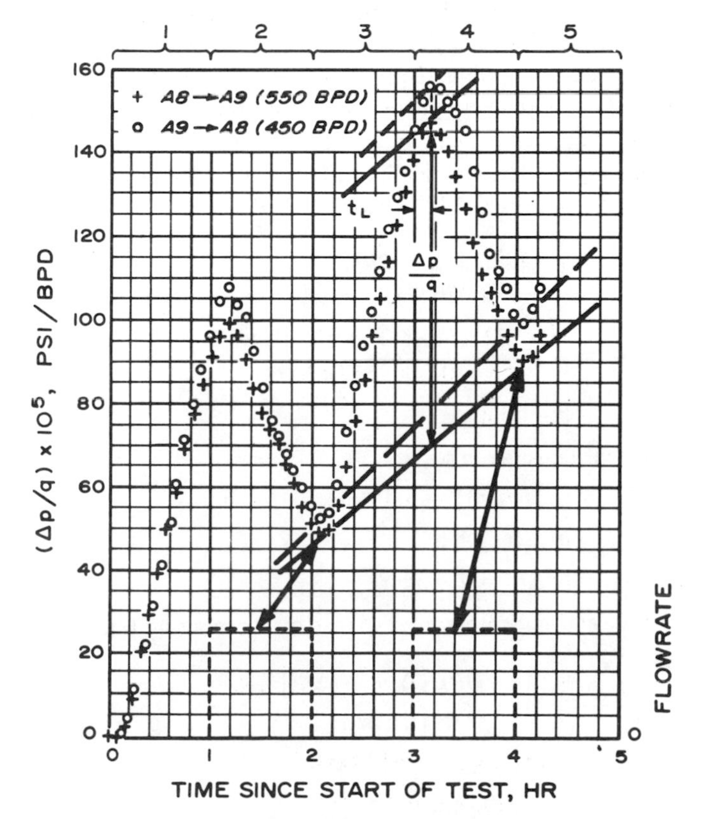

Fig. 9.23 Pulse-test data for a well pair tested in both directions, Example 9.4. After McKinley, Vela, and Carlton.[7]

*A paper that briefly discusses this subject appeared just before publication of the monograph — Jargon, J. R.: "Effect of Wellbore Storage and Wellbore Damage at the Active Well on Interference Test Analysis," *J. Pet. Tech.* (Aug. 1976) 851-858. That paper states that wellbore storage at the pulsing well reduces pulse amplitude and increases the time lag.

To estimate $\phi c_t h$ we use a rearranged form of Eq. 9.18 and $\{(t_L)_D/r_D^2\}_{\text{Fig.}} = 0.0876$ from Fig. 9.21 for all later odd pulses.

$$\phi c_t h = \frac{(0.0002637)(42{,}300)(0.15)}{(1{,}320)^2(0.0876)}$$

$$= 11.0 \times 10^{-6} \text{ ft/psi.}$$

Using their analysis technique, McKinley, Vela, and Carlton[7] estimated $kh/\mu = 40{,}000 \pm 2{,}000$ and $\phi c_t h = 11.0 \times 10^{-6} \pm 1.1 \times 10^{-6}$ for Well A8 pulsing and Well A9 responding. For Well A9 pulsing and Well A8 responding, they estimated $kh/\mu = 46{,}000 \pm 7{,}000$ and $\phi c_t h = 12.7 \times 10^{-6} \pm 1.3 \times 10^{-6}$.

For best results, pulse tests should be run with a good combination of pulse length, pulse amplitude, and pressure-measuring equipment. To accomplish this, we recommend designing pulse tests using estimated reservoir properties. Test design utilizes Figs. 9.15 through 9.22, as illustrated in the following example.

Example 9.5 Pulse-Test Design

We wish to design a pulse test for a reservoir with the following approximate properties:

$q = 100$ B/D $r = 660$ ft
$k = 200$ md $c_t = 10 \times 10^{-6}$ psi^{-1}
$\mu = 3$ cp $\phi = 0.18$
$h = 25$ ft $B = 1.1$ RB/STB.

The pulsing well is a producer, so to minimize shut-in time we choose a short shut-in pulse, such as $F' = 0.3$. For initial design calculations, we choose the *maximum* $\Delta p_D[t_L/\Delta t_C]^2$ points from Figs. 9.15 and 9.16 for the *first* odd and even pulses. For the first *even* pulse and $F' = 0.3$, $\{\Delta p_D[t_L/\Delta t_C]^2\}_{\text{Fig.}} \simeq 0.0042$ at $t_L/\Delta t_C \simeq 0.33$ from Fig. 9.16. Using Fig. 9.20 for that value of $t_L/\Delta t_C$, $\{(t_L)_D/r_D^2\}_{\text{Fig.}} = 0.122$. Rearranging Eq. 9.18,

$$t_L \simeq \frac{\{(t_L)_D/r_D^2\}_{\text{Fig.}}\, r^2 \phi\mu c_t}{(0.0002637)\,k}$$

$$\simeq \frac{\{0.122\}(660)^2(0.18)(3)(10 \times 10^{-6})}{(0.0002637)(200)}$$

$$\simeq 5.4 \text{ hours.}$$

Cycle time is given by

$$\Delta t_C = t_L/[t_L/\Delta t_C] = 5.4/0.33 = 16.4 \text{ hours,}$$

and the pulse length is

$$\Delta t_P = F'\Delta t_C = 4.9 \text{ hours.}$$

We estimate pressure response by rearranging Eq. 9.17:

$$\Delta p \simeq \frac{(141.2)qB\mu\{\Delta p_D[t_L/\Delta t_C]^2\}_{\text{Fig.}}}{kh[t_L/\Delta t_C]^2}$$

$$\simeq \frac{(141.2)(1.1)(3)\{0.0042\}}{(200)(25)[0.33]^2}\, q$$

$$\simeq 3.6 \times 10^{-3} q \text{ psi.}$$

Thus, $\Delta p = 3.6 \times 10^{-3}(100) = 0.36$ psi is the expected response amplitude for *even-pulse* analysis. We would shut in the well for 5 hours, produce for 11 hours, and so forth.

If we wish to analyze the first odd-pulse response, we design in a similar way:

$$\{\Delta p_D[t_L/\Delta t_C]^2\}_{\text{Fig.}} \simeq 0.0014$$

$$\text{at } t_L/\Delta t_C \simeq 0.27 \text{ from Fig. 9.15,}$$

and

$$\{(t_L)_D/r_D^2\}_{\text{Fig.}} = 0.074 \text{ from Fig. 9.19.}$$

Estimating as above, $t_L \simeq 3.3$ hours, $\Delta t_C \simeq 12.2$ hours, $\Delta t_P \simeq 3.7$ hours, and $\Delta p \simeq 0.12$ psi at a 100-B/D rate; we would shut in for 4 hours, produce for 8 hours, etc. We would require pressure-measuring instrumentation that will provide reliable pressure-change data when $\Delta p \simeq 0.1$ psi. We should analyze the even pulses, since they will show the greatest pressure response.

Alternatively, we could have chosen a convenient pulse length and calculated $\Delta p/q$. We could then use $\Delta p/q$ to choose pulse height (q) or the pressure-measuring instruments. Of course, for many wells, we may not be able to change the pulse magnitude (rate) significantly.

We could get a larger response amplitude with a longer test. For example, if we choose $\{\Delta p_D[t_L/\Delta t_C]^2\}_{\text{Fig. 9.16}} = 0.00156$ at $[t_L/\Delta t_C] = 0.07$, then $\{(t_L)_D/r_D^2\}_{\text{Fig. 9.20}} = 0.109$ and we get $t_L = 4.86$ hours, $\Delta t_C = 69.4$ hours, and $\Delta p = 2.96 \times 10^{-2} q$. Thus, by increasing cycle (and test) time by a factor of 4.23, we can increase the response amplitude by over eight times. However, a main advantage of pulse testing is the short test time, so it may be better to accept a smaller pressure response to keep test time short.

9.4 Heterogeneous and Anisotropic Reservoirs

Most multiple-well testing techniques are based on the assumptions that the reservoir system is isotropic and homogeneous in the influence region. If those restrictions are not satisfied, the analysis techniques presented in this chapter generally provide results that are some type of an average of the properties within the influence region (Fig. 9.3).[5] Vela and McKinley[5] propose a method for correcting values of kh/μ and $\phi c_t h$ determined from a *series* of pulse tests to reservoir average properties. This method should also apply to normal interference testing when sufficient data are available.

If enough observation wells are used, interference-test data sometimes can be analyzed by computer methods to give a description of the variation of reservoir properties with location. Jahns[6] describes a technique that applies to heterogeneous and anisotropic reservoirs. The technique presented by Earlougher and Kersch[10] applies to anisotropic but otherwise homogeneous, infinite-acting reservoirs. Other techniques also have been suggested.[8,9]

Pierce, Vela, and Koonce[24] describe a method for estimating the orientation and length of hydraulic fractures by pulse testing. Although that method requires computer simulation, it is potentially useful for estimating the existence and direction of permeability anisotropies. The method is qualitative unless the computer simulation is used, so it is not discussed further here.

Based on work by Papadopulos,[25] Ramey[11] presents a method for estimating anisotropic reservoir properties from interference data. Although more complex than single-well interference analyses, the method does not require computer assistance. Fig. 9.24 defines the necessary nomenclature. The major permeability axis, k_{max}, is rotated from the axis used for measuring well coordinates by the angle θ. The minor permeability axis, k_{min}, is oriented at 90° to the major permeability axis. The active well is located at the origin of the coordinate system and the observation wells are each located at coordinates indicated as (x,y). The anisotropic analysis requires pressure data from at least three observation wells located on different rays extending from the active well. It assumes that the active-well/observation-well system is infinite-acting and homogeneous (with the exception of having anisotropic permeability).

Ramey shows that the pressure at an observation well is

$$p(t,x,y) = p_i - \frac{141.2\, qB\mu}{\sqrt{k_{max}k_{min}}\, h}\, p_D\left(\left[t_D/r_D^2\right]_{dir}\right) \quad (9.20)$$

Eq. 9.20 has the familiar form of Eq. 2.2, except that the permeability has been replaced by $\sqrt{k_{max}k_{min}}$ and the definition

$$\left(\frac{t_D}{r_D^2}\right)_{dir} = \frac{0.0002637t}{\phi\mu c_t}\left[\frac{k_{max}k_{min}}{(k_x y^2 + k_y x^2 - 2k_{xy}xy)}\right], \quad (9.21)$$

is used. In Eq. 9.21, k_x, k_y, and k_{xy} are components of the symmetrical permeability tensor aligned with the coordinate system.[11] To estimate k_{max}, k_{min}, and θ, it is necessary to first estimate k_x, k_y, and k_{xy} as described below.

Type-curve matching is the first step of the analysis technique. Observed pressure data from at least three wells are plotted and matched to the type curve of Fig. C.2, as described in Section 9.2. Each of the three data sets must be matched so the *pressure match point* $[\Delta p_M,(p_D)_M]$ *is the same for all three observation-well responses*. The time match point $[t_M,(t_D/r_D^2)_M]$ will be different for each set of observation-well data. By making the match this way, and by obtaining the best match for all three data sets, the directional permeability characteristics and ϕc_t may be estimated.

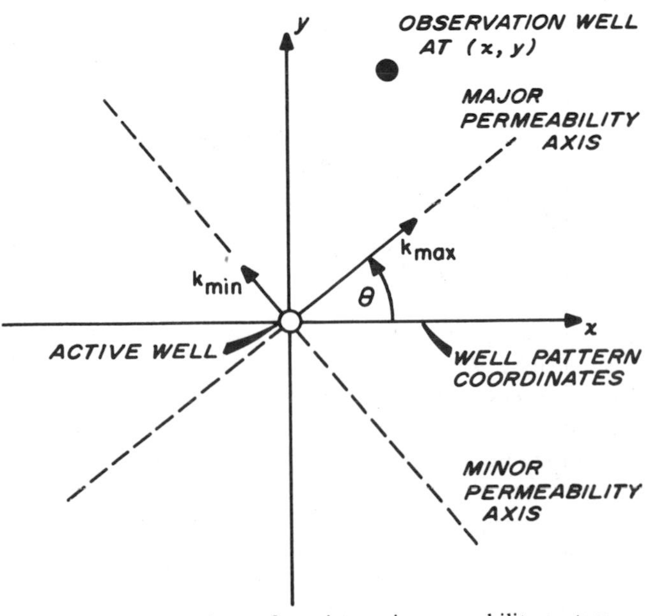

Fig. 9.24 Nomenclature for anisotropic permeability system. After Ramey.[11]

Once observed pressure responses are matched, average system permeability is estimated from

$$\bar{k} = \sqrt{k_{max}k_{min}} = \frac{141.2\, qB\mu(p_D)_M}{h\Delta p_M}, \quad (9.22)$$

where the pressure match points are the same for all pressure observation responses. Each of the three time match points, $[t_M,(t_D/r_D^2)_M]$, is used with a rearranged version of Eq. 9.21:

$$y^2k_x + x^2k_y - 2xy\,k_{xy} = \frac{(0.0002637)\,k_{max}k_{min}}{\phi\mu c_t}\,\frac{t_M}{(t_D/r_D^2)_M}. \quad (9.23)$$

We write Eq. 9.23 three times, once for each observation-well match. That gives three equations in four unknowns, k_x, k_y, k_{xy}, and $\phi\mu c_t$. They may be solved simultaneously to obtain k_x, k_y, and k_{xy}, each in terms of the unknown $\phi\mu c_t$. Then k_x, k_y, and k_{xy} (in terms of $\phi\mu c_t$) are substituted into

$$k_xk_y - k_{xy}^2 = k_{max}k_{min} = \bar{k}^2. \quad (9.24)$$

Since the right side of Eq. 9.24 is known (from Eq. 9.22), it can be solved to estimate $\phi\mu c_t$. Then we estimate k_x, k_y, and k_{xy} from their relationships to $\phi\mu c_t$. To complete the analysis and estimate the maximum and minimum directional permeabilities and the angle of orientation, we use

$$k_{max} = 0.5\left\{(k_x + k_y) + \left[(k_x - k_y)^2 + 4k_{xy}^2\right]^{1/2}\right\}, \quad (9.25)$$

$$k_{min} = 0.5\left\{(k_x + k_y) - \left[(k_x - k_y)^2 + 4k_{xy}^2\right]^{1/2}\right\}, \quad (9.26)$$

and

$$\theta = \arctan\left(\frac{k_{max} - k_x}{k_{xy}}\right) \quad (9.27)$$

Example 9.6 Interference-Test Analysis in an Anisotropic Reservoir

Ramey[11] gives data for an interference test in a nine-spot pattern at the end of a waterflood. Before testing, all wells were shut in. The test was run by injecting at −115 STB/D and observing the fluid levels in eight of the shut-in production wells, both during injection and during the subsequent falloff period. Ramey[11] tabulates complete pressure data during injection and falloff for all eight observation wells and falloff data for the injection well. Only a portion of the data is used here to illustrate the use of Eqs. 9.22 through 9.27.

Fig. 9.25 shows the well locations. Table 9.4 gives observed pressure changes during the injection period. Other data are

TABLE 9.4—PRESSURE DATA FOR INTERFERENCE TEST OF EXAMPLE 9.6.
After Ramey.[11]

Well 1-D		Well 5-E		Well 1-E	
t (hours)	Δp (psi)	t (hours)	Δp (psi)	t (hours)	Δp (psi)
23.5	−6.7	21	−4	27.5	−3
28.5	−7.2	47	−11	47	−5
51	−15	72	−16.3	72	−11
77	−20	94	−21.2	95	−13
95	−25	115	−22	115	−16
		122	−25		

$q = -115$ STB/D $\quad B = 1.0$ RB/STB
$h = 25$ ft $\quad \phi = 0.20$
$\mu = 1.0$ cp $\quad p_i = 240$ psi.

Fig. 9.26 shows the match of the data in Table 9.4 to the type curve of Fig. C.2. The match was made so the pressure match point $[p_M,(p_D)_M]$ is the same for all three responses, while the time match points vary.

Match-point data are $p_M = -10$ psi and $(p_D)_M = 0.26$.

	Well 1-D	Well 5-E	Well 1-E
t_M, hours	72	92	150
$(t_D/r_D^2)_M$	1.0	1.0	1.0

Average permeability is estimated from Eq. 9.22:

$$\bar{k} = \sqrt{k_{max}k_{min}} = \frac{(141.2)(-115)(1)(1)}{25}\,\frac{(0.26)}{(-10)}$$

$$= 16.89 \text{ md.}$$

$$k_{max}k_{min} = (16.89)^2 = 285.3.$$

We now write Eq. 9.23 for *each* time match point:

For Well 1-D,

$$(475)^2 k_x + (0)^2 k_y - 2(0)(475)\,k_{xy} = \frac{(0.0002637)(285.3)}{\phi\mu c_t}\,\frac{(72)}{(1.0)}.$$

For Well 5-E,

$$(0)^2 k_x + (475)^2 k_y - 2(475)(0)\,k_{xy} = \frac{(0.0002637)(285.3)}{\phi\mu c_t}\,\frac{(92)}{(1.0)}.$$

For Well 1-E,

$$(514)^2 k_x + (475)^2 k_y - 2(475)(514)\,k_{xy} = \frac{(0.0002637)(285.3)}{\phi\mu c_t}\,\frac{(150)}{(1.0)}.$$

Simplifying and normalizing, these equations become

$$k_x = \frac{2.401\times10^{-5}}{\phi\mu c_t} \quad \text{. . . (A)}$$

$$k_y = \frac{3.068\times10^{-5}}{\phi\mu c_t} \quad \text{. . . (B)}$$

$$0.5411k_x + 0.4621\,k_y - k_{xy} = \frac{2.311\times10^{-5}}{\phi\mu c_t} \quad \text{. . . (C)}$$

Combining Eqs. A, B, and C gives

$$k_{xy} = (0.5411)\,\frac{2.401\times10^{-5}}{\phi\mu c_t} + (0.4621)\,\frac{3.068\times10^{-5}}{\phi\mu c_t} - \frac{2.311\times10^{-5}}{\phi\mu c_t}$$

$$= \frac{4.059\times10^{-6}}{\phi\mu c_t} \quad \text{. . . (D)}$$

Using Eqs. A, B, and D in Eq. 9.24 results in

$$\frac{(2.401\times10^{-5})(3.068\times10^{-5})}{(\phi\mu c_t)(\phi\mu c_t)} - \left(\frac{4.059\times10^{-6}}{\phi\mu c_t}\right)^2$$

$$= 285.3, \quad \text{. . . (E)}$$

so

$$\phi\mu c_t = \sqrt{\frac{(2.401\times10^{-5})(3.068\times10^{-5}) - (4.059\times10^{-6})^2}{285.3}}$$

$$= 1.589\times10^{-6} \text{ cp/psi.}$$

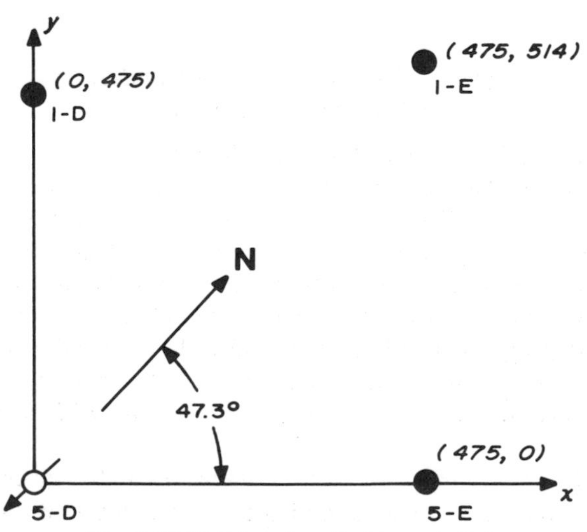

Fig. 9.25 Well locations for Example 9.6.

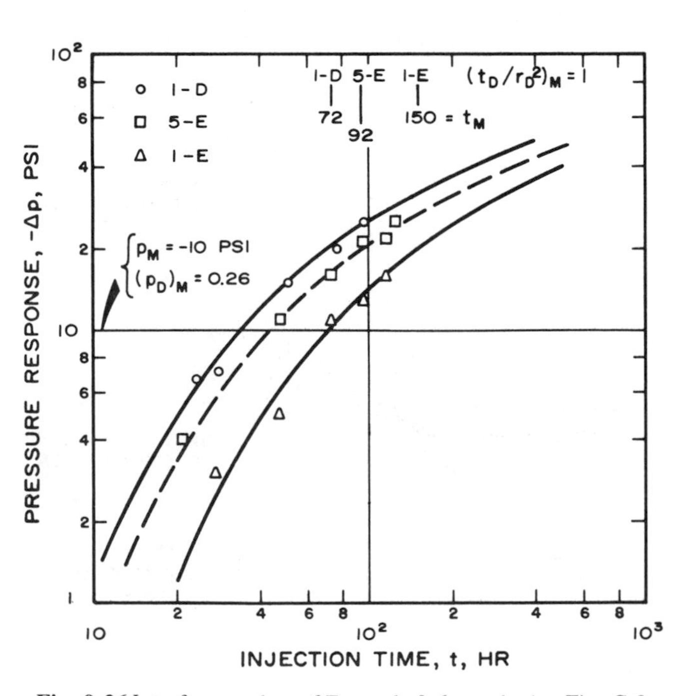

Fig. 9.26 Interference data of Example 9.6 matched to Fig. C.2. Pressure match is the same for all curves.

Thus,

$$c_t = \frac{1.589 \times 10^{-6}}{(0.20)(1)} = 7.95 \times 10^{-6}\ \text{psi}^{-1}.$$

Now, Eqs. A, B, and D are solved using the computed $\phi\mu c_t$:

$$k_x = \frac{2.401 \times 10^{-5}}{1.589 \times 10^{-6}} = 15.11\ \text{md},$$

$$k_y = \frac{3.068 \times 10^{-5}}{1.589 \times 10^{-6}} = 19.31\ \text{md},$$

$$k_{xy} = \frac{4.059 \times 10^{-6}}{1.589 \times 10^{-6}} = 2.55\ \text{md}.$$

We use Eq. 9.25 to estimate the major permeability value,

$$k_{max} = 0.5\left[(15.11 + 19.31) + \sqrt{(15.11 - 19.31)^2 + 4(2.55)^2}\right]$$

$$- 20.5\ \text{md},$$

and Eq. 9.26 to estimate the minor permeability value,

$$k_{min} = 0.5\left[(15.11 + 19.31) - \sqrt{(15.11 - 19.31)^2 + 4(2.55)^2}\right]$$

$$= 13.9\ \text{md}.$$

We know $\sqrt{k_{max}k_{min}} = 16.89$ from Eq. 9.22, so we can check the computations:

$$\sqrt{k_{max}k_{min}} = \sqrt{(20.5)(13.9)}$$

$$= 16.88$$

Finally, we estimate the direction of k_{max} from Eq. 9.27:

$$\theta = \arctan\left(\frac{20.5 - 15.11}{2.55}\right)$$

$$= 64.7° \text{ from the } x \text{ axis}.$$

Correcting for the orientation of the axes, the maximum permeability direction is

$$\theta = 64.7 - 47.3 = \text{N}17.4°\text{W}.$$

Example 9.6 indicates that although estimating anisotropic reservoir parameters may be a lengthy procedure, it is not particularly difficult, and does not necessarily require a computer. The most difficult part of the analysis is the simultaneous solution of the equations indicated as Eqs. A, B, and C in Example 9.6. In that example, solution was simple since two of the equations solved themselves. That is often the case, if the coordinate system is chosen judiciously. As indicated in the example, it is not necessarily easiest to choose the coordinate system so it is aligned with north and south.

Swift and Brown[26] propose a computationally different approach to estimating directional properties that is still based on the exponential integral. When applied in a manner similar to that described above, it will give the same results as Ramey's[11] technique, although solution of the simultaneous equations may be more difficult. Swift and Brown[26] suggest that, if enough data are available, a regression approach could be used to estimate k_{max}, k_{min}, and θ. They also state that, if the reservoir is heterogeneous, permeability values estimated by conventional interference-test analysis for homogeneous, anisotropic reservoirs (for example, by Eq. 9.2) may imply directional properties directly contrary to the true situation.

References

1. Matthews, C. S. and Russell, D. G.: *Pressure Buildup and Flow Tests in Wells,* Society of Petroleum Engineers of AIME, Dallas (1967) **1,** Chap. 7.
2. Driscoll, Vance J.: "Use of Well Interference and Build-Up Data for Early Quantitative Determination of Reserves, Permeability and Water Influx," *J. Pet. Tech.* (Oct. 1963) 1127-1136; *Trans.,* AIME, **228.**
3. Matthies, E. Peter: "Practical Application of Interference Tests," *J. Pet. Tech.* (March 1964) 249-252. Also *Reprint Series, No. 9 — Pressure Analysis Methods,* Society of Petroleum Engineers of AIME, Dallas (1967) 145-148.
4. Warren, J. E. and Hartsock, J. H.: "Well Interference," *Trans.,* AIME (1960) **219,** 89-91. Also *Reprint Series, No. 9 — Pressure Analysis Methods,* Society of Petroleum Engineers of AIME, Dallas (1967) 93-95.
5. Vela, Saul and McKinley, R. M.: "How Areal Heterogeneities Affect Pulse-Test Results," *Soc. Pet. Eng. J.* (June 1970) 181-191; *Trans.,* AIME, **249.**
6. Jahns, Hans O.: "A Rapid Method for Obtaining a Two-Dimensional Reservoir Description From Well Pressure Response Data," *Soc. Pet. Eng. J.* (Dec. 1966) 315-327; *Trans.,* AIME, **237.**
7. McKinley, R. M., Vela, Saul, and Carlton, L. A.: "A Field Application of Pulse-Testing for Detailed Reservoir Description," *J. Pet. Tech.* (March 1968) 313-321; *Trans.,* AIME, **243.**
8. Woods, E. G.: "Pulse-Test Response of a Two-Zone Reservoir," *Soc. Pet. Eng. J.* (Sept. 1970) 245-256; *Trans.,* AIME, **249.**
9. Elkins, Lincoln F. and Skov, Arlie M.: "Determination of Fracture Orientation From Pressure Interference," *Trans.,* AIME (1960) **219,** 301-304. Also *Reprint Series, No. 9 — Pressure Analysis Methods,* Society of Petroleum Engineers of AIME, Dallas (1967) 97-100.
10. Earlougher, Robert C., Jr., and Kersch, Keith M.: "Field Examples of Automatic Transient Test Analysis," *J. Pet. Tech.* (Oct. 1972) 1271-1277.
11. Ramey, Henry J., Jr.: "Interference Analysis for Anisotropic Formations — A Case History," *J. Pet. Tech.* (Oct. 1975) 1290-1298; *Trans.,* AIME, **259.**
12. Gringarten, A. C. and Witherspoon, P. A.: "A Method of Analyzing Pump Test Data From Fractured Aquifers," *Proc.,* Symposium on Percolation Through Fissured Rock, International Society for Rock Mechanics, Stuttgart (Sept. 18-19, 1972).
13. Earlougher, R. C., Jr., and Ramey, H. J., Jr.: "Interference Analysis in Bounded Systems," *J. Cdn. Pet. Tech.* (Oct.-Dec. 1973) 33-45.
14. Johnson, C. R., Greenkorn, R. A., and Woods, E. G.: "Pulse-Testing: A New Method for Describing Reservoir Flow Properties Between Wells," *J. Pet. Tech.* (Dec. 1966) 1599-1604; *Trans.,* AIME, **237.**

15. Greenkorn, R. A. and Johnson, C. R.: ''Method for Defining Reservoir Heterogeneities,'' U.S. Patent No. 3,285,064 (Nov. 15, 1966).

16. Johnson, C. R.: ''Portable 'Radar' for Testing Reservoirs Developed by Esso Production Research,'' *Oil and Gas J.* (Nov. 20, 1967) 162-164.

17. Johnson, C. R. and Raynor, R.: ''System for Measuring Low Level Pressure Differential,'' U.S. Patent No. 3,247,712 (April 26, 1966).

18. Miller, G. B., Seeds, R. W. S., and Shira, H. W.: ''A New, Surface Recording, Down-Hole Pressure Gauge,'' paper SPE 4125 presented at the SPE-AIME 47th Annual Fall Meeting, San Antonio, Tex., Oct. 8-11, 1972.

19. Culham, W. E.: ''Amplification of Pulse-Testing Theory,'' *J. Pet. Tech.* (Oct. 1969) 1245-1247.

20. Startzman, R. A.: ''A Further Note on Pulse-Test Interpretation,'' *J. Pet. Tech.* (Sept. 1971) 1143-1144.

21. Brigham, W. E.: ''Planning and Analysis of Pulse-Tests,'' *J. Pet. Tech.* (May 1970) 618-624; *Trans.*, AIME, **249.**

22. Kamal, Medhat and Brigham, William E.: ''Pulse-Testing Response for Unequal Pulse and Shut-In Periods,'' *Soc. Pet. Eng. J.* (Oct. 1975) 399-410; *Trans.*, AIME, **259.**

23. Prats, M. and Scott, J. B.: ''Effect of Wellbore Storage on Pulse-Test Pressure Response,'' *J. Pet. Tech.* (June 1975) 707-709.

24. Pierce, A. E., Vela, Saul, and Koonce, K. T.: ''Determination of the Compass Orientation and Length of Hydraulic Fractures by Pulse Testing,'' *J. Pet. Tech.* (Dec. 1975) 1433-1438.

25. Papadopulos, Istavros S.: ''Nonsteady Flow to a Well in an Infinite Anisotropic Aquifer,'' *Proc.*, 1965 Dubrovnik Symposium on Hydrology of Fractured Rocks, Int'l. Assoc. of Sci. Hydrology (1965) **I,** 21-31.

26. Swift, S. C. and Brown, L. P.: ''Interference Testing for Reservoir Definition — The State of the Art,'' paper SPE 5809 presented at the SPE-AIME Fourth Symposium on Improved Oil Recovery, Tulsa, March 22-24, 1976.

Chapter 10

Effect of Reservoir Heterogeneities on Pressure Behavior

10.1 Introduction

Most material elsewhere in this monograph assumes homogeneous and isotropic reservoirs. This chapter discusses the effects of some common reservoir heterogeneities on pressure transient behavior. Heterogeneities — variations in rock and fluid properties from one location to another — may result from deposition, folding and faulting, post-depositional changes in reservoir lithology, and changes in fluid type or properties.[1] Heterogeneities may be small-scale, as in carbonate reservoirs where the rock has two constituents, matrix and fractures, vugs, or solution cavities. They also may be large-scale, such as physical barriers, faults, fluid-fluid contacts, thickness changes, lithology changes, several layers with different properties in each layer, etc.

A related characteristic is permeability anisotropy. An anisotropic reservoir has permeability that varies with flow direction. Anisotropy can be caused by sedimentary processes (channel fill deposits, for example) or by tectonics (parallel fracture orientations). Anisotropy can occur in both homogeneous and heterogeneous reservoirs. As a result, anisotropy does not necessarily imply heterogeneity in the sense defined in the previous paragraph. Most reservoir rocks have a lower vertical permeability than horizontal permeability, and so are anisotropic in that regard.

All the above heterogeneities usually exist in the virgin state. In addition, man may induce heterogeneities in the reservoir. Man-made heterogeneities include changes near the wellbore from mud invasion during drilling, hydraulic fracturing, acidizing, or fluid injection. Chapter 11 treats man-induced changes such as hydraulic fracturing.

Since heterogeneities exist in varying degrees in many reservoirs, it is important to know how much of the information presented elsewhere in this monograph is usable in heterogeneous systems. Fortunately, many of the single-well transient testing techniques for homogeneous reservoirs may be applied to heterogeneous systems with useful results. Heterogeneities affect multiple-well tests more severely.

Jahns[2] and Vela and McKinley[3] propose techniques for estimating heterogeneous reservoir properties from a series of interference and pulse tests. Such techniques generally require a computer and some type of regression analysis. Even when using the computer and regression techniques, it is difficult (if not impossible) to delineate specific heterogeneities from well tests. The difficulty occurs because *many different conditions can cause the same or similar well-test response*. If we have an idea of the type of heterogeneity, it may be possible to determine some of the properties involved by pressure transient testing. However, it is dangerous — and poor engineering practice — to infer heterogeneous reservoir properties based *solely* on transient testing. Geological, seismic, fluid flow, and performance data should be considered before hypotheses are formed about the type and location of heterogeneities. It may be possible to design a specific transient test to investigate the possibility of a particular type of heterogeneity. For instance, if each layer in a layered reservoir is tested with a straddle packer arrangement, permeability, skin factor, and average pressure of that layer may be estimated from the test data. However, if layers are not isolated in the wellbore during such testing, meaningful individual-layer data cannot be obtained with current technology.

Fig. 10.1 illustrates one of many difficulties that arise in attempting to infer heterogeneities from transient data. The reciprocity principle[4] (Section 9.2) indicates that both situations shown in Fig. 10.1 will have *precisely the same interference response*. Thus, interference testing the two wells cannot indicate which of the two situations in Fig. 10.1 exists. The difficulty might be resolved by interference testing with other wells or by using other data. The importance of Fig. 10.1, however, is the indication that even very simple conditions cannot necessarily be resolved by transient testing.

In this chapter, we discuss permeability anisotropy, classify reservoir heterogeneities, and describe how these heterogeneities affect transient testing. We show that several types of heterogeneities can cause similar transient-test pressure response. We provide some analysis techniques, but again, we caution that results should be supported by other data. The main goal here is to illustrate a variety of situations — a comprehensive treatment would require a monograph itself.

10.2 Linear Discontinuities — Faults and Barriers

Linear discontinuities, particularly single sealing faults, have been a popular topic in the transient-testing literature.[1,5-9] Fig. 10.2 shows a single well, Well A, near a linear sealing fault in an otherwise infinite-acting reservoir. Horner[5] considers pressure buildup and Russell[7] discusses two-rate flow testing in that system. Regardless of test type, the linear flow barrier affects the test in about the same way. To obtain the effect of the linear fault, we use the method of images[1,5] (see Appendix B) and add an image well, Well A′, as shown in Fig. 10.2. Then the pressure drop anywhere on the left side of the fault is the sum of the pressure drops caused by Wells A and A′. The following illustrates the process for drawdown testing.

If Well A, near a linear fault, is put on production at $t = 0$, the pressure change at the well is estimated by using the superposition principle (Section 2.9) and the two wells shown in Fig. 10.2. Thus, the pressure drop at Well A is the Δp at Well A caused by production at Well A plus the Δp at Well A caused by production at Well A′:

$$\Delta p = p_i - p_{wf}$$

$$= \Delta p_{\mathrm{A,A}} + \Delta p_{\mathrm{A,A'}}.$$

Using Eq. 2.2,

$$\Delta p = \frac{141.2}{kh}\, qB\mu \left\{\left[p_D(t_D, r_D = 1) + s\right] + p_D(t_D, 2L/r_w)\right\}, \quad \ldots\ldots (10.1)$$

since q is the same at both wells. For infinite-acting systems, the dimensionless pressure is given by the exponential-integral solution, Eq. 2.5a. At relatively short times, the log approximation applies *at Well A* and p_D is given by Eq. 2.5b. Furthermore, at short times t_D/r_D^2 for the image well is small, so the dimensionless-pressure contribution from the image well is essentially zero (see Fig. C.2). As a result, at short times the flowing bottom-hole pressure at Well A is given by Eq. 3.5, the familiar drawdown relationship:[1,5]

$$p_{wf} = m \log t + p_{1\mathrm{hr}}\, . \quad \ldots\ldots (10.2)$$

Eq. 10.2 indicates that a plot of short-time drawdown bottom-hole pressure vs log t should have a straight-line portion with slope m given by Eq. 3.6,

$$m = \frac{-162.6\, qB\mu}{kh}, \quad \ldots\ldots (10.3)$$

and intercept $p_{1\mathrm{hr}}$ given by Eq. 3.7. Thus, for a well near a linear fault, drawdown (buildup, two-rate, etc.) testing can be used to estimate reservoir permeability and skin factor in the usual fashion, as long as wellbore storage effects do not mask the initial straight-line section. If the well is very close to the fault, the initial straight line may end so quickly that it is masked by wellbore storage.

As the drawdown proceeds, the dimensionless-pressure contribution from Image Well A′ becomes significant, and the pressure at the producing well falls below the initial semilog straight line. After a long-enough production time, p_D for the image well is given by Eq. 2.5b, so[9]

$$p_{wf} = 2(m \log t + p_{1\mathrm{hr}}) + \left\{p_i + m\left[0.86859\, s + \log\left(\frac{4L^2}{r_w^2}\right)\right]\right\}. \quad \ldots\ldots (10.4)$$

Eq. 10.4 indicates that the p_{wf} vs log t plot will have a second straight-line portion with a slope double that of the initial straight line. The double slope also occurs in two-rate testing, pressure buildup testing, injectivity testing, and pressure falloff testing.[1,5,7] Average drainage-region pressure is estimated in the manner normally used for buildup testing using the *second* straight line.

The simple occurrence of a doubling of slope in a transient test *does not guarantee* the existence of a linear boundary

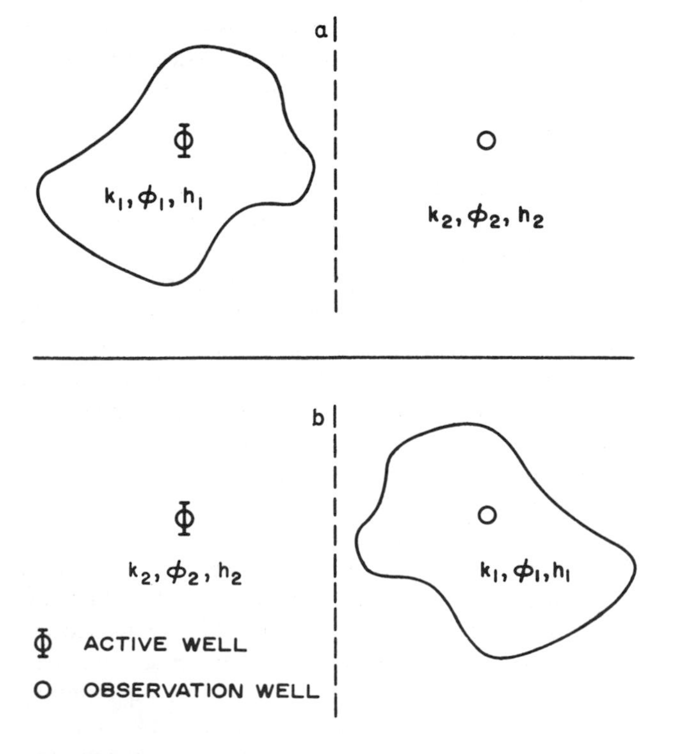

Fig. 10.1 Nonuniqueness because of reciprocity. Mirror-image models give identical multiple-well test responses. After Jahns.[2]

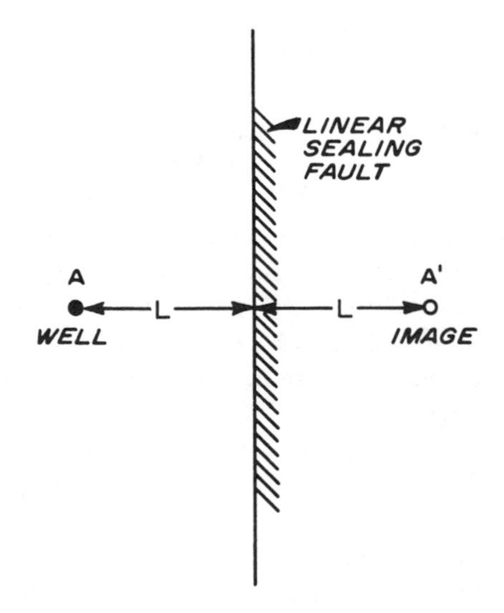

Fig. 10.2 Linear sealing fault near a producing well.

near the well. Pressure data taken during wellbore storage domination can cause two apparent semilog straight lines with a slope increase (for examples, see Figs. 5.2 and 7.4). In such cases, the apparent semilog straight lines are caused completely by wellbore effects and have nothing to do with reservoir characteristics. Thus, it is important to construct the log-log plot of transient test data to determine when wellbore storage effects are no longer important. That is particularly true when a slope increase is *expected* from a transient test.

To estimate distance to a linear discontinuity, we use the intersection time, t_x, of the two straight-line segments of the drawdown curve:[9]

$$L = 0.01217\sqrt{\frac{kt_x}{\phi\mu c_t}} \quad \text{. . . . . (10.5)}$$

Eq. 10.5 applies for drawdown testing. Russell[7] gives an equation for two-rate testing. For pressure buildup testing, the intersection point of the two straight lines is related to the dimensionless pressure at the intersection time by

$$p_D\left[t_D/(2L/r_w)^2\right] = \frac{1}{2}\ \ln\left(\frac{t_p + \Delta t}{\Delta t}\right)_x \quad \text{. . . . . (10.6)}$$

Thus, to estimate the distance to a linear fault from a pressure buildup test, we find $\left[(t_p + \Delta t)/\Delta t\right]$ when the semilog straight lines intersect and calculate p_D from Eq. 10.6. Then we enter Fig. C.2 with that value of p_D and determine $\left[t_D/(2L/r_w)^2\right]_{\text{Fig. C.2}}$. Finally,

$$L = \sqrt{\frac{0.0002637\, kt_p}{4\phi\mu c_t\left[t_D/(2L/r_w)^2\right]_{\text{Fig. C.2}}}} \quad \text{. . . . . . (10.7)}$$

When $t_p >> \Delta t$, a useful estimate of L may be made from a pressure buildup test by using Eq. 10.5 with Δt_x in place of t_x.

Example 10.1 Distance to a Fault From a Pressure Buildup Test

Pressure buildup data are shown in Figs. 10.3 and 10.4. The log-log plot, Fig. 10.3, indicates that wellbore storage effects are not important, so the increase in slope in Fig. 10.4 is probably caused by reservoir heterogeneity. The ratio of the two slopes is 1.79. Since the absolute value of the slopes is *increasing* with shut-in time, and since the slope ratio is about 2, a linear fault is suspected. Seismic data have verified the existence of a fault near this well.

Test data are

$$q = 10{,}250 \text{ STB/D} \qquad B = 1.55 \text{ RB/STB}$$
$$t_p = 530 \text{ hours} \qquad \phi = 0.09$$
$$p_{wf}(\Delta t = 0) = 3{,}666 \text{ psig} \qquad c_t = 22.6 \times 10^{-6} \text{ psi}^{-1}$$
$$h = 524 \text{ ft} \qquad r_w = 0.354 \text{ ft.}$$
$$\mu = 0.20 \text{ cp}$$

Formation permeability is estimated from the first straight line using Eq. 5.6. Recall that for a Horner plot the slope is $-m$, so m = 24.3 psig/cycle.

$$k = \frac{(162.6)(10{,}250)(1.55)(0.20)}{(524)(24.3)}$$
$$= 40.6 \text{ md.}$$

To estimate the distance to the fault, we determine $\left[(t_p + \Delta t)/\Delta t\right]_x$ = 285 from Fig. 10.4. Then using Eq. 10.6, $p_D\left[t_D/(2L/r_w)^2\right]$ = (1/2) ln(285) = 2.83. Referring to Fig. C.2, we see that $\left[t_D/(2L/r_w)^2\right]_{\text{Fig. C.2}}$ = 135 when p_D = 2.83. Next, we use Eq. 10.7:

$$L = \sqrt{\frac{(0.0002637)(40.6)(530)}{(4)(0.09)(0.20)(22.6 \times 10^{-6})(135)}}$$
$$= 161 \text{ ft.}$$

Since $t_p >> \Delta t$, we could use Eq. 10.5 with Δt_x = 1.87 hours to estimate L = 166 ft — quite good agreement.

If we wish, we may use Eq. 5.7 and data from the first straight line to estimate s = 4.4.

Multiple faults near a well may cause several different transient-test characteristics. For example, two faults intersecting at a right angle near a well may cause the slope to double, then redouble, or may simply cause a fourfold slope increase, depending on well location. It is not safe, however, to assume that additional boundaries continue to double transient-test response slopes. For example, a single well producing from the center of a closed square has an increase

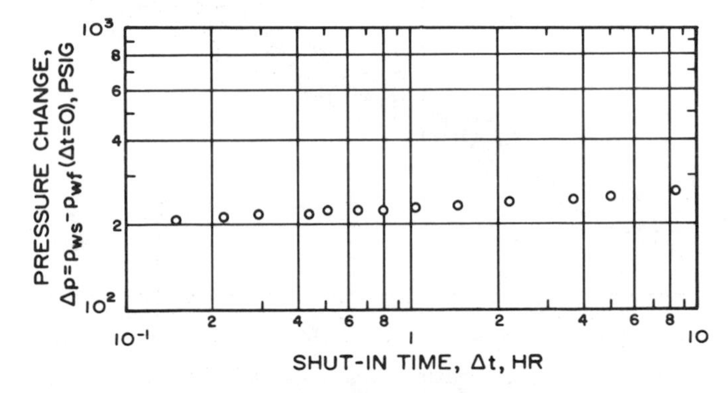

Fig. 10.3 Log-log data plot for Example 10.1.

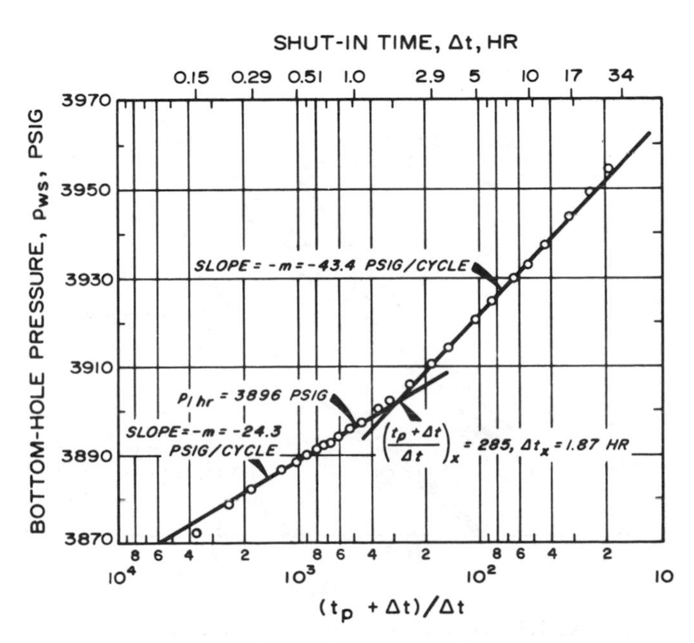

Fig. 10.4 Horner plot for pressure buildup data of Example 10.1.

in slope as the system reaches pseudosteady state; but during a pressure buildup test the slope decreases from the initial semilog straight line as the pressure approaches average reservoir pressure. Ramey and Earlougher[10] illustrate pressure buildup curves for several closed reservoir situations. Figs. 10.5 and 10.6 show such pressure buildup curves for closed-square and rectangular systems. Some of those buildup curves have characteristics commonly attributed to various kinds of heterogeneities. For example, Curve 3 in Fig. 10.5 has a peculiar bend that could be interpreted as the result of natural fracturing, as studied by Warren and Root[11] (Section 10.6). The curve shapes in Fig. 10.6 all show some upward bending that might be misinterpreted as indications of faulting, stratification, or some other reservoir heterogeneity. Yet, in all cases, the curve shapes are caused only by the shape of the closed system. In particular, there is no definite indication of multiple boundaries from any of the curves. Ramey and Earlougher[10] and

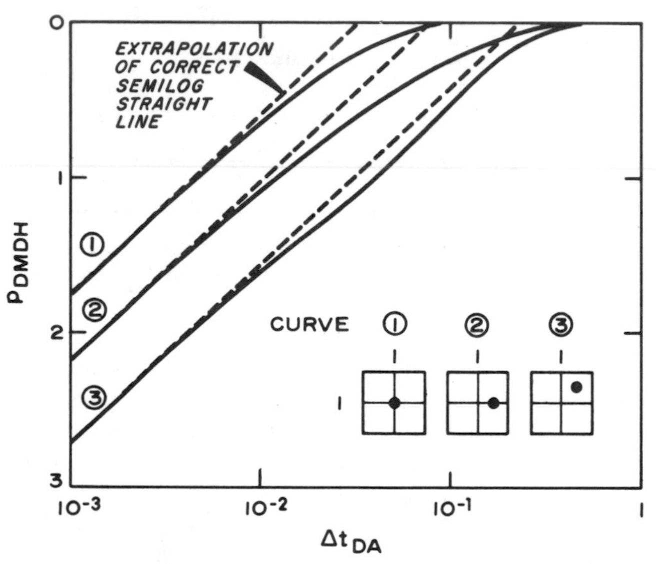

Fig. 10.5 Miller-Dyes-Hutchinson-type pressure buildup curves for square shapes. After Ramey and Earlougher.[10]

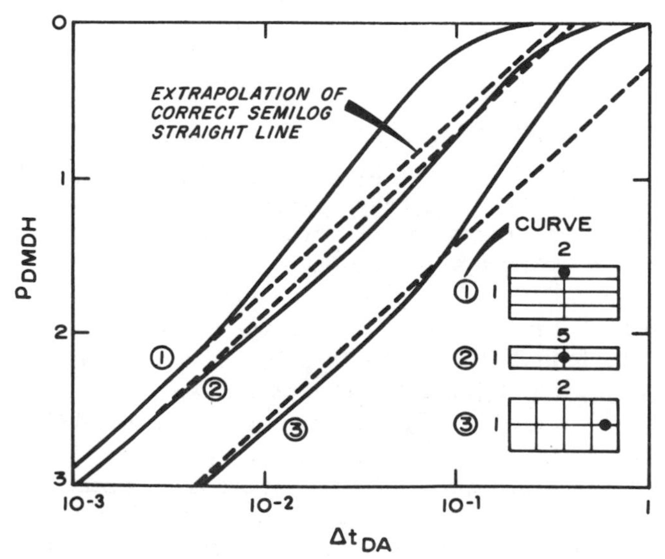

Fig. 10.6 Miller-Dyes-Hutchinson-type pressure buildup curves for rectangular shapes. After Ramey and Earlougher.[10]

Earlougher *et al.*[12] show several other pressure buildup curves with shapes that might be interpreted as reservoir heterogeneities. Generally, reservoir simulation must be used to estimate pressure transient response for complex, multiple-fault situations.

Fig. 10.7 illustrates three possible physical models for linear discontinuities in reservoirs. The upper case corresponds to the linear sealing fault already described. The lower two cases correspond to situations with a linear change in reservoir or fluid properties, but with flow still occurring. Bixel, Larkin, and van Poollen[8] discuss the third situation in detail for drawdown and less thoroughly for pressure buildup. They show that in both drawdown and buildup, the semilog straight-line slope may either *increase* or *decrease* depending on the contrast in properties between two reservoir regions. The slope ratio on the semilog plot equals the k/μ ratio *only* if ϕc_t does not vary appreciably across the discontinuity, and if the distance from the well to the reservoir boundary is much greater than the distance between the well and the discontinuity. (A similar situation is discussed in Section 7.5.) Bixel, Larkin, and van Poollen[8] suggest a curve-matching procedure to analyze for reservoir properties on both sides of the discontinuity and to estimate the distance to the discontinuity. Their procedure applies to drawdown testing if sufficient long-time, constant-rate data are available. Although the method could be applied to pressure buildup testing, Ref. 8 does not give the necessary curves.

10.3 Permeability Anisotropy

In some porous materials, permeability varies in different flow directions. In such materials, the permeability may be described mathematically by a symmetric tensor[13,14] or by maximum and minimum permeabilities oriented 90° apart, and by a direction for the maximum permeability. Those quantities are called the principal permeabilities and prin-

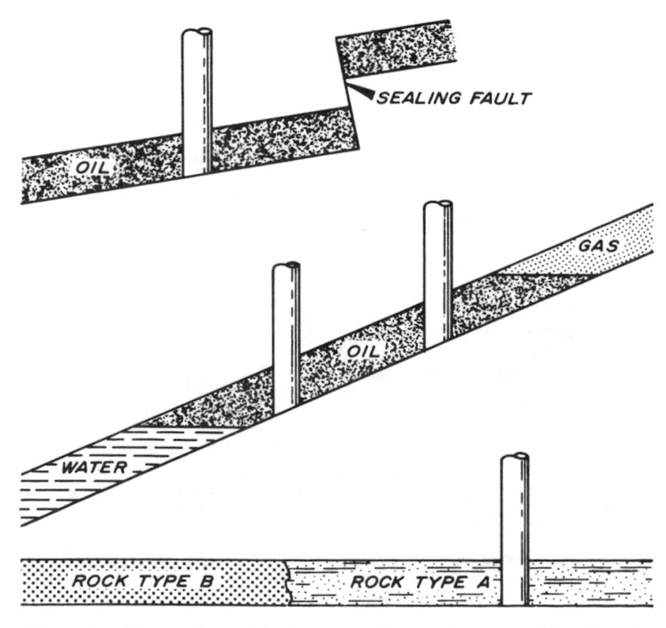

Fig. 10.7 Physical models for linear discontinuities. After Bixel, Larkin, and van Poollen.[8]

cipal direction. Refs. 13 and 14 give a complete treatment of permeability anisotropy. This section briefly describes fluid-flow equations for horizontal homogeneous anisotropic systems and shows the similarity of the equations to those for isotropic systems.

Consider a single well in an infinite, horizontal reservoir containing single-phase fluid of constant and small compressibility. The reservoir is homogeneous but anisotropic with k_{max} and k_{min} oriented in the x and y directions, respectively. (Note that we choose the coordinates for the flow equations to coincide with the principal permeability direction.) We transform the x-y coordinate system by changing the scale along each axis:

$$x' = x\sqrt{\bar{k}/k_{max}}\ , \quad \ldots\ldots \quad (10.8a)$$

$$y' = y\sqrt{\bar{k}/k_{min}}\ , \quad \ldots\ldots \quad (10.8b)$$

where

$$\bar{k} = \sqrt{k_{max}/k_{min}}\ . \quad \ldots\ldots \quad (10.9)$$

The diffusivity equation, Eq. 2.1, retains its normal form when expressed in x'-y' coordinates if $\bar{k}$ is used in place of k. Fig. 10.8 illustrates the transformation of Eq. 10.8. In particular, a circular region such as a wellbore is transformed into an elliptical region. The choice of k (Eq. 10.9) causes the area of the well to be the same in both the physical and transformed systems. The transformation allows dimensionless-pressure solutions for isotropic systems to be used for anisotropic systems, as long as the shape and well location of the *transformed system* matches that of the p_D chosen. For an infinite-acting anisotropic reservoir, the flowing wellbore pressure is given by

$$p_{wf} = p_i + m\left[\log t + \log\left(\frac{\bar{k}}{\phi\mu c_t r_w^2}\right) - 3.2275 + 0.86859\,(s - s'')\right]. \quad \ldots\ldots \quad (10.10)$$

The similarity of Eq. 10.10 to Eq. 3.4 for pressure drawdown in an isotropic system is important. In Eq. 10.10, m is the slope of a p_{wf} vs $\log t$ plot and is given by the anisotropic equivalent of Eq. 3.6:

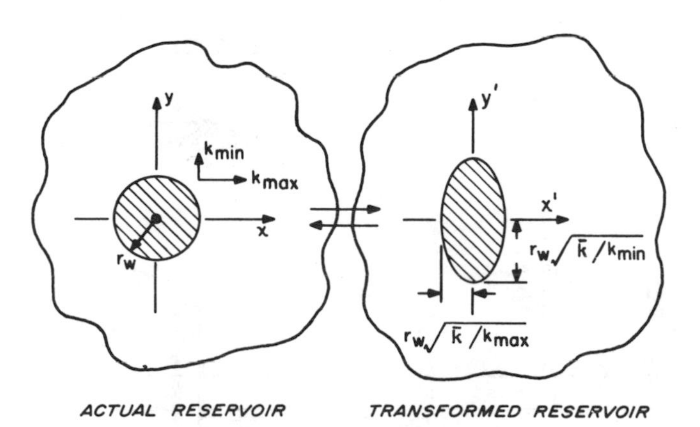

Fig. 10.8 Transformation of a homogeneous, anisotropic, infinite reservoir with a circular well to an equivalent homogeneous, isotropic reservoir.

$$m = -\frac{162.6\,qB\mu}{\bar{k}h}\ . \quad \ldots\ldots \quad (10.11)$$

The effective permeability may be estimated in the usual manner by solving Eq. 10.11 for $\bar{k}$ to obtain the analog of Eq. 3.9. The analogy applies to pressure buildup, injectivity, and falloff testing as well as to drawdown testing. It also carries over to multiple-rate testing, so all single-well testing techniques described elsewhere in this monograph may be used in anisotropic systems. The skin factor, estimated in the usual way, will be $(s - s'')$ in Eq. 10.10, where[11]

$$s'' = 2.303\log\left(\frac{\sqrt{k_{max}} + \sqrt{k_{min}}}{2\sqrt{\bar{k}}}\right)\ . \quad \ldots\ldots \quad (10.12)$$

The analogy between anisotropic-system and isotropic-system flow equations indicates the following for *single-well* transient testing in anisotropic reservoirs:

1. The permeability estimated by normal methods is $\bar{k}$ defined by Eq. 10.9.

2. The skin factor estimated by normal methods is $s - s''$. Eq. 10.12 shows that $s'' > 0$ always, so the estimated skin factor from a transient test in an anisotropic system will be somewhat low. Nonetheless, Eq. 10.12 does not provide a way to estimate k_{max}/k_{min} because, in general, there is no way to estimate s'' in single-well transient testing.

3. It is impossible to recognize that the system is anisotropic from a single-well test because the appearance of the pressure response curve is the same as for isotropic reservoirs.

Anisotropic reservoirs can be recognized by geological studies, oriented cores, reservoir performance (waterflood behavior, for example), etc. Correctly conducted multiple-well transient tests may be used to recognize and quantify anisotropic reservoir properties.[15-18] Section 9.4 describes a method for estimating anisotropic reservoir characteristics from interference testing. Refs. 2 and 16 describe other methods.

In most petroleum reservoirs, vertical permeability is significantly less than horizontal permeability. Methods for estimating vertical permeability are described in Section 10.8.

10.4 Composite Systems

In a composite system, fluid or rock properties vary in a steplike fashion in the radial direction away from a well (Fig. 10.9). (Transient pressure behavior of systems with linear discontinuities is discussed in Section 10.2.) Radial composite systems are generally man-induced. Examples are formation damage or improvement near the wellbore; the bubble formed when natural gas is injected into an aquifer; the burned zone in an in-situ combustion project; the water bank in a waterflood project (see Section 7.5); and the rubble and fracture zones in a nuclear stimulation project.

Both k/μ and ϕc_t may vary from one annular zone to another, as illustrated in Fig. 10.9. Analytical[19-23] and numerical[24-26] solutions have both been used to study transient behavior in composite systems. Refs. 19 through 26 indicate that the k/μ change in the radial direction is reflected as a slope change on the semilog plot, as shown in Fig. 7.14. The first straight line may be used to estimate k/μ for the inner region and the skin factor — if wellbore storage effects do not mask that straight line. Equations in Chapters 3 through 7 apply.

Because of the interplay between k/μ and ϕc_t in the two zones, the ratio of the late-time and early-time slopes, m_2/m_1, is not necessarily the same as the ratio of the mobilities $(k/\mu)_1/(k/\mu)_2$. Odeh[23] indicates that the ratios are the same for drawdown (and injection) only if ϕc_t is constant everywhere and if the outer radius or the outer reservoir boundary is more than 10 times the radius of the boundary between the two zones. Merrill, Kazemi, and Gogarty[26] provide similar results for falloff (or buildup). In some cases,[26] there may be no slope change at all.

The second straight-line portion of a falloff or buildup semilog plot may be used to estimate k/μ of the second zone, as indicated in Section 7.5. An alternative approach for pressure drawdown (injectivity) analysis is the curve-matching procedure proposed by Bixel and van Poollen.[24]

When there is an extreme variation in k/μ from Zone 1 to Zone 2, the pressure buildup (falloff) behavior may be similar to that shown by Carter,[22] Fig. 10.10. The figure is for a bounded, two-zone reservoir with the properties given in the caption. The semilog straight line for the first zone ends after about 0.4 day; then the pressure levels off before the slope increases toward that for the second zone. In the case shown in Fig. 10.10, the second straight line never completely develops before the pressure builds to average reservoir pressure. It is significant that the buildup-curve shape in Fig. 10.10 is commonly attributed to pressure buildup behavior in noncommunicating, layered reservoirs (see Section 10.5, Figs. 10.14 through 10.16). This illustrates the difficulty of analyzing transient pressure behavior in heterogeneous systems; many situations can cause

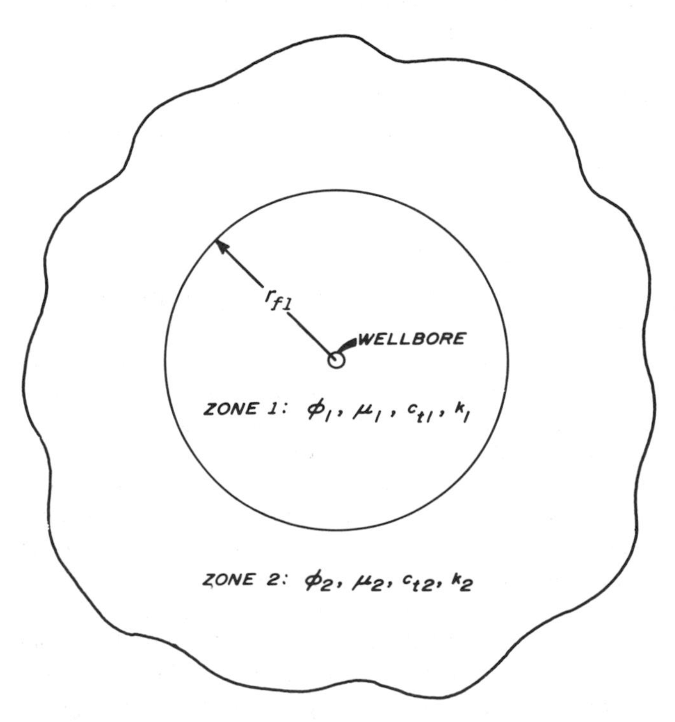

Fig. 10.9 Composite reservoir.

pressure transient tests to have similar characteristics, and the characteristics are not always the same for a particular heterogeneity.

The distance to the boundary between the inner and outer zones, r_{f1}, may be estimated from Eq. 2.41 for a *drawdown* or *injectivity* test,

$$r_{f1} = 0.029\sqrt{\frac{kt^*}{\phi\mu c_t}}, \quad \ldots\ldots\ldots\ldots\ldots\ldots (10.13)$$

where t^* is the time at the *end* of the first semilog straight line. Eq. 10.13 provides an order-of-magnitude estimate only. Section 7.5 gives a method for estimating r_{f1} for buildup and falloff. Eq. 7.16 or Eq. 7.17 and Fig. 7.15 are used. Odeh[23] proposes a trial-and-error method for estimating r_{f1} and $(k/\mu)_2$.

Bixel and van Poollen[24] and Carter[22] state that it may be difficult, or impossible, to estimate average reservoir pressure from pressure buildup (or falloff) tests in composite systems unless very long shut-in times occur. Carter[22] does show that if a drawdown test is continued long enough, the total reservoir volume communicating with the well may be estimated using the analysis techniques of Section 3.5.

10.5 Layered Reservoir Systems

Layered reservoirs can be divided into two groups: *layered reservoirs with crossflow,* in which layers are hydrodynamically communicating at the contact planes (Fig. 10.11), and *layered reservoirs without crossflow,* in which layers communicate only through the wellbore (Fig. 10.12). The latter type of system also has been called a ''commingled system''.

Layered Reservoirs With Crossflow

Fig. 10.11 schematically shows a three-layer reservoir with crossflow allowed between the layers. Many papers discuss pressure transient testing in such reservoirs; Russell and Prats[27] summarize the practical aspects of those papers. They conclude that pressure transient behavior of a layered reservoir is the same as the behavior of the *equivalent homogeneous system.* Thus, the layered system with crossflow behaves like a homogeneous system with an arithmetic total permeability-thickness product,

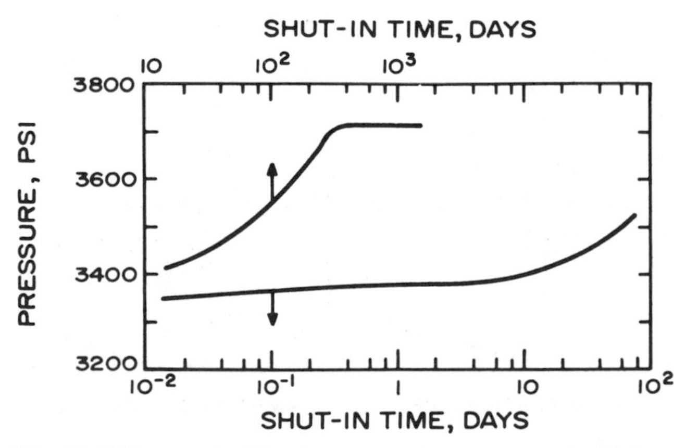

Fig. 10.10 Pressure buildup for a composite gas reservoir with k_1 = 10 md, k_2 = 0.01 md, r_{f1} = 500 ft, r_e = 1,000 ft, $\phi_1 = \phi_2$, $\mu_1 = \mu_2$, and $c_{t1} = c_{t2}$. After Carter.[22]

$$(kh)_t = \sum_{j=1}^{n} (kh)_j, \quad \text{.......... (10.14)}$$

substituted for kh and an arithmetic total porosity-compressibility-thickness product,

$$(\phi c_t h)_t = \sum_{j=1}^{n} (\phi c_t h)_j, \quad \text{.......... (10.15)}$$

substituted for $\phi c_t h$. The total number of layers is n. As a result, the appropriate semilog plot for any pressure transient test can be analyzed just like it can for homogeneous systems.

Kazemi and Seth[28] show how a series of drillstem tests on sequential intervals can be used to estimate the average permeability of the tested intervals and, thus, provide a gross picture of layering. A flow profile, such as a spinner survey, may also indicate gross reservoir stratification. If $(kh)_t$ is known from a well test, individual layer permeabilities may be approximated from[29]

$$k_j = (q_j/q)\left[(kh)_t/h_j\right], \quad j = 1, 2, \ldots, n. \quad \text{.......... (10.16)}$$

Layered Reservoirs Without Crossflow

Fig. 10.12 schematically illustrates a two-layer reservoir with the layers separated by a flow barrier. Production is commingled at the well so layers communicate only through the well. Early-time pressure drawdown in such a system yields a straight line on the semilog plot,[30-35] as illustrated in Fig. 10.13. Nothing distinguishes the drawdown curve from that for a single-layer, homogeneous reservoir. The slope of the semilog straight line may be used to estimate $(kh)_t$ and average skin factor with normal drawdown equations. The drawdown curves shown in Fig. 10.13 are for a layered system with no skin damage or with equal skin damage in each layer. If the damage varies from layer to layer, the behavior may differ from that shown. Unfortunately, there is little information available about that situation.

The upward bending in Fig. 10.13 is caused by boundary effects. After a long enough production time, pseudosteady-state conditions prevail and the pressure behavior will be linear with time. Pseudosteady-state flow generally begins much later in a commingled system than in the equivalent single-layer system because of the complex variation in flow contribution of each layer and the different times required for boundary effects to be felt in each layer. Cobb, Ramey, and Miller[33] indicate that pseudosteady-state flow begins at approximately

$$(t_{DA})_{pss} \simeq 23.5\,(k_1/k_2), \quad k_1 > k_2, \quad \text{.......... (10.17)}$$

for a single well in the center of a closed, circular, two-layered reservoir with porosity, compressibility, viscosity, and thickness equal in each layer. (Recall that $(t_{DA})_{pss} \simeq 0.1$ for a single-layer, closed, circular system.) Earlougher, Kersch, and Kunzman[35] indicate that the time to the beginning of pseudosteady state also depends on the relationship between the porosity, thickness, and compressibility in the various layers. We should also expect that time to depend on reservoir shape, number of layers, and well location.

A Horner or Miller-Dyes-Hutchinson plot of pressure buildup data for a single-well, closed, layered, no-crossflow system has an initial straight-line section with slope proportional to $(kh)_t$.[1,30-35] It has been stated that, after the initial semilog straight line, the buildup curve flattens, then steepens, and finally flattens toward static pressure,[1,30,33] as

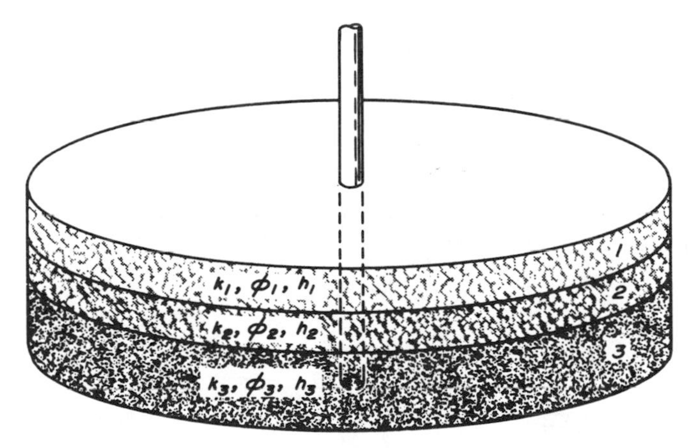

Fig. 10.11 Three-layer reservoir with crossflow.

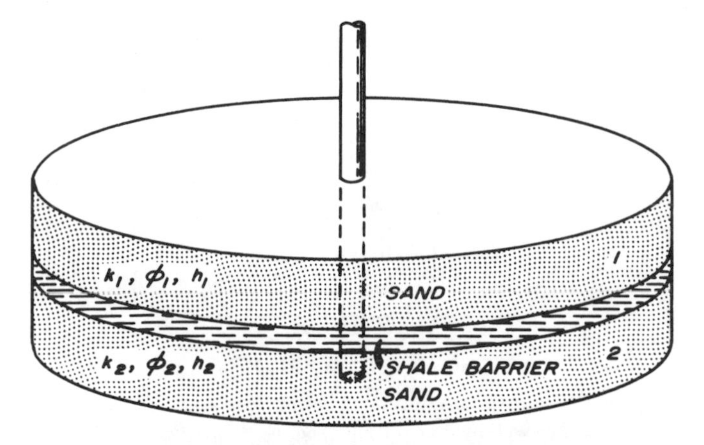

Fig. 10.12 Two-layer reservoir without crossflow.

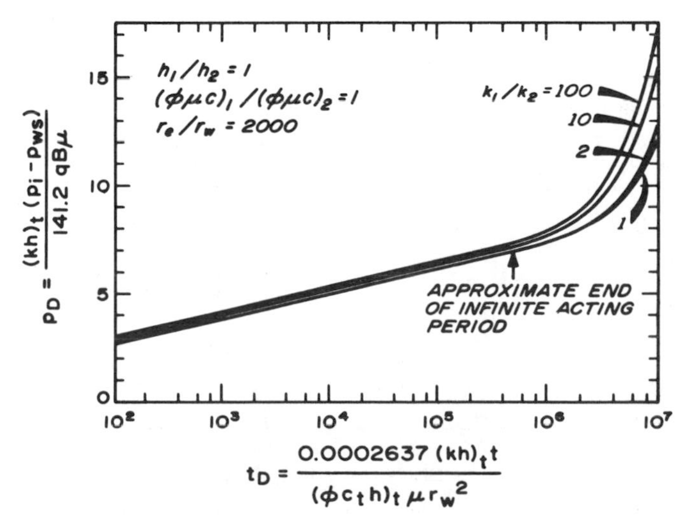

Fig. 10.13 Dimensionless drawdown behavior for a well in the center of a closed, circular, commingled, layered reservoir. After Cobb, Ramey, and Miller.[33]

indicated schematically in Fig. 10.14. *This is not always correct*. Several studies [32-35] show that the first flattening of the buildup curve, F-G in Fig. 10.14, can be insignificant for some systems. That is particularly true for large contrasts in thickness or porosity, for more than two layers, or for nonsymmetrical systems. Figs. 10.15 and 10.16 illustrate flattening and its absence. In Fig. 10.15, the flattening and secondary pressure rise are definitely apparent for the two lower curves. However, in the upper curve, where the *kh* ratio is the result of a large thickness ratio rather than a high permeability ratio, the buildup curve is indistinguishable from that for a single-layer system. Fig. 10.16 shows that the flattening does not occur for a 4:1 rectangle. Rather, the buildup curves take on a shape dominated by the drainage-area shape and the well location. Nevertheless, in all cases, the initial semilog straight-line segment may be used to estimate the total permeability-thickness product.[35] The second (steep) slope is not analyzable by known methods.

An additional danger lies in attributing the shape shown in Fig. 10.14 to layered reservoirs. Fig. 10.10, for a composite reservoir system, has the shape shown in Fig. 10.14. (There is no wellbore storage effect in Figs. 10.10, 10.15, and 10.16.) Changing wellbore storage can also cause the early part of a pressure buildup curve to be similar to the curve shown in Fig. 10.14; for examples, see Fig. 2.12 or Fig. 11.2, and Ref. 36.

Almost all the published material about layered systems is for a single well in the center of a symmetrical bounded system. If there are other wells in the reservoir, and if they continue producing while one well is shut in for a buildup test, none of the pressure response characteristics described in Refs. 30 through 35 and illustrated in Figs. 10.14 through 10.16 necessarily occur. Fig. 10.17 shows pressure behavior when a single well in a developed, two-layer, commingled reservoir is shut in for pressure buildup testing. The initial semilog straight line is observed and may be used to estimate $(kh)_t$. However, neither the flattening nor the secondary pressure rise and eventual leveling to average reservoir pressure occur. Rather, the influence of adjacent producing wells causes the pressure to decline at the shut-in well, completely obscuring evidence of possible layered behavior.

Raghavan *et al*.[34] propose a technique for estimating the *kh* ratio between layers for a single well in the center of a circular, two-layer, commingled system. The method re-

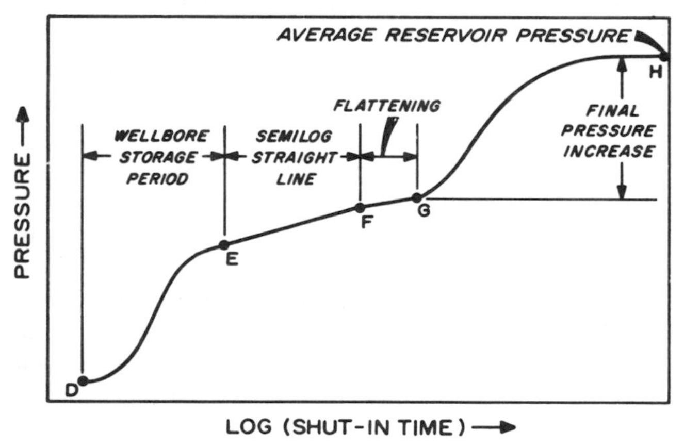

Fig. 10.14 Hypothetical pressure buildup curve for an ideal single-well, multiple-layer, bounded reservoir. After Earlougher, Kersch, and Kunzman.[35]

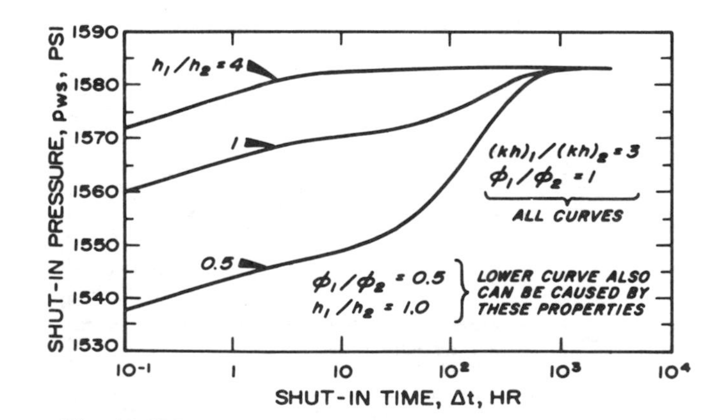

Fig. 10.15 Pressure buildup behavior as a function of thickness ratio for a well in the center of a square, two-layer reservoir; $(kh)_1/(kh)_2 = 3$, $\phi_1/\phi_2 = 1$. After Earlougher, Kersch and Kunzman.[35]

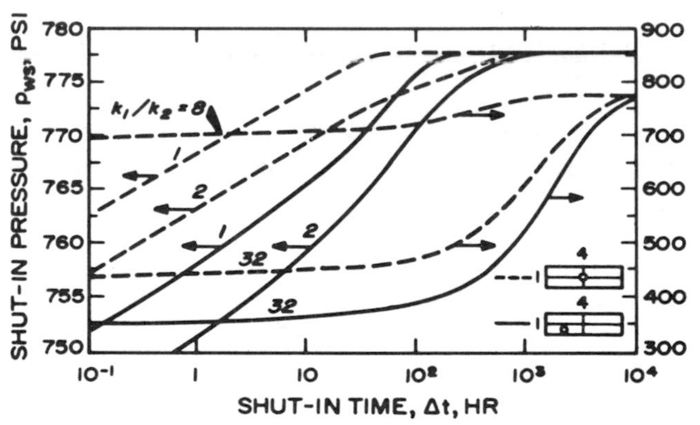

Fig. 10.16 Pressure buildup behavior as a function of permeability ratio for two different well locations in a two-layer, 4:1, rectangular reservoir; $\phi_1/\phi_2 = 1$, $h_1/h_2 = 1$. After Earlougher, Kersch, and Kunzman.[35]

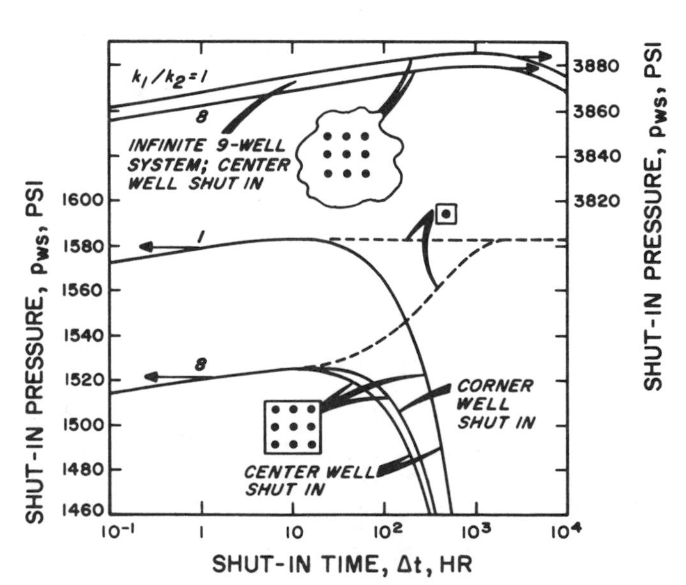

Fig. 10.17 Comparison of pressure buildup behavior in single-well and multiple-well layered systems; $\phi_1/\phi_2 = 1$, $h_1/h_2 = 1$. After Earlougher, Kersch, and Kunzman.[35]

quires a relatively long production time and requires that buildup data be taken *through* the final pressure rise.

Refs. 33 through 35 may be used to estimate average reservoir pressure from buildup tests in commingled systems. The techniques do require some knowledge of the layer properties and utilize correlations for specific systems.

Woods[29] has studied pulse-test behavior in a two-layer reservoir system. His studies include the commingled case, the full-communication case, and intermediate situations. He shows how a combination of single-well tests, pulse tests, and flowmeter surveys may be used to estimate individual-zone properties for two-layer reservoirs with communication only at the wellbores. He points out that the wellbores must be undamaged or uniformly damaged. He extensively studied *pulse-test* behavior in such reservoirs, with the following conclusions:

1. Apparent kh/μ is always equal to or greater than the actual total kh/μ for the reservoir.
2. Apparent $\phi c_t h$ is always equal to or less than total $\phi c_t h$ for the reservoir.
3. The deviation of apparent values from actual total values depends on the pulse duration (see Section 9.3).
4. When wells are undamaged or have uniform damage, the ratio of flow rates into the zones is a good estimator of kh/μ of the zones for noncommunicating systems. The estimate is usually valid within ±15 percent when the zones are partially communicating.

10.6 Naturally Fractured Reservoirs

Naturally fractured reservoirs are among the commonly encountered heterogeneous systems. Two basic behaviors may occur in such reservoirs. If the existing fractures dominantly trend in a single direction, the reservoir may appear to have anisotropic permeability and methods of Sections 9.4 and 10.3 apply. Earlougher and Kersch[16] show interference data from such a reservoir. The second class of naturally fractured reservoirs exhibits two distinct porosity types as shown on the left side of Fig. 10.18. The matrix region contains fine pores and often has low permeability. The remaining region — a set of interconnecting fractures, fissures, and vugs — has a significant porosity and a high permeability compared with the matrix. Ideally, we would like to be able to estimate the permeability and porosity of each region from transient-test data. Much theoretical information is available[11,37-45] about transient testing in fractured reservoirs. Unfortunately, published field data are scarce.

Pollard[37] and Pirson and Pirson[38] suggest analysis techniques for estimating average pressure, wellbore damage, fracture volume, and matrix volume. Warren and Root[11] and Kazemi[44] show that the Pollard-Pirson-Pirson method may be considerably in error for both infinite-acting and finite fractured reservoirs. In addition, unfractured reservoirs may be analyzed by this method and may be reported to be fractured. Thus, we recommend that the Pollard-Pirson-Pirson method not be used.

Warren and Root[11] assume that a fractured reservoir can be represented by the system shown on the right side of Fig. 10.18. The blocks represent the matrix; the space between blocks represents fractures. Warren and Root assume that formation fluid flows from the blocks into the fractures under pseudosteady-state conditions. The fractures carry all fluid to the wellbore. Kazemi[44] uses a similar model but does not require pseudosteady-state flow from the matrix to the fractures. His results verify those presented by Warren and Root.

Warren and Root define two characteristics of the fractured system, the ratio of the porosity-compressibility product of the fractures to the total system porosity-compressibility product,

$$F_{ft} = \frac{(\phi c_t)_f}{(\phi c_t)_f + (\phi c_t)_{ma}}, \quad \ldots\ldots\ldots\ldots (10.18)$$

and the interporosity flow parameter,

$$\epsilon = \frac{\alpha k_{ma} r_w^2}{k_f}, \quad \ldots\ldots\ldots\ldots\ldots\ldots (10.19)$$

In Eqs. 10.18 and 10.19 the subscript f indicates a fracture property, while ma indicates a matrix property. In Eq. 10.19 α is a matrix-to-fracture geometric factor with dimension of length^{-2}. Warren and Root[11] and Kazemi[44] indicate how this factor may be estimated. The effective fracture permeability is $\bar{k}_f$; Warren and Root consider fractures with different permeabilities in the principal directions.

Fig. 10.19 shows computed drawdown behavior for the Kazemi and Warren and Root analytical models; note the two parallel semilog straight-line portions. The slope of either straight line indicates the total system permeability-

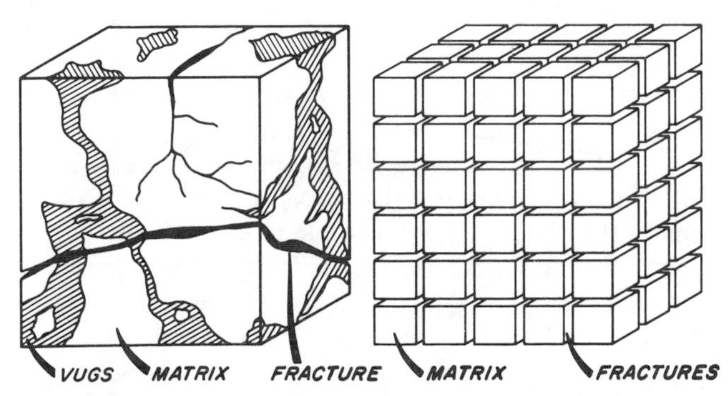

Fig. 10.18 Schematic illustration of a naturally fractured reservoir and its idealization. After Warren and Root.[11]

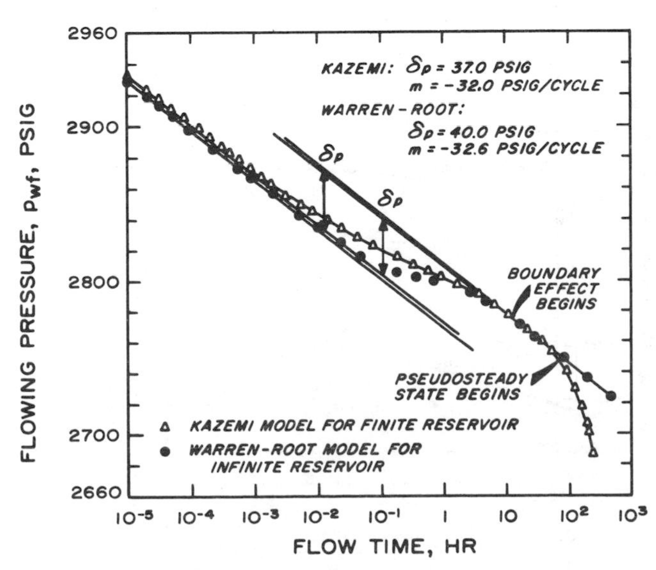

Fig. 10.19 Drawdown in a naturally fractured reservoir; comparing models of Kazemi and Warren and Root. After Kazemi.[44]

thickness product. The first semilog straight line may be obscured by wellbore storage effects, since it occurs at very short shut-in times. Thus, to successfully see both semilog straight lines, it may be necessary to minimize wellbore storage by assuring that the wellbore is full of liquid or by using a bottom-hole shut-in device.

Results from the two models in Fig. 10.19 differ because of the assumptions made in the models. The difference in location between the semilog straight lines results because Warren and Root assume pseudosteady-state flow from the matrix to the fracture, while Kazemi allows transient flow. The late-time deviation is caused by the boundary in the Kazemi model. Kazemi shows that the pseudosteady-state behavior for a naturally fractured system begins at a dimensionless time given by

$$(t_{DA})_{pss} = \frac{0.0002637\,(kh)_t\,t_{pss}}{\left[(\phi c_t)_f + (\phi c_t)_{ma}\right]h\mu A} \simeq 0.13. \quad \quad (10.20)$$

Eq. 10.20 applies for a single well in the center of a closed, circular drainage region. It also can be expected to apply to square and triangular regions. It is interesting that the time to beginning of pseudosteady state is about the same as for a homogeneous reservoir — when total-system properties are used in computing t_{DA}.

Fig. 10.20 shows actual pressure buildup data for a naturally fractured reservoir.[40] Theory[11,44] predicts such a pressure-buildup curve shape. As for drawdown, wellbore storage effects may obscure the initial semilog straight line. If both semilog straight lines develop, as in the test in Fig. 10.20, analysis for total permeability-thickness product is by normal methods. Skin factor is estimated using normal equations, taking p_{1hr} from the *second* straight line. Average reservoir pressure is estimated by extrapolating the *second* straight line to infinite shut-in time to obtain p^* and then using the methods of Chapter 6. The vertical distance between the two semilog straight lines, identified as δp in Figs. 10.19 and 10.20, may be used to estimate the ratio of the porosity-compressibility product in the fracture to that for the total system for isotropic or anisotropic systems:[11,44]

$$F_{ft} = \text{antilog}\,(-\delta p/m). \quad \quad (10.21)$$

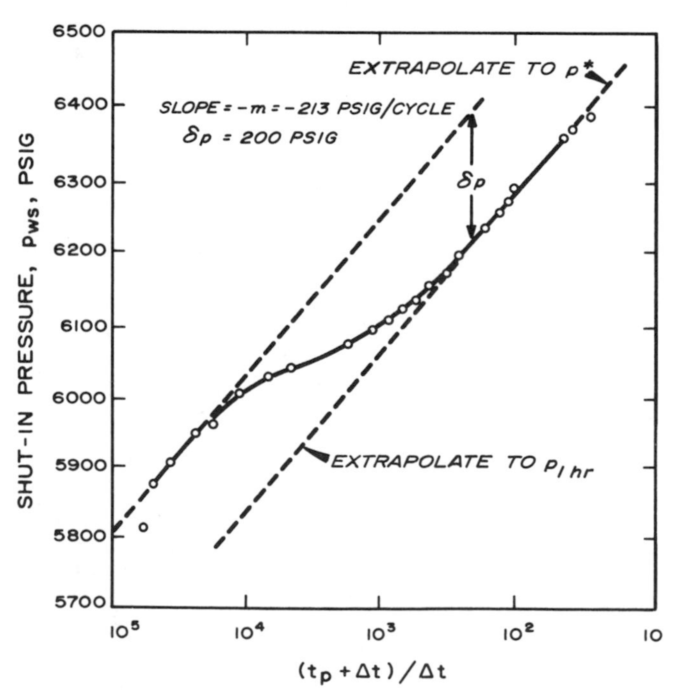

Fig. 10.20 Buildup curve from a fractured reservoir. After Warren and Root.[40]

The pressure buildup curve shown in Fig. 10.20 has a semilog straight line, then flattens and then steepens again. That is precisely the characteristic in Fig. 10.14 that is often attributed to reservoirs with noncommunicating layers. Thus, we again see that several types of heterogeneous systems may have transient response curves with similar shapes. Note also that Curve 3 in Fig. 10.5 has the same general shape as the curves in Figs. 10.19 and 10.20.

Example 10.2 Estimating the Ratio of Fracture to Total ϕc_t

Fig. 10.20 is an example pressure buildup test presented by Warren and Root.[40] From that figure, m = 213 psig/cycle, and δp = 200 psig. From Eq. 10.21,

$$F_{ft} = \text{antilog}(-200/213) = 0.12.$$

This may be interpreted that the fracture pore volume is 12 percent of the total pore volume, provided $(c_t)_{ma} = (c_t)_f$.

Odeh,[39] Warren and Root,[40] and Kazemi[44] show that when ϵ (Eq. 10.19) is relatively large (approaching 1), the Warren-Root model degenerates to the normal model for a homogeneous reservoir. For practical purposes, this may happen when the block dimensions are small, on the order of 3 ft, and the matrix permeability exceeds 0.01 md. In support of the Warren-Root model, fractured reservoirs in southwestern Iran have matrix permeabilities in the range of 0.00005 to 0.5 md, and have huge fracture blocks.[46]

Kazemi, Seth, and Thomas[44,45] show that observation-well behavior in interference testing in naturally fractured reservoirs can be substantially different from that of homogeneous reservoirs. Fig. 10.21 is a comparison of calculated interference pressures using a fracture model and a homogeneous model with equivalent permeability-thickness and porosity-compressibility characteristics. At short times, the pressures are significantly different in the two systems. However, at longer times, say for $t_D/r_D{}^2 > 5$ ($t_D > 1.6 \times 10^7$ in Fig. 10.21), based on the Kazemi-Seth-Thomas[45] data, normal analysis techniques should apply. Thus, it should be possible to analyze interference tests in naturally fractured reservoirs by the semilog methods of Section 9.2. *Type-curve methods and pulse testing may not provide correct results.* Kazemi[44] suggests a technique for using both pressure drawdown (or buildup) data and interference data to estimate properties of both fracture and matrix.

10.7 Effect of Pressure-Dependent Rock Properties

In some reservoirs, particularly low-permeability, deep gas reservoirs, rock properties can change significantly as reservoir fluid pressure declines. Although both permeability and porosity tend to decline with declining reservoir pressure, the permeability change is most frequently studied.[47-50] Vairogs *et al.*[47] and Thomas and Ward[48] studied the effect of declining permeability on gas production. Raghavan, Scorer, and Miller[50] provide some information on flow of liquids when permeability and porosity are pressure dependent.

Vairogs and Rhoades[49] studied pressure drawdown and buildup in low-permeability gas reservoirs with pressure-dependent permeability. Fig. 10.22 shows simulated drawdown curves for such a situation. The upper three curves are for real gas at different production rates; the lower curve is for an ideal gas at a relatively high production rate. If drawdown data for such a system could be analyzed using techniques like those given in Chapter 3, all four curves in Fig. 10.22 would be identical. Not only do the curves differ, but they have different slopes. The slopes taken at t_D = 1,000 for the upper three curves decrease as the rate decreases, but are greater than the theoretically expected value, 1.15, that is observed for the ideal-gas, constant-permeability case. Based on analysis of such data, Vairogs and Rhoades conclude that *neither* permeability nor skin factor should be estimated from *drawdown* tests in formations with pressure-dependent permeability. Wattenbarger and Ramey[51,52] reach a similar conclusion for testing gas wells with damage and wellbore storage in *constant permeability* formations.

Fig. 10.23 shows simulated pressure buildup curves taken from Ref. 49. Table 10.1 shows the results of analyzing 12 such pressure buildup curves for pressure-dependent permeability. In all cases, the permeability at initial pressure was 0.061 md. For correct analysis by methods of Chapter 5, we would expect all curves in Fig. 10.23 to have m = 1.15. That is the case for the short production period, but m exceeds 1.15 for longer production periods. Vairogs and Rhoades conclude that the permeability value estimated from a *buildup* test in such a reservoir is a good estimate of the permeability of that reservoir. For short production times, the initial permeability is calculated. For longer production times, the reservoir pressure has decreased and the actual reservoir permeability decreases as a result. Thus, the increase in slope with increasing production time shown in Fig. 10.23 indicates that the buildup-test analysis could give useful estimates of average reservoir permeability *at the time of the buildup test*.

The calculated skin factors for the first six cases in Table 10.1 are not a result of physical damage, but are caused by low permeability around the wellbore owing to pressure drawdown. Thus, a positive skin factor is estimated from buildup tests in formations with pressure-dependent permeability when no actual wellbore damage is present. The same is true if turbulence exists.[51] Vairogs and Rhoades point out that it may be impossible to determine whether an excessive pressure drop (positive skin factor) is a result of pressure-dependent permeability, skin damage, or turbulent

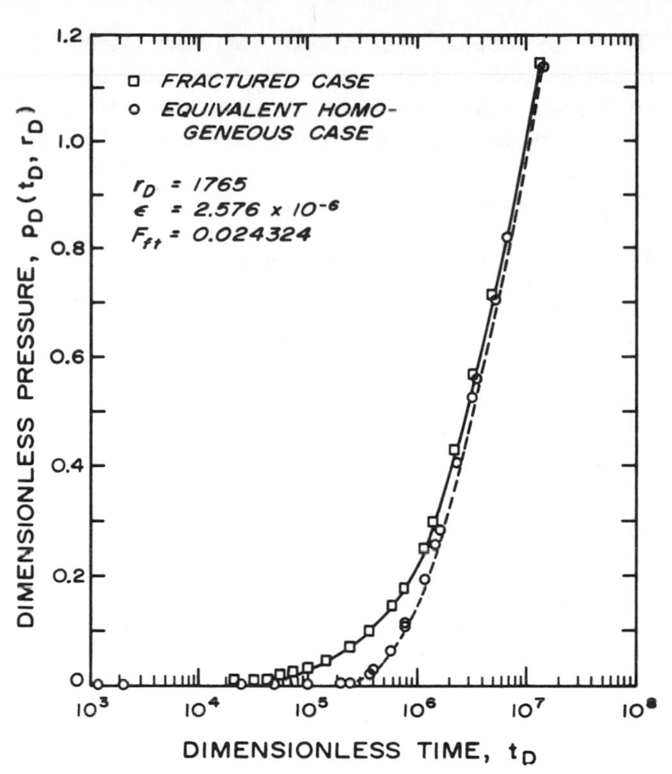

Fig. 10.21 Comparison of interference effect for a fractured reservoir with that for the equivalent homogeneous reservoir. After Kazemi, Seth, and Thomas.[45]

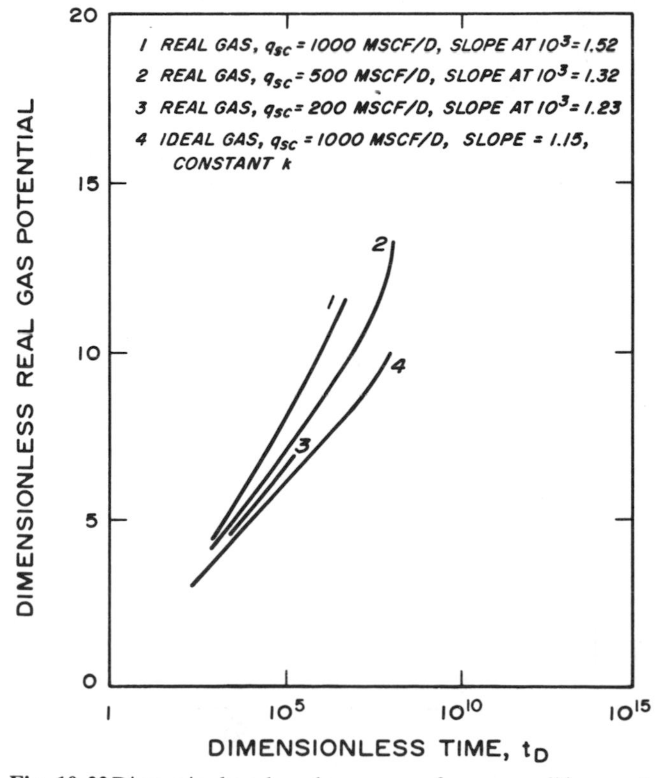

Fig. 10.22 Dimensionless drawdown curves for a gas well in a sand with pressure-dependent permeability. Well in the center of a closed circular reservoir. After Vairogs and Rhoades.[49]

flow at the wellbore. They also point out that when the well is mechanically damaged, the skin factor estimated from buildup-test analysis will be large for short production times and will decrease for significantly longer production times. A stimulated well will show a negative skin factor of about the right magnitude for all production times (Table 10.1).

10.8 Well Tests for Vertical Permeability

The vertical permeability in a formation is normally different from horizontal permeability, even when the system is otherwise homogeneous. Such vertical anisotropic effects are generally a result of depositional environment and post-depositional compaction history. In addition to microscopic characteristics that may cause the vertical permeability to be lower on a pore-to-pore basis, macroscopic features such as shale-sand interbedding can also create that effect. All transient testing and analysis methods described elsewhere in this monograph provide information about only the horizontal (radial or x and y direction) permeability. This section discusses techniques for estimating vertical permeability, k_z.

The transient tests proposed for estimating vertical permeability generally may be classified as vertical interference testing[53,54] or vertical pulse testing.[55-57] In such tests the well must be completed so that part may be used for production or injection (that is, be active) and part may be used for pressure observation (Fig. 10.24). Although Fig. 10.24 shows injection in the upper perforations and observation in the lower perforations, the opposite situation is equally acceptable theoretically. Operational considerations generally favor the situation illustrated in Fig. 10.24. A single-interval test for estimating vertical permeability also has been proposed.[58]

Vertical inferference testing is similar to normal interference testing. Of course, equipment and operational requirements are more demanding, since only one well is used. Burns[53] gives a good discussion of test design, well selection, and test operation. Vertical interference tests use

TABLE 10.1—SUMMARY OF SIMULATED BUILDUP TEST ANALYSES FOR RESERVOIRS WITH PRESSURE-DEPENDENT PERMEABILITY.
After Vairogs and Rhoades.[49]

Test Parameters			Buildup-Curve Analysis Results	
q (Mscf/D)	t_p (days)	s	k (md)	Skin Factor
500	2	0	0.061	1.19
1,000	2	0	0.055	2.22
1,250	2	0	0.053	4.86
500	3,000	0	0.054	2.15
500	6,000	0	0.050	2.98
500	9,000	0	0.050	5.13
1,000	2	−3	0.061	−2.60
1,000	2	1	0.061	4.72
500	3,000	−3	0.047	−2.82
500	3,000	1	0.047	1.98
500	6,000	−3	0.041	−2.91
500	6,000	1	0.041	1.85

relatively short-time pressure response data, so wellbore storage may have a significant effect on the observed pressure response. Unfortunately, that problem has not been studied to date, so no quantitative information can be provided here. Based on the behavior of other kinds of well tests, it is recommended that all possible efforts be made to minimize the effects of wellbore storage. If possible, testing in wells with a changing fluid level should be avoided. If that is not possible, consideration should be given to down-hole shut-in devices or to loading the wellbore with a low-density liquid to minimize the wellbore storage coefficient. We also recommend using a tubing packer (see Fig. 10.25) to minimize any fluid entry or exit from the observation perforations.

It should be clear that fluid communication through the wellbore behind the pipe or through vertical fractures in the formation cannot be tolerated. If such conditions are expected, an alternate well should be chosen for the test. Such conditions will cause a rapid pressure response at the observation perforations and will preclude analysis for vertical permeability. The test interval should not contain fluid-fluid contacts, since they will affect pressure response in a currently unknown way. It is recommended that both the active and test perforations be over a short interval. Because of that requirement and the requirement for a relatively long unperforated interval in the center of the well, it is expected that

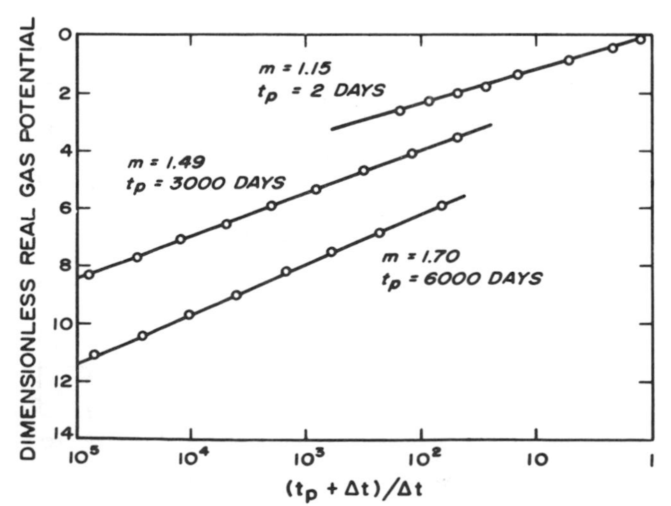

Fig. 10.23 Dimensionless pressure-buildup curves for a gas well in a sand with pressure-dependent permeability. Well in the center of a closed circular system. After Vairogs and Rhoades.[49]

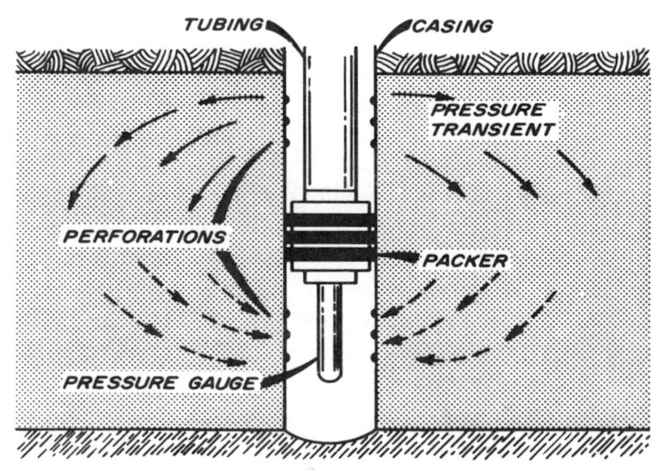

Fig. 10.24 Vertical interference or pulse test. After Burns.[53]

most existing wells would not be suitable for the testing described in this section. The possibility of using long packers or a straddle packer setup exists, but normally a special completion of a new well is necessary for vertical permeability testing.

As in any transient well test, good data collection is important. Rate control is essential for all the analysis techniques described in this section. If the rate varies significantly, computer analysis beyond the scope of this monograph is required.[53]

Vertical Pulse Testing

Hirasaki[55] proposes using vertical pulse testing to estimate k_z. He describes a testing and analysis technique based on arrival time of the first pressure peak. He considers the situation with perforations at the upper and lower reservoir boundaries. Because of that and because the very short pulse periods used can be expected to be influenced significantly by wellbore storage, we prefer the vertical pulse-testing technique described by Falade and Brigham.[56,57]

Fig. 10.25 shows the well configuration and the nomenclature used for vertical pulse testing and for vertical interference testing. Although flow perforations are shown at the top and observation perforations are shown at the bottom of the well, the arrangement may be reversed without affecting the analysis techniques presented here.

The techniques of this section apply only for wells far from a boundary that limits the reservoir laterally, such as a pinchout. Fortunately, vertical well tests are of short duration, so boundaries should not have much effect unless they are closer than given by Eq. 2.41 for the longest test time.

Fig. 10.26 is a schematic representation of rate history and pressure response for a vertical pulse test. The method presented here requires equal pulse times. Cycle time (Δt_C), pulse time (Δt_P), time lag (t_L), and pressure response amplitude (Δp) are identified. The pressure response amplitude for vertical pulse testing is determined just as it is for normal pulse testing. A tangent is drawn between the two peaks (or valleys) of the pressure response and a line parallel to that is drawn tangent to the valley (or peak) being analyzed. Pressure response amplitude is the vertical distance between those two lines.

Falade and Brigham[56,57] base dimensionless pulse time on the vertical permeability, k_z, and the vertical distance between the flow and observation perforations, ΔZ_R:

$$\Delta t_{PDV} = \frac{0.0002637\, k_z\, \Delta t_P}{\phi \mu c_t \Delta Z_R{}^2} \quad \text{. . . . . (10.22)}$$

Since this dimensionless time differs significantly from dimensionless times used elsewhere in this monograph, we have used subscript V to identify it as applying to vertical testing. Falade and Brigham[56,57] also define a dimensionless pulse response amplitude:

$$\Delta p_{DV} = \frac{k_r\, \Delta Z_R\, \Delta p}{141.2\, qB\mu} \quad \text{. . . . . (10.23)}$$

The vertical dimensionless pressure is based on the horizontal permeability, k_r, and the distance between the two sets of perforations, ΔZ_R. Again, because of the difference from the dimensionless pressures used elsewhere in this monograph, subscript V is added.

Once the test well is chosen, it is important to design the pulse length and magnitude to be compatible with formation characteristics and the pressure instrument resolution. The pulse duration must be long enough so the pressure instrument detects the pulses, but short enough so the pulses are easily identified. Analysis techniques are simpler for short pulses because the upper and lower boundary effects are less important for short pulses. Fig. 10.27 is a design curve for vertical pulse testing. The middle of the operational range, near the maximum of the upper curve in Fig. 10.27, should be chosen. That corresponds to a dimensionless pulse length of about 0.05 and a value of $(\Delta p_{DV})_\infty \left[(t_L)_\infty/\Delta t_P\right] = 0.0175$. Although those precise values need not be used, the values chosen should be close to those. Once the values of the parameters in Fig. 10.27 have been chosen, the pulse length is designed using a rearranged form of Eq. 10.22:

$$\Delta t_P = \frac{\phi \mu c_t \Delta Z_R{}^2\, (\Delta t_{PDV})_{\text{Fig. 10.27}}}{0.0002637\, k_z} \quad \text{. . . . . (10.24)}$$

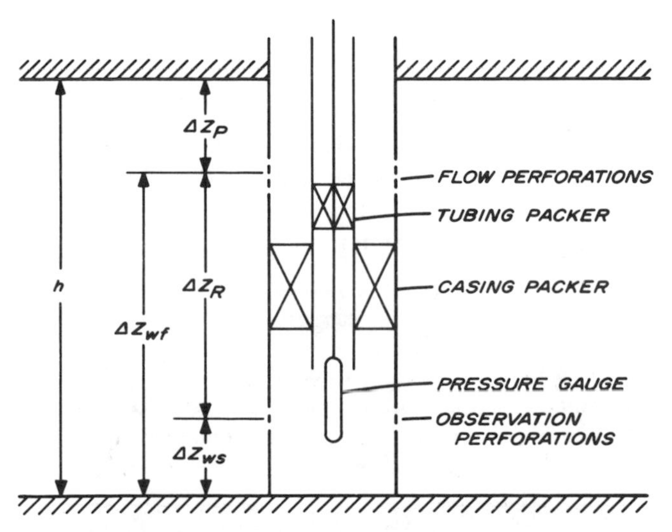

Fig. 10.25 Vertical interference and pulse test nomenclature.

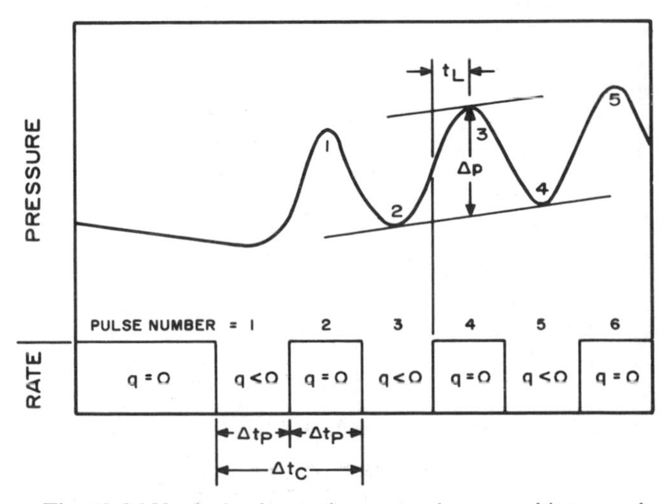

Fig. 10.26 Vertical pulse-testing rate and pressure history and nomenclature. Analysis procedure requires equal pulse lengths.

The expected response amplitude is estimated from a form of Eq. 10.23:

$$\Delta p = \frac{141.2\, qB\mu \left\{(\Delta p_{DV})_\infty\left[(t_L)_\infty/\Delta t_P\right]\right\}_{\text{Fig. 10.27}}}{k_r \Delta Z_R \left[(t_L)_\infty/\Delta t_P\right]_{\text{Fig. 10.27}}}, \quad \ldots\ldots (10.25)$$

and the expected time lag is estimated from

$$t_L = \left[(t_L)_\infty/\Delta t_P\right]_{\text{Fig. 10.27}} \Delta t_P \quad \ldots\ldots (10.26)$$

The expected response amplitude may be used to choose the combination of rate and pressure instrument for good resolution.

Example 10.3 Vertical Pulse Testing: Test Design

Falade and Brigham[57] give the following estimated parameters for a formation to be vertically pulse tested:

$k_z \simeq 1.0$ md $c_t = 1.0 \times 10^{-5}$ psi^{-1}
$k_r \simeq 20.0$ md $\Delta Z_R = 50$ ft
$\phi = 0.20$ $B = 1.0$ RB/STB.
$\mu = 2.0$ cp

We wish to design a pulse test. From Fig. 10.27 we choose

$$(\Delta t_{PDV})_{\text{Fig. 10.27}} = 0.05,$$

$$\left\{(\Delta p_{DV})_\infty\left[(t_L)_\infty/\Delta t_P\right]\right\}_{\text{Fig. 10.27}} = 0.0175,$$

and

$$\left[(t_L)_\infty/\Delta t_P\right]_{\text{Fig. 10.27}} = 0.37.$$

We estimate pulse length from Eq. 10.24:

$$\Delta t_P = \frac{(0.20)(2.0)(1.0\times10^{-5})(50)^2(0.05)}{(0.0002637)(1.0)}$$

$$\simeq 1.9 \text{ hours}.$$

The time lag is estimated from Eq. 10.26:

$$t_L \simeq (0.37)(1.9) \simeq 0.7 \text{ hour}.$$

Finally, we estimate the response amplitude from Eq. 10.25:

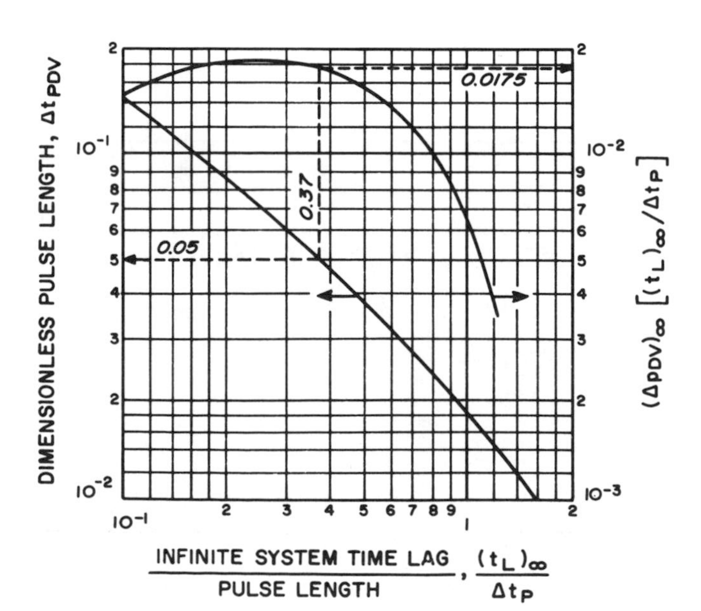

Fig. 10.27 Vertical pulse testing: design curve. After Falade and Brigham.[57]

$$\Delta p \simeq \frac{(141.2)(1.0)(2.0)(0.0175)}{(20.0)(50)(0.37)} q$$

$$\simeq 0.0134\, q \text{ psi}.$$

Thus, for an injection (or production) rate of 500 STB/D we would expect a pressure change of about

$$\Delta p = (0.0134)(500) = 6.7 \text{ psi}.$$

The pressure gauge should be chosen appropriately.

Vertical pulse-test data analysis is more complex than horizontal test analysis. That is because of the influence of the upper and lower formation boundaries on the test. For thick formations or short pulses, the system is infinite-acting and the upper and lower boundaries do not influence the observed pressures. For longer pulses, thinner formations, or when the perforations tested are close to one of the boundaries, either one or both of the boundaries influence the test response. In the latter cases, an iterative analysis procedure is required.

Regardless of the analysis procedure used (infinite-acting, single boundary, or double boundary), the first step in vertical pulse-test analysis is to calculate the geometric fraction (factor) needed for the correlation curves. The *primal geometric factor* is

$$G_P = \frac{\Delta Z_P}{\Delta Z_R}, \quad \ldots\ldots (10.27a)$$

and the *reciprocal geometric factor* is

$$G_R = \frac{h}{\Delta Z_R} - G_P - 1. \quad \ldots\ldots (10.27b)$$

By a clever application of the reciprocity principle, Falade and Brigham[56] show that all pulse-test responses can be correlated with only the two geometric factors. Once the two factors are calculated, if $G_P > G_R$, the two values are interchanged. Thus, the analysis techniques in this section and in Ref. 57 *require* that $G_P < G_R$. The values of G_P and G_R determine the type of analysis used.

Infinite-Acting System Analysis
($G_P > 2$, $G_R > 2$)

When G_P and G_R are each greater than 2, the system is infinite-acting and analysis is noniterative. The measured ratio of time lag to pulse length, $(t_L)_\infty/\Delta t_P$, is used with Fig. 10.28 to determine the dimensionless pulse length, Δt_{PDV}. Fig. 10.28 has curves for the first and second pulses. All later pulses fall between the curves shown. The vertical permeability is estimated from

$$k_z = \frac{\phi\mu c_t \Delta Z_R^2}{0.0002637\, \Delta t_P} (\Delta t_{PDV})_{\text{Fig. 10.28}}. \quad \ldots\ldots (10.28)$$

To estimate horizontal permeability, the dimensionless pulse time from Fig. 10.28 is used with Fig. 10.29 to obtain a dimensionless pulse response amplitude for the infinite system, $(\Delta p_{DV})_\infty$. Horizontal permeability is estimated from

$$k_r = \frac{141.2\, qB\mu}{\Delta Z_R\, \Delta p}\left[(\Delta p_{DV})_\infty\right]_{\text{Fig. 10.29}}, \quad \ldots\ldots (10.29)$$

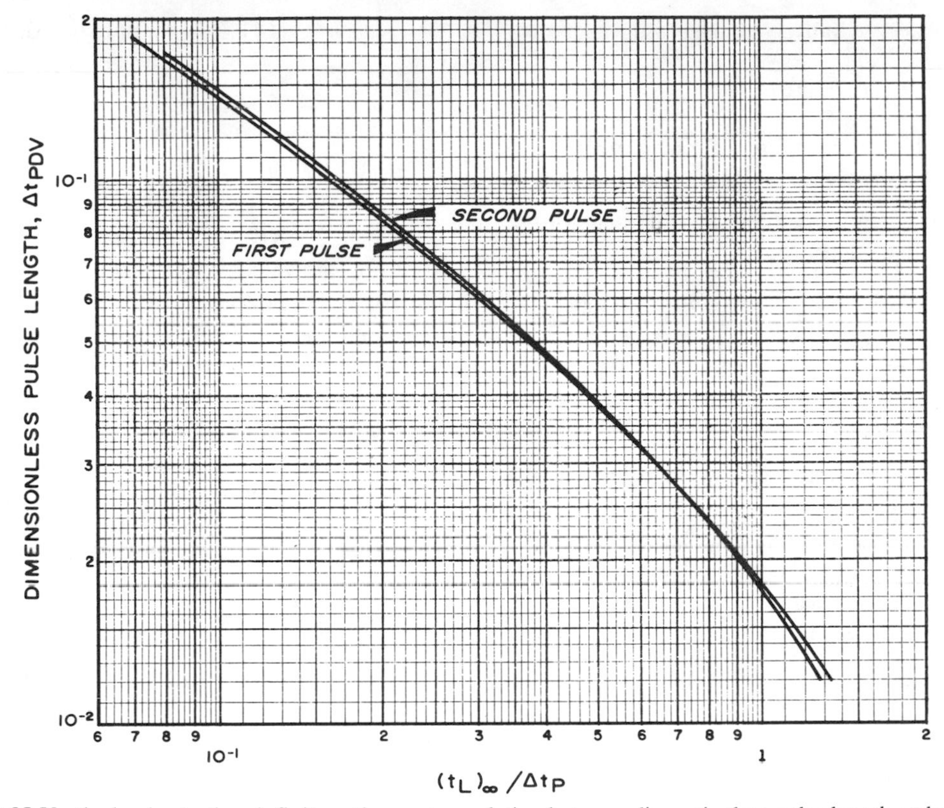

Fig. 10.28 Vertical pulse testing: infinite-acting system relation between dimensionless pulse length and time lag. After Falade and Brigham.[57]

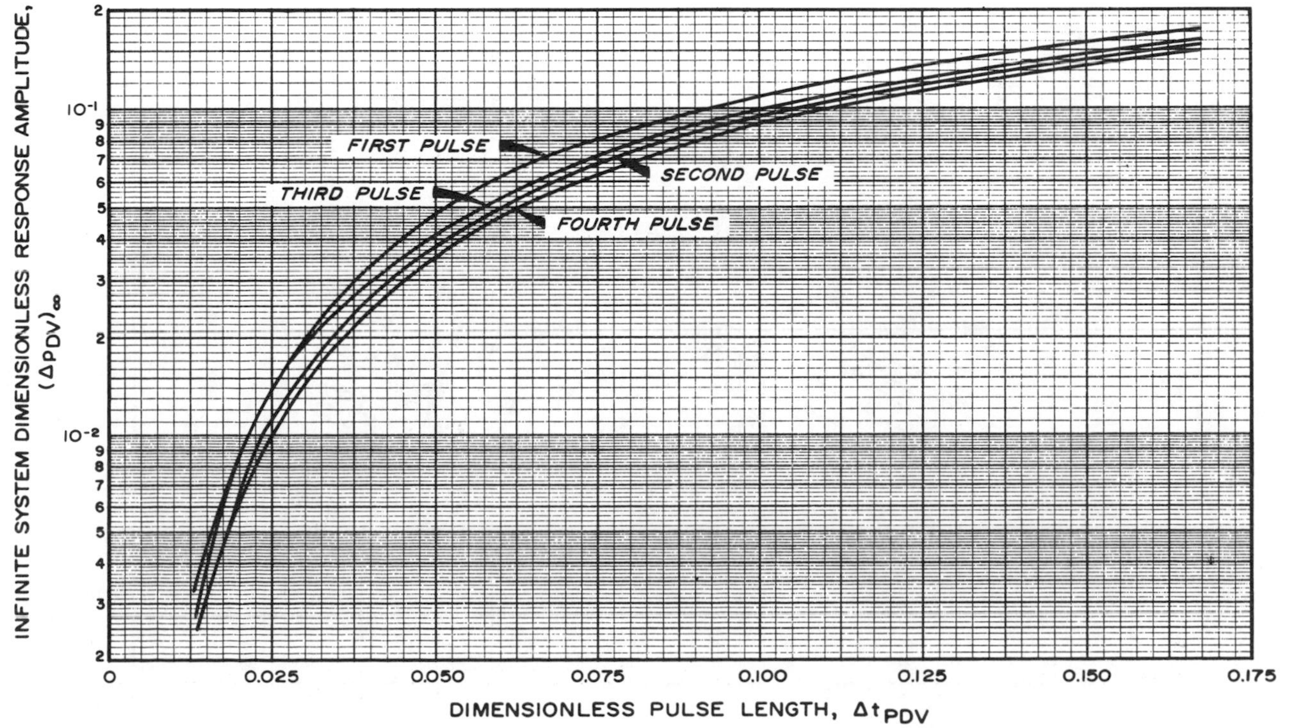

Fig. 10.29 Vertical pulse testing: infinite-acting system relationship between pulse length and response amplitude. After Falade and Brigham.[57]

where Δp is the observed pulse response amplitude. No iteration is required.

Example 10.4 Vertical Pulse Testing in an Infinite-Acting System

Falade and Brigham[57] provide the following data for the *first* pulse of a vertical pulse test:

$\phi = 0.20$ | $\Delta t_P = 1.9$ hours
$\mu = 2.0$ cp | $t_L = 0.837$ hour
$c_t = 1.0 \times 10^{-5}$ psi^{-1} | $\Delta p = -1.0$ psi
$\Delta Z_R = 50$ ft | $B = 1.0$ RB/STB.
$q = -100$ B/D

(Note that $\Delta p/q$ is always positive.) We wish to estimate the horizontal and vertical permeabilities.

We assume that the system is infinite-acting (Examples 10.5 and 10.6 show the analysis when one and two boundaries influence the test) and use Fig. 10.28. Because the system is infinite-acting, we take

$$(t_L)_\infty = t_L = 0.837 \text{ hour,}$$

so

$$(t_L)_\infty/\Delta t_P = 0.837/1.9 = 0.441.$$

From the first-pulse curve of Fig. 10.28, $(\Delta t_{PDV})_{\text{Fig. 10.28}} = 0.0425$, and from Fig. 10.29, $\left[(\Delta p_{DV})_\infty\right]_{\text{Fig. 10.29}} = 0.037$.

We estimate the vertical permeability from Eq. 10.28,

$$k_z = \frac{(0.20)(2.0)(1.0 \times 10^{-5})(50)^2(0.0425)}{(0.0002637)(1.9)}$$

$$= 0.85 \text{ md,}$$

and the horizontal permeability from Eq. 10.29,

$$k_r = \frac{(141.2)(-100)(1.0)(2.0)(0.037)}{(50)(-1.0)}$$

$$= 20.9 \text{ md.}$$

Analysis When One Boundary Affects Pulse Response ($G_P < 2$, $G_R > 2$)

If $G_P < 2$ and $G_R > 2$ (recall that we require that $G_P < G_R$ — if not, the values are interchanged), then one of the two horizontal boundaries is affecting the pulse-test response and an iterative analysis is required. Fig. 10.30, Fig. 10.31, or Fig. 10.32 is entered with G_P (after it has been interchanged with G_R, if necessary) and $(\Delta t_{PDV})_{\text{Fig. 10.28}}$. From the figure for the pulse being analyzed, $\left[t_L/(t_L)_\infty\right]_{\text{Fig.}}$ is obtained. Since Δt_{PDV} is a parameter in Figs. 10.30 through 10.32, a cross-plot (Fig. 10.33) for the appropriate geometric factor is usually helpful. Once $\left[t_L/(t_L)_\infty\right]_{\text{Fig.}}$ is estimated from Fig. 10.30, Fig. 10.31, or Fig. 10.32, we calculate

$$\left[(t_L)_\infty/\Delta t_P\right]_{\text{new}} = \frac{(t_L/\Delta t_P)}{\left[t_L/(t_L)_\infty\right]_{\text{Fig.}}} \quad \ldots\ldots\ldots\ldots (10.30)$$

The new value of $(t_L)_\infty/\Delta t_P$ is used in Fig. 10.28 to estimate a new value of $(\Delta t_{PDV})_{\text{Fig. 10.28}}$. If that value does not agree with the previous value, Fig. 10.30, Fig. 10.31, or Fig. 10.32 is used again and Eq. 10.30 is applied again. This process continues until two successive values of $(\Delta t_{PDV})_{\text{Fig. 10.28}}$ are the same. Then the vertical permeability is estimated from Eq. 10.28 using the final $(\Delta t_{PDV})_{\text{Fig. 10.28}}$ value.

Horizontal permeability is estimated from

$$k_r = \frac{141.2\, qB\mu}{\Delta Z_R\, \Delta p} \left[\frac{\Delta p_{DV}}{(\Delta p_{DV})_\infty}\right]_{\text{Fig.}} \left[(\Delta p_{DV})_\infty\right]_{\text{Fig. 10.29}}, \quad \ldots\ldots\ldots\ldots (10.31)$$

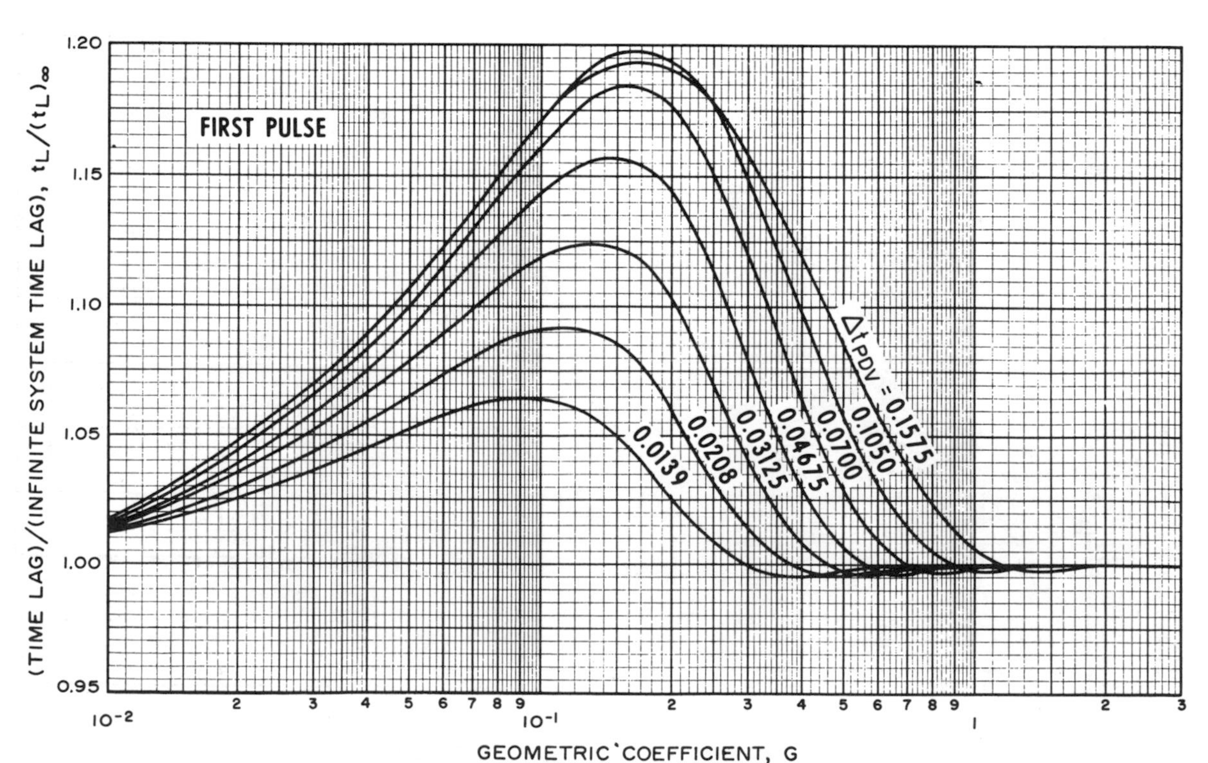

Fig. 10.30 Vertical pulse testing: correlation curves for the first-pulse time lag. After Falade and Brigham.[57]

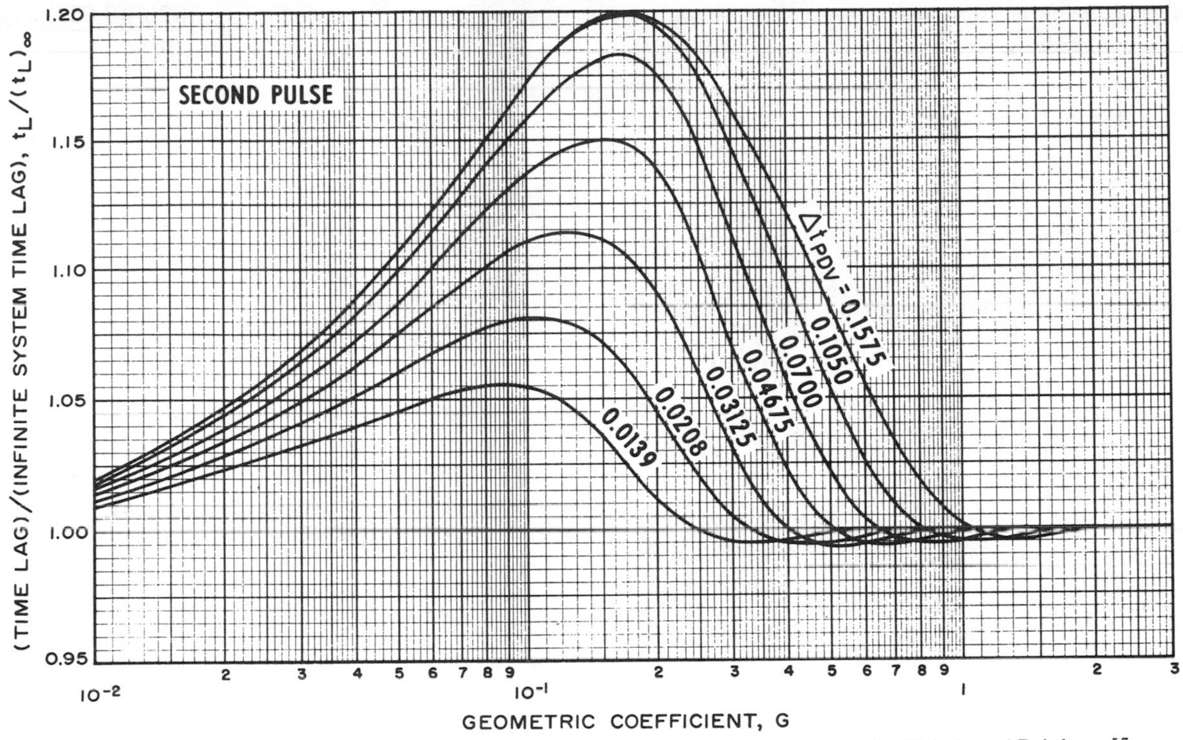

Fig. 10.31 Vertical pulse testing: correlation curves for the second-pulse time lag. After Falade and Brigham.[57]

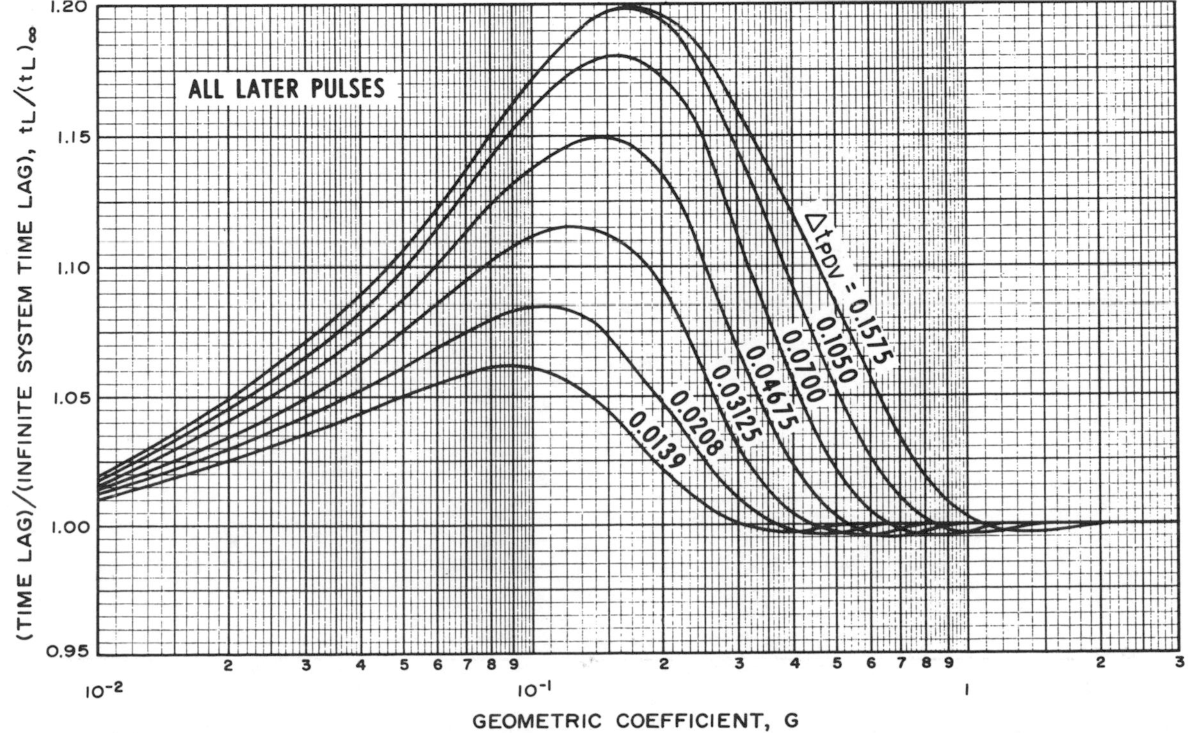

Fig. 10.32 Vertical pulse testing: correlation curves for all-later-pulses time lag. After Falade and Brigham.[57]

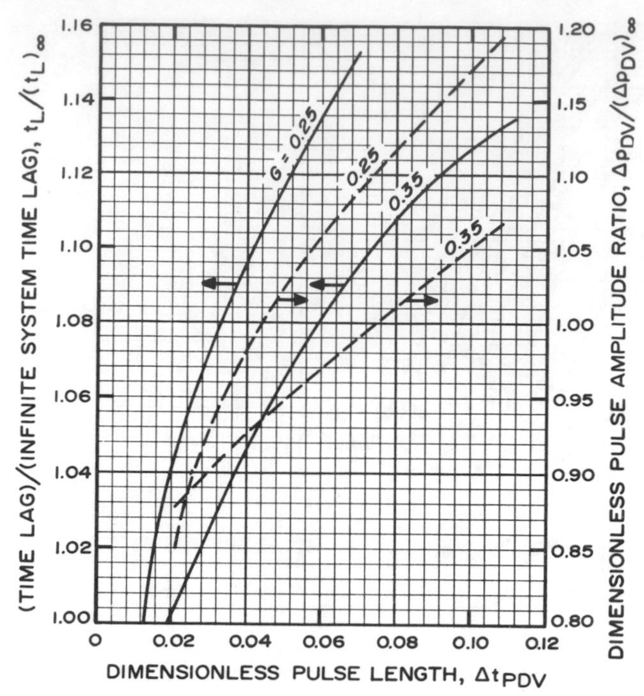

Fig. 10.33 Cross-plot of data in Figs. 10.30 and 10.34 for Examples 10.5 and 10.6.

where Fig. 10.34, Fig. 10.35, Fig. 10.36, or Fig. 10.37 is used to estimate $\left[\Delta p_{DV}/(\Delta p_{DV})_\infty\right]_{\text{Fig.}}$. $\left[(\Delta p_{DV})_\infty\right]_{\text{Fig. 10.29}}$ is estimated from Fig. 10.29 using the final $(\Delta t_{PDV})_{\text{Fig. 10.28}}$ value. No iteration is required to estimate the horizontal permeability once $(\Delta t_{PDV})_{\text{Fig. 10.28}}$ is estimated by the iterative technique described above. A cross-plot of material from Fig. 10.34, Fig. 10.35, Fig. 10.36, or Fig. 10.37 is usually helpful in this analysis (see Fig. 10.33).

Example 10.5 Vertical Pulse Testing in a System With One Nearby Boundary

We continue with the example of Falade and Brigham[57] using the data of Example 10.4 plus $\Delta Z_P = 12.5$ ft and $h = 250$ ft.

We use Eq. 10.27 to calculate the geometric factors

$$G_P = \frac{12.5}{50} = 0.25,$$

and

$$G_R = \left(\frac{250}{50} - 0.25 - 1\right) = 3.75.$$

We must use the smallest of G_P and G_R as G_P, in this case 0.25. Since $G_R = 3.75 > 2$, we analyze for a system with a single influencing boundary. The analysis starts as if the system were infinite-acting, so it is like that in Example 10.4. For the first iteration, $(t_L)_\infty/\Delta t_P = 0.837/1.9 = 0.441$ and

$$(\Delta t_{PDV})_{\text{Fig. 10.28}} = 0.0425.$$

Fig. 10.30 provides the first-iteration estimate of $t_L/(t_L)_\infty$; since Δt_{PVD} is a parameter in that figure, we use the cross-plot shown in Fig. 10.33 to determine $\left[t_L/(t_L)_\infty\right]_{\text{Fig.}} = 1.102$ for $G_P = 0.25$. Using Eq. 10.30,

$$\left[(t_L)_\infty/\Delta t_P\right]_{\text{new}} = \frac{(0.837/1.9)}{1.102} = 0.400.$$

Using Fig. 10.28 again,

$$(\Delta t_{PDV})_{\text{Fig. 10.28}} = 0.0465.$$

This completes the first step. Since the new value of $(\Delta t_{PDV})_{\text{Fig. 10.28}}$ does not agree with the previous value, we must iterate. All iterations are summarized in Table 10.2.

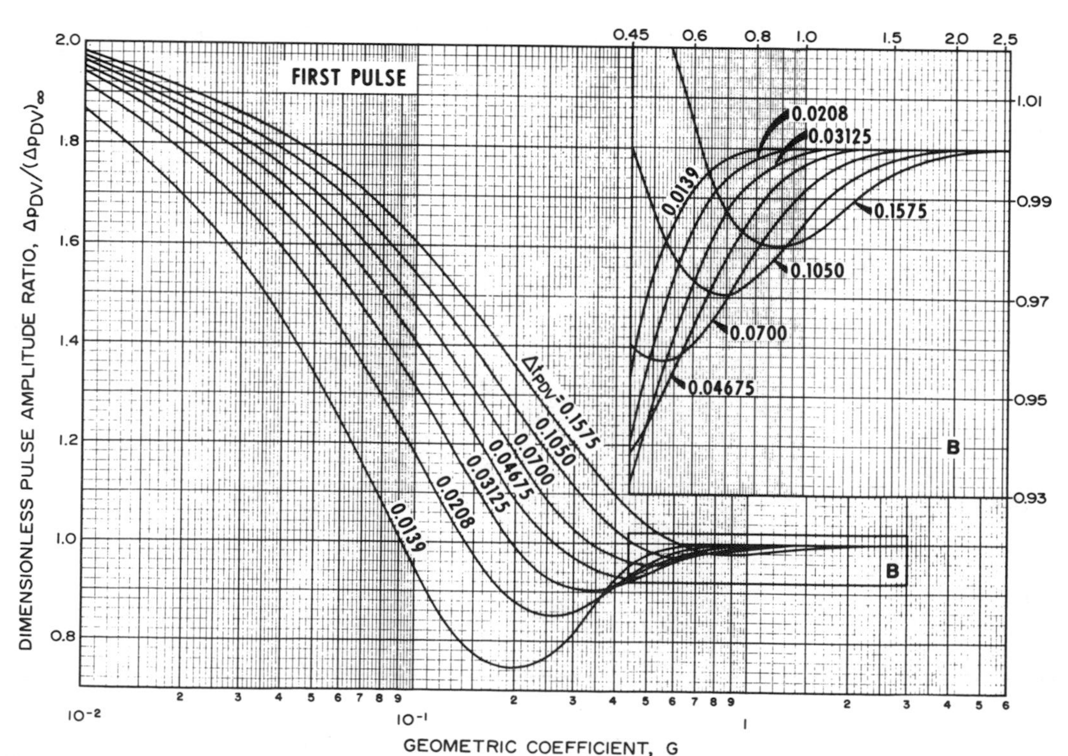

Fig. 10.34 Vertical pulse testing: correlation curves for the first-pulse response amplitude. After Falade and Brigham.[57]

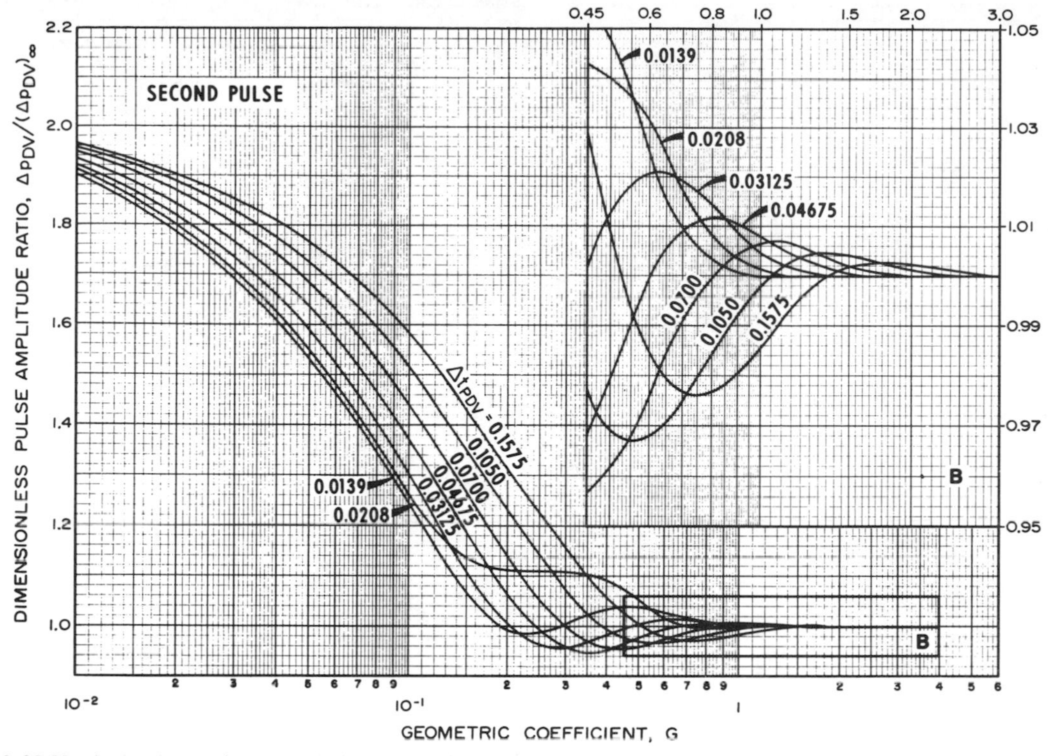

Fig. 10.35 Vertical pulse testing: correlation curves for the second-pulse response amplitude. After Falade and Brigham.[57]

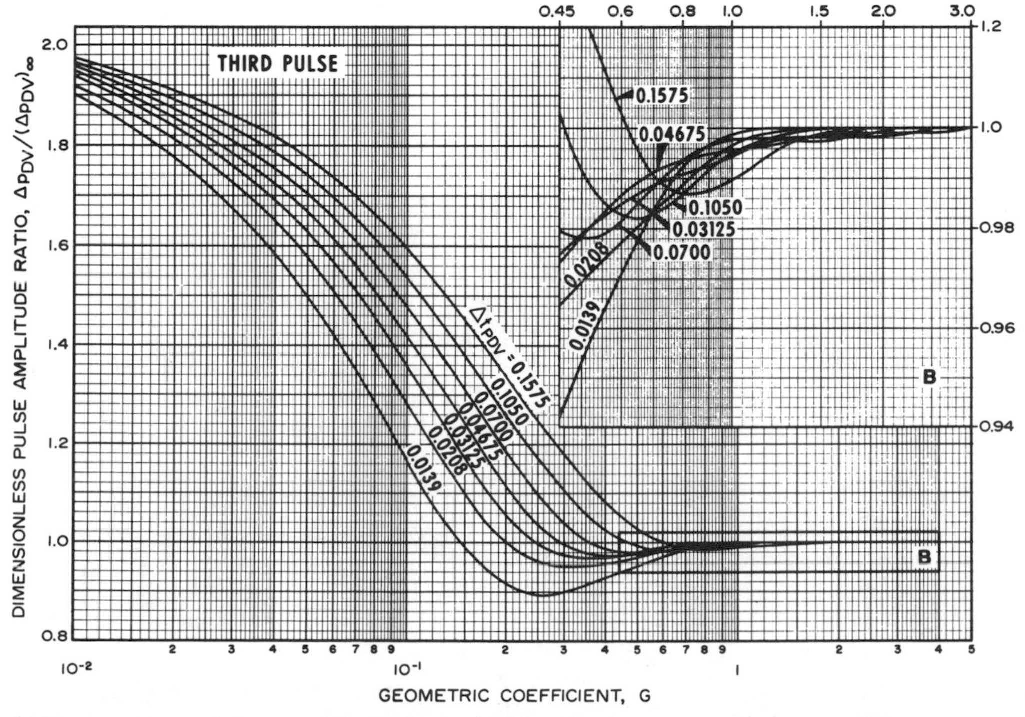

Fig. 10.36 Vertical pulse testing: correlation curves for the third-pulse response amplitude. After Falade and Brigham.[57]

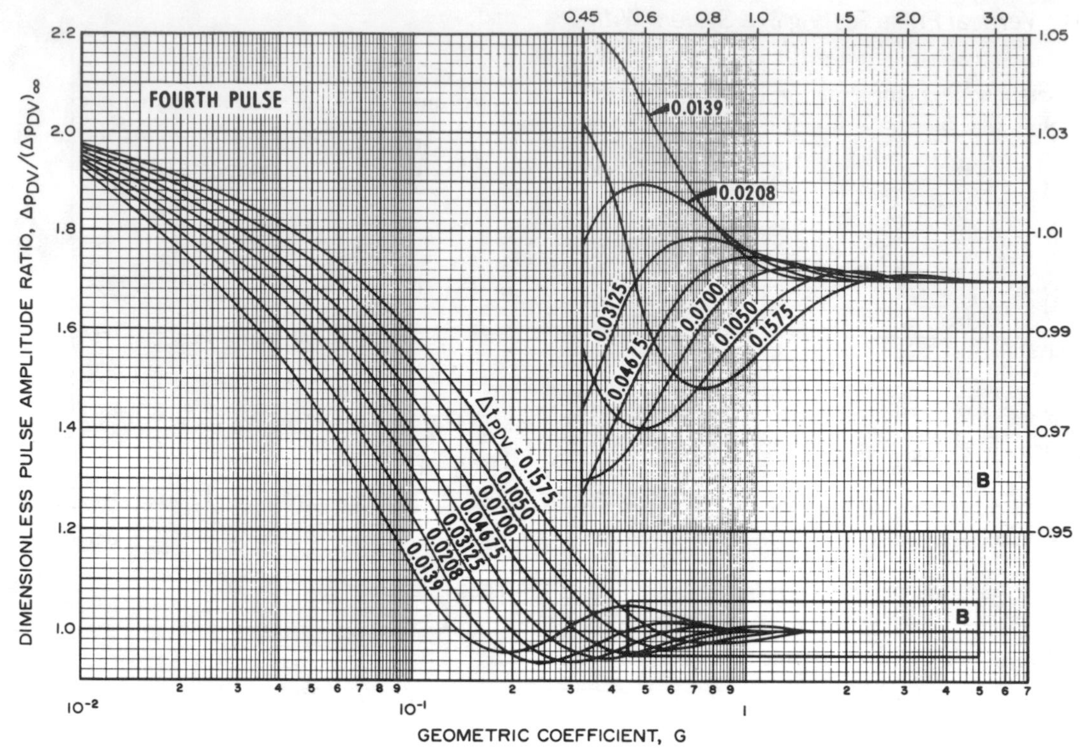

Fig. 10.37 Vertical pulse testing: correlation curves for the fourth-pulse response amplitude. After Falade and Brigham.[57]

The final result is

$$(\Delta t_{PDV})_{\text{Fig. 10.28}} = 0.0468,$$

and

$$\left[t_L/(t_L)_\infty\right]_{\text{Fig.}} = 1.110.$$

The vertical permeability is estimated using Eq. 10.28:

$$k_z = \frac{(0.20)(2.0)(1.0 \times 10^{-5})(50)^2(0.0468)}{(0.0002637)(1.9)}$$

$$= 0.93 \text{ md}.$$

To estimate k_r we use Figs. 10.29 and 10.34 for the first pulse to determine

$$\left[(\Delta p_{DV})_\infty\right]_{\text{Fig. 10.29}} = 0.0430,$$

and

$$\left[\frac{\Delta p_{DV}}{(\Delta p_{DV})_\infty}\right]_{\text{Fig.}} = 1.02.$$

Then, from Eq. 10.31,

$$k_r = \frac{(141.2)(-100)(1.0)(2.0)\left[1.02\right]\left[0.0430\right]}{(50)(-1.0)}$$

$$= 24.8 \text{ md}.$$

TABLE 10.2—ITERATIVE CALCULATION FOR VERTICAL PULSE TEST ANALYSIS IN A SYSTEM INFLUENCED BY ONE HORIZONTAL BOUNDARY; EXAMPLE 10.5.
After Falade and Brigham.[57]

Parameter	Iteration 1	Iteration 2	Iteration 3	Use
Δt_{PDV}	0.0425	0.0465	0.0468	Fig. 10.28
$t_L/(t_L)_\infty$	1.102	1.110	1.110	Fig. 10.30
$(t_L)_\infty/\Delta t_P$	0.400	0.397	0.397	Eq. 10.30
$(\Delta t_{PDV})_{\text{new}}$	0.0465	0.0468	0.0468	Fig. 10.28

Analysis When Both Boundaries Affect Pulse Response ($G_P < 2$, $G_R < 2$)

When G_P and G_R are each less than 2, both the upper and lower formation boundaries affect pulse-test response and an iterative analysis is required. We start by estimating Δt_{PDV} from Fig. 10.28, then use Fig. 10.30, Fig. 10.31, or Fig. 10.32 to estimate values of $t_L/(t_L)_\infty$ for both G_P and G_R. Next, a corrected value of $t_L/(t_L)_\infty$ is estimated from

$$\left[\frac{t_L}{(t_L)_\infty}\right]_{\text{Fig.}} = \left[\frac{t_L}{(t_L)_\infty}\right]_{\text{Fig.}, G_P} \left[\frac{t_L}{(t_L)_\infty}\right]_{\text{Fig.}, G_R} \quad \ldots\ldots (10.32)$$

Eq. 10.30 is used to estimate a new value of $(t_L)_\infty/\Delta t_P$. That $\left[(t_L)_\infty/\Delta t_P\right]_{\text{new}}$ is used with Fig. 10.28 to estimate the next value of Δt_{PDV}. If that value does not agree with the previous value, the computation is repeated until two successive values of Δt_{PDV} are the same. Finally, Eq. 10.28 is used to estimate k_z.

To estimate horizontal permeability, Fig. 10.34, Fig. 10.35, Fig. 10.36, or Fig. 10.37 is used for both G_P and G_R. Then we apply

$$\left[\frac{\Delta p_{DV}}{(\Delta p_{DV})_\infty}\right]_{\text{Fig.}} = \left[\frac{\Delta p_{DV}}{(\Delta p_{DV})_\infty}\right]_{\text{Fig.}, G_P} \times \left[\frac{\Delta p_{DV}}{(\Delta p_{DV})_\infty}\right]_{\text{Fig.}, G_R} \quad \ldots\ldots (10.33)$$

Finally, the horizontal permeability is estimated from Eq. 10.31 without iteration.

Example 10.6 Vertical Pulse Testing in a System With Two Nearby Boundaries

The Falade-Brigham[57] example can be carried one step further by assuming that the test geometric configuration is $\Delta Z_P = 12.5$ ft and $h = 80$ ft. Other data are the same.

From Eq. 10.27,

$$G_P = \frac{12.5}{50} = 0.25,$$

and

$$G_R = \frac{80}{50} - 0.25 - 1 = 0.35.$$

Since $G_P < G_R$ we need not exchange the values. Both values are less than 2, so a finite-system analysis is required. As in Example 10.5, we start by using an infinite-acting system analysis, then iterating. To start the first iteration we enter Fig. 10.28 with $(t_L)_\infty/\Delta t_P = 0.837/1.9 = 0.441$ to get

$$(\Delta t_{PDV})_{\text{Fig. 10.28}} = 0.0425.$$

From Fig. 10.33, the cross-plot of data in Fig. 10.30, we get the *two* required values of $\{t_L/(t_L)_\infty\}$ for the first pulse:

$$\left[t_L/(t_L)_\infty\right]_{\text{Fig.}, G_P = 0.25} = 1.102,$$

and

$$\left[t_L/(t_L)_\infty\right]_{\text{Fig.}, G_R = 0.35} = 1.051.$$

Applying Eq. 10.32,

$$\left[t_L/(t_L)_\infty\right]_{\text{Fig.}} = (1.102)(1.051) = 1.158.$$

Then, Eq. 10.30 is used to get

$$\left[(t_L)_\infty/\Delta t_P\right]_{\text{new}} = \frac{(0.837)/(1.9)}{1.158}$$

$$= 0.3804.$$

Using Fig. 10.28 again,

$$\left[\Delta t_{PDV}\right]_{\text{Fig. 10.28}} = 0.0488.$$

The first iteration is complete. Since the new $\{\Delta t_{PDV}\}_{\text{Fig. 10.28}}$ does not agree with the initial value, we must continue. All iterations are summarized in Table 10.3. The final iteration gives

$$\left[\Delta t_{PDV}\right]_{\text{Fig. 10.28}} = 0.0492,$$

and

$$\left[t_L/(t_L)_\infty\right]_{\text{Fig.}} = 1.184.$$

We estimate vertical permeability from Eq. 10.28:

$$k_z = \frac{(0.20)(2.0)(1.0 \times 10^{-5})(50)^2(0.0492)}{(0.0002637)(1.9)}$$

$$= 0.98 \text{ md}.$$

To estimate k_r we use Figs. 10.29 and 10.33 (the cross-plot of Fig. 10.34) for both geometric factors:

$$\left[(\Delta p_{DV})_\infty\right]_{\text{Fig. 10.29}} = 0.0470,$$

$$\left[\Delta p_{DV}/(\Delta p_{DV})_\infty\right]_{\text{Fig.}, G_P = 0.25} = 1.030,$$

$$\left[\Delta p_{DV}/(\Delta p_{DV})_\infty\right]_{\text{Fig.}, G_R = 0.35} = 0.961.$$

From Eq. 10.33,

$$\left[\Delta p_{DV}/(\Delta p_{DV})_\infty\right]_{\text{Fig.}} = (1.030)(0.961)$$

$$= 0.990$$

Then k_r is from Eq. 10.31:

$$k_r = \frac{(141.2)(-100)(1.0)(2.0)(0.990)(0.0470)}{(50)(-1)}$$

$$= 26.3 \text{ md}.$$

TABLE 10.3—ITERATIVE CALCULATION FOR VERTICAL PULSE TEST ANALYSIS IN A SYSTEM INFLUENCED BY TWO HORIZONTAL BOUNDARIES; EXAMPLE 10.6. After Falade and Brigham.[57]

Parameter	Iteration 1	2	3	Use
Δt_{PDV}	0.0425	0.0488	0.0492	Fig. 10.28
$t_L/(t_L)_\infty$	1.102	1.113	1.114	Fig. 10.30; G_P=0.25
$t_L/(t_L)_\infty$	1.051	1.063	1.063	Fig. 10.30; G_R=0.35
$t_L/(t_L)_\infty$	1.158	1.183	1.184	Eq. 10.32
$(t_L)_\infty/\Delta t_P$	0.3804	0.3724	0.3721	Eq. 10.30
$(\Delta t_{PDV})_{\text{new}}$	0.0488	0.0492	0.0492	Fig. 10.28

TABLE 10.4—COMPARISON OF RESULTS FOR THREE VERTICAL PULSE TEST ANALYSIS METHODS.

Analysis Method	k_z	k_r
Infinite-acting	0.85	20.9
One boundary	0.93	24.8
Two boundaries	0.98	26.3

Table 10.4 compares the results from Examples 10.4 through 10.6. For the situations used in those examples, the estimated vertical permeability changes by about 13 percent, while the horizontal permeability changes by about 21 percent. Thus, for the conditions used in the examples, the results do not vary as widely as might have been anticipated. Nevertheless, we recommend using the full iterative analysis if it is indicated by the values of G_P and G_R.

Vertical Interference Testing

Burns[53] proposes a method for vertical interference testing that has considerable utility. Unfortunately, it requires the use of a computer for test analysis. Although Burns' technique can be used with type-curve matching, a computer program is still required to generate the type curves. Thus, the Burns approach is not discussed further here. Prats[54] suggests a vertical interference testing method that does not require computer solutions. He shows that if observed pressure, p_{ws}, is plotted vs the logarithm of time from the beginning of injection or production, a straight line should result with slope m and intercept at $t = 1$ hour of p_{1hr}. The horizontal permeability is estimated from the slope using

$$k_r = \frac{-162.6\, qB\mu}{mh} \quad \text{. . . . . (10.34)}$$

The vertical permeability is estimated from the slope and intercept using

$$k_z = \frac{\phi\mu c_t h^2}{0.0002637} \text{ antilog}\left(\frac{p_{1hr} - p_i}{m} - \frac{G^* + h/\left|\Delta Z_{wf} - \Delta Z_{ws}\right|}{2.3025}\right) \quad \text{. . . . . (10.35)}$$

G^* in Eq. 10.35 is a geometric factor (fraction) provided by Prats and shown in Fig. 10.38. The vertical distances used in Eq. 10.35 and Fig. 10.38 are defined in Fig. 10.25. As in all transient test analysis, p_{1hr} must be taken from the semilog straight line, which is extrapolated if necessary.

Example 10.7 Vertical Interference Test Analysis

Prats[54] presents example data from a vertical interference test. Fig. 10.39 is the plot of observed pressure vs log of injection time. Other data are

$$h = 50 \text{ ft} \qquad \mu = 1.0 \text{ cp}$$
$$\Delta Z_{wf} = 45 \text{ ft} \qquad c_t = 2.0 \times 10^{-5} \text{ psi}^{-1}$$
$$\Delta Z_{ws} = 10 \text{ ft} \qquad \phi = 0.10$$
$$q = -50 \text{ STB/D} \qquad p_i = 3{,}015 \text{ psi.}$$
$$B = 1.0 \text{ RB/STB}$$

From Fig. 10.39, $m = 22.5$ psi/cycle and $p_{1hr} = 3{,}022$ psi. Eq. 10.34 is used to estimate horizontal permeability:

$$k_r = \frac{(-162.6)(-50)(1.0)(1.0)}{(22.5)(50)}$$
$$= 7.2 \text{ md.}$$

To estimate the vertical permeability, we use Eq. 10.35 and Fig. 10.38. We enter that figure with

$$\Delta Z_{wf}/h = 45/50 = 0.9$$

and

$$\Delta Z_{ws}/h = 10/50 = 0.2$$

to read $G^* = 0.76$. Then, from Eq. 10.35,

$$k_z = \frac{(0.10)(1.0)(2.0 \times 10^{-5})(50)^2}{0.0002637}$$
$$\text{antilog}\left(\frac{3{,}022 - 3{,}015}{22.5} - \frac{0.76 + 50/|45 - 10|}{2.3025}\right)$$
$$= 4.3 \text{ md.}$$

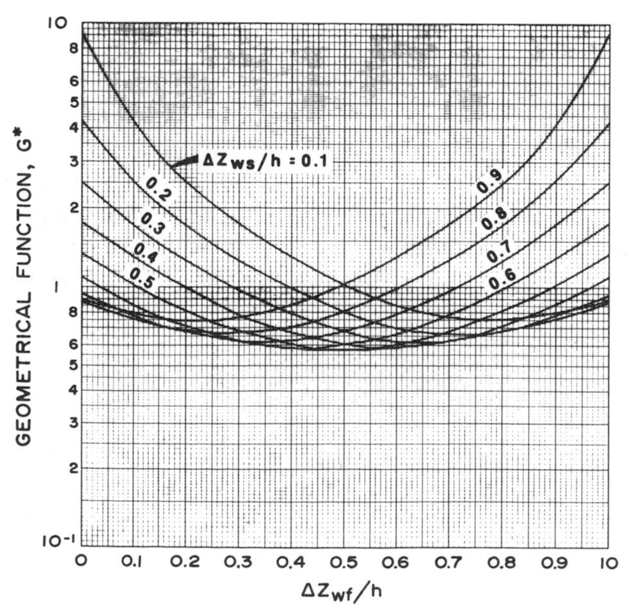

Fig. 10.38 Geometric factor for vertical interference testing. After Prats.[54]

The Prats analysis technique requires the well to be thoroughly stabilized before testing. The initial pressure in the region of the well at the time of the test is p_i. Although Prats does not show it, if the observation pressure at the well is changing according to a trend, that trend may be extrapolated for the test duration, and the pressure difference between the trend and observed pressures can be used in the plot of pressure vs log of time and in the analysis. (That approach is analogous to that given in Section 3.4.) The Prats testing and analysis technique is limited to perforated intervals that are short compared with the distance between the flow and observation perforations.

Because of the repetitive nature of pulse testing, we recommend that vertical pulse testing be used in preference to vertical interference testing. An added advantage of pulse testing is the larger range of allowed perforation locations with respect to the formation boundaries than for the interference method proposed by Prats.[54]

Single-Interval Testing

Raghavan and Clark[58] propose a method for estimating vertical permeability when only a single set of perforations is used. They use a spherical flow equation, so the perforated interval must be small compared with the formation thickness and it should be near the center of the formation interval. The analysis requires that a very restrictive portion of the data response be analyzed. It appears that wellbore storage could mask the data portion being analyzed and, thus, could render the technique ineffective. For this reason, we prefer interference or pulse testing techniques. However, the technique can be applied for many drillstem tests where only a small interval is tested and wellbore storage is suppressed.

10.9 Summary

The surface has just been scratched in the study of the effects of reservoir heterogeneities on well testing and well test analysis. As indicated in this chapter, many different physical situations can cause similar well-test pressure responses. Thus, it can be dangerous to infer a particular reservoir condition or heterogeneity from a single well-test response. Generally, much additional data from other types of tests, geology, cores, logs, etc., are required to verify the existence of reservoir heterogeneities. The nonexpert should

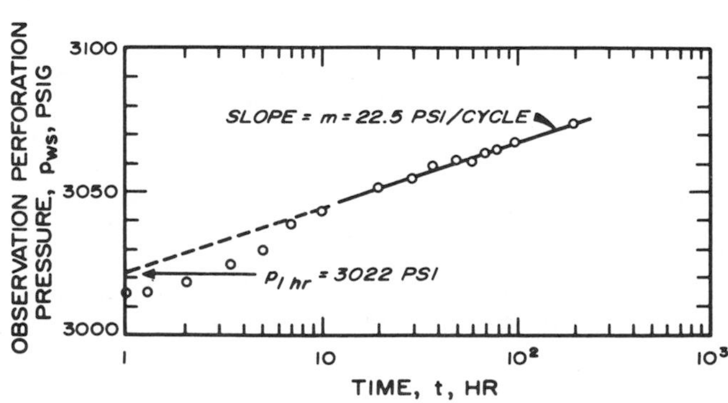

Fig. 10.39 Pressure response for vertical interference test with injection; Example 10.7. After Prats.[54]

be reluctant to make positive interpretations in heterogeneous systems. Perhaps the best thing to do is recognize that a test is unusual because of some peculiarity either in the test or in the reservoir and then call on the services of an expert. In many cases, it will not be possible to determine what the actual situation is to any degree of certainty. Often, even sophisticated computer analyses of single- or multiple-well test results cannot provide a definitive indication of the actual situation.

Vertical well testing is one area that is well enough developed for general application — when the test is performed with sufficient care. The analysis methods presented apply to specific restricted testing conditions but are sufficient for estimating vertical and horizontal permeabilities if other needed reservoir data can be estimated from other sources. Many other methods for vertical well testing are also available and should provide adequate results.

References

1. Matthews, C. S. and Russell, D. G.: *Pressure Buildup and Flow Tests in Wells,* Monograph Series, Society of Petroleum Engineers of AIME, Dallas (1967) **1,** Chap. 10.
2. Jahns, Hans O.: "A Rapid Method for Obtaining a Two-Dimensional Reservoir Description From Well Pressure Response Data," *Soc. Pet. Eng. J.* (Dec. 1966) 315-327; *Trans.,* AIME, **237.**
3. Vela, Saul and McKinley, R. M.: "How Areal Heterogeneities Affect Pulse-Test Results," *Soc. Pet. Eng. J.* (June 1970) 181-191; *Trans.,* AIME, **249.**
4. McKinley, R. M., Vela, Saul, and Carlton, L. A.: "Field Application of Pulse-Testing for Detailed Reservoir Description," *J. Pet. Tech.* (March 1968) 313-321; *Trans.,* AIME, **243.**
5. Horner, D. R.: "Pressure Build-Up in Wells," *Proc.,* Third World Pet. Cong., The Hague (1951) Sec. II, 503-523. Also *Reprint Series, No. 9 — Pressure Analysis Methods,* Society of Petroleum Engineers of AIME, Dallas (1967) 25-43.
6. Dolan, John P., Einarsen, Charles A., and Hill, Gilman A.: "Special Application of Drill-Stem Test Pressure Data," *Trans.,* AIME (1957) **210,** 318-324. Also *Reprint Series, No. 9 — Pressure Analysis Methods,* Society of Petroleum Engineers of AIME, Dallas (1967) 68-74.
7. Russell, D. G.: "Determination of Formation Characteristics From Two-Rate Flow Tests," *J. Pet. Tech.* (Dec. 1963) 1347-1355; *Trans.,* AIME, **228.** Also *Reprint Series, No. 9 — Pressure Analysis Methods,* Society of Petroleum Engineers of AIME, Dallas (1967) 136-144.
8. Bixel, H. C., Larkin, B. K., and van Poollen, H. K.: "Effect of Linear Discontinuities on Pressure Build-Up and Drawdown Behavior," *J. Pet. Tech.* (Aug. 1963) 885-895; *Trans.,* AIME, **228.**
9. Gray, K. E.: "Approximating Well-to-Fault Distance From Pressure Build-Up Tests," *J. Pet. Tech.* (July 1965) 761-767.
10. Ramey, H. J., Jr., and Earlougher, R. C., Jr.: "A Note on Pressure Build-Up Curves," *J. Pet. Tech.* (Feb. 1968) 119-120.
11. Warren, J. E. and Root, P. J.: "The Behavior of Naturally Fractured Reservoirs," *Soc. Pet. Eng. J.* (Sept. 1963) 245-255; *Trans.,* AIME, **228.**
12. Earlougher, Robert C., Jr., Ramey, H. J., Jr., Miller, F. G., and Mueller, T. D.: "Pressure Distributions in Rectangular Reservoirs," *J. Pet. Tech.* (Feb. 1968) 199-208; *Trans.,* AIME, **243.**
13. Collins, Royal Eugene: *Flow of Fluids Through Porous Materials,* Reinhold Publishing Corp., New York (1961) 115.
14. Polubarinova-Kochina, P. Ya.: *Theory of Ground Water Movement,* Princeton U. Press, Princeton, N.J. (1962) 343-369.
15. Elkins, Lincoln F. and Skov, Arlie M.: "Determination of Fracture Orientation From Pressure Interference," *Trans.,* AIME (1960) **219,** 301-304. Also *Reprint Series, No. 9 — Pressure Analysis Methods,* Society of Petroleum Engineers of AIME, Dallas (1967) 97-100.
16. Earlougher, Robert C., Jr., and Kersch, Keith M.: "Field Examples of Automatic Transient Test Analysis," *J. Pet. Tech.* (Oct. 1972) 1271-1277.
17. Ramey, Henry J., Jr.: "Interference Analysis for Anisotropic Formations — A Case History," *J. Pet. Tech.* (Oct. 1975) 1290-1298; *Trans.,* AIME, **259.**
18. Papadopulos, Istavros S.: "Nonsteady Flow to a Well in an Infinite Anisotropic Aquifer," *Proc.,* 1965 Dubrovnik Symposium on Hydrology of Fractured Rocks, Int'l. Assoc. of Sci. Hydrology (1965) **I,** 21-31.
19. Hurst, William: "Interference Between Oil Fields," *Trans.,* AIME (1960) **219,** 175-192.
20. Larkin, Bert K.: "Solutions to the Diffusion Equation for a Region Bounded by a Circular Discontinuity," *Soc. Pet. Eng. J.* (June 1963) 113-115; *Trans.,* AIME, **228.**
21. Loucks, T. L. and Guerrero, E. T.: "Pressure Drop in a Composite Reservoir," *Soc. Pet. Eng. J.* (Sept. 1961) 170-176; *Trans.,* AIME, **222.**
22. Carter, R. D.: "Pressure Behavior of a Limited Circular Composite Reservoir," *Soc. Pet. Eng. J.* (Dec. 1966) 328-334; *Trans.,* AIME, **237.**
23. Odeh, A. S.: "Flow Test Analysis for a Well With Radial Discontinuity," *J. Pet. Tech.* (Feb. 1969) 207-210; *Trans.,* AIME, **246.**
24. Bixel, H. C. and van Poollen, H. K.: "Pressure Drawdown and Buildup in the Presence of Radial Discontinuities," *Soc. Pet. Eng. J.* (Sept. 1967) 301-309; *Trans.,* AIME, **240.** Also *Reprint Series No. 9 — Pressure Analysis Methods,* Society of Petroleum Engineers of AIME, Dallas (1967) 188-196.
25. Kazemi, Hossein, Merrill, L. S., and Jargon, J. R.: "Problems in Interpretation of Pressure Fall-Off Tests in Reservoirs With and Without Fluid Banks," *J. Pet. Tech.* (Sept. 1972) 1147-1156.
26. Merrill, L. S., Jr., Kazemi, Hossein, and Gogarty, W. Barney: "Pressure Falloff Analysis in Reservoirs With Fluid Banks," *J. Pet. Tech.* (July 1974) 809-818; *Trans.,* AIME, **257.**
27. Russell, D. G. and Prats, M.: "The Practical Aspects of Interlayer Crossflow," *J. Pet. Tech.* (June 1962) 589-594. Also *Reprint Series, No. 9 — Pressure Analysis Methods,* Society of Petroleum Engineers of AIME, Dallas (1967) 120-125.
28. Kazemi, Hossein and Seth, Mohan S.: "Effect of Anisotropy and Stratification on Pressure Transient Analysis of Wells With Restricted Flow Entry," *J. Pet. Tech.* (May 1969) 639-647; *Trans.,* AIME, **246.**
29. Woods, E. G.: "Pulse-Test Response of a Two-Zone Reservoir," *Soc. Pet. Eng. J.* (Sept. 1970) 245-256; *Trans.,* AIME, **249.**

30. Lefkovits, H. C., Hazebroek, P., Allen, E. E., and Matthews, C. S.: "A Study of the Behavior of Bounded Reservoirs Composed of Stratified Layers," *Soc. Pet. Eng. J.* (March 1961) 43-58; *Trans.*, AIME, **222.**

31. Duvaut, G.: "Drainage des Systèmes Hétérogènes," *Revue IFP* (Oct. 1961) 1164-1181.

32. Kazemi, Hossein: "Pressure Buildup in Reservoir Limit Testing of Stratified Systems," *J. Pet. Tech.* (April 1970) 503-511; *Trans.*, AIME, **249.**

33. Cobb, William M., Ramey, H. J., Jr., and Miller, Frank G.: "Well-Test Analysis for Wells Producing Commingled Zones," *J. Pet. Tech.* (Jan. 1972) 27-37; *Trans.*, AIME, **253.**

34. Raghavan, R., Topaloglu, H. N., Cobb, W. M., and Ramey, H. J., Jr.: "Well-Test Analysis for Wells Producing From Two Commingled Zones of Unequal Thickness," *J. Pet. Tech.* (Sept. 1974) 1035-1043; *Trans.*, AIME, **257.**

35. Earlougher, Robert C., Jr., Kersch, K. M., and Kunzman, W. J.: "Some Characteristics of Pressure Buildup Behavior in Bounded Multiple-Layer Reservoirs Without Crossflow," *J. Pet. Tech.* (Oct. 1974) 1178-1186; *Trans.*, AIME, **257.**

36. Earlougher, Robert C., Jr., Kersch, K. M., and Ramey, H. J., Jr.: "Wellbore Effects in Injection Well Testing," *J. Pet. Tech.* (Nov. 1973) 1244-1250.

37. Pollard, P.: "Evaluation of Acid Treatments From Pressure Build-Up Analysis," *Trans.*, AIME (1959) **216,** 38-43.

38. Pirson, Richard S. and Pirson, Sylvain J.: "An Extension of the Pollard Analysis Method of Well Pressure Build-Up and Drawdown Tests," paper SPE 101 presented at the SPE-AIME 36th Annual Fall Meeting, Dallas, Oct. 8-11, 1961.

39. Odeh, A. S.: "Unsteady-State Behavior of Naturally Fractured Reservoirs," *Soc. Pet. Eng. J.* (March 1965) 60-64; *Trans.*, AIME, **234.**

40. Warren, J. E. and Root, P. J.: "Discussion of Unsteady-State Behavior of Naturally Fractured Reservoirs," *Soc. Pet. Eng. J.* (March 1965) 64-65; *Trans.*, AIME, **234.**

41. Morris, Earl E. and Tracy, G. W.: "Determination of Pore Volume in a Naturally Fractured Reservoir," paper SPE 1185 presented at the SPE-AIME 40th Annual Fall Meeting, Denver, Oct. 3-6, 1965.

42. Huskey, William L. and Crawford, Paul B.: "Performance of Petroleum Reservoirs Containing Vertical Fractures in the Matrix," *Soc. Pet. Eng. J.* (June 1967) 221-228; *Trans.*, AIME, **240.**

43. Adams, A. R., Ramey, H. J., Jr., and Burgess, R. J.: "Gas Well Testing in a Fractured Carbonate Reservoir," *J. Pet. Tech.* (Oct. 1968) 1187-1194; *Trans.*, AIME, **243.**

44. Kazemi, H.: "Pressure Transient Analysis of Naturally Fractured Reservoirs With Uniform Fracture Distribution," *Soc. Pet. Eng. J.* (Dec. 1969) 451-462; *Trans.*, AIME, **246.**

45. Kazemi, H., Seth, M. S., and Thomas, G. W.: "The Interpretation of Interference Tests in Naturally Fractured Reservoirs With Uniform Fracture Distribution," *Soc. Pet. Eng. J.* (Dec. 1969) 463-472; *Trans.*, AIME, **246.**

46. Levorsen, A. I.: *Geology of Petroleum,* 2nd ed., W. H. Freedman and Co., San Francisco (1967) 125.

47. Vairogs, Juris, Hearn, C. L., Dareing, Donald W., and Rhoades, V. W.: "Effect of Rock Stress on Gas Production From Low-Permeability Reservoirs," *J. Pet. Tech.* (Sept. 1971) 1161-1167; *Trans.*, AIME, **251.**

48. Thomas, Rex D. and Ward, Don C.: "Effect of Overburden Pressure and Water Saturation on the Gas Permeability of Tight Sandstone Cores," *J. Pet. Tech.* (Feb. 1972) 120-124.

49. Vairogs, Juris and Rhoades, Vaughan W.: "Pressure Transient Tests in Formations Having Stress-Sensitive Permeability," *J. Pet. Tech.* (Aug. 1973) 965-970; *Trans.*, AIME, **255.**

50. Raghavan, R., Scorer, J. D. T., and Miller, F. G.: "An Investigation by Numerical Methods of the Effect of Pressure-Dependent Rock and Fluid Properties on Well Flow Tests," *Soc. Pet. Eng. J.* (June 1972) 267-275; *Trans.*, AIME, **253.**

51. Ramey, H. J., Jr.: "Non-Darcy Flow and Wellbore Storage Effects in Pressure Build-Up and Drawdown of Gas Wells," *J. Pet. Tech.* (Feb. 1965) 223-233; *Trans.*, AIME, **234.** Also *Reprint Series, No. 9 — Pressure Analysis Methods,* Society of Petroleum Engineers of AIME, Dallas (1967) 233-243.

52. Wattenbarger, Robert A. and Ramey, H. J., Jr.: "Gas Well Testing With Turbulence, Damage, and Wellbore Storage," *J. Pet. Tech.* (Aug. 1968) 877-887; *Trans.*, AIME, **243.**

53. Burns, William A., Jr.: "New Single-Well Test for Determining Vertical Permeability," *J. Pet. Tech.* (June 1969) 743-752; *Trans.*, AIME, **246.**

54. Prats, Michael: "A Method for Determining the Net Vertical Permeability Near a Well From In-Situ Measurements," *J. Pet. Tech.* (May 1970) 637-643; *Trans.*, AIME, **249.**

55. Hirasaki, George J.: "Pulse Tests and Other Early Transient Pressure Analyses for In-Situ Estimation of Vertical Permeability," *Soc. Pet. Eng. J.* (Feb. 1974) 75-90; *Trans.*, AIME, **257.**

56. Falade, Gabriel K. and Brigham, William E.: "The Dynamics of Vertical Pulse Testing in a Slab Reservoir," paper SPE 5055A presented at the SPE-AIME 49th Annual Fall Meeting, Houston, Oct. 6-9, 1974.

57. Falade, Gabriel K. and Brigham, William E.: "The Analysis of Single-Well Pulse Tests in a Finite-Acting Slab Reservoir," paper SPE 5055B presented at the SPE-AIME 49th Annual Fall Meeting, Houston, Oct. 6-9, 1974.

58. Raghavan, R. and Clark, K. K.: "Vertical Permeability From Limited Entry Flow Tests in Thick Formations," *Soc. Pet. Eng. J.* (Feb. 1975) 65-73; *Trans.*, AIME, **259.**

Chapter 11

Effect of Wellbore Conditions on Pressure Behavior

11.1 Introduction

Chapter 2 introduced the concepts of wellbore storage (Section 2.6) and partial well completions (Section 2.5) and indicated how those phenomena affect transient test behavior. This chapter further discusses the effects of more complex wellbore storage conditions and partial completions, as well as the effects of hydraulic fracturing and slanted holes on transient pressure behavior. Such features usually must be considered when analyzing well test data — if they are not, incorrect formation permeabilities and skin factors may be estimated. Therefore, it is important to recognize test data that are influenced by special wellbore conditions.

11.2 Changing Wellbore Storage

Wellbore storage[1-4] influences pressure transient data as discussed in Section 2.6. The wellbore storage coefficient is defined as the change in total volume of wellbore fluids per unit change in bottom-hole pressure,

$$C=\frac{\Delta V}{\Delta p}, \quad \text{(11.1)}$$

where

C = wellbore storage constant (coefficient, factor), bbl/psi
ΔV = change in volume of fluid in the wellbore at wellbore conditions, bbl
Δp = change in bottom-hole pressure, psi.

Sometimes ΔV is defined to include the volume of fractures as well as of the wellbore.

When the relationship between ΔV and Δp does not change during a well test, the wellbore storage coefficient is constant and usually may be estimated from the well completion. Then,

$$C=\frac{V_u}{\left(\frac{\rho}{144}\frac{g}{g_c}\right)}, \quad \text{(11.2)}$$

for a changing liquid level and

$$C=c\,V_w, \quad \text{(11.3)}$$

for a completely liquid- or gas-filled wellbore. The dimensionless wellbore storage coefficient is

$$C_D=\frac{5.6146\,C}{2\pi\phi c_t h r_w^2}. \quad \text{(11.4)}$$

For a constant wellbore storage coefficient, we may estimate the time when wellbore storage stops influencing a transient well test. That time corresponds to the beginning of the semilog straight line on the normal semilog transient data plot (see Chapters 3 through 5 and Chapter 7). As indicated in Section 2.6, wellbore storage effects are essentially negligible for drawdown and injection when[4]

$$t_D>(60+3.5s)\,C_D, \quad \text{(11.5a)}$$

or

$$t>\frac{(200{,}000+12{,}000s)\,C}{(kh/\mu)}. \quad \text{(11.5b)}$$

For pressure buildup and falloff tests, the corresponding times are[5]

$$t_D>50\,C_D e^{0.14s}, \quad \text{(11.6a)}$$

or

$$t>\frac{170{,}000\,C\,e^{0.14s}}{(kh/\mu)}. \quad \text{(11.6b)}$$

The wellbore storage coefficient often is not constant throughout a well test.[6,7] Abrupt changes in the wellbore storage coefficient are easy to visualize and occur relatively frequently. Fig. 11.1 depicts a wellbore condition that can cause an *increasing* wellbore storage coefficient. When an injection well with a positive wellhead pressure is shut in for a falloff test, the wellhead pressure remains high immediately after shut-in. However, a few minutes (or hours) later, the bottom-hole pressure falls below the pressure required to maintain the liquid column to the surface and the liquid level begins to fall (the well "goes on vacuum"). When that happens, the wellbore storage coefficient increases from one for a compressible system (Eq. 11.3) to one for a falling liquid-level system (Eq. 11.2); the change can be by a factor of 100 or more. Fig. 11.2 shows both log-log and semilog falloff curves for such a step increase in wellbore storage coefficient; the light curves are for constant wellbore storage coefficients, and the heavy curves show what happens when the wellbore storage coefficient increases at shut-in time Δt_1. The log-log data plot initially shows a unit-slope straight-line pressure response; then the response flattens and finally steepens as the pressure approaches the response for the larger wellbore storage coefficient. The semilog data plot shows a flattening period, a steepening, and a final flattening. The correct semilog straight line is reached at Δt_2. The abrupt change in slope of the data plot (flattening period) normally corresponds to the wellhead pressure reaching atmospheric (going on vacuum). Fig. 11.3 shows data from a falloff test with increasing wellbore storage.[7] Example 11.1 discusses that test in more detail. If there is a pressurized gas cushion in the well, the wellbore storage coefficient still decreases, but not as abruptly as illustrated in Fig. 11.2. The flattening occurs, but it is muted considerably. That is actually the case for the test shown in Fig. 11.3.

For the wellbore storage coefficient to increase, the liquid level must begin to fall *during* (rather than at the start of) the well test. A common situation for that occurrence is a pressure falloff test in an injection well. However, similar behavior may occur during drawdown testing of pumping wells. The fluid level starts falling when pumping reduces bottom-hole pressure below that required to maintain a fluid column to the surface or to a packer, thus giving a larger storage coefficient.

It is important to recognize that the shape of the falloff curve shown in Figs. 11.2 and 11.3 is quite similar to the falloff-curve shape that might be expected from a layered reservoir without communication between the layers. Comparing Figs. 11.2 and 11.3 with Fig. 10.14 shows that many physical situations can result in a single characteristic well-test-curve shape.

Decreasing wellbore storage can occur during buildup testing in a production well or during injectivity testing. Fig. 11.4 illustrates a typical production-well completion that can cause a decreasing wellbore storage coefficient. While the well is pumping, the liquid level stands below the packer. The liquid level is low just after shut-in, but rises as pressure increases. Gas in the wellbore is either compressed or redissolved. When the liquid level reaches the packer (there may be a small gas cushion), the wellbore storage coefficient drops from the relatively large value for a rising liquid level with annular volume between the casing and tubing to the relatively small value for a compression-controlled situation. (Normally, in this situation, the tubing is held full of liquid by the standing valve in the pump and that volume does not play a part in test response.) If there is gas above the liquid, its compressibility must be considered when estimating the wellbore storage coefficient. In that case, it is generally best to use Eq. 11.1 and a careful analysis of the situation; Example 11.1 illustrates such a

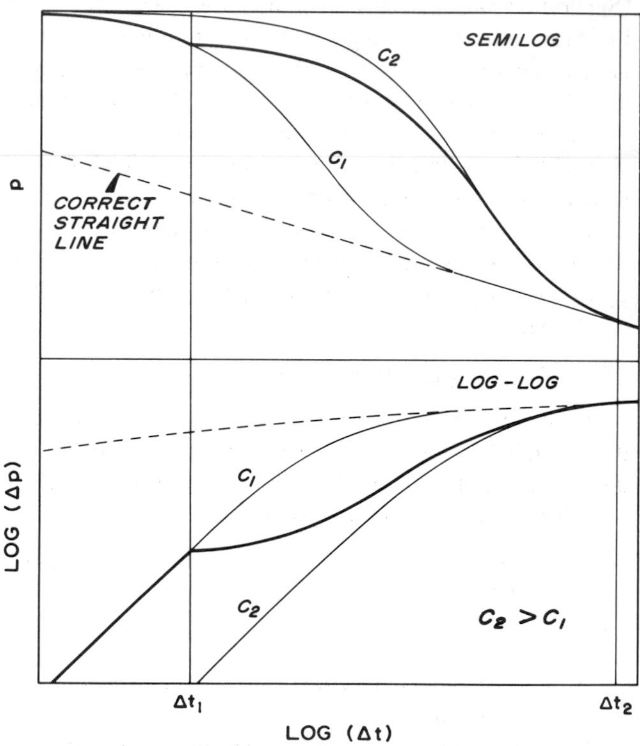

Fig. 11.2 Log-log and semilog theoretical falloff curves for a step increase in wellbore storage coefficient. After Earlougher, Kersch, and Ramey.[7]

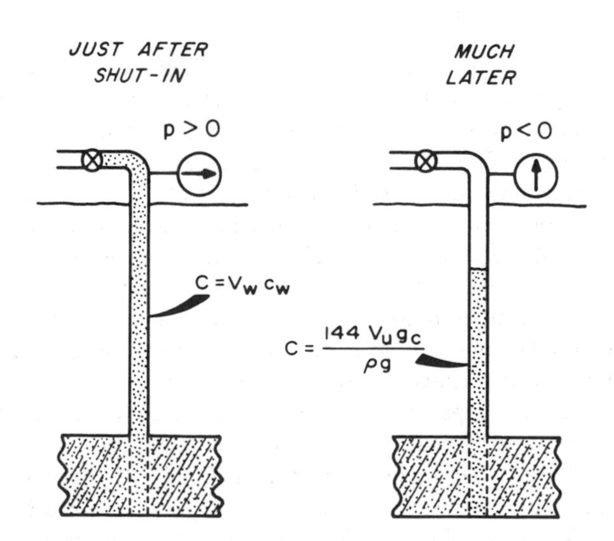

Fig. 11.1 Increasing wellbore storage — shut-in injection well going on vacuum.

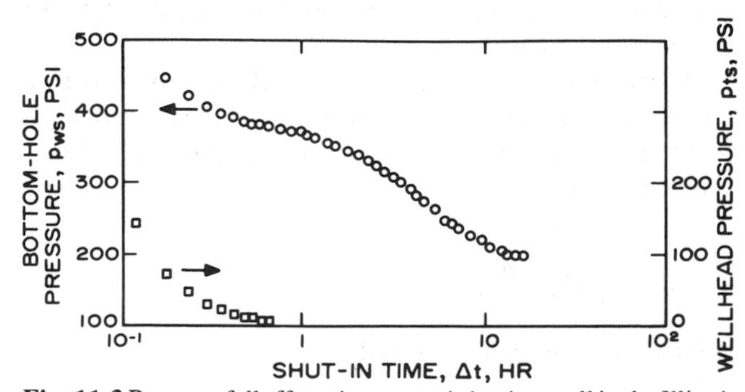

Fig. 11.3 Pressure falloff test in a water injection well in the Illinois basin; increasing wellbore storage. After Earlougher, Kersch, and Ramey.[7]

computation. Wellbore storage also may decrease during an injectivity test. To visualize that case, reverse the sequence shown in Fig. 11.1 and picture injection into a well with a liquid level standing below the surface. As the bottom-hole pressure increases, the liquid level rises until it reaches the surface. The result is a compression-controlled wellbore storage coefficient.

Fig. 11.5 shows theoretical data plots for decreasing wellbore storage.[7] In the log-log plot, the data initially fall on the line for the higher wellbore storage coefficient. When the coefficient decreases at time t_1, the pressure increases rapidly until it reaches the line for the lower wellbore storage coefficient. The semilog plot also shows a rapid increase in slope. Data reach the correct semilog straight line at t_2. Fig. 11.6 shows actual data for an injectivity test with a decreasing wellbore storage coefficient.[7] Note that the test data start with a relatively low slope that rapidly increases and finally flattens to the semilog straight line indicated.

When the wellbore storage coefficient changes during a transient well test, the *second* storage coefficient determines when the correct semilog straight line will begin. Thus, for the case in Fig. 11.2, wellbore storage coefficient C_2 would be used in Eq. 11.5 or Eq. 11.6 to estimate the starting time for the semilog straight line. For the situation in Fig. 11.5, wellbore storage coefficient C_1 would be used. An exception would be certain cases of increasing wellbore storage coefficient where the increase in wellbore storage did not occur until a test time exceeding the "die-out" time (Eq. 11.5 or Eq. 11.6) based on the smaller initial storage coefficient. In any event, afterflow should be negligible at the die-out time based on the larger final wellbore storage coefficient.

Changing wellbore storage is usually easy to detect in a transient well test — if the engineer is cognizant of its characteristics. Since storage usually changes at a bottom-hole pressure corresponding to the hydrostatic pressure of the fluid column in the well, verification is straightforward. In some cases, it is easier to recognize changing wellbore storage effects on the semilog plot than on the log-log plot, although both are recommended for diagnostic purposes. Ref. 7 illustrates when it may be difficult to recognize changing wellbore storage from the log-log plot.

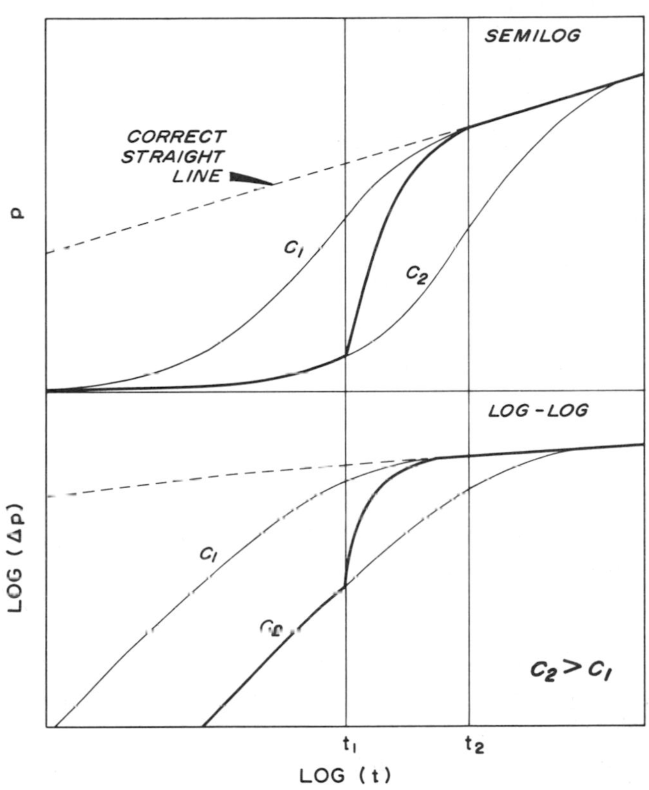

Fig. 11.5 Theoretical log-log and semilog plots for an injectivity test with a step decrease in wellbore storage. After Earlougher, Kersch, and Ramey.[7]

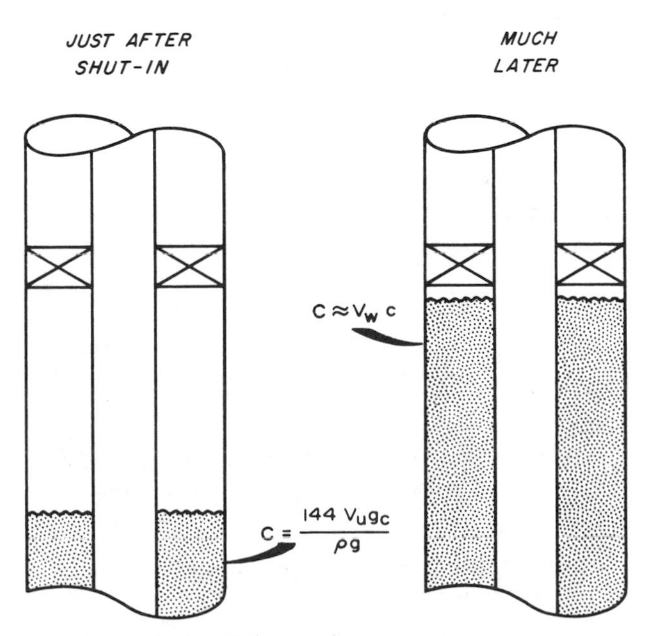

Fig. 11.4 Decreasing wellbore storage — fluid level in shut-in pumping well reaches packer. V_w = total annular volume below the packer.

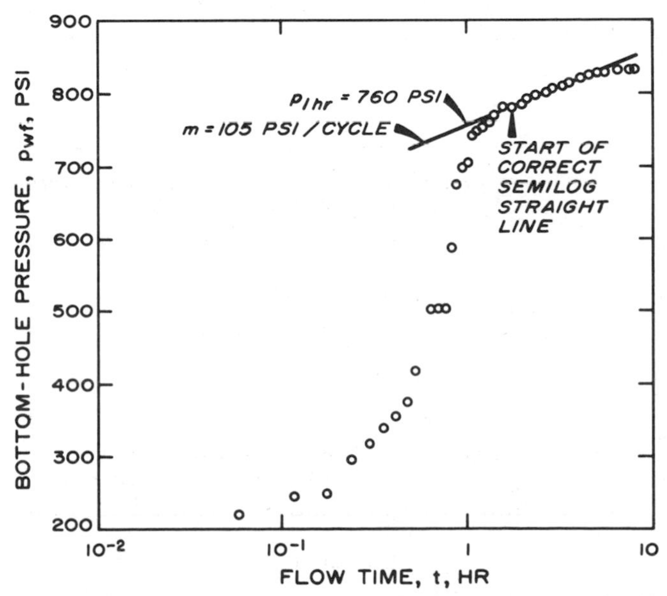

Fig. 11.6 Injectivity test for the well of Fig. 11.3; decreasing wellbore storage. After Earlougher, Kersch, and Ramey.[7]

Example 11.1 Effect of Changing Wellbore Storage on Falloff and Injectivity Tests

Earlougher, Kersch, and Ramey[7] present the falloff and injectivity test data shown in Figs. 11.3 and 11.6. Fig. 11.7 is the log-log data plot for the two tests. Known data are

$q_w = -100$ STB/D
$B_w = 1.0$ RB/STB
$\mu_w = 1.0$ cp
$h = 16$ ft
$\phi = 0.22$
$c_t = 7 \times 10^{-6}$ psi^{-1}
$r_w = 0.29$ ft
depth = 994 ft, and

2-in. EUE tubing with packer set at 979 ft. Also,

$$p_{wf}(\Delta t = 0) = 837 \text{ psi}$$

and

$$p_{tf}(\Delta t = 0) = 457 \text{ psi}$$

for the falloff, and

$$p_{ws}(t = 0) = 194 \text{ psi}$$

and

$$p_{ts}(t = 0) = 0 \text{ psi}$$

for the injectivity test.

The falloff data in Figs. 11.3 and 11.7 have the characteristic shape of an increasing wellbore storage coefficient. The increase starts at about 0.3 hour. That is not when the wellhead pressure reaches atmospheric, as we would expect if the wellbore were *full* of water at the start of the test. The data show that the difference between surface and bottom-hole pressure is only 837 − 457 = 380 psi. Using 0.433 psi/ft as the static gradient for water, 380 psi corresponds to 878 ft of water column. Thus, there must have been about 994 − 878 = 116 ft of gas above the water column, assuming no significant static pressure gradient in the gas.

Assuming that the gas column does exist and that it is 116 ft long, the wellbore storage coefficient at the start of the test can be estimated by applying Eq. 11.1. We assume ΔV = 0.1 bbl and estimate the resulting Δp. For 2-in. tubing, $V_u \simeq$ 0.004 bbl/ft, so 0.1 bbl corresponds to 25 ft of fluid-level change. Applying the static water gradient, and assuming the perfect gas law may be used for the gas column,

$$\Delta p = 25(0.433) + 457\left(\frac{116}{116-25} - \frac{116}{116}\right)$$

$$= 136 \text{ psi}.$$

Then,

$$C = 0.1/136 = 7.35 \times 10^{-4} \text{ bbl/psi}.$$

In a similar manner we find that $C = 2.3 \times 10^{-3}$ bbl/psi when the liquid level has fallen 100 ft, indicating a continual increase in C. The maximum value of C must occur when wellhead pressure reaches atmospheric; then Eq. 11.2 applies and

$$C_{\max} = \frac{0.004}{(62.4/144)(32.17/32.17)}$$

$$= 9.23 \times 10^{-3} \text{ bbl/psi}.$$

Since the beginning of the semilog straight line is determined by the final wellbore storage coefficient, Eq. 11.6b can be used with assumed values of $kh/\mu = 155$ md ft/cp and $s = 0.9$ to estimate

$$t_{bsl} > \frac{(170{,}000)(9.23 \times 10^{-3})\, e^{(0.14)(0.9)}}{155}$$

$$> 11.5 \text{ hours}.$$

Only the last four data points in Fig. 11.3 occur after 11.5 hours, and they are not adequate for analysis. Thus, we must conclude that we cannot analyze this increasing-wellbore-storage falloff test.

During the injectivity test the wellbore storage coefficient decreases, as indicated by Figs. 11.6 and 11.7. The correct semilog straight line, with $m = 105$ psi/cycle and $p_{1hr} = 760$ psi, is shown in Fig. 11.6. Using Eq. 7.4,

$$k = \frac{(-162.6)(-100)(1.0)(1.0)}{(105)(16)} = 9.7 \text{ md},$$

and

$$kh/\mu = (9.7)(16)/(1.0) = 155 \text{ md ft/cp}.$$

From Eq. 7.5,

$$s = 1.1513\left\{\frac{760-194}{105} - \log\left[\frac{9.7}{(0.22)(1)(7\times10^{-6})(0.29)^2}\right] + 3.2275\right\}$$

$$= 0.9.$$

From the unit-slope straight line in Fig. 11.7 for the injectivity test, $\Delta p = 41.5$ psi at $\Delta t = 0.1$ hour. Then, from Eq. 2.20,

$$C = \frac{(100)(1)(0.1)}{(24)(41.5)} = 0.0100 \text{ bbl/psi}.$$

C_D at the beginning of both the falloff and the injectivity tests can be estimated from the wellbore storage coefficients at the beginning of the tests and Eq. 11.4. For the falloff test,

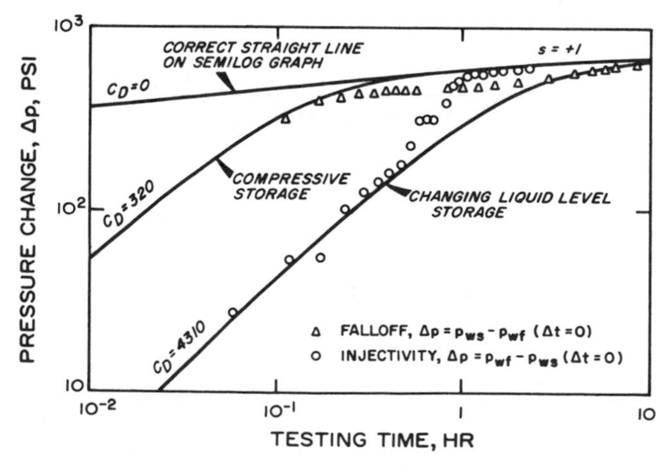

Fig. 11.7 Log-log plot of falloff and injectivity data of Figs. 11.3 and 11.6, Example 11.1. After Earlougher, Kersch, and Ramey.[7]

$$C_{Di} = \frac{(5.6146)(7.35 \times 10^{-4})}{2\pi(0.22)(7 \times 10^{-6})(16)(0.29)^2} \simeq 320,$$

and for the injectivity test,

$$C_{Di} = \frac{(5.6146)(0.0100)}{2\pi(0.22)(7 \times 10^{-6})(16)(0.29)^2} = 4{,}310.$$

The p_D vs t_D curves for those two values of C_D and for s = 1 have been matched to the observed data in Fig. 11.7. The curves verify the changing wellbore storage condition.

If a well test encounters a changing wellbore storage condition, it may be possible to analyze test data by being careful, or it may be necessary to devise a test that minimizes or eliminates the wellbore storage change. Frequently, tests with increasing wellbore storage are not interpretable. That is because of the unlikely appearance of the correct semilog straight line during the compressive wellbore storage period and the long duration of the falling-liquid-level wellbore storage period. The second factor often results in an inadequate semilog straight-line definition because of insufficient shut-in time or boundary or interference effects. In fortunate circumstances, the compressive storage may last long enough for the semilog straight line to develop and the test may be analyzed using data before the liquid level starts to fall. Tests with a decreasing storage coefficient (injectivity or buildup) have more potential for analysis than tests with increasing wellbore storage (falloff or drawdown). Figs. 11.2 and 11.5 indicate that the semilog straight line is reached sooner when the wellbore storage decreases than when it increases. In some instances, it may be worthwhile to use a two-rate test (Section 4.3), or some other multiple-rate test (Chapter 4), to attempt preventing a changing storage situation.

11.3 Artificially Fractured Wells

Since the beginning of intentional hydraulic fracturing of wells,[8] thousands of production and injection wells have been fractured hydraulically. Hydraulic fracturing has a definite effect on pressure transient response, so we should be aware of that effect when analyzing well test data. Although both horizontal and vertical fractures may be induced by the hydraulic fracturing process, it is believed that essentially all induced fractures at depths greater than 3,000 ft are vertical.[8] Thus, most studies of pressure transient behavior in fractured wells have been devoted to vertically fractured wells,[9-15] while horizontally fractured wells have been studied less thoroughly.[13,16-18] Gringarten, Ramey, and Raghavan[19] discuss several techniques for analyzing transient test data for fractured wells.

Vertically Fractured Wells

Both infinite and closed systems containing a vertically fractured well have been studied thoroughly. Additional information is still needed about vertically fractured wells in drainage areas with constant-pressure boundaries. Such information would be more applicable to wells in waterfloods after fillup and in strong water-drive reservoirs. Also, more data are needed for geometries other than currently discussed in the literature.

Fig. 11.8 defines nomenclature for a closed-square system with a vertically fractured well at its center. The half-fracture length, x_f, and the half-length of the side of the square, x_e, are commonly used to characterize the system. In all systems discussed in this chapter, we assume that the fracture fully penetrates the vertical extent of the formation and is the same length on both sides of the well; in closed systems the fracture is parallel to a boundary. Only one fracture is considered.

Figs. C.3, C.4, C.17, C.18, and C.19 show dimensionless pressure data for vertically fractured, infinite, and closed systems. The difference between an infinite-conductivity and a uniform-flux fracture is explained in Section C.2. Except for highly propped and conductive fractures, it is thought that the uniform-flux fracture better represents reality than the infinite-conductivity fracture.

In either infinite systems or in closed systems with relatively short vertical fractures ($x_e/x_f > 1.5$), the *early-time* flow behavior is linear* from the formation into the fracture. (If the wellbore storage coefficient is large, the linear flow portion may be obscured.) Eq. C.8 gives the dimensionless pressure as a function of dimensionless time for the linear flow period. By using Eqs. 2.2 and C.8, the pressure at the well during the linear flow period may be written as[10]

$$p_{ws} = p_i + m_{Vf}\sqrt{t}\,. \quad \ldots\ldots\ldots\ldots\ldots \quad (11.7)$$

Eq. 11.7, which applies for drawdown or injection, indicates that a plot of bottom-hole pressure vs the square root of time should have an early-time straight-line portion with intercept p_i. The slope is

$$m_{Vf} = \frac{-4.064\,qB}{h}\sqrt{\frac{\mu}{k\phi c_t x_f^2}}\,. \quad \ldots\ldots\ldots \quad (11.8)$$

The slope of the $\sqrt{t}$ plot may be used to estimate

$$kx_f^2 = \left(\frac{-4.064\,qB}{m_{Vf}h}\right)^2 \frac{\mu}{\phi c_t}\,. \quad \ldots\ldots\ldots \quad (11.9)$$

*But a recent paper indicates that systems with finite conductivity vertical fractures may not exhibit linear flow behavior. See Cinco, Heber, Samaniego-V., F., and Dominguez-A., N.: "Transient Pressure Behavior for a Well With a Finite-Conductivity Vertical Fracture," paper SPE 6014 presented at the SPE-AIME 51st Annual Fall Technical Conference and Exhibition, New Orleans, Oct. 3-6, 1976.

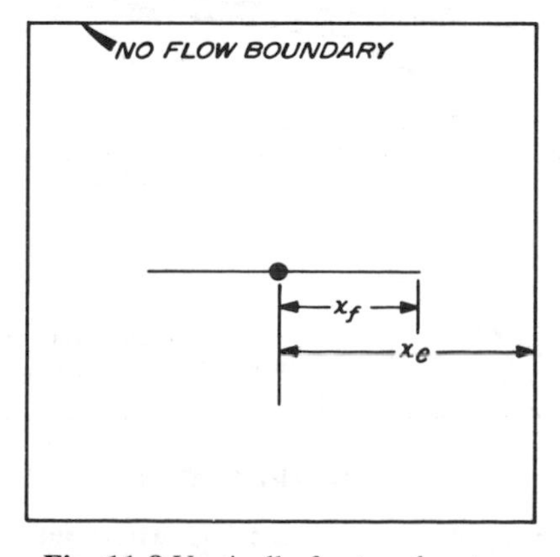

Fig. 11.8 Vertically fractured system.

For values of $x_e/x_f > 1$, the linear flow period ends at about $t_{Dxf} \simeq 0.016$ for an infinite-conductivity vertical fracture and at $t_{Dxf} \simeq 0.16$ for a uniform-flux vertical fracture.

After the initial linear flow period, there is a transition to an infinite-acting, pseudoradial flow period in which the normal semilog analysis (Chapters 3 through 7) applies. The pseudoradial period begins at $t_{Dxf} \simeq 3$ for the infinite-conductivity case and at $t_{Dxf} \simeq 2$ for the uniform-flux case. Eqs. C.7 and C.10 give the dimensionless pressure for that flow period, providing boundary effects are not encountered. In a *closed system,* the infinite-acting, pseudoradial behavior only develops completely if $x_e/x_f > 5$. When pseudoradial flow occurs, permeability may be estimated from the semilog straight-line slope with the familiar equation

$$k = \frac{\pm 162.6\, qB\mu}{mh}, \qquad (11.10)$$

where the appropriate sign is chosen depending on the test type, as indicated in Chapters 3 through 7 and Appendix E.

Wattenbarger and Ramey[12] have observed that there is an approximate relationship between the pressure change at the end of the linear flow period, Δp_{el}, and at the beginning of the semilog straight line, Δp_{bsl}. Since the linear flow period is a straight line on a p vs $\sqrt{t}$ plot (or a line of one-half slope on a log Δp vs log Δt plot), it is generally not too difficult to estimate Δp_{el}, the pressure change at the end of the linear flow period. If the semilog straight line develops, it is also possible to estimate Δp_{bsl}, the pressure change at the beginning of the semilog straight line. If the relationship

$$\Delta p_{bsl} \geq 2\, \Delta p_{el}, \qquad (11.11)$$

is not satisfied, it is likely that either an incorrect linear flow period or an incorrect radial flow period has been chosen. Fig. 11.9 shows dimensionless pressure for an infinite-acting, vertically fractured system during both linear and radial flow periods. Note that the data curve on the $\sqrt{t}$ scale appears to be straight at times after the end of the linear flow period. Both Fig. 11.9 and experience indicate it is not difficult to find an *apparent* $\sqrt{t}$ straight line in the transition period from linear to radial flow. The slope of that apparent straight line is not related to formation permeability and fracture length, as indicated by Eq. 11.9. Fig. 11.9 illustrates the importance of making the check indicated by Eq. 11.11.

Pressure buildup (falloff) testing in vertically fractured wells is similar to that in unfractured wells. However, when performing the superposition to get the effect of the shut-in period (Section 5.2), it must be understood that the linear flow period does not last long. Thus, if the flow period is long enough to deviate from linear flow, Eq. C.8 (or Eq. 11.7), which indicates pressure is a function of $\sqrt{t}$, cannot be used in the flow portion of the superposition. Instead, Eq. C.7 or Eq. C.10, which are longer-time solutions for the

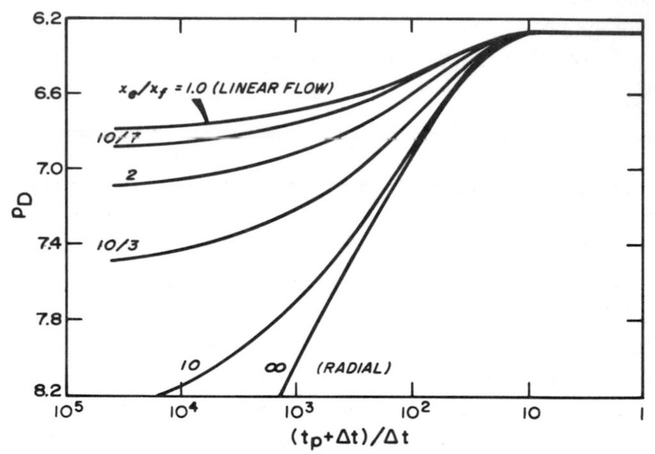

Fig. 11.10 Horner graph for a vertically fractured well in the center of a closed-square reservoir, $t_{pDA} = 1.0$. After Russell and Truitt.[9]

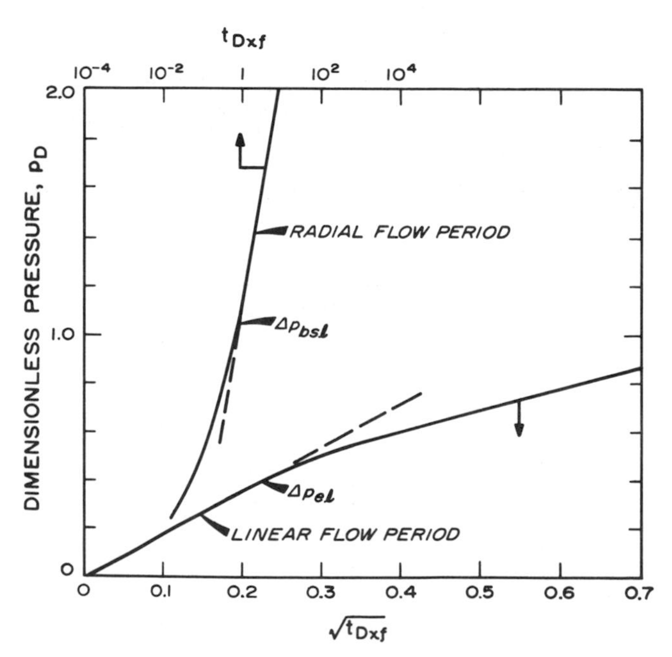

Fig. 11.9 Linear and radial flow periods for a fractured well. After Wattenbarger and Ramey;[12] data of Russell and Truitt.[9]

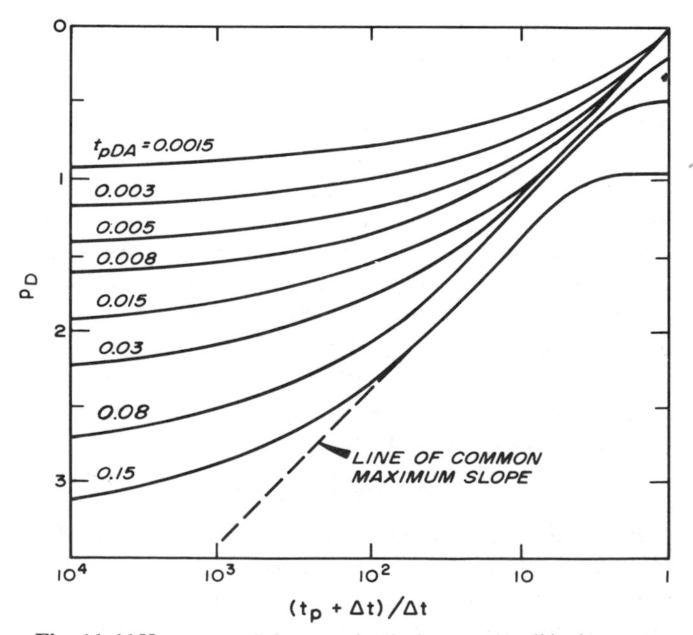

Fig. 11.11 Horner graph for a vertically fractured well in the center of a closed-square reservoir, $x_e/x_f = 10$. After Raghavan, Cady, and Ramey.[11]

pseudoradial flow period (that still assume no boundary effects) may have to be substituted. Care is required in devising an analysis technique under such circumstances.

Pressure data from the pseudoradial flow portion of a buildup (falloff) test is analyzed using the normal Horner plot, p_{ws} vs $\log[(t_p + \Delta t)/\Delta t]$. Figs. 11.10 and 11.11 show theoretical Horner plots for pressure buildup in vertically fractured systems. The two figures indicate the effect of fracture length and producing time on the Horner-plot slope. The maximum Horner-plot slope depends on fracture length (Fig. 11.10), but is independent of producing time for a fixed fracture length (Fig. 11.11) providing the buildup or falloff test is of sufficient duration to show the maximum slope. This indicates that permeabilities estimated from the measured Horner-plot slope must be modified as indicated below.

To correct permeability estimated from a Horner or Miller-Dyes-Hutchinson (MDH) plot, we use (when shut-in lasts long enough to observe the maximum slope)

$$k = k_C \left[\frac{(kh)_{tr}}{(kh)_a}\right]_{\text{Fig. 11.12}}$$

$$= k_C F_{cor}, \quad \ldots \quad (11.12)$$

where k_C is the k value calculated with Eq. 11.10; the ratio of true to apparent kh, F_{cor}, is taken from Fig. 11.12.* In Fig. 11.12 the upper solid line applies to the Horner data plot for any producing time. If the MDH data plot is used, the correction factor is a function of production time up to $t_{pDA} = 0.12$. Because of its independence of t_p and its lower

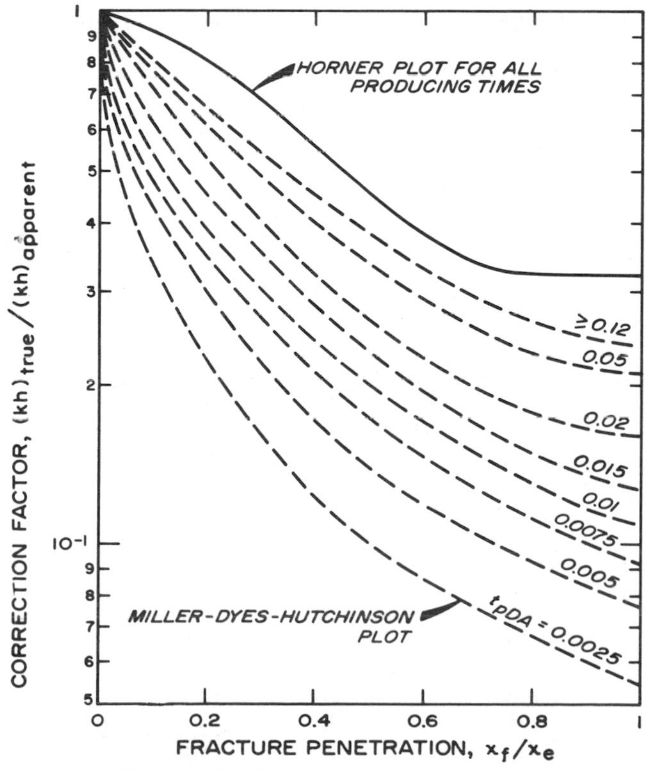

Fig. 11.12 Correction factor for kh estimated from pressure buildup tests in vertically fractured wells, assuming sufficient shut-in time to reach maximum slope. After Raghavan, Cady, and Ramey.[11]

sensitivity (correction factor nearer 1.0), the Horner graph is recommended for analyzing data from vertically fractured wells. To use Eq. 11.12, it is necessary to know x_f/x_e (note in Fig. 11.12 F_{cor} is plotted vs x_f/x_e rather than x_e/x_f used in most figures in this monograph). If both the linear ($\sqrt{t}$) and radial ($\log t$) flow periods develop, it is possible to estimate kx_f^2 from Eq. 11.9 and k from Eq. 11.12 using a reasonable value of x_f/x_e in Fig. 11.12. Then the estimated k may be used with results from Eq. 11.9 to estimate x_f. That x_f is used to compute a new x_f/x_e to be used in Fig. 11.12 and to improve the estimation of k. This process continues until two successive values are the same. If both flow periods do not develop, it is necessary to make an independent estimate of fracture length (from fracture-job parameters) or of permeability (from an unfractured nearby well). (Fig. 9.10 and Ref. 20 indicate that interference testing in a system with a vertically fractured well may provide incorrect results. A more sophisticated analysis is required in that instance.)

Eqs. 11.9, 11.10, and 11.12 also may be combined to give another useful form for estimating fracture length similar to that proposed by Clark:[10]

$$x_f = \frac{0.3187}{m_{Vf}} \sqrt{\frac{mqB}{\phi c_t h F_{cor}}}, \quad \ldots \quad (11.13)$$

where m is the apparent semilog straight-line slope and F_{cor} is the factor that corrects m to the true semilog slope, as in Eq. 11.12. F_{cor}, given in Fig. 11.12, has a minimum value of about 0.32 for a Horner plot and 0.23 for an MDH plot for a closed-square drainage area when production time is sufficient to reach pseudosteady state ($t_{pDA} > 0.12$). When flow times are less than required to reach pseudosteady state, much lower values of F_{cor} (larger corrections) can be applicable. The proper F_{cor} can be estimated by an iterative technique similar to the one described above.

An alternate approach to analyzing fractured-well transient test data is type-curve matching. Gringarten, Ramey, and Raghavan[19] provide a good illustration. Curve matching normally would be performed with Fig. C.3, Fig. C.18, or Fig. C.19 using the technique described in Section 3.3 and illustrated in Fig. 3.5. Once a match point is chosen, the dimensionless parameters on the axis of the type curve are used to estimate formation permeability and fracture length. For Figs. C.3, C.18, and C.19, permeability would be estimated from the pressure match using[19]

$$k = \frac{141.2\, qB\mu}{h} \frac{(p_D)_M}{(\Delta p)_M}, \quad \ldots \quad (11.14)$$

where the data-plot match point, $(\Delta p)_M$, falls on top of the type-curve match point, $(p_D)_M$. The fracture length is estimated from the time-axis match point:[19]

$$x_f = \sqrt{\frac{0.0002637\, k\, (\Delta t)_M}{\phi \mu c_t\, (t_{Dxf})_M}}. \quad \ldots \quad (11.15)$$

*Fig. 11.12 applies for a vertical fracture in a closed-square region. A recent paper (Raghavan, R.: "Analysis of Pressure Data for Fractured Wells: The Constant-Pressure Outer Boundary," paper SPE 6015 presented at the SPE-AIME 51st Annual Fall Technical Conference and Exhibition, New Orleans, Oct. 3-6, 1976) gives similar information for a vertically fractured well in a constant-pressure-boundary square. In that case, the Horner graph for pressure buildup analysis is also best, but the correction factors also depend on production time, t_p.

If all test data fall on the half-slope line on the log Δp vs log Δt plot (the straight line in the $\sqrt{t}$ plot), then permeability cannot be estimated by either type-curve matching or semilog plotting. This situation often occurs in tight gas wells, where the linear flow period may last for several hundred hours. However, the last data point on the half-slope line (on the $\sqrt{t}$ straight line) may be used to estimate an upper limit of the permeability:[19]

$$k \leqslant \frac{(0.215)(141.2)\,qB\mu}{h\,\Delta p}, \quad \ldots\ldots\ldots\ldots (11.16)$$

where Δp is the observed pressure change at the last point on the half-slope ($\sqrt{t}$) straight line. That permeability and the corresponding time value for the last point on the $\sqrt{t}$ straight line, t, may be used to estimate a minimum fracture length:[19]

$$x_f \geqslant \sqrt{\frac{0.0002637\,kt}{(0.016)\,\phi\mu c_t}}. \quad \ldots\ldots\ldots\ldots (11.17)$$

Eqs. 11.16 and 11.17 apply only for $x_e/x_f >> 1$ and for infinite-conductivity fractures. If the fracture is expected to be more like a uniform-flux fracture, 0.215 in Eq. 11.16 becomes 0.76 and 0.016 in Eq. 11.17 becomes 0.16.

Often in type-curve matching applications, it is also possible to estimate reservoir size. In Figs. C.18 and C.19 the dimensionless pressure deviates upward from the infinite-system solution as boundary effects become important. When observed data fall on one of the upward-deviating curves, the x_e/x_f parameter on that curve may be used to estimate x_e and, thus, drainage area, assuming that the square system shape applies (if it does not, it is unlikely that a good match will be obtained[19]). If the last data point is still on the infinite-acting solution curve, then a limiting estimate can be made concerning the drainage volume. In that case one observes the x_e/x_f parameter for the last upward-deviating curve that the matched data pass and uses that as an indication of the smallest drainage volume for the system.

As additional type curves become available for other fractured systems, capabilities for analyzing fractured-well transient pressure data by curve-matching techniques should expand significantly.

Pierce, Vela, and Koonce[15] propose a method for using pulse testing to estimate the orientation and length of a vertical fracture. Their method requires computer analysis of test data; however, a qualitative idea of fracture orientation may be obtained from pulse testing without computer assistance. See Ref. 15 for details.

Russell and Truitt[9] show that the vertical-fracture half-length can be related to an apparent wellbore radius for a single vertical fracture in a closed square by

$$r_{wa} \simeq 0.48\,x_f, \quad \ldots\ldots\ldots\ldots (11.18a)$$

when $x_e/x_f > 2$.

By using the skin factor equation and Eqs. C.7 and C.10, it is possible to write similar expressions for *infinite-acting systems*. For a uniform-flux fracture,

$$r_{wa} \simeq 0.37 x_f, \quad \ldots\ldots\ldots\ldots (11.18b)$$

and for an infinite-conductivity fracture,

$$r_{wa} \simeq 0.50 x_f. \quad \ldots\ldots\ldots\ldots (11.18c)$$

Eqs. 11.18a through 11.18c provide only a rough estimate, but may be helpful in estimating fracture length when only an apparent wellbore radius can be estimated from the skin factor using Eq. 2.11. That approach is not recommended.

Square, vertically fractured systems approach pseudosteady state after long enough production time[9] ($t_{pDA} >$ 0.12). Then, the dimensionless pressure is

$$p_D = 2\pi t_{DA} + \ln\left(\frac{x_e}{x_f}\right) + \frac{1}{2}\ln\left(\frac{2.2458}{C_A}\right), \quad \ldots\ldots (11.19)$$

where the shape factor, C_A, is given in Table C.1. Note the similarity of Eq. 11.19 to the pseudosteady-state equation for an unfractured well, Eq. 2.23. Russell and Truitt[9] show that reservoir-limit testing techniques (Section 3.5) apply to vertically fractured wells.

Example 11.2 Buildup Test Analysis for a Vertically Fractured Well

Gringarten, Ramey, and Raghavan[13] provide the pressure buildup data in Table 11.1. Other pertinent data are

$q_o = 2{,}750$ STB/D $h = 230$ ft
$\mu_o = 0.23$ cp $B_o = 1.76$ RB/STB
$\phi = 0.30$ depth = 9,500 ft.
$c_t = 30 \times 10^{-6}$ psi^{-1}

The usual log-log plot (not shown) has no unit slope, but has a slope of ½ from 5 to 75 minutes. Thus, we suspect a fractured well. For a formation at a depth of 9,500 ft, the fracture should be vertical.[8]

Fig. 11.13 is a plot of Δp vs $\sqrt{t}$, as suggested by Eq. 11.7. The graph has a straight line with $m_{Vf} = 97.3$ psi hr$^{-1/2}$ up to at least $\Delta p = 160$ psi. Eq. 11.9 can be used to estimate

$$kx_f^2 = \left[\frac{-4.064(2{,}750)(1.76)}{(97.3)(230)}\right]^2 \frac{0.23}{(0.30)(30 \times 10^{-6})}$$

$$= 19{,}700 \text{ md sq ft.}$$

According to Eq. 11.11, Δp at the beginning of the semilog straight line should be at least 2 × 160 = 320 psi. Since the test ended before that Δp, the semilog plot will not

TABLE 11.1—PRESSURE BUILDUP DATA FROM A WELL WITH A VERTICAL FRACTURE, EXAMPLE 11.2.
After Gringarten, Ramey, and Raghavan.[13]

Δt (min)	$p_{ws} - p_{wf}(\Delta t = 0)$ (psi)
0	0
5	31
10	43
15	54
20	66
25	66
30	72
35	78
40	83
45	89
50	100
55	100
60	100
75	114
120	136
150	159
240	181
285	206
480	218

be helpful. Thus, we must use type-curve matching for further analysis of the test. (This is not meant to imply that type-curve matching would not be useful if we could use the semilog plot; it could be useful, particularly to give limits on x_e/x_f.) Fig. 11.14 shows the data of Table 11.1 matched to Fig. C.19. Note that the data do not show any effects of a boundary. The match-point data are

$$(\Delta p)_M = 100 \text{ psi at } (p_D)_M = 0.77,$$

and

$$(\Delta t)_M = 100 \text{ minutes at } (t_{Dxf})_M = 0.36.$$

Using Eq. 11.14,

$$k = \frac{(141.2)(2{,}750)(1.76)(0.23)(0.77)}{(230)(100)}$$

$$= 5.26 \text{ md}.$$

From Eq. 11.15,

$$x_f = \sqrt{\frac{(0.0002637)(5.26)(100/60)}{(0.30)(0.23)(30 \times 10^{-6})(0.36)}}$$

$$= 55.7 \text{ ft}.$$

These results can be compared with the result of using Fig. 11.13 and Eq. 11.9. The value

$$kx_f^2 = (5.26)(55.7)^2 = 16{,}300 \text{ md sq ft},$$

compares with 19,700 md sq ft computed from Eq. 11.9 — about a 17-percent difference. If k is assumed to be correct, then

$$56 < x_f < 61 \text{ ft};$$

or if x_f is assumed to be correct,

$$5.3 < k < 6.3 \text{ md}.$$

The discrepancy provides an estimate of the accuracy of the type-curve matching method. When the well is retested, it would be advisable to get a longer buildup period in an attempt to define the semilog straight line.

Note from Fig. 11.14 that observed pressure data are still on the infinite-acting curve — at least to $x_e/x_f = 2$. Thus, we can say that x_e/x_f must be greater than 2 for the tested well. Using the value $x_f = 55.7$ ft, we estimate that $x_e > 111$ ft. Thus, we expect the drainage area for the test well to be larger than a 222-ft square, with an equivalent area of 1.13 acres. As we see, this test did not investigate a very large amount of reservoir.

Horizontal Fractures

Dimensionless pressure data for a single horizontal fracture located at the center of the productive interval in an infinite-acting system are given in Fig. C.5. No allowance for wellbore or fracture storage is included in that figure. At *very* short production times, the log-log plot of p_D vs t_D may have a unit slope resulting from storage effects within the fracture.[13,18] In most cases wellbore storage would obscure such fracture storage.

Fig. 11.15 also shows p_D vs t_D for a horizontally fractured well.[18] When plotted as shown in Fig. 11.15 (as contrasted to the log-log plot of p_D/h_D shown in Fig. C.5), the pressure-

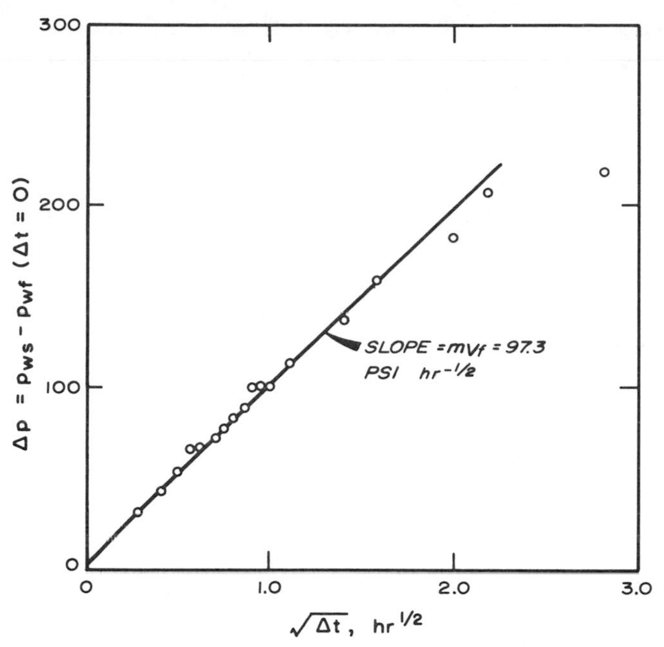

Fig. 11.13 Square-root data plot for buildup test of Example 11.2.

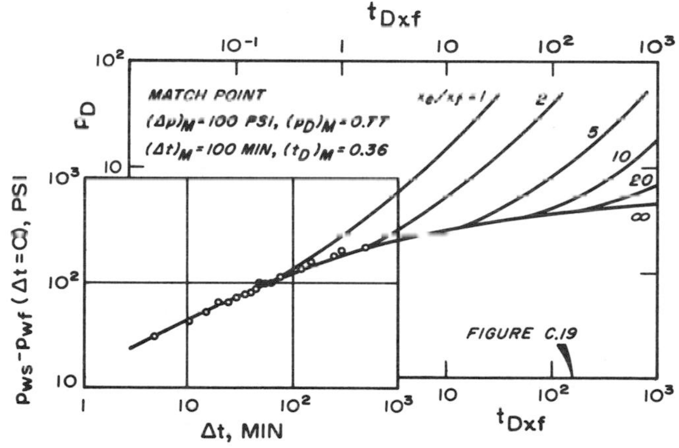

Fig. 11.14 Type-curve match for a uniform-flux vertical fracture, Example 11.2. After Gringarten, Ramey, and Raghavan.[13]

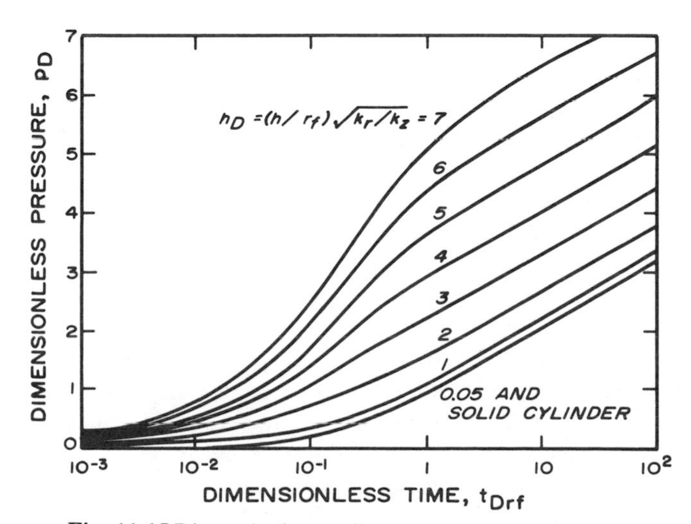

Fig. 11.15 Dimensionless well pressure for a single-plane, uniform-flux, horizontal fracture at center of formation. After Gringarten and Ramey.[18]

time curve has a peculiar S shape when $h_D > 3$ that distinguishes it from the behavior of a vertically fractured well. The log-log plot (Fig. C.5) has an S shape[13,18] for $h_D < 1$. Thus, for low and high h_D values, it may be possible to distinguish between vertical and horizontal fractures based on pressure behavior. However, the semilog curves of Fig. 11.15 for the larger h_D values have a shape typical of pressure data from wells with significant wellbore storage effects (compare Fig. 11.15 with Figs. 5.2 and 7.6).

At short flow times, there is a period of linear vertical flow from the formation to the horizontal fracture. During those times, dimensionless pressure is given by[19]

$$p_D = 2h_D \sqrt{\frac{t_{Drf}}{\pi}}, \quad \ldots\ldots\ldots\ldots\ldots\ldots (11.20)$$

where h_D is given by Eq. C.12 and t_{Drf} is given by Eq. C.11. The p_D in Eq. 11.20 may be used in Eq. 2.2 to obtain an expression for the flowing bottom-hole pressure during drawdown or injection — during the linear flow period;

$$p_{wf} = p_i + m_{Hf}\sqrt{t}\,. \quad \ldots\ldots\ldots\ldots\ldots\ldots (11.21)$$

Eq. 11.21 indicates that a plot of flowing bottom-hole pressure vs $\sqrt{t}$ should have an early-time straight-line portion with intercept p_i and slope

$$m_{Hf} = \frac{-2.587\,qB}{r_f^2}\sqrt{\frac{\mu}{k_z\,\phi c_t}}\,. \quad \ldots\ldots\ldots\ldots (11.22)$$

Note in Eq. 11.22 that the slope is controlled by the formation *vertical* permeability and the fracture radius.

Gringarten, Ramey, and Raghavan[13,19] show that the long-time pressure behavior for an infinite-acting, horizontally fractured well is the same as that for an unfractured well with an additional (negative) skin effect; that is, there is a pseudoradial flow period. Thus, if the fracture is not too long for the test duration, drawdown or injectivity test analysis by normal semilog methods can be used to estimate k_r, the radial direction permeability. Fig. C.5 can be used for curve matching for shorter tests. To date, there are no thorough studies of pressure buildup (falloff) behavior in horizontally fractured wells, so there are no published data for slope or k_r corrections similar to those in Fig. 11.12 for vertically fractured wells. Thus, we would preferably estimate k_r from a drawdown or injectivity test. When we estimate it from a buildup test, we can only assume that the long-time semilog slope gives a reasonable estimate of the true radial permeability when boundary and interference effects are not important.

Type-curve matching with Fig. C.5 may be used to estimate k_r/r_f^2. By using k_r/r_f^2 from type-curve matching and k_r estimated from the semilog straight line (Eq. 11.10), it is possible to estimate r_f. That r_f may be used in a rearranged form of Eq. 11.22 to estimate k_z. Thus, it may be possible to estimate k_r, k_z, and r_f from a single-well test for a well with a horizontal fracture. Of course, the fracture must be at the center of the formation to use Fig. C.5 and Eq. 11.22; the system must be infinite-acting; and the pseudoradial flow period must be fully developed. In addition, wellbore storage effects must not mask the initial $\sqrt{t}$ straight line. It should also be remembered that no data are presented for correcting pressure buildup (falloff) analyses in a horizontally fractured well, so the accuracy of k_r may be in doubt. Eq. 11.20 may be used with superposition to devise a technique for pressure buildup plotting for the linear flow period. Normally, that would be done by having a long production period before the shut-in and using a long-term equation for that period such as one based on a p_D given by[19]

$$p_D = \frac{1}{2}\left(\ln t_{Drf} + 1.80907 + \frac{h_D^2}{6}\right), \quad \ldots\ldots (11.23)$$

where t_{Drf} is given by Eq. C.11. Then the linear-flow portion of the shut-in period would be represented by Eq. 11.20. If Eq. 11.20 is used for the drawdown period as well as the shut-in period, the linear flow period must last *throughout* the *drawdown* portion of the test or the derived equation will be incorrect, as will be the subsequent analysis.

Gringarten, Ramey, and Raghavan[19] also indicate how type-curve matching may be used to estimate k_r, k_z, and r_f for a horizontally fractured well. In that situation pressure drawdown or injection data would be matched to the type curve of Fig. C.5. Sufficient data must be available so one of the h_D curves is clearly matched. The pressure match is used to estimate

$$\sqrt{k_r k_z}\, r_f = \frac{141.2\,qB\mu\,(p_D/h_D)_M}{(\Delta p)_M}\,. \quad \ldots\ldots\ldots (11.24)$$

The time-scale match is used to compute

$$\frac{k_r}{r_f^2} = \frac{\phi\mu c_t\,(t_D)_M}{0.0002637\,(t)_M}\,. \quad \ldots\ldots\ldots\ldots (11.25)$$

Then the value computed in Eq. 11.25 is used with the matched h_D curve to estimate vertical permeability:

$$k_z = \frac{k_r}{r_f^2}\left[\frac{h}{(h_D)_M}\right]^2. \quad \ldots\ldots\ldots\ldots\ldots\ldots (11.26)$$

The three results from Eqs. 11.24 through 11.26 are then used simultaneously to obtain k_r and r_f. Gringarten, Ramey, and Raghavan[19] show an example calculation using the procedure.

11.4 Partial Penetration and Partial Completion

Most material presented in this monograph assumes that a single, vertical well completely penetrates a horizontal formation. When a well penetrates (or is completed in) only a portion of the formation, normal transient analysis may still be used to estimate formation kh/μ and $\bar{p}$, but the skin factor estimated with the normal equations will reflect the partial well completion. Fig. 11.16, based on information presented by Kazemi and Seth,[21] indicates that the pressure transient behavior of a partially completed well may show two semilog straight-line portions. The first straight line represents kh/μ of the perforated interval, while the second represents kh/μ of the full formation interval. The initial straight line may not appear because of wellbore storage or other effects; but under ideal conditions, it does exist. Even when it does occur, we do not recommend trying to estimate the total pay thickness based on the interval open in the well and the two values of kh/μ from a semilog plot unless extraordinary steps have been taken to suppress wellbore storage.[22] Straight lines often seem to appear in data that is dominated completely by wellbore storage effects. As stated elsewhere, the log Δp-log Δt data plot will often clearly identify wellbore-storage-dominated data. Culham[23] shows that the transition between the two semilog straight lines is often characteristic of spherical flow. As such, those data may provide an estimate of k.

The apparent skin factor estimated from normal transient-test analysis methods for a well with restricted flow entry or for a well that is not vertical in the formation is[24]

$$s_a = s_{tr} + s_p + s_{swp} + \ldots, \quad \ldots\ldots\ldots (11.27)$$

where s_{tr} is the true skin factor caused by damage to the completed portion of the well; s_p is the pseudoskin factor resulting from restricted flow entry; and s_{swp} is a pseudoskin factor resulting from a slanted well. Brons and Marting[25] presented pseudoskin factors resulting from partial penetration — wells not drilled completely through the section (Fig. 2.7). Odeh[24] gives information for estimating the pseudoskin factor for wells that are either partially penetrating or are only partially completed (entire productive interval not perforated). His method may be used to estimate the pseudoskin factor for a large variety of completion conditions.

Jones and Watts[26] propose an equation for estimating the actual skin factor resulting from a combination of a partial completion and a change in permeability in an annular zone around the well over the perforated interval, such as might result from a sand consolidation treatment. The skin so estimated is caused only by the effect of the short damaged interval. The pseudoskin resulting from the partial completion must be added to it. The skin factor resulting from the permeability change near the wellbore is

$$s_{cp} = \frac{h}{\Delta Z_P}\left[1 - 0.2\left(\frac{r_s - r_w}{\Delta Z_P}\right)\right]\left(\frac{k - k_s}{k_s}\right)\ln\left(\frac{r_s}{r_w}\right). \quad \ldots\ldots\ldots (11.28)$$

The equation applies when the perforated interval, ΔZ_P, is much smaller than the formation thickness, h, but that is not thought to be a serious restriction.[26]

Fig. 11.17 schematically shows a well penetrating a formation at an angle α from the line perpendicular to the formation top and bottom. Cinco, Miller, and Ramey[27] show that the pseudoskin factor resulting from a slanted, completely perforated well can be approximated by

$$s_{swp} = -(\alpha/41)^{2.06} - (\alpha/56)^{1.865}\log\left(\frac{h}{100\, r_w}\right), \quad \ldots\ldots\ldots (11.29)$$

when $0° \leq \alpha \leq 75°$, $h/r_w > 40$, and $t_D > 100$. Fig. 11.18 shows the pseudoskin factor for a slanted well. The effect of the slanted well is to provide more wellbore area and, thus, a negative pseudoskin factor.

Skin factors estimated from transient testing include all features that affect the efficiency of fluid flow into the wellbore. The material presented in this section may be useful for either predicting the pressure rate behavior of a well, or for attempting to estimate the skin actually caused by damage. Clearly, a large skin factor resulting from a sand consolidation treatment or from partial completion probably cannot be reduced significantly by most stimulation treatments. Thus, it is important that the analyst recognize such situations.

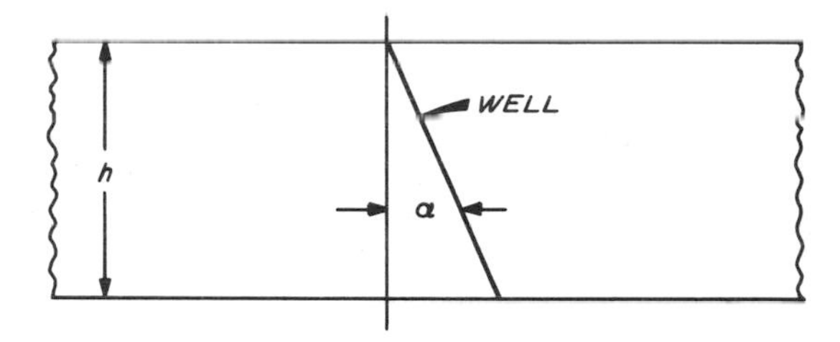

Fig. 11.17 Definition of terms for slanted wells.

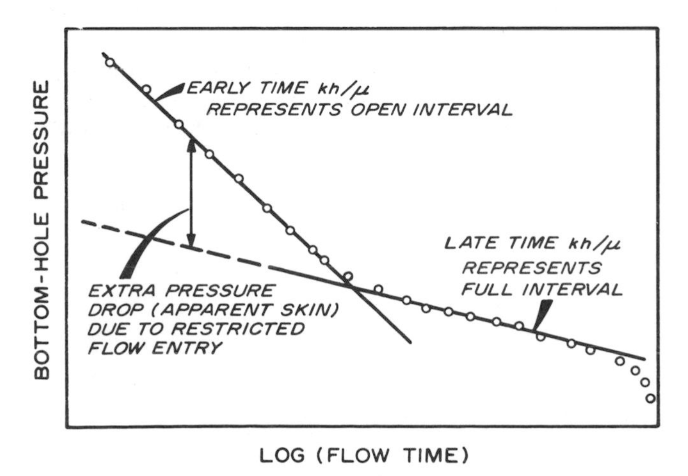

Fig. 11.16 Schematic drawdown behavior for a well with restricted flow entry.

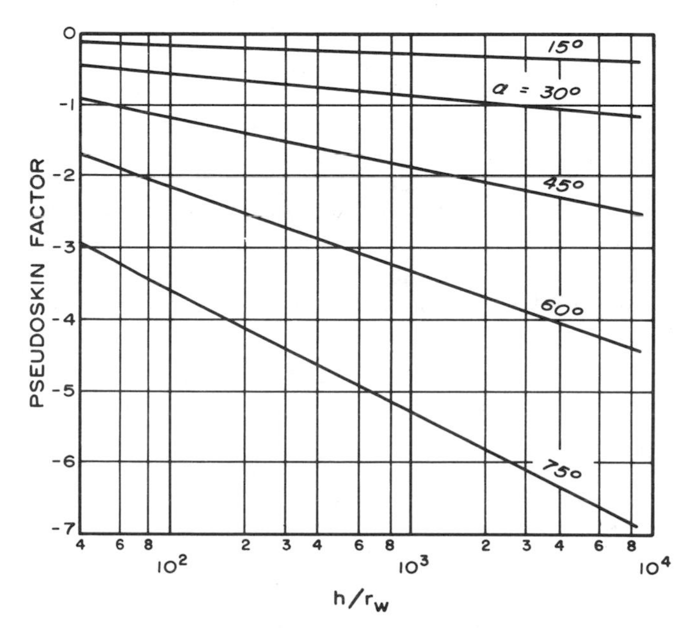

Fig. 11.18 Pseudoskin factor for slanted wells. After Cinco, Miller, and Ramey.[27]

References

1. Matthews, C. S. and Russell, D. G.: *Pressure Buildup and Flow Tests in Wells,* Monograph Series, Society of Petroleum Engineers of AIME, Dallas (1967) **1.**
2. Ramey, H. J., Jr.: "Non-Darcy Flow and Wellbore Storage Effects in Pressure Build-Up and Drawdown of Gas Wells," *J. Pet. Tech.* (Feb. 1965) 223-233; *Trans.,* AIME, **234.** Also *Reprint Series, No. 9 — Pressure Analysis Methods,* Society of Petroleum Engineers of AIME, Dallas (1967) 233-243.
3. Ramey, H. J., Jr.: "Short-Time Well Test Data Interpretation in the Presence of Skin Effect and Wellbore Storage," *J. Pet. Tech.* (Jan. 1970) 97-104; *Trans.,* AIME, **249.**
4. Agarwal, Ram G., Al-Hussainy, Rafi, and Ramey, H. J., Jr.: "An Investigation of Wellbore Storage and Skin Effect in Unsteady Liquid Flow: I. Analytical Treatment," *Soc. Pet. Eng. J.* (Sept. 1970) 279-290; *Trans.,* AIME, **249.**
5. Chen, Hsiu-Kuo and Brigham, W. E.: "Pressure Buildup for a Well With Storage and Skin in a Closed Square," paper SPE 4890 presented at the SPE-AIME 44th Annual California Regional Meeting, San Francisco, April 4-5, 1974.
6. Ramey, Henry J., Jr., and Agarwal, Ram G.: "Annulus Unloading Rates as Influenced by Wellbore Storage and Skin Effect," *Soc. Pet. Eng. J.* (Oct. 1972) 453-462; *Trans.,* AIME, **253.**
7. Earlougher, Robert C., Jr., Kersch, K. M., and Ramey, H. J., Jr.: "Wellbore Effects in Injection Well Testing," *J. Pet. Tech.* (Nov. 1973) 1244-1250.
8. Howard, G. C. and Fast, C. R.: *Hydraulic Fracturing,* Monograph Series, Society of Petroleum Engineers of AIME, Dallas (1970) **2.**
9. Russell, D. G. and Truitt, N. E.: "Transient Pressure Behavior in Vertically Fractured Reservoirs," *J. Pet. Tech.* (Oct. 1964) 1159-1170; *Trans.,* AIME, **231.** Also *Reprint Series, No. 9 — Pressure Analysis Methods,* Society of Petroleum Engineers of AIME, Dallas (1967) 149-160.
10. Clark, K. K.: "Transient Pressure Testing of Fractured Water Injection Wells," *J. Pet. Tech.* (June 1968) 639-643; *Trans.,* AIME, **243.**
11. Raghavan, R., Cady, Gilbert V., and Ramey, Henry J., Jr.: "Well-Test Analysis for Vertically Fractured Wells," *J. Pet. Tech.* (Aug. 1972) 1014-1020; *Trans.,* AIME, **253.**
12. Wattenbarger, Robert A. and Ramey, Henry J., Jr.: "Well Test Interpretation of Vertically Fractured Gas Wells," *J. Pet. Tech.* (May 1969) 625-632; *Trans.,* AIME, **246.**
13. Gringarten, Alain C., Ramey, Henry J., Jr., and Raghavan, R.: "Pressure Analysis for Fractured Wells," paper SPE 4051 presented at the SPE-AIME 47th Annual Fall Meeting, San Antonio, Tex., Oct. 8-11, 1972.
14. Gringarten, Alain C., Ramey, Henry J., Jr., and Raghavan, R.: "Unsteady-State Pressure Distributions Created by a Well With a Single Infinite-Conductivity Vertical Fracture," *Soc. Pet. Eng. J.* (Aug. 1974) 347-360.
15. Pierce, A. E., Vela, Saul, and Koonce, K. T.: "Determination of the Compass Orientation and Length of Hydraulic Fractures by Pulse Testing," *J. Pet. Tech.* (Dec. 1975) 1433-1438.
16. Hartsock, J. H. and Warren, J. E.: "The Effect of Horizontal Hydraulic Fracturing on Well Performance," *J. Pet. Tech.* (Oct. 1961) 1050-1056; *Trans.,* AIME, **222.**
17. Gringarten, Alain C.: "Unsteady-State Pressure Distributions Created by a Well With a Single Horizontal Fracture, Partial Penetration, or Restricted Flow Entry," PhD dissertation, Stanford U., Stanford, Calif. (1971) 106. (Order No. 71-23,512, University Microfilms, P.O. Box 1764, Ann Arbor, Mich. 48106.)
18. Gringarten, Alain C. and Ramey, Henry J., Jr.: "Unsteady-State Pressure Distributions Created by a Well With a Single Horizontal Fracture, Partial Penetration, or Restricted Entry," *Soc. Pet. Eng. J.* (Aug. 1974) 413-426; *Trans.,* AIME, **257.**
19. Gringarten, A. C., Ramey, H. J., Jr., and Raghavan, R.: "Applied Pressure Analysis for Fractured Wells," *J. Pet. Tech.* (July 1975) 887-892; *Trans.,* AIME, **259.**
20. Gringarten, A. C. and Witherspoon, P. A.: "A Method of Analyzing Pump Test Data from Fractured Aquifers," *Proc.,* Symposium on Percolation Through Fissured Rock, Int'l. Society for Rock Mechanics, Stuttgart (Sept. 18-19, 1972).
21. Kazemi, Hossein and Seth, Mohan S.: "Effect of Anisotropy and Stratification on Pressure Transient Analysis of Wells With Restricted Flow Entry," *J. Pet. Tech.* (May 1969) 639-647; *Trans.,* AIME, **246.**
22. Lescarboura, Jaime A.: "New Downhole Shut-in Tool Boosts BPH Test Accuracy," *World Oil* (Nov. 1974) 71-73.
23. Culham, W. E.: "Pressure Buildup Equations for Spherical Flow Regime Problems," *Soc. Pet. Eng. J.* (Dec. 1974) 545-555.
24. Odeh, A. S.: "Steady-State Flow Capacity of Wells With Limited Entry to Flow," *Soc. Pet. Eng. J.* (March 1968) 43-51; *Trans.,* AIME, **243.**
25. Brons, F. and Marting, V. E.: "The Effect of Restricted Fluid Entry on Well Productivity," *J. Pet. Tech.* (Feb. 1961) 172-174; *Trans.,* AIME, **222.** Also *Reprint Series, No. 9 — Pressure Analysis Methods,* Society of Petroleum Engineers of AIME, Dallas (1967) 101-103.
26. Jones, L. G. and Watts, J. W.: "Estimating Skin Effect in a Partially Completed Damaged Well," *J. Pet. Tech.* (Feb. 1971) 249-252; *Trans.,* AIME, **251.**
27. Cinco, H., Miller, F. G., and Ramey, H. J., Jr.: "Unsteady-State Pressure Distribution Created by a Directionally Drilled Well," *J. Pet. Tech.* (Nov. 1975) 1392-1400; *Trans.,* AIME, **259.**

Chapter 12

Application of Computers to Well Testing

12.1 Introduction

Most material in this monograph deals with analysis techniques that do not require the aid of a computer. We have avoided the necessity of using the computer by considering simple transient testing procedures and using simplified dimensionless pressures for analysis. Most transient well tests we have considered use no more than two flow rates plus, perhaps, a shut-in period. All the test analysis techniques have been based on some simple form of a combination of Eqs. 2.30 and 2.31:

$$\Delta p = \frac{141.2\,\mu}{kh} \sum_{j} \sum_{i=1}^{N} \Big[\left(q_i B_i - q_{i-1} B_{i-1} \right) \left\{ p_D\left([t - t_{i-1}]_D, r_D \right) + s \right\} \Big], \quad \ldots\ldots\ldots\ (12.1)$$

where the j sum is taken over all wells and r_D refers to the distance between the point of interest and the active well. In most cases we have assumed infinite-acting behavior and have used a logarithmic approximation to the exponential-integral dimensionless pressure expression, Eq. 2.5b.

Sometimes testing situations use so many wells and rate changes that Eq. 12.1 is difficult to use by hand, even though hand calculations are possible with any form of that equation. Earlougher *et al.*[1] provide a means for simplying calculations in bounded systems; but even that technique is burdensome if the system is complex. The approach presented by Jargon and van Poollen[2] is feasible for hand calculations — for a single well with multiple rate changes. Regardless of the number of wells or rate changes involved, it is almost essential to use a digital computer when a system is so complex that tabulated dimensionless pressures do not exist. Examples of such systems are fractured reservoirs;[3,4] stratified reservoirs;[5-9] reservoirs with various kinds of discontinuities;[10-12] reservoirs containing non-Newtonian fluids;[13] and reservoirs with peculiar conditions around the wellbore.[14,15] In such cases, a computer and some sort of mathematical model of the reservoir are required just to simulate* the behavior of the system. Analyzing transient tests in such complex systems generally also requires computer assistance.[10-23] Sometimes simulations lead to empirical hand-analysis techniques; for examples, see Sections 7.5, 9.3, 10.8, and 11.3.

A major advantage of using the computer is that we are able to study systems for which dimensionless pressure data are not available. A second advantage is that computer simulation and analysis is fast and *computationally* error-free. That, unfortunately, does not guarantee correct results, since the mathematical model or the computer program may not be *conceptually* error-free or since the data may be incorrect or supplied to the program incorrectly. The major disadvantages in applying the computer to transient testing are that both a computer and a computer program must be available. Surmounting the first problem does not necessarily overcome the second. The engineer may have to write a new program, modify an existing one, or find someone to do that for him. Thus, using computers in well testing can create a major time bottleneck if the required computer programs are not readily available. Fortunately, most large oil companies and many consulting organizations have both computers and libraries of programs that the engineer can use.

This chapter briefly discusses computer-aided transient test analysis, computer-aided transient test design, and reservoir simulation. The material is limited to results without detailed descriptions of the calculation procedures used, since the intent is to demonstrate possibilities rather than provide computational methods. The cited references give computational details.

12.2 Computer-Aided Well Test Analysis

When using the computer for analyzing transient tests, it is important to remember that it is just a tool. The computer does not perform the analysis, even though it may print answers. The engineer is responsible for the analysis — he must ascertain that the data were correctly supplied and that the computed results are consistent with the physical situation. If the results are unreasonable, the engineer must determine the problems and decide how to correct them, whether that involves rerunning the test, reanalyzing the data, or rewriting the computer program.

*As used in this monograph, a reservoir simulator is a computer program that solves fluid-flow equations to compute (simulate) reservoir behavior.

In the simplest type of computer-aided transient test analysis, the computer does the bulk of the time-consuming computations, but the engineer makes all the decisions related to test analysis. For example, suppose we wish to analyze data from a multiple-rate transient test using the technique of Section 4.2. We might use a computer program to do the computationally tedious superposition calculation. Then we could plot the pressure data vs the time-superposition function, draw the appropriate straight line, and estimate formation characteristics. Jargon and van Poollen[2] describe a technique of that nature. Such a general approach may be applied to almost any testing situation.

A more sophisticated approach to computer-aided test analysis has the computer both calculate and make decisions regarding the analysis. In the automated-analysis approach, the computer prints the answer and gives some indication of how reliable the results are. The engineer must then evaluate the results and accept or reject them.

Any computer program that automatically analyzes transient test data consists of two basic parts: (1) a part to compute pressure behavior for a given set of reservoir properties (for example, superposition in time and space using an appropriate p_D), and (2) a part that judges how well the computed behavior matches observed behavior and then changes the reservoir parameters used in Part 1 until the match of computed and observed data is acceptable.

The first part of the process may be accomplished using techniques of Chapter 2, some variation of those techniques, or a reservoir simulator.[13,19,24] To accomplish the second part of the process, the program may use a regression analysis technique[19,20] or some other technique to minimize the error.[22,23] The following examples indicate the kinds of results that can be obtained from computer-aided transient test analysis.

Example 12.1 Interference Test Analysis by Computer Methods

Examples 9.1 and 9.3 present interference test data and show two analysis techniques. Example 9.1 analysis results are

$$k = 5.1 \text{ md}$$

and

$$\phi c_t = 1.01 \times 10^{-6} \text{ psi}^{-1}.$$

We also have analyzed the data by three different computer-aided techniques. The first method, a trial-and-error approach, compares observed pressure response with pressure calculated by a computer program that uses Eq. 12.1 with given reservoir parameters. After several computer runs (sometimes the technique requires 12 to 15 trials), the match shown in Fig. 12.1 was obtained. The parameters used to calculate the line in that figure are

$$k = 6.0 \text{ md}$$

and

$$\phi c_t = 0.8 \times 10^{-6} \text{ psi}^{-1}.$$

The second computer-analysis method uses the regression technique described by Earlougher and Kersch.[20] Fig. 12.2 is a plot of the match obtained by that method. In that case, the computed results are

$$k = 5.7 \text{ md}$$

and

$$\phi c_t = 0.99 \times 10^{-6} \text{ psi}^{-1}.$$

A third computer-analysis technique, similar to that described by Coats, Dempsey, and Henderson,[22] was also used. Although not shown here, it gave essentially the same results as the regression technique.

All these analysis techniques give essentially the same results. If we had to choose one set of results, it would probably be best to use 5.7 md and 0.99×10^{-6} psi^{-1}. We do this because the regression technique provides less chance for an error in subjective judgment during analysis than do the other techniques. This example is a simple one that is analyzed by hand with relative ease. Many transient tests do not provide an option; they can be analyzed within reasonable time only by using computer methods.

Example 12.2 Interference Test Analysis — Heterogeneous Reservoir Case

Jahns[19] describes an interference test in a five-spot pattern with the center well being produced for several hours while the pressure decline in the four surrounding wells was ob-

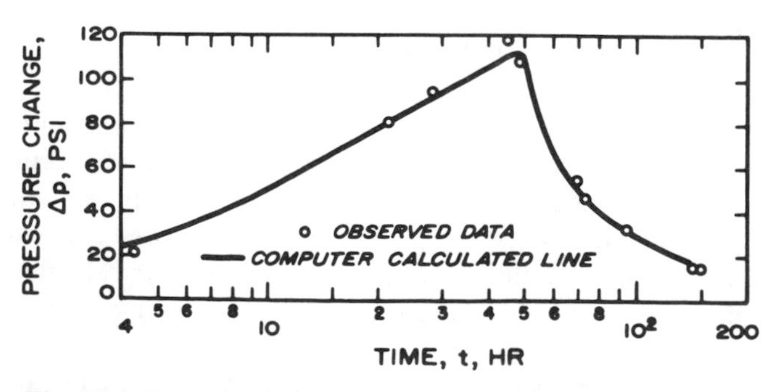

Fig. 12.1 Example 12.1, trial-and-error computer analysis of the interference test of Example 9.1.

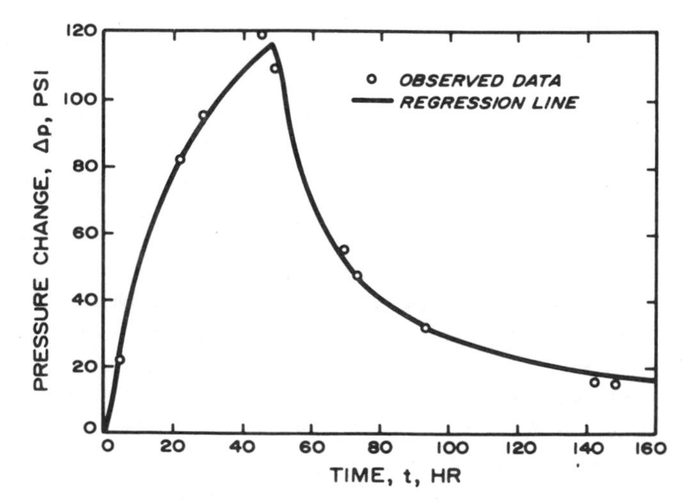

Fig. 12.2 Example 12.1, regression-type computer analysis of the interference test of Example 9.1.

TABLE 12.1—EXAMPLE 12.3: COMPUTER ANALYSIS RESULTS FOR INTERFERENCE TEST OF EXAMPLE 9.6.

Analysis Method and Data Used	Analysis Results					
	k_{max} (md)	k_{min} (md)	$\bar{k}$ (md)	θ (degrees)	ϕc_t (psi^{-1})	Standard Deviation (psi)
Type-Curve Matching						
Three wells, injection only	20.5	13.9	16.9	64.7	1.59×10^{-6}	——
Regression						
Three wells, injection only	17.6	12.1	14.6	60.3	1.61×10^{-6}	1.11
Three wells, injection and falloff	17.2	12.7	14.8	59.3	1.61×10^{-6}	1.86
Seven wells,* injection and falloff	15.8	10.1	12.6	53.7	1.67×10^{-6}	2.07
Eight wells, injection and falloff	14.5	11.1	12.7	63.6	1.60×10^{-6}	2.69

*Omit data from well with poorest match in eight-well analysis.

served. Using a grid-type mathematical simulator and a regression-type analysis technique, Jahns analyzed the data for pattern properties. Fig. 12.3 shows how he segmented the reservoir for analysis of test data. He estimated kh and $\phi c_t h$ in each block from the pressure data at the four observation wells. Jahns gives considerable information about this test, including plots of observed and calculated response data at each observation well. Fig. 12.4 shows results at Observation Well 2. The computed response matches the observed response closely.

Example 12.3 Interference Test Analysis — Anisotropic Reservoir Case

In Example 9.6 we analyzed injection-period pressure response data from three observation wells in an anisotropic reservoir. Ramey[25] gives both injection and shut-in period data for eight observation wells for that test. We have applied the regression analysis described by Earlougher and Kersch[20] to those data in several ways. Table 12.1 summarizes the results. Fig. 12.5 shows the range of data-match results for the analysis corresponding to the next-to-last line in Table 12.1. Clearly, the results depend on the amount of data used, and may be changed by excluding portions of the available data. If a different reservoir model had been chosen (such as one composed of many heterogeneous blocks, each with its own properties, as illustrated in Example 12.2), the estimated reservoir properties might have been quite different.

As seen from Table 12.1, the degree of anisotropy in this example is small ($k_{max}/k_{min} < 2$). Thus, it is possible that reservoir properties estimated from some other type of res-

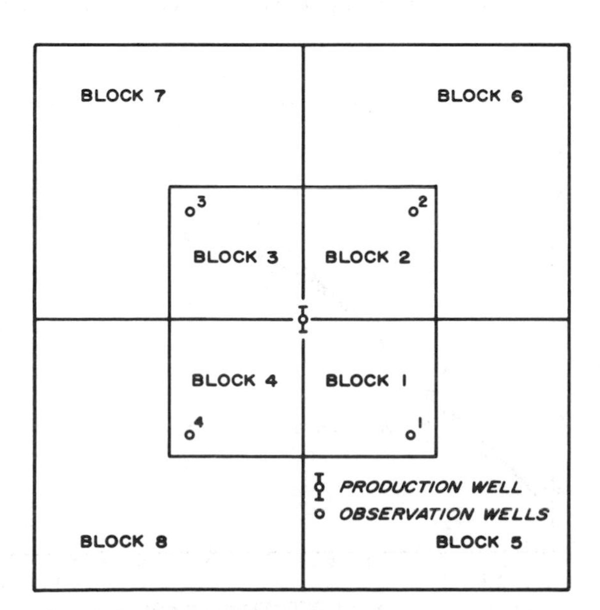

Fig. 12.3 Example 12.2, eight-region model used for a computer-aided analysis of an interference test in a heterogeneous reservoir. After Jahns.[19]

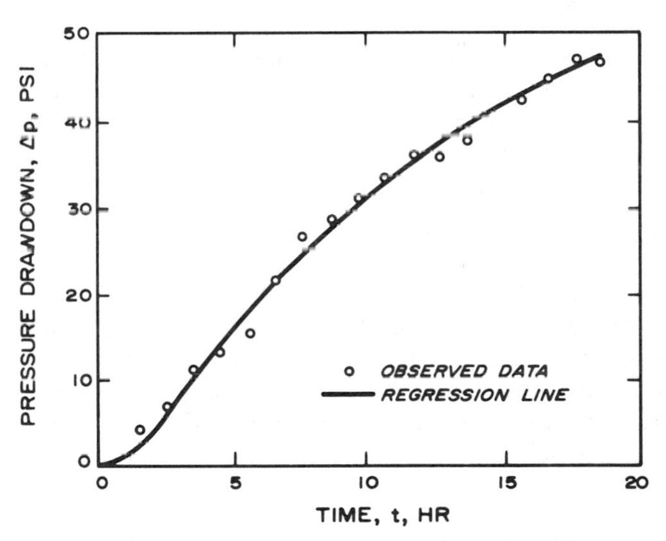

Fig. 12.4 Example 12.2, results of a computer-aided analysis of an interference test in a heterogeneous reservoir. Results for Well 2 of Fig. 12.3. After Jahns.[19]

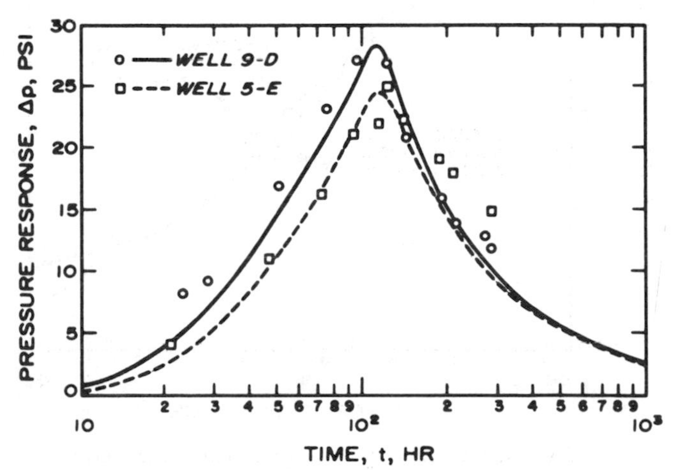

Fig. 12.5 Regression-analysis data fit for seven-well interference-test data analysis of Example 12.3 and Table 12.1.

ervoir model also might be applicable. Nevertheless, there are situations where other data available to the engineer would strongly suggest that extreme anisotropy must exist owing to natural fracturing, depositional trends, etc. Thus, it is important that data other than that of the test be considered during test analysis. In fact, for the data of this example, the sandstone is a channel deposit and directional permeability can be expected as a result of the depositional environment.

Example 12.4 Computer Analysis of Falloff Data With Significant Wellbore Storage

Earlougher and Kersch[20] present an analysis of the falloff test shown in Fig. 12.6. The S shape of the falloff curve is indicative of wellbore storage. The log-log data plot in Fig. 12.7 verifies that wellbore storage is dominant for about 5 hours.

The final straight-line part of the data in Fig. 12.6 can be analyzed using the techniques in Section 7.3 to obtain

$$k = 12.7 \text{ md}$$

and

$$s = 1.7.$$

Earlougher and Kersch used completion data to estimate an afterflow schedule. When they included that schedule in the automated regression-type analysis, the solid curve in Fig. 12.6 resulted. The best match occurred when

$$k = 13.7 \text{ md}$$

and

$$s = 1.7.$$

The data were also analyzed using the regression program, but assuming no wellbore storage. The best-match calculated response for that case is the dashed curve in Fig. 12.6. Clearly, it is not a good representation of observed data. Actually, the analysis shown by the dashed curve was the first one made. In that case, engineering judgment indicated the analysis should be repeated using more realistic conditions. Since the data (especially the log-log data plot) indicated significant wellbore storage, wellbore storage was included in the next analysis, resulting in the solid line in Fig. 12.6.

12.3 Computer-Aided Test Design

Chapter 13 provides guidelines for designing transient well tests. One approach includes computing the expected pressure response for the well test. By making such a computation, one can evaluate the consequences of wellbore storage, the type of pressure-measuring equipment necessary, the magnitude of flow rate required, the required test time, or any combination of those factors. If the reservoir system or the flow history is complex, such calculations tend to be time-consuming or must be based on over-idealized, simplified models. Thus, it is sensible to use the computer for test design, both to save labor and to extend the scope of situations the engineer can handle. Such design calculations estimate the expected pressure responses based on assumed reservoir properties, or estimate a range of responses based on a range of possible reservoir properties. Such computations can help insure that one gathers adequate test data.

12.4 Reservoir Simulation

As indicated in Section 12.2, one of the basic chores of any computer program for analyzing transient tests is calculating pressure response caused by a specific flow-rate history. Such computations are referred to by the general term "reservoir simulation." All dimensionless pressure functions in Appendix C, when coupled with the superposition techniques of Section 2.9, result in simple reservoir simulators. However, such simple simulators do not apply to many complex situations. For example, we might be interested in extremely heterogeneous systems, layered systems, systems with two or three phases flowing, systems with fluid-fluid contacts, systems with water or gas coning,

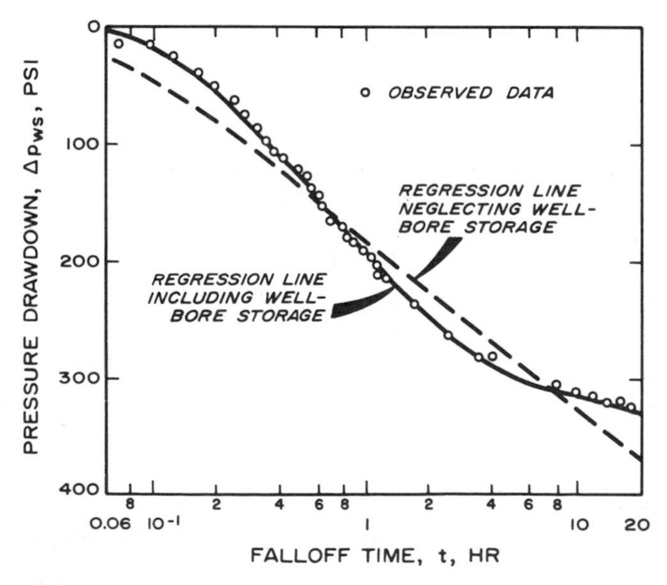

Fig. 12.6 Example 12.4, computer-aided falloff test analysis when test shows significant wellbore storage. After Earlougher and Kersch.[20]

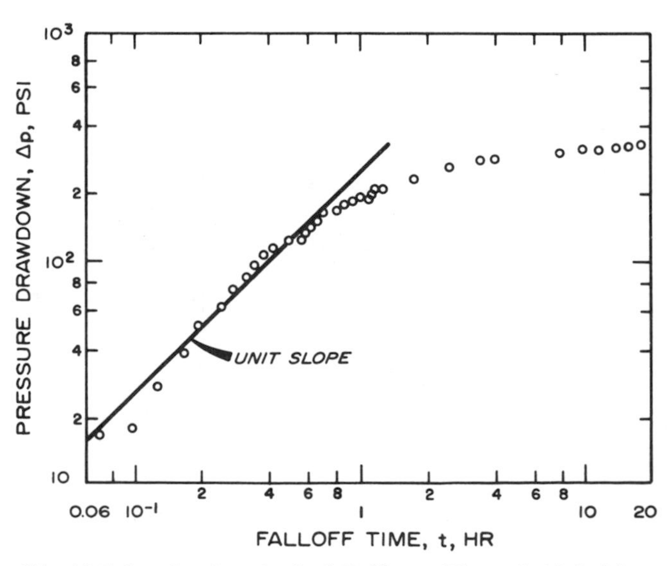

Fig. 12.7 Log-log data plot for falloff test of Example 12.4. After Earlougher and Kersch.[20]

or systems with significant gravity drainage. During the past several years, many papers appearing in the petroleum literature have discussed various kinds of reservoir simulators.[26] Three of the classic references are Aronofsky and Jenkins,[27] Bruce, Peaceman, Rachford, and Rice,[28] and West, Garvin, and Sheldon.[29] Many of the facets of reservoir simulation were summarized by van Poollen, Bixel, and Jargon[30-42] in a series of articles.

Most reservoir simulators are not written for direct application to well testing; rather, they sacrifice description of near-wellbore effects so they can predict over-all reservoir behavior using an economical amount of computer time. Reservoir simulation is used mainly for reservoir analysis — not well analysis — in the petroleum industry. However, special reservoir simulators are being used more frequently for well-testing applications; such simulators commonly include radial flow geometry and wellbore storage effects as well as other special effects.[4-23] The inherent flexibility of the well simulators enables the engineer to tackle difficult, real situations without either a high degree of mathematical skill or programming expertise. It is this flexibility that makes reservoir simulators and computer methods valuable tools in well testing.

References

1. Earlougher, Robert C., Jr., Ramey, H. J., Jr., Miller, F. G., and Mueller, T. D.: "Pressure Distributions in Rectangular Reservoirs," *J. Pet. Tech.* (Feb. 1968) 199-208; *Trans.*, AIME, **243.**

2. Jargon, J. R. and van Poollen, H. K.: "Unit Response Function From Varying-Rate Data," *J. Pet. Tech.* (Aug. 1965) 965-969; *Trans.*, AIME, **234.**

3. Russell, D. G. and Truitt, N. E.: "Transient Pressure Behavior in Vertically Fractured Reservoirs," *J. Pet. Tech.* (Oct. 1964) 1159-1170; *Trans.*, AIME, **231.** Also *Reprint Series, No. 9 — Pressure Analysis Methods,* Society of Petroleum Engineers of AIME, Dallas (1967) 149-160.

4. Kazemi, H.: "Pressure Transient Analysis of Naturally Fractured Reservoirs With Uniform Fracture Distribution," *Soc. Pet. Eng. J.* (Dec. 1969) 451-462.

5. Kazemi, Hossein and Seth, Mohan S.: "Effect of Anisotropy and Stratification on Pressure Transient Analysis of Wells With Restricted Flow Entry," *J. Pet. Tech.* (May 1969) 639-647; *Trans.*, AIME, **246.**

6. Cobb, William M., Ramey, H. J., Jr., and Miller, Frank G.: "Well-Test Analysis for Wells Producing Commingled Zones," *J. Pet. Tech.* (Jan. 1972) 27-37; *Trans.*, AIME, **253.**

7. Kazemi, Hossein: "Pressure Buildup in Reservoir Limit Testing of Stratified Systems," *J. Pet. Tech.* (April 1970) 503-511; *Trans.*, AIME, **249.**

8. Raghavan, R., Topaloglu, H. N., Cobb, W. M., and Ramey, H. J., Jr.: "Well-Test Analysis for Wells Producing From Two Commingled Zones of Unequal Thickness," *J. Pet. Tech.* (Sept. 1974) 1035-1043; *Trans.*, AIME, **257.**

9. Earlougher, Robert C., Jr., Kersch, K. M., and Kunzman, W. J.: "Some Characteristics of Pressure Buildup Behavior in Bounded Multiple-Layer Reservoirs Without Crossflow," *J. Pet. Tech.* (Oct. 1974) 1178-1186; *Trans.*, AIME, **257.**

10. Bixel, H. C., Larkin, B. K., and van Poollen, H. K.: "Effect of Linear Discontinuities on Pressure Build-Up and Drawdown Behavior," *J. Pet. Tech.* (Aug. 1963) 885-895; *Trans.*, AIME, **228.**

11. Kazemi, Hossein: "Locating a Burning Front by Pressure Transient Measurements," *J. Pet. Tech.* (Feb. 1966) 227-232; *Trans.*, AIME, **237.**

12. Bixel, H. C. and van Poollen, H. K.: "Pressure Drawdown and Buildup in the Presence of Radial Discontinuities," *Soc. Pet. Eng. J.* (Sept. 1967) 301-309; *Trans.*, AIME, **240.**

13. van Poollen, H. K. and Jargon, J. R.: "Steady-State and Unsteady-State Flow of Non-Newtonian Fluids Through Porous Media," *Soc. Pet. Eng. J.* (March 1969) 80-88; *Trans.*, AIME, **246.**

14. Kazemi, H.: "A Reservoir Simulator for Studying Productivity Variation and Transient Behavior of a Well in a Reservoir Undergoing Gas Evolution," *J. Pet. Tech.* (Nov. 1975) 1401-1412; *Trans.*, AIME, **259.**

15. Jones, L. G. and Watts, J. W.: "Estimating Skin Effect in a Partially Completed Damaged Well," *J. Pet. Tech.* (Feb. 1971) 249-252; *Trans.*, AIME, **251.**

16. Kazemi, H., Seth, M. S., and Thomas, G. W.: "The Interpretation of Interference Tests in Naturally Fractured Reservoirs With Uniform Fracture Distribution," *Soc. Pet. Eng. J.* (Dec. 1969) 463-472; *Trans.*, AIME, **246.**

17. Brill, J. P., Bourgoyne, A. T., and Dixon, T. N.: "Numerical Simulation of Drillstem Tests as an Interpretation Technique," *J. Pet. Tech.* (Nov. 1969) 1413-1420.

18. Dixon, Thomas N., Seinfeld, John H., Startzman, Richard A., and Chen, W. H.: "Reliability of Reservoir Parameters From History-Matched Drillstem Tests," paper SPE 4282 presented at the SPE-AIME Third Symposium on Numerical Simulation of Reservoir Performance, Houston, Jan. 10-12, 1973.

19. Jahns, Hans O.: "A Rapid Method for Obtaining a Two-Dimensional Reservoir Description From Well Pressure-Response Data," *Soc. Pet. Eng. J.* (Dec. 1966) 315-327; *Trans.*, AIME, **237.**

20. Earlougher, Robert C., Jr., and Kersch, Keith M.: "Field Examples of Automatic Transient Test Analysis," *J. Pet. Tech.* (Oct. 1972) 1271-1277.

21. Pierce, A. E., Vela, Saul, and Koonce, K. T.: "Determination of the Compass Orientation and Length of Hydraulic Fractures by Pulse Testing," *J. Pet. Tech.* (Dec. 1975) 1433-1438.

22. Coats, K. H., Dempsey, J. R., and Henderson, J. H.: "A New Technique for Determining Reservoir Description From Field Performance Data," *Soc. Pet. Eng. J.* (March 1970) 66-74; *Trans.*, AIME, **249.**

23. Hernandez, Victor M. and Swift, George W.: "A Method for Determining Reservoir Parameters From Early Drawdown Data," paper SPE 3982 presented at the SPE-AIME 47th Annual Fall Meeting, San Antonio, Tex., Oct. 8-11, 1972.

24. Breitenbach, E. A., Thurnau, D. H., and van Poollen, H. K.: "Solution of the Immiscible Fluid Flow Simulation Equations," *Soc. Pet. Eng. J.* (June 1969) 155-169. Also *Reprint Series, No. 11 — Numerical Simulation,* Society of Petroleum Engineers of AIME, Dallas (1973) 16-30.

25. Ramey, Henry J., Jr.: "Interference Analysis for Anisotropic Formations — A Case History," *J. Pet. Tech.* (Oct. 1975) 1290-1298; *Trans.*, AIME, **259.**

26. *Reprint Series, No. 11 — Numerical Simulation,* Society of Petroleum Engineers of AIME, Dallas (1973).

27. Aronofsky, J. A. and Jenkins, R.: "A Simplified Analysis of Unsteady Radial Gas Flow," *Trans.*, AIME (1954) **201,** 149-154. Also *Reprint Series, No. 9 — Pressure Analysis Methods,* Society of Petroleum Engineers of AIME, Dallas (1967) 197-202.

28. Bruce, G. H., Peaceman, D. W., Rachford, H. H., Jr., and Rice, J. D.: "Calculations of Unsteady-State Gas Flow Through Porous Media," *Trans.*, AIME (1953) **198,** 79-92.

29. West, W. J., Garvin, W. W., and Sheldon, J. W.: "Solution of the Equations of Unsteady-State Two-Phase Flow in Oil Reservoirs," *Trans.*, AIME (1954) **201,** 217-229.

30. van Poollen, H. K., Bixel, H. C., and Jargon, J. R.: "Reservoir Modeling — 1: What It Is, What It Does," *Oil and Gas J.* (July 28, 1969) 158-160.

31. van Poollen, H. K., Bixel, H. C., and Jargon, J. R.: "Reservoir Modeling — 2: Single-Phase Fluid-Flow Equations," *Oil and Gas J.* (Aug. 18, 1969) 94-96.

32. van Poollen, H. K., Bixel, H. C., and Jargon, J. R.: "Reservoir Modeling — 3: Finite Differences," *Oil and Gas J.* (Sept. 15, 1969) 120-121.

33. van Poollen, H. K., Bixel, H. C., and Jargon, J. R.: "Reservoir Modeling — 4: Explicit Finite-Difference Technique," *Oil and Gas J.* (Nov. 3, 1969) 81-87.

34. van Poollen, H. K., Bixel, H. C., and Jargon, J. R.: "Reservoir Modeling — 5: Implicit Finite-Difference Approximation," *Oil and Gas J.* (Jan. 5, 1970) 88-92.

35. van Poollen, H. K., Bixel, H. C., and Jargon, J. R.: "Reservoir Modeling — 6: General Form of Finite-Difference Approximations," *Oil and Gas J.* (Jan. 19, 1970) 84-86.

36. van Poollen, H. K., Bixel, H. C., and Jargon, J. R.: "Reservoir Modeling — 7: Single-Phase Reservoir Models," *Oil and Gas J.* (March 2, 1970) 77-80.

37. van Poollen, H. K., Bixel, H. C., and Jargon, J. R.: "Reservoir Modeling — 8: Single-Phase Gas Flow," *Oil and Gas J.* (March 30, 1970) 106-107.

38. van Poollen, H. K., Bixel, H. C., and Jargon, J. R.: "Reservoir Modeling — 9: Here are Fundamental Equations for Multiphase Fluid Flow," *Oil and Gas J.* (May 11, 1970) 72-78.

39. van Poollen, H. K., Bixel, H. C., and Jargon, J. R.: "Reservoir Modeling — 10: Applications of Multiphase Immiscible Fluid-Flow Simulator," *Oil and Gas J.* (June 29, 1970) 58-63.

40. van Poollen, H. K., Bixel, H. C., and Jargon, J. R.: "Reservoir Modeling — 11: Comparison of Multiphase Models," *Oil and Gas J.* (July 27, 1970) 124-130.

41. van Poollen, H. K., Bixel, H. C., and Jargon, J. R.: "Reservoir Modeling — 12: Individual Well Pressures in Reservoir Modeling," *Oil and Gas J.* (Oct. 26, 1970) 78-80.

42. van Poollen, H. K., Bixel, H. C., and Jargon, J. R.: "Reservoir Modeling — 13: (Conclusion) A Review — and A Look Ahead," *Oil and Gas J.* (March 1, 1971) 78-79.

Chapter 13

Test Design and Instrumentation

13.1 Introduction

Each transient-test analysis technique described in this monograph requires specific test data. Complete and adequate data are essential for satisfactory transient-test results. Thus, an important part of preparation for a transient well test is deciding which data are needed and how they will be obtained. This chapter discusses the design of transient tests, from the choice of test type to determining data required, and describes characteristics of suitable equipment.

The first step in designing a transient test is to choose the appropriate test for the existing situation: buildup, drawdown, multiple rate, interference, etc. When we desire specific information about a reservoir (for example, an indication of a mobility change or a boundary), test design is critical since many things can mask the response we seek, or can cause a response that is misleading because it merely resembles the behavior expected. (Sections 7.5, 10.1 through 10.7, and 11.2 through 11.4 illustrate a variety of situations that have similar test responses.) Once the test is chosen, the test duration and expected pressure response should be estimated so appropriate pressure-measuring equipment may be used. We also must decide what other data are required, determine how those data will be obtained, and consider how the testing plan will fit into the work schedules of the individuals who will perform the test. Occasionally, this part of test design indicates that a different kind of test than originally chosen should be used. If that happens, the entire design process is repeated.

Test design should minimize problems such as those caused by excessive wellbore storage, unintentional variation in rates, rate changes in nearby wells, etc. This chapter presents a broad approach to test design, emphasizing general design concepts rather than details. The design material in this chapter is limited to conventional transient testing. Pulse test design is discussed in Sections 9.3 and 10.8.

A discussion of pressure- and rate-measuring instruments indicates the kinds of instruments available, and what factors should be considered in instrument choice. We make no attempt to consider all available instruments nor to evaluate the relative merits of the instruments considered.

13.2 Choice of Test Type

When deciding what kind of transient well test to use, the foremost considerations are the type and status of the well: injection or production, active or shut-in. We may choose a single-well or a multiple-well test, depending on what we wish to learn about the reservoir.

When planning a production-well transient test, the engineer can choose between drawdown, buildup, and multiple-rate testing. He also must determine how he will make pressure measurements in artificially lifted wells. It is particularly difficult to measure down-hole pressure in rod-pumped wells unless the well is equipped with a permanent bottom-hole pressure gauge. Although it is possible to run some pressure gauges in the tubing-casing annulus, that can be risky and, generally, is not recommended. Pulling the pump and then running the pressure gauge seldom solves the problem; when the pump is pulled, the fluid in the tubing is dumped in the hole, creating an injection transient. It is possible to pull the pump, run the gauge in tubing below the pump, rerun the pump, continue production for several days, and then test. That approach requires a gauge with a long time span; it also involves considerable expense for the necessary well-service work. One common way to obtain pressure buildup data in a rod-pumped well completed without a packer is to measure the fluid level in the tubing-casing annulus with an acoustic sounder. Pressure measurement in flowing wells, in gas-lift wells, and in some wells with hydraulic or submersible electric pumps is not so difficult. But even in those situations, mechanical problems such as gas-lift valves that open suddenly during pressure-buildup surveys must be considered and avoided in test design.

The composition and rate of the fluid produced is important, since multiple-phase effects may be significant. The system may have to be treated as an oil well or as a gas well, depending on the gas-liquid ratio. As indicated in Section 2.6, phase segregation in the tubing string may create anomalies that make analysis difficult or impossible. Such anomalies should be anticipated and avoided, if possible.

Test duration may be a problem in producing wells; generally, one does not like to shut in a producing well for a long time since delayed production can be a major cost in a test. Deferred income often may be reduced by using a two-rate test.

The choice of test type is less complicated for injection

wells than for production wells — because the difficulties associated with artificial lift are not present. Normally, an injectivity test or falloff test will provide usable results. The pressure falloff is preferred, since it is easier to perform than an injectivity test and since minor rate variations have less influence on falloff-test response. It is good practice to run an injectivity test after the falloff test, since the cost is low and additional information may be obtained. Injection wells that take fluid on vacuum are difficult to test because of high wellbore storage coefficients associated with the free liquid level in the injection string. Usually, it is advisable to test such wells by increasing the injection rate enough to obtain wellhead pressure and then performing either an injectivity test at high rate or a two-rate injection test with positive wellhead pressure maintained during both rates. Changing wellbore storage (Section 11.2) tends to be more of a problem in injection wells than in production wells.

Ideally, pressures should be recorded continuously during a transient test. Best results are obtained when the bottom-hole pressure is measured, although surface pressure often can be converted to bottom-hole values if adequate information is available about the wellbore system. If possible, one should avoid changing pressure gauges during the test because of offsets that usually occur when a gauge is changed — even when the same gauge is removed and rerun with a new chart. Remember, we are often concerned with relatively subtle pressure trends in transient pressure analyses. It is often possible to avoid changing charts in a gauge and still get good short-time and long-time data by running two gauges in tandem with different-speed clocks.

It is usually beneficial to record bottom-hole, tubing-head, and casing-head pressures during a well test. That combination of data can provide information about wellbore effects — such as fluid redistribution, wellbore storage, and leaking packers or tubing — and may allow analysis of a test that could not be analyzed adequately based on bottom-hole pressure data alone. Such surface pressure data may be valuable in verifying correct operation of the down-hole pressure gauge.

Some well tests may require bottom-hole shut-in; some may even require extra packers or drillstem test equipment. Such requirements must be considered in the test design so that all important data are obtained.

13.3 Design Calculations

There are three general approaches to designing a transient well test:

1. Estimate the complete expected pressure response using assumed formation properties.
2. Estimate key factors in test response, such as the end of wellbore storage effects, the end of the semilog straight line, the semilog straight-line slope, and the general magnitude of the pressure response.
3. Just run the test without design calculations.

Option 3 is generally a poor one except in wells or reservoirs that have been tested frequently enough so their behavior is well known. Estimating the entire pressure response for a test can be a time-consuming task and may require computer assistance. (In complex situations, the subsequent analysis also may require the use of a computer.) Nevertheless, in some cases that is the only reliable way to design a test. To estimate pressure response for relatively simple systems, we use superposition and Eq. 2.2:

$$p_{wf} = p_i - 141.2 \frac{qB\mu}{kh} \left[p_D(t_D, r_D, \ldots) + s \right]. \qquad (13.1)$$

For complex systems, or to reduce the labor required for test design, we normally use a computer to estimate expected pressure response. Once the pressure response has been estimated, the data may be analyzed by normal methods to determine potential analysis problems. Example 13.2 illustrates that approach.

In most transient well tests, we need not know the complete pressure response for design purposes. It is normally sufficient to estimate the beginning time of the correct semilog straight line by using Eq. 2.21,

$$t > \frac{(200{,}000 + 12{,}000\, s)\, C}{(kh/\mu)}, \qquad (13.2a)$$

for pressure drawdown or injectivity, or Eq. 2.22,

$$\Delta t > \frac{170{,}000\, C\, e^{0.14s}}{(kh/\mu)}, \qquad (13.2b)$$

for falloff or buildup. The wellbore storage coefficient, C, is estimated from completion details using the techniques of Section 2.6. Formation kh/μ and skin factor must be assumed to use Eq. 13.2. If $s < 0$, we recommend using $s = 0$ in Eqs. 13.2a and 13.2b to get conservative results.

The next step is to estimate the end time of the semilog straight line. For drawdown and injectivity tests, we estimate the time when the system no longer acts infinite by using Eq. 2.8:

$$t \simeq \frac{\phi\mu c_t A (t_{DA})_{eia}}{0.0002637\, k}. \qquad (13.3a)$$

t_{DA} at the end of the infinite-acting period is taken from the "Use Infinite System Solution With Less Than 1% Error for t_{DA} <" column of Table C.1. The end of the semilog straight line for buildup and falloff tests may be estimated from Eq. 5.16:

$$\Delta t = \frac{\phi\mu c_t A}{0.0002637\, k} (\Delta t_{DA})_{esl}. \qquad (13.3b)$$

$(\Delta t_{DA})_{esl}$ is from Fig. 5.6 or Fig. 5.7.

Finally, the slope of the semilog straight line is estimated from

$$m = \frac{\pm 162.6\, qB\mu}{kh}, \qquad (13.4)$$

where the sign used depends on the test type. It may be necessary to consider the general pressure-level decline in developed reservoirs when computing m. See Sections 3.4, 4.5, and 5.3 to determine how that would be done.

Once the slope is estimated, the pressure change expected between two times on the semilog straight line, t_1 and t_2, can be estimated from

$$\Delta p = \pm m \log(t_2/t_1). \qquad (13.5)$$

Again, the sign chosen depends on test type. The pressure instrument chosen must be sensitive enough to detect the expected pressure change during the test period. The normal skin equation for the test used may be used to estimate p_{1hr} (after assuming an s value and a value for bottom-hole pressure at the start of the test); thus, we may estimate the pressure at any time on the semilog straight line by appropriate application of Eq. 13.5. The range of the pressure instrument used must be chosen so that the pressure can be measured with acceptable accuracy but without exceeding the instrument upper pressure limit.

Test design also may be important in reservoir limit testing — if it is possible to estimate reservoir parameters well enough to make the design calculations. When that can be done, the time to the start of the straight line on a plot of pressure vs time (arithmetic or linear coordinates) is estimated from

$$t_{pss} \simeq \frac{\phi \mu c_t A}{0.0002637\, k} (t_{DA})_{pss}\,, \quad \ldots\ldots\ldots\ldots\ldots (13.6)$$

where $(t_{DA})_{pss}$ is taken from the "Exact for t_{DA} >" column of Table C.1. By using Eq. 13.6, even when estimates of reservoir parameters are incorrect by a factor of 2 to 3, it is possible to get a reasonable idea of the time to the beginning of analyzable data for a reservoir limit test. In many cases, that time may be impractically long, indicating the inadvisability of attempting to run the test. The slope of the straight-line portion of the Cartesian pressure-time plot is estimated from Eq. 3.33:

$$m^* = -\frac{0.23395\, qB}{\phi c_t h A}\,. \quad \ldots\ldots\ldots\ldots\ldots\ldots\ldots (13.7)$$

The slope estimate may be used to indicate the sensitivity required of the pressure gauge and to get an idea of how long the test need be run after the straight-line portion starts. Pressure levels generally will be below the initial pressure, so it usually suffices to choose a gauge with a maximum range equivalent to initial reservoir pressure. If there has been a substantial pressure decline, then a lower-range gauge would be sufficient; but reservoir limit testing is not necessarily applicable to wells that have experienced much pressure decline, unless they have been shut in long enough to stabilize at average reservoir pressure.

When designing an interference test, it is best to estimate the pressure response at the observation well as a function of time. That may be done by using Eq. 13.1 with p_D taken from Fig. C.2 (Eq. 2.5a) or Fig. C.12. Such design is particularly important because observation-well response may be small and may occur only after a long time. For a reservoir with several noncommunicating layers, the most rapid response generally will be associated with the most permeable layer. This time of response can be much shorter than the time corresponding to the average permeability. In such cases, we recommend that interference-test design and analysis be backed up with a reservoir simulator, since little has been published about pressure behavior away from the active well in such layered systems.

Pulse-test design is discussed in Sections 9.3 and 10.8.

Example 13.1 Pressure-Buildup Test Design

We wish to run a pressure buildup test in a well in an undersaturated reservoir developed on 40-acre spacing. The field is suspected to be operating at pseudosteady-state conditions; that suspicion is confirmed later in the example. The well is currently producing 132 BOPD and 23 BWPD at a bottom-hole pressure of about 2,450 psi. Known data from production operations, laboratory tests, and log analyses are

$q_o = 132$ BOPD
$q_w = 23$ BWPD
$\mu_o = 2.30$ cp
$\mu_w = 0.940$ cp
$c_o = 14.6 \times 10^{-6}$ psi^{-1}
$c_w = 3.20 \times 10^{-6}$ psi^{-1}
$c_f = 3.40 \times 10^{-6}$ psi^{-1}
$B_o = 1.21$ RB/STB
$B_w = 1.00$ RB/STB
$A = 40$ acres $= 1{,}742{,}400$ sq ft
$h = 63$ ft
$\phi = 16.3$ percent
$r_w = 0.26$ ft
depth $= 3{,}600$ ft
tubing $= 2\frac{3}{8}$ in. OD
$V_u = 0.00387$ bbl/ft.

Estimated data are

$p_{wf} = 2{,}450$ psi
$k = 135$ md
$s = 2$.

Based on observed flow rates, known fluid properties, and relative permeability data, we estimate

$S_w = 0.29$
$S_o = 0.71$
$k_{rw} = 0.02$
$k_{ro} = 0.2$.

Thus, composite properties may be estimated. Using Eq. 2.38,

$$\begin{aligned} c_t &= (0.71)(14.6 \times 10^{-6}) \\ &\quad + (0.29)(3.20 \times 10^{-6}) + 3.40 \times 10^{-6} \\ &= 14.7 \times 10^{-6} \text{ psi}^{-1} \end{aligned}$$

for the reservoir.

To estimate c_t for the wellbore, we weight the compressibilities by the relative volume of wellbore fluids,

$$\begin{aligned} q_t B_t &= (132)(1.21) + (23)(1.00) \\ &= 182.7, \end{aligned}$$

so

$$\begin{aligned} c_{twb} &= c_o \frac{q_o B_o}{q_t B_t} + c_w \frac{q_w B_w}{q_t B_t} \\ &= \frac{(14.6 \times 10^{-6})(132)(1.21) + (3.2 \times 10^{-6})(23)(1.00)}{182.7} \\ &= 13.2 \times 10^{-6} \text{ psi}^{-1} \end{aligned}$$

for the wellbore.

We also must estimate total mobility of the fluids flowing

in the formation using Eq. 2.37:

$$\left(\frac{k}{\mu}\right)_t = k\left(\frac{0.2}{2.3} + \frac{0.02}{0.94}\right)$$
$$= 0.11\,k.$$

The semilog straight line will be bounded on the low-time end by wellbore storage effects and on the upper end as indicated by Eq. 13.3b. To estimate the start of the semilog straight line, we determine the wellbore storage coefficient for the liquid-filled wellbore. We use the compressibility estimated above for the wellbore and the total mobility based on an estimated permeability of 135 md. Then we use Eq. 13.2b for pressure buildup with an estimated $s = 2.0$. First we compute the wellbore storage coefficient from Eq. 2.17:

$$C = (0.00387)(3{,}600)(13.2 \times 10^{-6})$$
$$= 1.84 \times 10^{-4}\text{ bbl/psi}.$$

Then we apply Eq. 13.2b:

$$\Delta t > \frac{(170{,}000)(1.84 \times 10^{-4})\,e^{(0.14)(2)}}{(0.11)(135)(63)}$$
$$> 0.044\text{ hour} = 2.6\text{ minutes}.$$

This indicates that wellbore storage should not be a problem.

We may now check the assertion that the well was producing at pseudosteady state. For a square system with a well in the center, $(t_{DA})_{pss} = 0.1$. Using Eq. 2.24,

$$t_{pss} = \frac{(0.163)(14.7 \times 10^{-6})(1{,}742{,}400)(0.1)}{(0.0002637)(0.11)(135)}$$
$$= 107\text{ hours}$$
$$\simeq 4.5\text{ days}.$$

Since the well has been operating for many weeks, it can be treated like it is operating at pseudosteady state.

To estimate the end of the semilog straight line, we use Eq. 13.3b and the appropriate figure from Chapter 5. Since the reservoir is producing at pseudosteady state, we assume that the pattern is a closed square with the well in the center. We use $t_{pDA} \simeq 0.1$ (see Sections 5.2, 6.3, and Table C.1) and Curve 1 of Figs. 5.6 and 5.7. From Fig. 5.6 $(\Delta t_{DA})_{esl} \simeq$ 0.013 for a Horner plot, and from Fig. 5.7 $(\Delta t_{DA})_{esl} \simeq$ 0.0038 for a Miller-Dyes-Hutchinson plot. Then, applying Eq. 13.3b,

$$\Delta t = \frac{(0.163)(14.7 \times 10^{-6})(1{,}742{,}400)}{(0.0002637)(0.11)(135)}(\Delta t_{DA})_{esl}$$
$$= 1{,}070\,(\Delta t_{DA})_{esl}.$$

The semilog straight line should end at about

$$\Delta t = (1{,}070)(0.013) = 14\text{ hours}$$

for the Horner plot, and

$$\Delta t = (1{,}070)(0.0038) = 4.1\text{ hours}$$

for the Miller-Dyes-Hutchinson graph. These times are probably conservative since the other wells will not be shut in, and thus, the shut-in well will not really be at the center of a closed 40-acre square. Nevertheless, the Horner plot should have a longer semilog straight line.

We estimate the slope of the semilog straight line from Eq. 13.4:

$$m = \frac{(162.6)\left[(132)(1.21) + (23)(1.00)\right]}{(0.11)(135)(63)}$$
$$= 31.8\text{ psi/cycle}.$$

The relatively small m value indicates a slowly increasing pressure so we need a sensitive pressure gauge. We might consider stabilizing the well at a higher production rate before the buildup test to create a larger pressure response.

We may estimate the pressure level expected during the buildup test. A simple way to do that is to solve the skin-factor equation, Eq. 5.7, for p_{1hr}. We may then use p_{1hr} and m to estimate the pressure at later times. Rearranging Eq. 5.7,

$$p_{1hr} = p_{wf}(\Delta t = 0) + m\left[\log\left(\frac{k}{\phi\mu c_t r_w^2}\right) - 3.2275 + 0.86859\,s\right]$$
$$= 2{,}450 + 31.8\left[\log\left(\frac{(0.11)(135)}{(0.163)(14.7 \times 10^{-6})(0.26)^2}\right) - 3.2275 + 0.86859\,s\right]$$
$$= 2{,}600 + 27.6\,s.$$

Then the pressure at any time on the semilog straight line can be estimated using Eq. 13.5:

$$p(\Delta t) \simeq 2{,}600 + 27.6\,s + 31.8\log\Delta t.$$

If we expect $s = 2$ and a 24-hour test, $p \simeq 2{,}700$ psi. The pressure gauge we choose probably should be in the 3,300-psi range and still be capable of detecting a 30-psi change over a period of several hours.

The average pressure to be expected likewise can be estimated by the Dietz method, using $C_A = 30.88$ for the square drainage area. The time when the semilog straight line should reach $\bar{p}$ is given by Eq. 6.7a:

$$(\Delta t)_{\bar{p}} = \frac{(0.163)(14.7 \times 10^{-6})(1{,}742{,}400)}{(0.0002637)(0.11)(135)(30.88)}$$
$$= 34.5\text{ hours}.$$

Using the equation above,

$$\bar{p} \simeq 2{,}600 + 27.6(2) + 31.8\log(34.5)$$
$$\simeq 2{,}704\text{ psi}.$$

Since the reservoir pressure is declining, we may have to

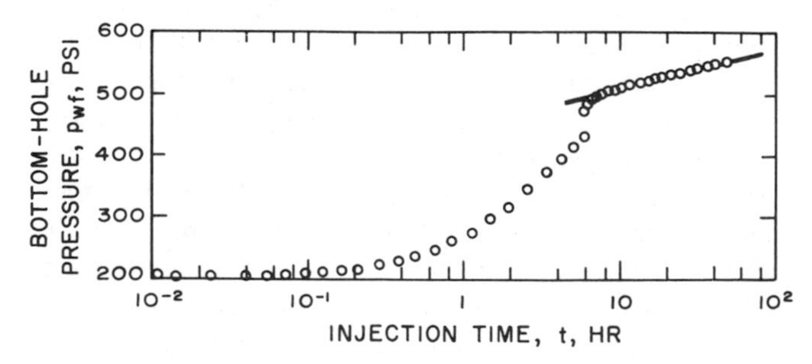

Fig. 13.1 Calculated injectivity pressure response for Example 13.2

consider that decline in the analysis, as indicated in Section 5.3. Nevertheless, the information estimated in this example indicates correct times; the pressure changes are in addition to the established trend at the well before testing. It would be worthwhile to try to measure that trend before doing the test, so it could be included in the analysis if necessary.

The pressure decline before shut-in may be estimated from Eq. 13.7:

$$m^* = \frac{dp}{dt}$$

$$= -\frac{(0.23395)\left[(132)(1.21) + (23)(1.0)\right]}{(0.163)(14.7 \times 10^{-6})(63)(1,742,400)}$$

$$= -0.16 \text{ psi/hr.}$$

Example 13.2 Injection-Well Test Design

An injection-well transient testing program was proposed for several input wells in a waterflooded reservoir before starting fluid injection for a tertiary recovery project. The reservoir was liquid-filled with water flowing at residual oil saturation. Reservoir pressure was low with the static liquid level standing about 600 ft below the surface. Because of the importance of the tests, and since changing wellbore storage could be expected as a result of the low liquid level, we decided to compute the expected pressure response by using a reservoir simulator.

We supplied estimated reservoir properties and computed the injectivity-test response shown in Fig. 13.1. As expected, the liquid level rose in the well during injection until it reached the surface about 5.9 hours after starting injection. The rapid increase in pressure in Fig. 13.1 is a result of the wellbore storage coefficient decreasing abruptly from a value corresponding to a rising liquid level to one for compression only. Note the similarity of the response in Fig. 13.1 to Figs. 11.5 and 11.6. We analyzed the apparent semilog straight line starting at about 10 hours to estimate a permeability about 15 percent lower than the input value and a skin factor that is *low* by 1.1. The discrepancy is due to choosing a semilog straight line with too steep a slope — apparently a result of the rapid wellbore storage decrease. The fact that the analysis of simulated data does not return the values supplied to the simulator indicates that the assumptions used in the analysis technique are not completely correct (assuming the simulator is working correctly). This says that if we wish to analyze data by standard techniques, we should seek a type of test for which conditions more closely approach the ideality of the analysis equations. Alternatively, we could use a more sophisticated analysis, perhaps one based on a reservoir simulator. Practically speaking, that is not often required.

Fig. 13.2 shows a simulated pressure falloff test following 48 hours of injection. The well goes on vacuum about 40 seconds after shut-in, so wellbore storage *increases* from compression to falling liquid level. To analyze data from a 24-hour falloff, it would be necessary to draw the straight line through the last five data points shown — a risky approach at best. We conclude the falloff test is essentially worthless. Although a computer program similar to the one used to generate the data could be used with regression analysis to analyze such falloff data, such an approach is not often practical. Even when that is done, the margin for error is significant because the analyst must make some assumptions about the wellbore storage characteristics (constant, step increase, gradual increase because of a gas cushion in the well, etc.). Thus, we prefer to find a type of test that is relatively insensitive to wellbore storage effects.

We also simulated a two-rate injection test, with the injection rate increasing by 73 percent after 48 hours. The second rate continues for 24 hours. Since the wellbore storage was from liquid compression only (wellhead pressure is positive), wellbore storage had little effect on the pressure response to the rate increase. Fig. 13.3 shows the data plotted as suggested by Eq. 4.6. The straight line can be analyzed to give k within 3 percent of the input value and a skin factor within 0.2 of the input value.

As a result of the design work shown in Figs. 13.1 through 13.3, the wells were tested with a two-rate injection followed by a falloff. The injection-pressure data were analyzed successfully; standard analysis was not possible for the falloff data; no attempt was made to use computer analysis.

13.4 Test Data and Operation Requirements

An important part of test planning and execution is complete data acquisition and safe and correct test operation. The important parts of test operation include good and complete rate stabilization (or rate control during tests requiring the well to be active), placement of the pressure instrument before the test begins, and careful documentation of what happens during the test, both at the test well and at nearby operating wells.

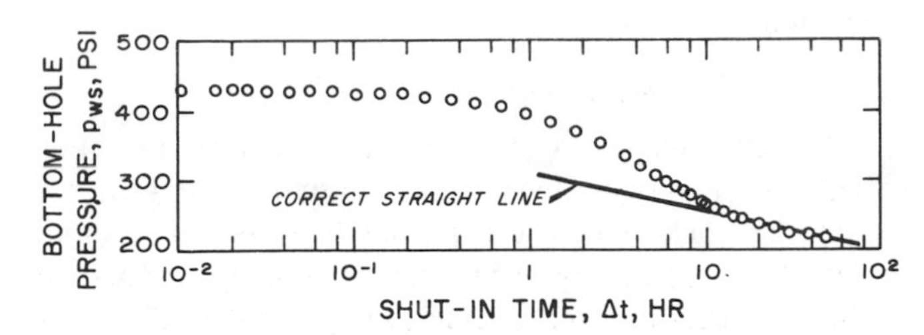

Fig. 13.2 Calculated falloff pressure response for Example 13.2.

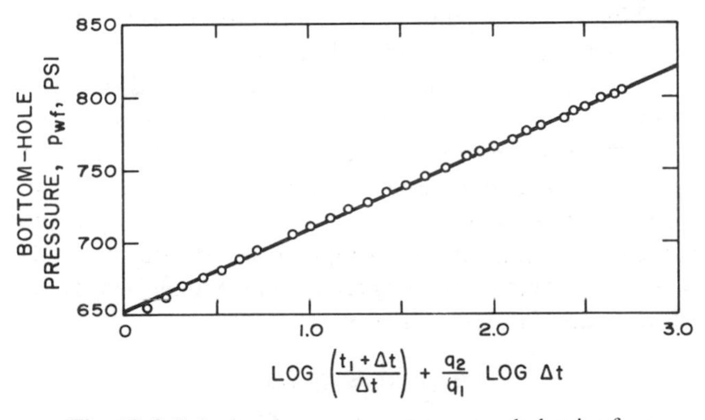

Fig. 13.3 Calculated two-rate test pressure behavior for Example 13.2.

The following general data check list is an aid to complete data acquisition. Depending on the testing situation and the information desired, it may require modification. When testing in wells or reservoirs that have been thoroughly tested before, much of the data may not be necessary based on past experience.

1. Well-Completion Data

Pipe in the Hole and Packers — Size and location of casing and tubing, location of any packers, and an indication of what pipe strings they separate should be recorded. A sketch of the well completion should accompany test data. Such information is important for running the gauge and determining the effects of wellbore storage. It also should assure that the gauge can be run to the depth where pressure is to be measured.

Type of Completion in the Producing Interval — Data should include information about whether the well is open hole, cased, perforated, uses a liner or a gravel pack, type of completion fluid, etc. If there is a dual completion, give details. Data should include location and number of perforations and indication of partial penetration, if applicable.

Stimulation — Has the well been stimulated by shooting, acidizing, fracturing, etc?

2. Pattern Data

Well Pattern — Data should include pattern size and shape, and information about location of other wells. Usually a map suffices.

Rate Information at Other Wells — When testing developed patterns, it may be important to know how the rates behave at other wells before and during the test. We must know of any major rate changes during the test since these may influence the test response considerably.

3. Rate Data

Stabilized Rate Before Testing — It is best to have spot checks on the rate for several days before testing so any problems can be isolated. Severe fluctuations may dictate postponing the test or changing the analysis technique. It is especially important to know when an active well might have been shut in before testing, if only for a few minutes.

Detailed Test-Rate Data — For producing, injecting, multiple-rate, and multiple-well tests, it is advantageous to have a detailed description of the rate behavior during testing. Ideally, one should use a recording flowmeter; if that is not possible, rate checks should be made frequently.

Fluid Type — It is important to know rates and properties of all fluids flowing at the well. It is advantageous to know the composite density and compressibility of the fluids for wellbore storage considerations and for correcting pressure measurements to some other datum. A separate pressure survey consisting of short stops at several depths in the wellbore will provide information relative to the density and distribution of fluids in the well.

4. Pressure Data

Bottom-Hole Measurements — Continuously recorded bottom-hole measurements are usually essential to good well test analysis. However, care in obtaining pressure data and additional data is also important, as stated below.

Trends Before Testing — We should know the pressure trend before testing since it may affect the analysis technique. Generally, such information is not important in undeveloped reservoirs or in fluid-injection projects with injection and production approximately balanced.

Short-Time Data — It is advisable to take pressure data at short intervals while wellbore storage is important. That allows isolation of the storage-affected part of the response data and aids in correct analysis. Design calculations give some indication of how rapidly such data are required. In the absence of other information, we recommend taking data as frequently as every 15 seconds for the first few minutes of the test. Data should be taken at least every 15 minutes until wellbore storage effects have essentially ended.

Pressure Just Before Testing — Record the pressure observed just before the test is started. Skin-factor calculations and log-log plots depend on this information.

Wellhead Pressures — In wells that have not been tested before and whose characteristic response is not well known, it is good practice to periodically record surface tubing *and* casing pressures. Such data are usually recorded manually at intervals of 1 to 4 hours from pressure gauges on the wellhead. It is sometimes particularly useful to have such data at short test times, so that unusual wellbore effects may be more thoroughly understood. Usually by comparing surface pressures with bottom-hole pressures, it is possible to estimate how much fluid is accumulating in or leaving the wellbore.

5. Other Data

Surface Piping — A diagram of wellhead and surface piping should accompany test data. That information often helps explain anomalous test behavior.

Chronology — A complete list of how things happen during the test is frequently the only key to unraveling unusual test response. Thus, the engineer should keep a log of times at which various events occur.

13.5 Pressure-Measurement Instruments

Accurate pressure data are an essential part of transient well testing. For best results, pressure should be measured near the sand-face. If that is impossible, useful data usually may be obtained by correcting wellhead-pressure or fluid-level measurements to bottom-hole conditions.[1]

Three basic types of bottom-hole pressure gauges are available:[2] self-contained wireline gauges; permanently installed surface-recording gauges; and retrievable surface-recording gauges.

Self-Contained Wireline Gauges

The self-contained wireline gauge, the gauge most often used in the petroleum industry, is lowered into a well on a solid wire (slick-line). The gauge has three essential components: (1) a pressure-sensing device, usually a Bourdon tube; (2) a pressure-time recorder; and (3) a clock. The clock is designed to run for a specific length of time; if data are desired after that time, the gauge must be retrieved from the well, prepared for another period of recording, and rerun into the well.

Section 1 of Table 13.1 summarizes characteristics of several self-contained wireline gauges. The table also presents data for surface-recording gauges. Information in Table 13.1 is meant to be illustrative, not all-encompassing. Data shown are included because they are readily available and because the gauges are generally well known. Other gauges are available; we do not intend to imply that they are less useful than those in Table 13.1.

The Amerada RPG-3 is probably the most commonly used self-contained wireline gauge. It is typical of many such gauges. Fig. 13.4 schematically shows the important parts of the Amerada RPG-3 gauge. The clock, at the top of the instrument, is connected to a recording section that houses a metal chart covered with a black coating. The clock moves the chart vertically down across a stylus as time passes. The stylus is connected to a shaft that is twisted as the Bourdon tube coils and uncoils in response to pressure changes. Thus, pressure is recorded as a function of time by the stylus, which scratches a very fine line on the black coating of the metal chart. When the chart is removed from the instrument and is laid flat, the time scale is 5 in. long (for the RPG-3) and the pressure scale is 2 in. long. Because the size of the recording surface is fixed, most gauge manufacturers offer pressure-sensing elements (Bourdon tubes) in a variety of ranges; the lowest may be 0 to 500 or 1,000 psi; the highest ranges are 0 to the value given in Table 13.1. Similarly, a series of clocks is available.

In the Amerada RPG-3 gauge shown in Fig. 13.4, the pressure element is a helically wound Bourdon tube. The tube is anchored at the bottom and is free to rotate at the top. Fluid enters the lower end of the gauge and transmits pressure to the Bourdon tube, causing the tube to uncoil and rotate at the free end. The Bourdon tube is filled with oil by the manufacturer. The oil in the tube is generally protected from the fluid in the well by a bellows or filter arrangement.

Most self-contained gauges have provisions for recording bottom-hole temperature with a maximum recording thermometer, as indicated in Fig. 13.4. Temperature measurement is important, since the gauge calibration is somewhat temperature-dependent. Most gauges are temperature-compensated to about 175 to 200 °F; they must be specially calibrated at higher temperatures.

Self-contained (Amerada-type) gauges have a stated accuracy (ability to correctly indicate the pressure) of about ±0.25 percent of full scale. A 2,000-psi gauge with accuracy of ±0.25 percent can be considered to be correct within ±5 psi when properly calibrated and when operated at temperatures within the calibration range. The sensitivity of Bourdon-tube gauges to detect small pressure changes is much greater than the absolute accuracy of the gauge. For most gauges, the sensitivity is in the range of ±0.05 percent of the full-scale reading, or 1 psi for a 2,000-psi pressure element.

Fig. 13.5 shows the Leutert Precision Subsurface Pressure Recorder, which has a stated accuracy of ±0.025 percent of full scale and a precision of ±0.005 percent of full

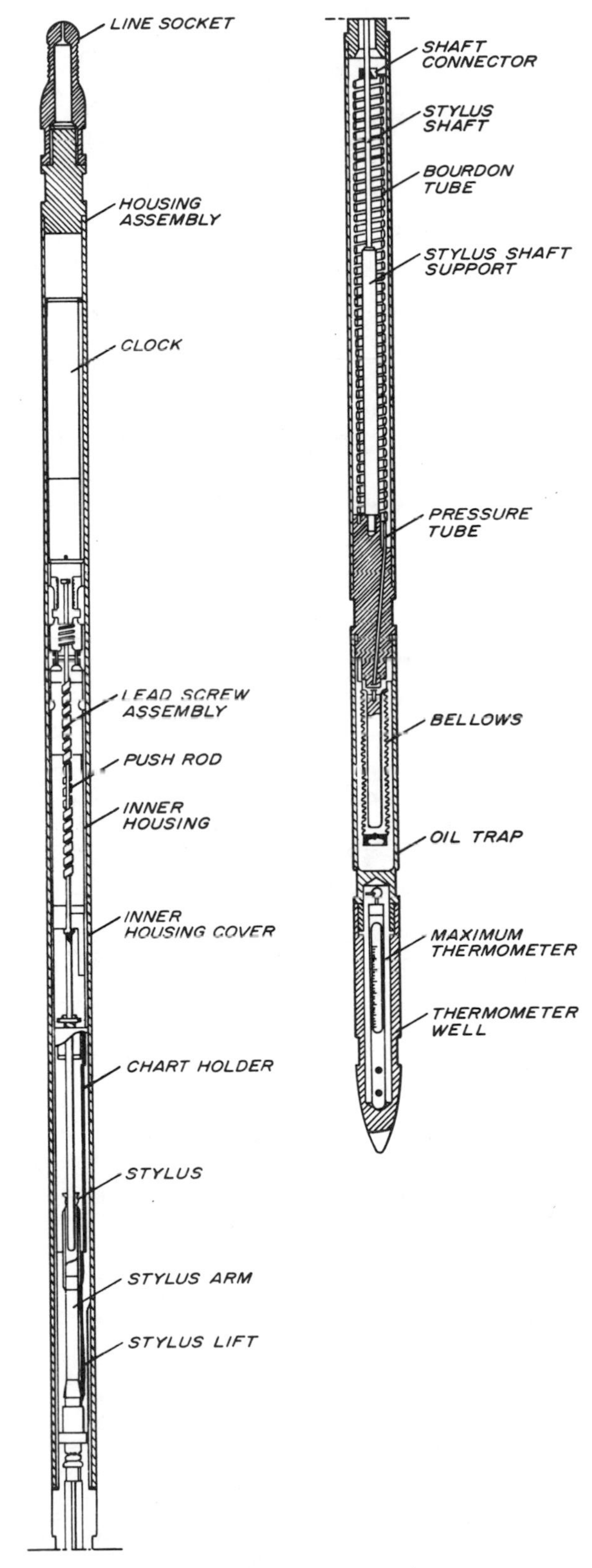

Fig. 13.4 Amerada RPG-3 gauge. Courtesy Geophysical Research Corp.

TABLE 13.1—DOWN-HOLE PRESSURE GAUGES.*

Section 1: Self-Contained Wireline Gauges

Gauge	Maximum Pressure[1] (psi)	Sensitivity, Percent of Full Scale	Accuracy, Percent of Full Scale	OD (in.)	Approximate Length[2] (in.)	Maximum Service Temperature[3] (°F)	Type Pressure Element[4]	Maximum Time Down Hole[5] (hours)	Approximate Chart Size, $p \times t$ (in.)
Amerada RPG-3	25,000	0.05	0.2	1.25	77	650	B	360	2 × 5
Amerada RPG-4	25,000	0.056	0.2	1	76	650	B	144	1.8 × 5
Amerada RPG-5	20,000	0.05	0.25	1.5	20	450	B	120	2 × 5
Kuster KPG	25,000	0.05	0.2	1.25	66	700	B	360	2 × 5
Kuster K-2	20,000	0.05	0.25	1	41	500	B	120	2 × 3
Kuster K-3	20,000	0.042	0.25	1.25	43	500	B	120	2.4 × 4
Kuster K-4	12,000	0.067	0.25	0.75	42	450	B	72	1.5 × 2.5
Leutert Precision Subsurface Pressure Recorder	6,400	0.005	0.025	1.25	139	300	P	360	9.8 × 3.1
Leutert Precision Subsurface Pressure Recorder	10,000	0.005	0.025	1.42	139	300	P	360	9.8 × 3.1
Sperry-Sun Precision Subsurface Gauge	16,000	0.005	0.05	1.5	108	300	B	672[6]	2.3 × 7.1

Section 2. Permanently Installed, Surface-Recording Gauges

Gauge	Maximum Pressure[1] (psi)	Sensitivity, Percent of Full Scale	Accuracy, Percent of Full Scale	OD (in.)	Approximate Length[2] (in.)	Maximum Service Temperature[3] (°F)	Type Pressure Element[4]	Type Signal[7]	Type Conductor[8]
Amerada EPG-512[9]	10,000	0.002	0.02	1.25	13	300	D	F	S
Amerada SPG-3	25,000	0.04	0.2	1.25	49	350	B	R	S
Flopetrol	10,000	0.001	0.06	1.42	29	257	S	F	S
Lynes Pressure Sentry MK-9PES	10,000	0.2	0.2	1.5	33	300	B	B	S
Maihak SG-2	5,700	0.1	1.0	3.54	11.54	176	D	F	S
Maihak SG-5	5,700	0.1	1.0	1.65	8.43	176	D	F	S
Sperry-Sun Permagauge	10,000	0.005	0.05	1.66	120 or 240	no max.	G	G	T
BJ Centrilift-PHD System[10]	3,500		3[11]	N/A[12]	N/A[12]		B	C	P

Section 3. Retrievable Surface-Recording Gauges

Gauge	Maximum Pressure[1] (psi)	Sensitivity, Percent of Full Scale	Accuracy, Percent of Full Scale	OD (in.)	Approximate Length[2] (in.)	Maximum Service Temperature[3] (°F)	Type Pressure Element[4]	Type Signal[7]	Type Conductor[8]
Amerada EPG-512[9]	10,000	0.002	0.02	1.25	13	300	D	F	S
Amerada SPG-3	25,000	0.04	0.2	1.25	49	350	B	R	S
Flopetrol[13]	10,000	0.001	0.06	1.42	29	257	S	F	S
Hewlett Packard HP-2811B	12,000	0.00009[14]	0.025[15]	1.44	39	302	Q	F	S
Kuster PSR	5,000	0.04	0.02	1.38	36	212	Q	F	S
Lynes Sentry MK-9PES	10,000	0.2	0.2	1.5	33	300	B	B	S
Maihak SG-3	5,700	0.1	1.0			176	D	F	S
Sperry-Sun Surface Recording	15,000	0.006	0.05	1.5	72	300	B	D	S

*Other gauges are available — no endorsement is implied by inclusion in this table. Data are from information supplied by the manufacturer and other sources believed to be reliable. Although we have been careful in assembling this table, neither the author nor SPE-AIME can guarantee accuracy of the data supplied. The reader should contact the manufacturer for specifics. Blank values could not be obtained by the author.

1. Normally, elements are available in several ranges, with the lowest being about 0 to 500 or 0 to 1,000 psi.
2. Length may vary depending on tool configuration; value is approximate normal length without weight sections.
3. Normally, temperature above which gauge cannot be used, not maximum temperature for normal calibration.
4. B — Bourdon tube.
   D — Diaphragm.
   G — Gas chamber with transducer at surface.
   P — Rotating piston.
   Q — Oscillating quartz crystal.
   S — Strain gauge.
5. Time depends on clock chosen. Clocks normally come in several ranges, starting as low as about 3 hours.
6. Clock is electronic without mechanical linkage to recorder.
7. B — Binary signal.
   C — Current.
   D — Digital.
   F — Frequency.
   G — Gas column to surface.
   R — Resistance.
8. P — Normal power cable for pump, no special conductor.
   S — Single-conductor armored cable, ground return.
   T — 3/32-in.-OD steel tubing.
9. Also measures temperature to an accuracy of 0.1 °F and a sensitivity of 0.01 °F.
10. Part of the BJ Centrilift submersible pump. Gauge is an integral part of the motor assembly.
11. Approximately 3 percent of reading.
12. Imbedded in pump motor assembly.
13. Flopetrol has under development a slick-line retrievable surface-recording gauge. The gauge is set in a side pocket mandrel; a conductor cable goes from the mandrel to the surface on the outside of the tubing.
14. Sensitivity is constant across the entire pressure range: 0.01 psi with nominal 1-second count time, 0.001 psi with nominal 10-second count time.
15. Accuracy, if temperature is known within 1° C: ±0.5 psi to 2,000 psi, ±0.025 percent of *reading* above 2,000 psi.

scale. Wellbore pressure causes a piston to operate against the tension of a helical spring. Piston extension, recorded on a metal chart by a stylus, is converted to pressure with calibration tables. The high accuracy of this gauge is obtained by rotating the measuring piston continuously to minimize frictional resistance in the instrument. Piston rotation is provided by a specially designed clock that simultaneously provides time displacement of the stylus on the chart.

The Sperry-Sun Well Survey Co. manufactures a high-precision wireline pressure gauge that is especially useful for long-time transient tests. Stated accuracy is ±0.05 percent of full scale and precision is ±0.005 percent of full scale. The gauge uses a Bourdon tube, but there is no physical connection between the Bourdon tube and the recording section, which is controlled by a battery-powered electronic programmer. Pressure recordings are made on a programmed time schedule rather than continuously. The programmer minimum is one recording every 15 seconds (giving a chart life of 5 hours) and the maximum is a recording every 32 minutes (for a chart life of 28 days).

Since pressure and time data are recorded as a finely scribed line on a metal chart in self-contained wireline gauges, a precision chart reader is needed to read the gauge charts. Chart readers should be capable of measurements at least five times as accurate as the accuracy of the pressure element. In most cases, that requires a reading accuracy of at least 0.001 in. in the pressure direction; slightly less accuracy is usually acceptable in the time direction. Although most manufacturers provide time scales for use with their chart scanners, our experience indicates that it is usually best to use two or more pressure events that occur at known times to determine a linear scale for the particular clock used.

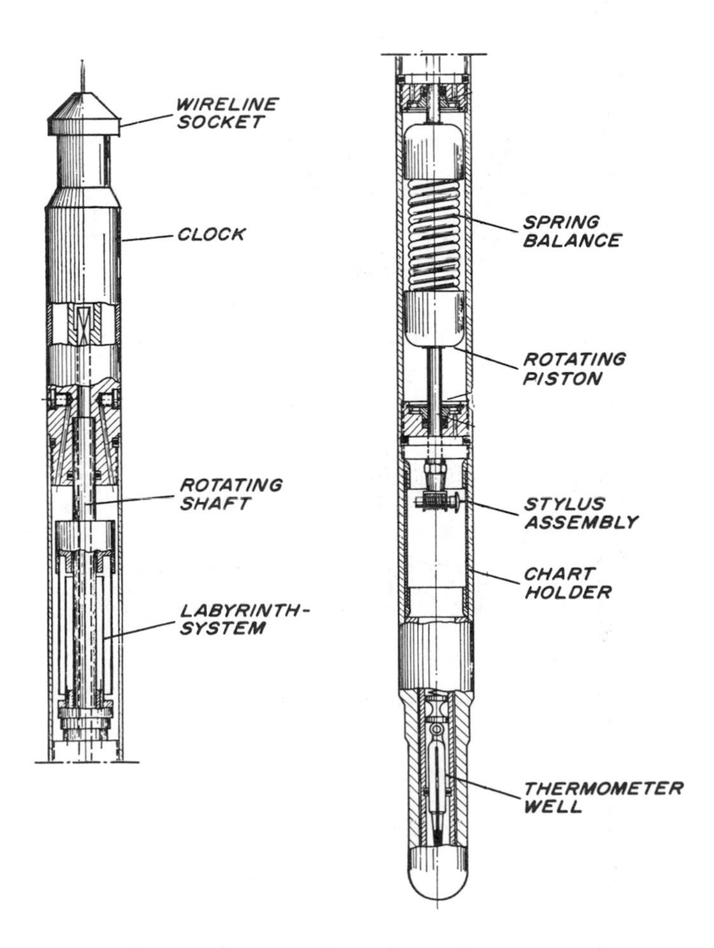

Fig. 13.5 Leutert Precision Subsurface Pressure Recorder. Courtesy Friedrich Leutert.

The factors influencing the accuracy of wireline gauges can be divided into two categories:[3-7]

1. Inherent errors owing to reproducibility of the gauge and to changes in the gauge characteristics with use.

2. Avoidable errors such as use of poor charts and stylus points, failure to reach thermal or mechanical equilibrium, neglect of temperature effects on the gauge, hysteresis effects, pressure shock during calibration, change in zero-pressure base line with temperature, etc. Ref. 3 lists several procedures that should help eliminate avoidable errors. It is important to check the calibration of the gauge frequently to verify that the calibration being used is still acceptably accurate. Generally, calibration checks should be made using a dead-weight tester. Most manufacturers will make calibration checks for a reasonable charge; all will recalibrate gauges when necessary. Nevertheless, the engineer should be aware of what constitutes good calibration. Ref. 3 presents a good explanation.

Permanently Installed Surface-Recording Gauges

Permanently installed surface-recording gauges are generally attached to the tubing string. Fig. 13.6 illustrates a common installation. These gauges are especially useful for performing transient tests on pumping wells. The instrument includes a means for measuring bottom-hole pressure and a way to transmit the measurements to the surface for recording as a function of time. Permanently installed surface-recording gauges may be used to provide continuous pressure data or occasional pressure data. Section 2 of Table 13.1 provides information about several permanently installed surface-recording gauges.

Most such gauges use a single-conductor armored cable to transmit the signal from the sensor to the recorder at the surface. Normally, the cable is strapped to the outside of the production tubing. Care must be exercised to prevent damage to the cable and the cable-to-instrument splice when the tubing is run. Formed steel protectors are normally used to protect the wire as it passes over tubing collars. To prevent damage to the cable resulting from tubing movement, a tubing anchor should be used for installations in rod-pumped wells. Hydraulic tubing anchors that can be set without turning the tubing are preferred.

One well known permanently installed surface-recording gauge is the Lynes Pressure Sentry[8] shown in Fig. 13.7. (This instrument is also available as a retrievable gauge.) The pressure element is a helically wound Bourdon tube similar in operation to that in an Amerada-type gauge. A code wheel attached to the shaft of the Bourdon tube rotates as pressure increases; the degree of rotation is an indication of the pressure applied. An electronic gating mechanism scans the code wheel and records either a 0 (the white portions of the code wheel in Fig. 13.7) or a 1 (the black portions). The resulting surface readout is a series of 1's and 0's inscribed on a chart paper as illustrated in Fig. 13.8. Tables provided with the instrument are used to convert the binary output into a deflection number, which is then converted to pressure.*

The Maihak gauge listed in Table 13.1 uses a pressure sensor based on the displacement of a diaphragm. A frequency signal is measured by a tuned circuit in the receiver. The receiver frequency reading is converted to pressure using calibration tables. Kolb[9] reports that the Maihak gauge exhibits both temperature and elastic effects from 85 to 205 °F. He found that it is possible to obtain an accuracy of about 0.25 percent by calibrating the gauge under elastic equilibrium at operating temperatures. However, Kolb found that if this is not done errors of as much as 4 percent of full scale could be incurred.

*Some newer models indicate pressure directly.

Sperry-Sun offers a permanently installed surface-recording pressure instrument they call the Permagauge. Fig. 13.9 shows one possible installation of that gauge.

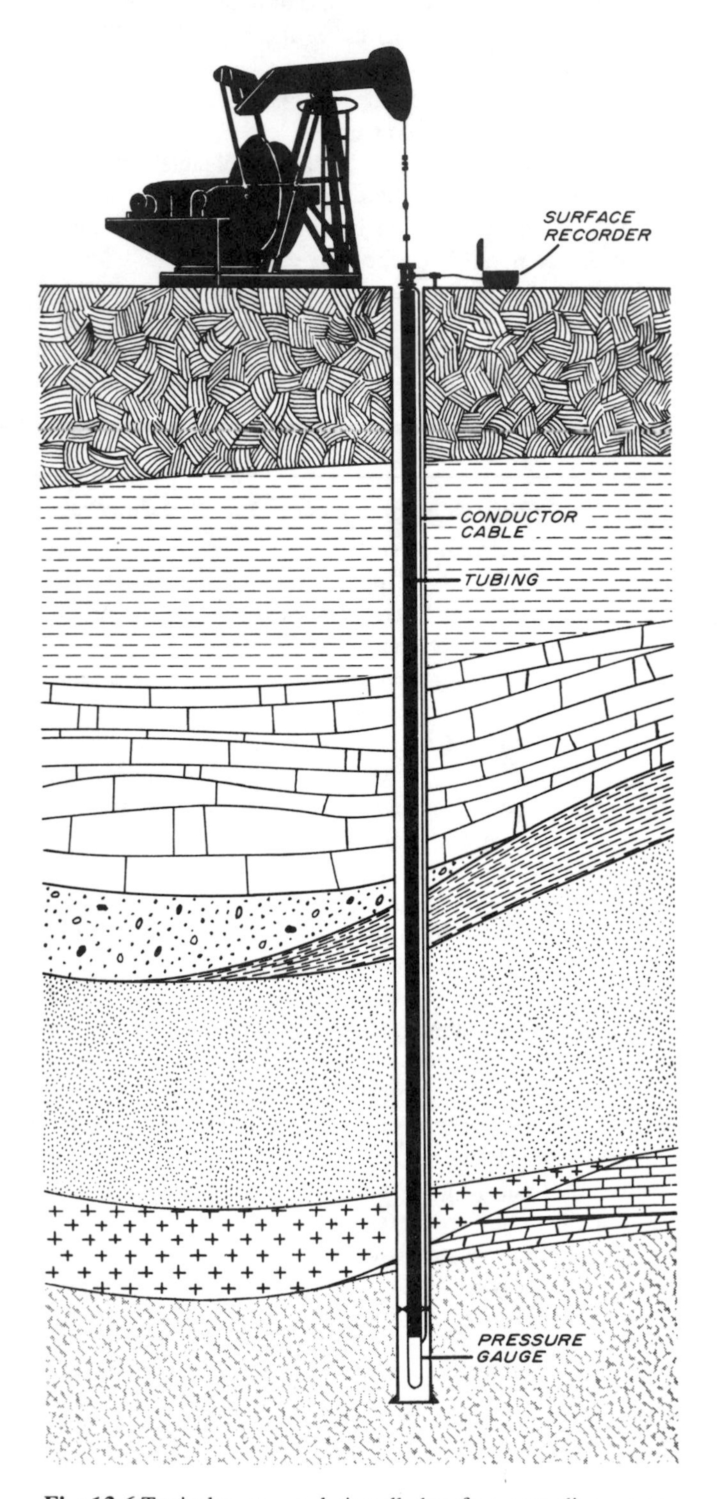

Fig. 13.6 Typical permanently installed surface-recording gauge installation. Courtesy Lynes, Inc.

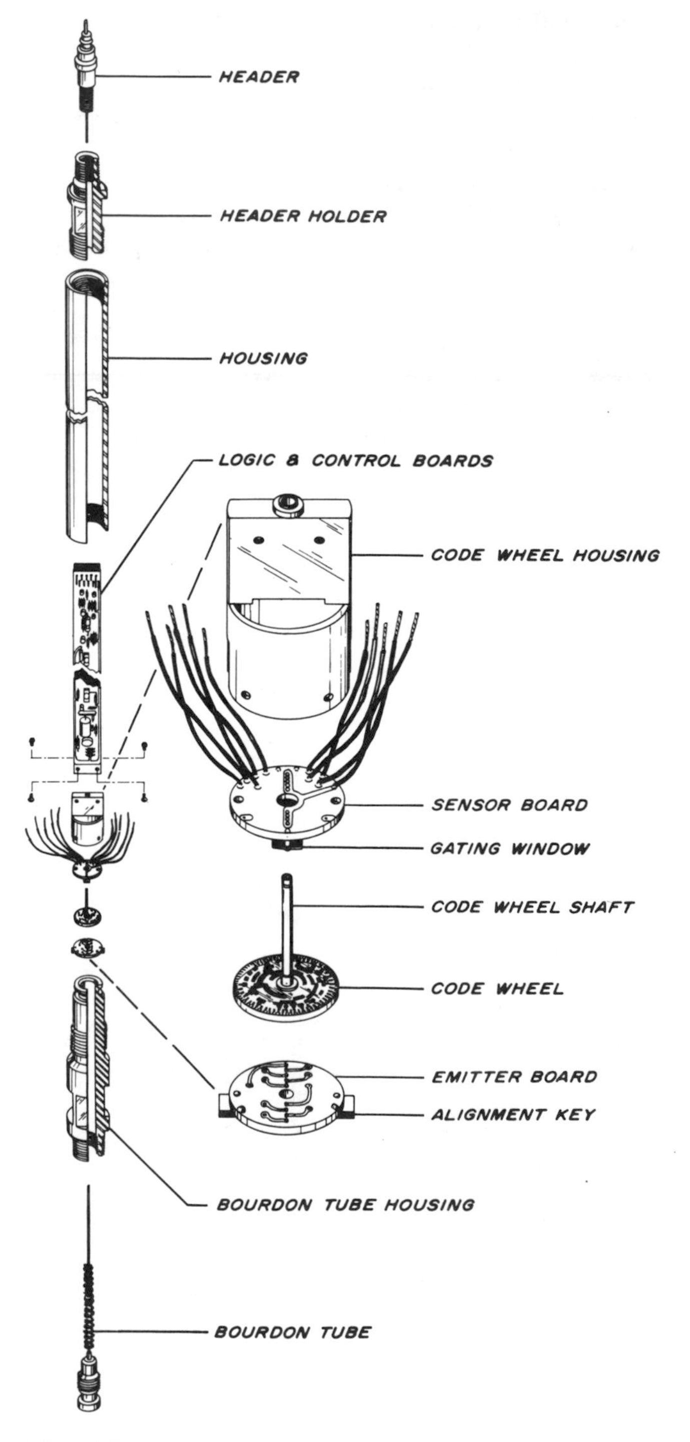

Fig. 13.7 Lynes Electronic Sentry System MK-9PES. Courtesy Lynes, Inc.

Concentric (with tubing) and suspension (like a sonde) expansion chambers are also available. The down-hole instrument is just an expansion chamber (pipe) connected to the surface by 3/32-in.-diameter stainless steel tubing. The expansion chamber and tube are charged with an inert gas and the pressure is measured at the surface by an accurate pressure transducer. Occasionally, it is necessary to purge the system with additional gas, but otherwise it requires no close supervision. The surface recorder provides pressure as a function of time. Sperry Sun provides down-hole expansion chambers in 10- and 20-ft lengths; they come in saddle-mount installations as shown in Fig. 13.9 or concentric installations that completely surround the tubing. The surface pressure detection and recording equipment need be connected to the well only when pressure data are desired. The Sperry-Sun Permagauge system requires no down-hole power and has no apparent temperature, pressure, or life limitations. The 3/32-in.-OD stainless steel tubing is strapped to the exterior of the tubing just as is the single-conductor cable used for most permanently installed surface-recording gauges.

Permanently installed down-hole pressure gauges are especially valuable for observation wells in enhanced recovery tests. In the course of normal operation, many useful transient pressure effects are generated and can be routinely and inexpensively measured by such installations.

Retrievable Surface-Recording Gauges

Section 3 of Table 13.1 presents data on several retrievable surface-recording gauges. Many are similar in operation to the permanently installed surface-recording gauges, except that they are run on a single-conductor armored wireline. Most use Bourdon tubes for pressure measurement.

The Hewlett-Packard Corp. manufactures a gauge that uses a quartz crystal[10] as a pressure-sensing device. The quartz crystal changes vibrational frequency as the imposed pressure changes. That vibrational frequency is compared with the frequency of a reference crystal and then a frequency signal is transmitted to surface-monitoring equipment. Frequency output is converted to pressure at the appropriate temperature by use of calibration equations supplied by the manufacturer. The *accuracy* of the Hewlett-Packard gauge is stated to be ±0.5 psi or ±0.025 percent of the *reading* obtained, whichever is greater (for example, ±0.75 psi at 3,000 psi). The sensitivity is stated to be better

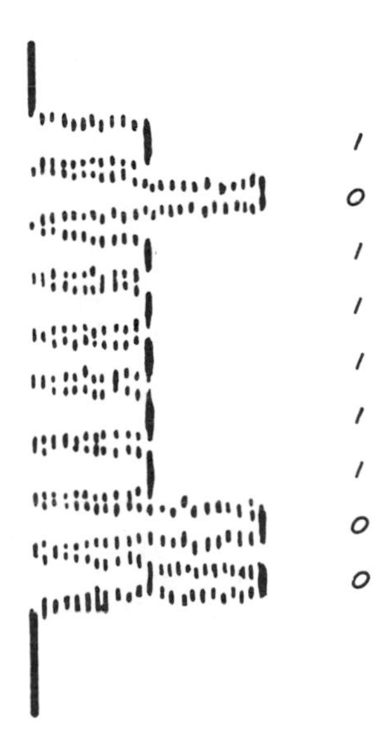

Fig. 13.8 Example of recording trace for Lynes Electronic Sentry MK-9PES. Courtesy Lynes, Inc.

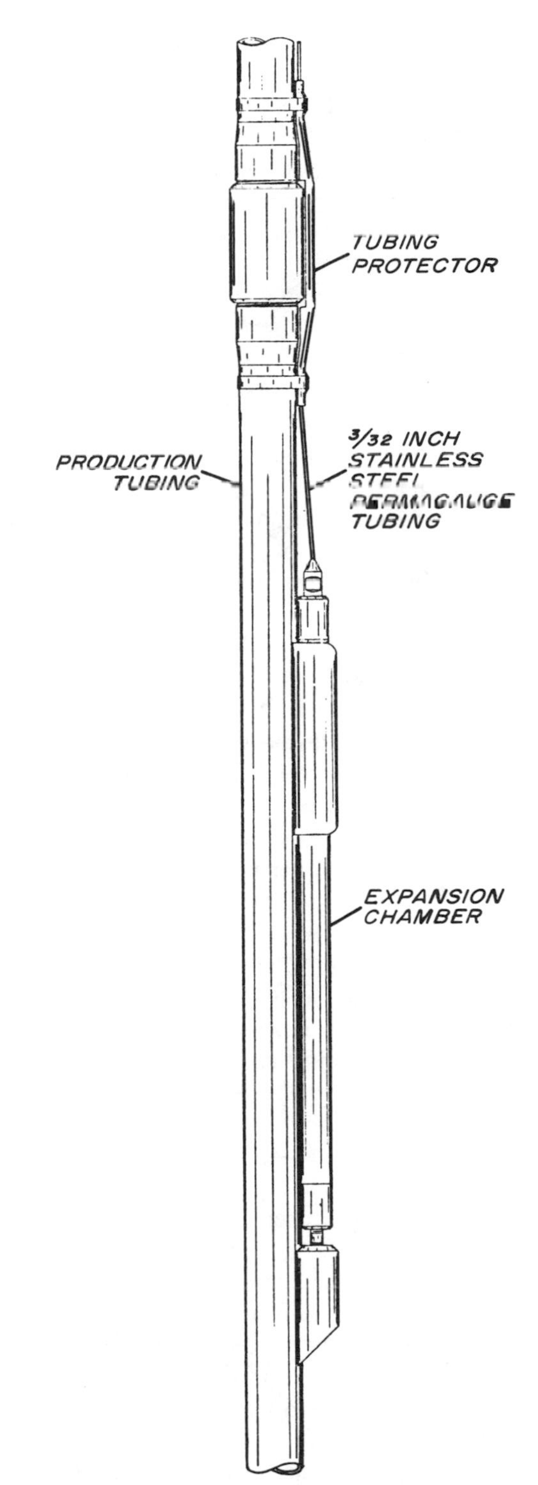

Fig. 13.9 Sperry-Sun Permagauge in saddle-mount configuration. Courtesy Sperry-Sun.

than ±0.01 psi at any pressure. Only one pressure range is offered. Although the gauge is temperature-compensated, it does have a definite response to temperature *changes* and cannot be used in changing temperature environments. About 30 minutes are required for temperature equalization of the Hewlett-Packard instrument.

Choice of Instrument

Gauge choice and operation must be considered during transient test design. We could devote dozens of pages to practical considerations. Instead, we point out only a few important items, leaving the practical aspects to be gained by experience. Many gauge manufacturers provide short instructive sessions in gauge maintenance and trouble-shooting for their customers.

It is advisable to choose the gauge pressure range so that the maximum observed pressure falls between 60 and 80 percent of the upper limit of the gauge. (However, Flopetrol and Hewlett-Packard state there is no need to limit the maximum pressure to the 80-percent value for their gauges.) If a gauge with too high a pressure range is chosen, the accuracy and sensitivity obtained may not be adequate.

All pressure gauges require periodic calibration checks and recalibration. When pressure is to be measured in high-temperature environments, it is usually necessary to specially calibrate the gauge at the temperature of interest. Bourdon-tube gauges demonstrate a hysteresis effect, so calibration must be done with a specific sequence of flexing and relaxing the Bourdon tube. The manufacturers' instructions should be followed.

When running a self-contained gauge, it is important to choose the clock so that most of the length of the chart is used during the test. It is also advisable to choose the clock so that the gauge need be run only once during the test, if that is possible. The times that the gauge is run into and out of the hole provide definite events on the pressure chart. It is recommended that those times be used to calibrate the time scale (by calculating hours per inch based on the distance between the events and the known elapsed time) to verify that the clock is running at a constant known rate and that it did not stop during the test. If there is doubt that a clock is running throughout a test, small pressure events may be put on the chart at known times simply by raising the gauge several feet and then lowering it back to its original position. The hours per inch calculated between each event should be the same.

During a test, a clock may stop for several reasons. It may be defective or damaged and not capable of running for its full term; or it may not have been wound completely. A tight recorder can stop an otherwise sound clock. Therefore, it is recommended that the recorder section of the gauge be checked before running the gauge to make sure that the stylus moves freely when the recorder is not connected to the clock or to the pressure element.

If the chart obtained from a gauge shows a stair-stepping pattern, a recorder malfunction is indicated. That pattern indicates that the pressure must change by a certain amount before the stylus moves. This difficulty must be corrected for good pressure results by freeing the recorder so it does not stick. Most manufacturers offer routine maintenance service for their gauges. In addition, many provide hints in the instruction manuals for detecting and correcting problems.

Operational problems with permanently installed gauges generally are not great. As mentioned previously, it is important to protect the cable so it is not damaged when the tubing is run in. That is usually adequately done by strapping the cable to the tubing, using protectors across the collars, and running the tubing without rotation. Readout method for permanent gauges varies depending on the application. In most situations, a portable readout device is adequate since only occasional pressures are required. However, if continuous pressure monitoring is desired, it may be necessary to provide readout devices at each wellhead, or to manifold the cables or tubing to a central location where one or more readout devices can be programmed to scan a series of wells and record the results. Most surface-recording pressure instruments require an electrical power supply (although some are battery powered). When a power supply is required, it should be determined that the voltage and frequency available fall within the range specified by the instrument manufacturer. Frequently, surface electronics are intolerant of extreme environments and must be kept in closed, heated (and possibly cooled) buildings. Such requirements should be investigated before acquiring instrumentation.

Retrievable surface-recording gauges generally are used for short-term applications. Therefore, readout is normally on a continuous basis while the gauge is in the hole. The comments above about power supply and environment apply. Normally, these gauges are used with a single-conductor armored cable.

13.6 Flow-Rate Measurement

Accurate production (or injection) rate measurement is as important as accurate pressure measurement for successful transient test analysis, since the flow rate is an integral part of all analysis equations.

Liquid flow rate is measured either by determining the time required to fill a calibrated container or by some type of flowmeter. Any suitable container can be used. For a low-rate well producing liquid with little or no gas, a rate measurement can be made by diverting the well stream into a barrel and measuring the time required to collect a given quantity of liquid. The water and oil rates are then determined from the amount of those two liquids collected during the test. The same kind of measurement can be made with a calibrated test tank.

Flowmeters are either direct or inferential. Direct flowmeters measure the total volume of fluid passing the meter. The stream is normally divided into segments of known volume and the volume segments are totaled and recorded on the meter dial. To measure flow rate with this type of meter, the time required for a given volume of fluid to pass through the meter is measured.

Inferential flowmeters measure instantaneous flow rate. Such instruments determine flow rate by measuring a related variable such as differential pressure, induced voltage, etc.

One example is a turbine flowmeter in which a precision turbine is rotated by liquid movement through the body of the meter. Rotational speed is proportional to the flow rate. As the turbine blade spins it creates measurable electrical impulses that can be converted to flow rate. Fig. 13.10 shows a turbine meter manufactured by the Halliburton Co.

All flowmeters require frequent calibration. Before using any meter on a transient test, it should be calibrated using a fluid similar to that to be measured during the test.

Gas flow-rate measurements are usually made with an orifice meter or a critical flow prover. (Orifice meters also may be used for high-rate liquid measurements.[11]) The orifice meter allows rate computation from the pressure drop occurring across an orifice placed concentrically in the flowline. Pressure taps on each side of the orifice plate allow the pressure difference across the orifice to be measured. Then, flow rate is estimated from

$$q = 0.024\, C_o \sqrt{h_w p}\ , \qquad (13.8)$$

where

q = flow rate, Mcf/D
C_o = orifice flow constant = rate of flow in cu ft/hr at base conditions when $\sqrt{h_w p} = 1$
h_w = pressure differential, inches of water
p = static pressure upstream from the orifice, psia.

The orifice constant depends on orifice size, pressure tap location, gas gravity, and temperature. It is commonly expressed as*

$$C_o = (F_b)(F_{pb})(F_{tb})(F_g)(F_{tf})(F_{pv})(F_r)\ , \qquad (13.9)$$

where

F_b = basic orifice flowmeter factor, cu ft/hr
F_{pb} = pressure base factor = $14.6955/p_b$
F_{tb} = temperature base factor = $T_b/520$
F_g = specific gravity factor = $\sqrt{1/\gamma}$
F_{tf} = flowing temperature factor = $\sqrt{520/T}$
F_{pv} = supercompressibility factor = $\sqrt{1/z}$
F_r = Reynolds number factor
T = temperature, °R.

*The nomenclature is taken from Ref. 11.

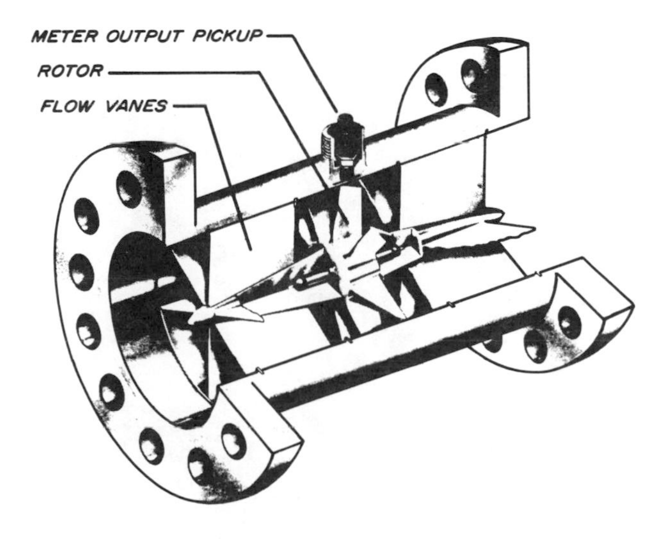

Fig. 13.10 Halliburton turbine flowmeter. Courtesy Halliburton Co.

Tables of orifice factors[11,12] are available for use in calculating C from Eq. 13.9. Precautions that should be observed in orifice-meter installation are summarized in Refs. 11 and 12.

Gas flow-rate measurement with a critical flow prover depends on gas flow through an orifice under critical conditions.[12] When gas is in critical flow, the velocity remains constant at a maximum value (sonic velocity) and the delivery rate depends only on gas density. Therefore, the flow rate is proportional to the pressure upstream of the orifice and does not change with variations in downstream pressure. A gas is normally in critical flow as long as the upstream pressure is about twice the downstream pressure.

Gas flow rate through an orifice under critical flow conditions is estimated from

$$q = \frac{0.024\, C_o p}{\sqrt{\gamma_g T}}\ , \qquad (13.10)$$

where

q = flow rate, Mcf/D
C_o = orifice coefficient, cu ft/hr
p = upstream pressure, psia
T = upstream temperature, °R
γ_g = specific gravity of the gas.

Rawlins and Schellhardt[13] give C_o values for orifices in 2- and 4-in. pipe. They neglect deviations from the ideal gas law.

References

1. "Guide for Calculating Static Bottom-Hole Pressures Using Fluid-Level Recording Devices," ERCB Report 74-S, Energy Resources Conservation Board, Calgary, Alta., Canada (Nov. 1974).

2. Matthews, C. S. and Russell, D. G.: *Pressure Buildup and Flow Tests in Wells,* Monograph Series, Society of Petroleum Engineers of AIME, Dallas (1967) **1,** Chap. 11.

3. "Guide for the Planning, Conducting, and Reporting of Subsurface Pressure Tests," ERCB Report 74-T, Energy Resources Conservation Board, Calgary, Alta., Canada (Nov. 1974).

4. Brownscombe, E. R. and Conlon, D. R.: "Precision in Bottom-Hole Pressure Measurements," *Trans.,* AIME (1946) **165,** 159-174.

5. Laird, A. and Birks, J.: "Performance and Accuracy of Amerada Bottom-Hole Pressure Recorder With Special Reference to Use in Drill Stem Formation Tests and Repeatability of Reservoir Pressures Obtained Therein," *J. Inst. Pet.* (1951) **37,** 678-695.

6. Smith, R. V. and Dewees, E. J.: "Sources of Error in Subsurface-Pressure-Gage Calibration and Usage," *Oil and Gas J.* (Dec. 9, 1948) 85-98.

7. Brownscombe, E. R.: "A Field Calibration Technique for Bottom-Hole Pressure Measurement," *Pet. Eng.* (Aug. 1947) 84-88.

8. Nestlerode, W. A.: "Permanently Installed Bottom-Hole Pressure Gauge," Paper 875-16-L presented at the API Div. of Production meeting, Denver, April 11-13, 1962.

9. Kolb, R. H.: "Two Bottom-Hole Pressure Instruments Providing Automatic Surface Recording," *Trans.*, AIME (1960) **219,** 346-349.

10. Miller, G. B., Seeds, R. W. S., and Shira, H. W.: "A New, Surface-Recording, Down-Hole Pressure Gauge," paper SPE 4125 presented at the SPE-AIME 47th Annual Fall Meeting, San Antonio, Tex., Oct. 8-11, 1972.

11. *Engineering Data Book,* 9th ed., Gas Processors Suppliers Assn., Tulsa (1972) Sec. 1.

12. Katz, Donald L., Cornell, Donald, Kobayashi, Riki, Poettmann, Fred H., Vary, John A., Elenbaas, John R., and Weinaug, Charles F.: *Handbook of Natural Gas Engineering,* McGraw-Hill Book Co., Inc., New York (1959) Chap. 8, 761-763.

13. Rawlins, E. L. and Schellhardt, M. A.: *Back-Pressure Data on Natural-Gas Wells and Their Application to Production Practices,* Monograph 7, USBM (1936).

APPENDICES

Appendix A

Units Systems and Conversions

A.1 Introduction

In any book of this nature, it is worthwhile to include a comprehensive list of units conversion factors, since data are often reported in units different from those used in the equations. Such factors are presented in this appendix. Because of the possibility of eventual conversion of engineering calculations to a metric standard, we also include information about the "SI" system of weights and measures.[1] Finally, we compare some important units and equations in five different unit systems.

A.2 The International (SI) Metric System

"SI" is the official abbreviation, *in all languages,* for the International System of Units (les Systéme International d'Unités). That system is neither the centimetre-gram-second (cgs) system nor the metre-kilogram-second (mks) system. Rather, it is a modernized version of mks. A complete description of SI is presented by Hopkins.[1] The American Petroleum Institute has proposed a set of metric standards for use in the petroleum industry.[2]

TABLE A.1—SI SYSTEM UNITS.

Base SI Units Used in Well Test Analysis

Quantity	Name	Symbol
length	metre	m
time	second	s
mass	kilogram	kg
temperature	kelvin	K
amount of substance	mole	mol

Units That Are Multiples or Submultiples of SI Base Units Given Special Names

Quantity	Name of Unit	Symbol	Definition	SI Term
mass	tonne	t	$1 t = 10^3 kg$	Mg
volume	litre*	l	$1 l = 1 dm^3$	dm^3

SI-Derived Units With Special Names Used in Well Test Analysis

Quantity	Name	Symbol	Expression in Terms of Other Units	Expression in Terms of SI Base Units
force	newton	N	—	$m \cdot kg \cdot s^{-2}$
pressure	pascal	Pa	N/m^2	$m^{-1} \cdot kg \cdot s^{-2}$
energy, work, quantity of heat	joule	J	$N \cdot m$	$m^2 \cdot kg \cdot s^{-2}$

*In 1964, the 12th Conférence Générale des Poids et Mesures (CGPM) redefined the litre to be $1 dm^3 = 0.001 m^3$. At the same time, it abrogated the 1901 definition of the litre given by the third CGPM.

Table A.1 lists the five *base* SI units encountered in well test analysis. The approved spelling is the French spelling. Names of units are never capitalized, although some of the abbreviations are. Most units are abbreviated with a single symbol. Table A.1 also lists two SI units that have been given special names and three derived units with special SI names. No units are normally used in well test analysis other than those presented in Table A.1.

SI allows prefixes to indicate multiples of the base units. The prefixes are summarized in Table A.2. Compound prefixes, such as micro-micro, are not allowed; the correct prefix, in this case pico, should always be used.

A.3 Constants and Conversion Factors

Table A.3 presents values for several physical constants useful in petroleum engineering, in several sets of units. Table A.4 summarizes useful conversion factors. SI units are indicated by boldface type. To use Table A.4, multiply the quantity given in the left-hand column by the number given in the "multiply by" column to obtain units in the second column. For simplicity, the "inverse" column may be used to reverse this procedure. Thus, to convert from square feet to acres, one would multiply the number of square feet by 2.296×10^{-5}.

Some permeability units are given under the heading "area" in Table A.4, since permeability has the units of area. Table A.5 is also supplied to simplify other conver-

TABLE A.2—SI PREFIXES.

Factor	Prefix	Symbol*
10^{12}	tera	T
10^9	giga	G
10^6	mega	M
10^3	kilo	k
10^2	hecto	h
10	deka	da
10^{-1}	deci	d
10^{-2}	centi	c
10^{-3}	milli	m
10^{-6}	micro	μ
10^{-9}	nano	n
10^{-12}	pico	p
10^{-15}	femto	f
10^{-18}	atto	a

*Only the symbols T (tera), G (giga), and M (mega) are capital letters. Compound prefixes are not allowed — for example, use nm (nano metre) rather than mμm (milli micro metre).

TABLE A. 3—PHYSICAL CONSTANTS AND VALUES.*

Quantity	Magnitude	Unit
Triple point of water	273.16 exactly	**K**
	0.01 exactly	°C
	491.688 exactly	°R
	32.018 exactly	°F
Absolute zero	0.00 exactly	**K**
	−273.15 exactly	°C
	0.00 exactly	°R
	−459.67 exactly	°F
Gas constant (R)	8.3143	**J·mol^{-1}·K^{-1}**
	8.3143 E + 07	erg·(gm mole)$^{-1}$·°K^{-1}
	10.732	psi·ft^{3}·(lb mole)$^{-1}$·°R^{-1}
Maximum density of water	999.973	**kg·m^{-3}**
	0.999 973	g·cm^{-3}
	62.426 1	lb$_{m}$·ft^{-3}
Density of water at 60 °F (15.56 °C, 288.71 K)	999.014	**kg·m^{-3}**
	0.999 014	g·cm^{-3}
	62.366 4	lb$_{m}$·ft^{-3}
Water gradient at 60 °F (15.56 °C, 288.71 K)	9,796.98	**Pa·m^{-1}**
	979.698	dyne·cm^{-3}
	0.433 100	psi·ft^{-1}
Standard atmosphere	1.013 25 E + 05	**Pa**
	1.013 25 E + 06	dyne·cm^{-2}
	14.695 9	psi
Density of air at 1 atm, 60 °F (15.56 °C, 288.71 K)	1.223 2	**kg·m^{-3}**
	1.223 2 E − 03	g·cm^{-3}
	0.076 362	lb$_{m}$·ft^{-3}
Earth's gravitational acceleration,	9.806 650	**m·s^{-2}**
g	980.665 0	cm·s^{-2}
	32.174 05	ft·s^{-2}
g_c	1.000 000	**kg·m·N^{-1}·sec^{-2}**
	1.000 000	g·cm·dyne^{-1}·sec^{-2}
	32.174 05	lb$_{m}$·ft·lb$_{f}^{-1}$·s^{-2}
π	3.141 593	
e	2.718 282	
1n (10)	2.302 585	
γ (Euler's constant)	0.577 215 66	
°API	$\dfrac{141.5}{\gamma(60\ °F)} - 131.5$	

*SI values are in boldface type. All quantities are consistent with conversion factors for the current SI system.

sions using permeability. That table is similar to data given by Amyx, Bass, and Whiting[3] but there is some variation in numbers. Differences occur because numerical values are shown only to the significance possible (in this case, limited by the SI agreed-on accuracy of atmospheric pressure and the density of water and mercury), and are based on conversion factors derived from the SI Standards. Some of those factors are slightly different from those used previously because of more precise definitions for certain quantities.

Table A.6 provides conversions for various temperature scales. The SI standard temperature unit is the kelvin; that unit is neither capitalized nor associated with the word "degree." One kelvin is equivalent to 1 degree Celsius (the SI system drops the use of the term Centigrade). The triple point of water is *defined* as 273.16 kelvin exactly. All other temperatures are derived from that. The normal Celsius and Fahrenheit scales are the same, as are their conversions to the other scales.

Table A.7 compares units and equations from five systems of units. The oilfield units are used exclusively throughout this monograph. The column for SI units is a coherent system (that is, one in which basic equations contain no units conversion factors). The preferred API standard SI unit system[2] is not a coherent system, but has the advantages of providing reasonable size values for most physical quantities. The cgs units column is the standard cgs system used for many years in petroleum engineering. Finally, the column for groundwater units is provided for those who practice in that field. The symbols used in groundwater hydrology vary from reference to reference, so the reader should check carefully when using groundwater literature. Table A.8 shows the correspondence between some groundwater quantities and oilfield quantities for any consistent unit system.

Many of the conversion factors provided in Tables A.3 through A.7 have been calculated from SI-stated factors for other unit conversions. Depending on the approach used in performing such a calculation, the seventh significant digit may vary by a few units. The reader should be aware of that when attempting to verify the values given, or to use them in precision computations.

References

1. Hopkins, Robert A.: *The International (SI) Metric System and How It Works,* Polymetric Services, Inc., Tarzana, Calif. (1974).
2. "Conversion of Operational and Process Measurement Units to the Metric (SI) System," *Manual of Petroleum Measurement Standards,* Pub. API 2564, American Petroleum Institute (March 1974) Chap. 15, Sec. 2.
3. Amyx, James W., Bass, Daniel, M., Jr., and Whiting, Robert L.: *Petroleum Reservoir Engineering: Physical Properties,* McGraw-Hill Book Co., Inc., New York (1960) 79.

TABLE A.4—CONVERSION FACTORS USEFUL IN WELL TEST ANALYSIS.

SI conversions are in boldface type. All quantities are current to SI standards as of 1974. An asterisk (*) after the sixth decimal indicates the conversion factor is exact and all following digits are zero. All other conversion factors have been rounded. The notation E+03 is used in place of 10^3, and so on.

To Convert From	To	Multiply by	Inverse
AREA			
acre	**metre² (m²)**	4.046 856 E+03	2.471 054 E−04
	foot²	4.356 000* E+04	2.295 684 E−05
darcy	**metre² (m²)**	9.869 23 E−13	1.013 25 E+12
	centimetre² (cm²)	9.869 23 E−09	1.013 25 E+08
	micrometre² (μm²)	9.869 23 E−01	1.013 25 E+00
	millidarcy	1.000 000* E+03	1.000 000*E−03
	cm²·cp·sec⁻¹·atm⁻¹	1.000 000* E+00	1.000 000*E+00
foot²	**metre² (m²)**	9.290 304* E−02	1.076 391 E+01
	centimetre²	9.290 304* E+02	1.076 391 E−03
	inch²	1.440 000* E+02	6.944 444 E−03
hectare	**metre² (m²)**	1.000 000* E+04	1.000 000*E−04
	acre	2.471 054 E+00	4.046 856 E−01
mile²	**metre² (m²)**	2.589 988 E+06	3.861 022 E−07
	acre	6.400 000* E+02	1.562 500*E−03
DENSITY			
gram/centimetre³	**kilogram/metre³ (kg·m⁻³)**	1.000 000* E+03	1.000 000*E−03
	pound-mass/foot³	6.242 797 E+01	1.601 846 E−02
	pound-mass/gallon	8.345 405 E+00	1.198 264 E−01
	pound-mass/barrel	3.505 070 E+02	2.853 010 E−03
pound-mass/foot³	**kilogram/metre³ (kg·m⁻³)**	1.601 846 E+01	6.242 797 E−02
	pound-mass/gallon	1.336 805 E−01	7.480 520 E+00
	pound-mass/barrel	5.614 583 E+00	1.781 076 E−01
pound-mass/gallon	**kilogram/metre³ (kg·m⁻³)**	1.198 264 E+02	8.345 406 E−03
	pound-mass/barrel	4.200 000 E+01	2.380 952 E−02
FORCE			
dyne	**newton (N)**	1.000 000* E−05	1.000 000*E+05
	pound-force	2.248 089 E−06	4.448 222 E+05
kilogram-force	**newton (N)**	9.806 650* E+00	1.019 716 E−01
	pound-force	2.204 622 E+00	4.535 924 E−01
pound-force	**newton (N)**	4.448 222 E+00	2.248 089 E−01
LENGTH			
angstrom	**metre (m)**	1.000 000* E−10	1.000 000*E+10
centimetre	**metre (m)**	1.000 000* E−02	1.000 000*E+02
foot	**metre (m)**	3.048 000* E−01	3.280 840 E+00
	centimetre	3.048 000* E+01	3.280 840 E−02
inch	**metre (m)**	2.540 000* E−02	3.937 008 E+01
	centimetre	2.540 000* E+00	3.937 008 E−01
micron	**metre (m)**	1.000 000* E−06	1.000 000*E+06
mile (U.S. statute)	**metre (m)**	1.609 344* E+03	6.213 712 E−04
	foot	5.280 000* E+03	1.893 939 E−04
MASS			
gram-mass	**kilogram (kg)**	1.000 000* E−03	1.000 000*E+03
ounce-mass (av)	**kilogram (kg)**	2.834 952 E−02	3.527 397 E+01
	gram	2.834 952 E+01	3.527 397 E−02
pound-mass	**kilogram (kg)**	4.535 923 7*E−01	2.204 623 E+00
	ounce-mass	1.600 000* E+01	6.250 000*E−02
slug	**kilogram (kg)**	1.459 390 E+01	6.852 178 E−02
	pound-mass	3.217 405 E+01	3.108 095 E−02
ton (U.S. short)	**kilogram (kg)**	9.071 847 E+02	1.102 311 E−03
	pound-mass	2.000 000* E+03	5.000 000*E−04
ton (U.S. long)	**kilogram (kg)**	1.016 047 E+03	9.842 064 E−04
	pound-mass	2.240 000* E+03	4.464 286 E−04
ton (metric)	**kilogram (kg)**	1.000 000* E+03	1.000 000*E−03
tonne	**kilogram (kg)**	1.000 000* E+03	1.000 000*E−03

TABLE A.4—CONT'D.

To Convert From	To	Multiply by	Inverse
PRESSURE			
atmosphere (normal—760 mm Hg)	**pascal (Pa)**	1.013 25 E+05	9.869 23 E−06
	mm Hg (0 °C)	7.600 000*E+02	1.315 789 E−03
	feet water (4 °C)	3.389 95 E+01	2.949 90 E−02
	psi	1.469 60 E+01	6.804 60 E−02
	bar	1.013 25 E+00	9.869 23 E−01
bar	**pascal (Pa)**	1.000 000*E+05	1.000 000*E−05
	psi	1.450 377 E+01	6.894 757 E−02
centimetre of Hg (0 °C)	**pascal (Pa)**	1.333 22 E+03	7.500 64 E−04
	psi	1.933 67 E−01	5.171 51 E+00
dyne/centimetre²	**pascal (Pa)**	1.000 000*E−01	1.000 000*E+01
	psi	1.450 377 E−05	6.894 757 E+04
feet of water (4 °C)	**pascal (Pa)**	2.988 98 E+03	3.345 62 E−04
	psi	4.335 15 E−01	2.306 73 E+00
kilogram-force/centimetre²	**pascal (Pa)**	9.806 650*E+04	1.019 716 E−05
	bar	9.806 650*E−01	1.019 716 E+00
	psi	1.422 334 E+01	7.030 695 E−02
psi	**pascal (Pa)**	6.894 757 E+03	1.450 377 E−04
TIME			
day	**second (s)**	8.640 000*E+04	1.157 407 E−05
	minute	1.440 000*E+03	6.944 444 E−04
	hour	2.400 000*E+01	4.166 667 E−02
hour	**second (s)**	3.600 000*E+03	2.777 778 E−04
	minute	6.000 000*E+01	1.666 667 E−02
minute	**second (s)**	6.000 000*E+01	1.666 667 E−02
VISCOSITY			
centipoise	**pascal-second (Pa·s)**	1.000 000*E−03	1.000 000*E+03
	dyne-second/centimetre²	1.000 000*E−02	1.000 000*E+02
	pound-mass/(foot-second)	6.719 689 E−04	1.488 164 E+03
	pound-force-second/foot²	2.088 543 E−05	4.788 026 E+04
	pound-mass/(foot-hour)	2.419 088 E+00	4.133 789 E−01
centistoke	**metre²/second (m²/s)**	1.000 000*E−06	1.000 000*E+06
	centipoise/(gram/centimetre³)	1.000 000*E+00	1.000 000*E+00
poise	**pascal-second (Pa·s)**	1.000 000*E−01	1.000 000*E+01
pound-mass/(foot-second)	**pascal-second (Pa·s)**	1.488 164 E+00	6.719 689 E−01
pound-mass/(foot-hour)	**pascal-second (Pa·s)**	4.133 789 E−04	2.419 088 E+03
pound-force-second/foot²	**pascal-second (Pa·s)**	4.788 026 E+01	2.088 543 E−02
VOLUME			
acre-foot	**metre³ (m³)**	1.233 482 E+03	8.107 131 E−04
	foot³	4.356 000*E+04	2.295 684 E−05
	barrel	7.758 368 E+03	1.288 931 E−04
barrel	**metre³ (m³)**	1.589 873 E−01	6.289 811 E+00
	foot³	5.614 583 E+00	1.781 076 E−01
	gallon	4.200 000*E+01	2.380 952 E−02
foot³	**metre³ (m³)**	2.831 685 E−02	3.531 466 E+01
	inch³	1.728 000 E+03	5.787 037 E−04
	gallon	7.480 520 E+00	1.336 805 E−01
gallon	**metre³ (m³)**	3.785 412 E−03	2.641 720 E+02
	inch³	2.310 001 E+02	4.329 003 E−03
litre	**metre³ (m³)**	1.000 000*E−03	1.000 000*E+03
VOLUMETRIC RATE			
barrel/day	**metre³/sec (m³/s)**	1.840 131 E−06	5.434 396 E+05
	metre³/hour (m³/h)	6.624 472 E−03	1.509 554 E+02
	metre³/day (m³/d)	1.589 873 E−01	6.289 810 E+00
	centimetre³/second	1.840 131 E+00	5.434 396 E−01
	foot³/minute	3.899 016 E−03	2.564 750 E+02
	gallon/minute	2.916 667 E−02	3.428 571 E+01
foot³/minute	**metre³/sec (m³/s)**	4.719 474 E−04	2.118 880 E+03
foot³/second	**metre³/sec (m³/s)**	2.831 685 E−02	3.531 466 E+01
gallon/minute	**metre³/sec (m³/s)**	6.309 020 E−05	1.585 032 E+04

TABLE A.5—AUXILIARY PERMEABILITY CONVERSIONS.

To Convert From	To	Multiply by		Inverse	
md	darcy	1.000 000*	E−03	1.000 000*	E+03
	metre² (m²)	9.869 23	E−16	1.013 25	E+15
	centimetre² (cm²)	9.869 23	E−12	1.013 25	E+11
	micrometre² (μm²)	9.869 23	E−04	1.013 25	E+03
	$\frac{(cm^3/s)\ cp}{cm^2\ (atm/cm)}$	1.000 000*	E−03	1.000 000*	E+03
	$\frac{(cm^3/s)\ cp}{cm^2[(dyne/cm^2)/cm]}$	9.869 23	E−10	1.013 25	E+09
	$\frac{(ft^3/s)\ cp}{ft^2\ (psi/ft)}$	7.324 41	E−08	1.365 30	E+07
	$\frac{(ft^3/s)\ cp}{cm^2[(cm\ water)/cm]}$	3.417 80	E−11	2.925 85	E+10
	$\frac{(B/D)\ cp}{ft^2\ (psi/ft)}$	1.127 12	E−03	8.872 17	E+02
	$\frac{(gal/min)\ cp}{ft^2\ [(ft\ water)/ft]}$	1.425 15	E−05	7.016 81	E+04
	ft²	1.062 32	E−14	9.413 40	E+13

*Conversion factor is exact; all following digits are zero.

TABLE A.6—TEMPERATURE SCALE CONVERSIONS.*

To Convert	To	Solve
degree Fahrenheit	kelvin	$T_K = (T_F + 459.67)/1.8$
degree Rankine	kelvin	$T_K = T_R/1.8$
degree Fahrenheit	degree Rankine	$T_R = T_F + 459.67$
degree Fahrenheit	degree Celsius	$T_C = (T_F - 32)/1.8$
degree Celsius	kelvin	$T_K = T_C + 273.15$

*The SI standard, the kelvin (K), is *defined* so the triple point of water is 273.16 K *exactly*. The SI temperature symbol is written K, without a degree symbol. The cgs (and common) temperature unit is the degree Celsius, °C; the common oilfield unit is the degree Fahrenheit, °F.

TABLE A.8—RELATIONSHIP OF COMMON GROUNDWATER AND OILFIELD QUANTITIES.

A consistent-unit system is assumed. Variable definitions for each system are given in Table A.7.

Groundwater Quantity		Oilfield Quantity
Coefficient of permeability	$= P = K$	$= \frac{k}{\mu}\left(\frac{\rho g}{g_c}\right)$
Transmissivity	$= T = Km$	$= \frac{kh}{\mu}\left(\frac{\rho g}{g_c}\right)$
Coefficient of storage	$= S$	$= \phi c_t h \left(\frac{\rho g}{g_c}\right)$
Drawdown	$= s$	$= \frac{p_i - p}{(\rho g/g_c)}$
Head	$= h$	$= \frac{p}{(\rho g/g_c)}$
Dimensionless drawdown	$= W(1/4\alpha)$	$= 2p_D(t_D)$

TABLE A.7—COMPARISON OF UNITS AND EQUATIONS IN VARIOUS UNIT SYSTEMS.*

Oilfield Units	SI Units	Preferred API Standard SI Units	cgs Units*	Groundwater Units
q — production rate, STB/D	m^3/s	dm^3/s	cm^3/s	Q — production rate, gal/min
h — formation thickness, ft	m	m	cm	m — formation thickness, ft
k — permeability, md	m^2	μm^2	darcy	—
μ — viscosity, cp	Pa·s	Pa·s	cp	—
k/μ — mobility, md/cp	$m^2/(Pa\cdot s)$	$\mu m^2/(Pa\cdot s)$	darcy/cp	P or K — coefficient of permeability, gal/day ft²**
kh/μ — mobility-thickness product, md ft/cp	$m^3/(Pa\cdot s)$	$m(\mu m^2)/(Pa\cdot s)$	darcy·cm/cp	T — coefficient of transmissivity, gal/(day ft)**
Δp — pressure difference, psi	Pa	kPa	atm	s — drawdown, ft of water, >0 for pressure drawdown**
p — pressure, psi	Pa	kPa	atm	h — head of water, ft of water
r — radius, ft	m	m	cm	r — radius, ft
t — time, hours	s	h	s	t — time, days
ϕ — porosity, fraction				—
c_t — total system compressibility, psi⁻¹	Pa^{-1}	kPa^{-1}	atm^{-1}	—
$\phi c_t h$ — porosity-compressibility-thickness product, ft psi⁻¹	$m\cdot Pa^{-1}$	$m\cdot kPa^{-1}$	$cm\cdot atm^{-1}$	S — coefficient of storage, fraction**

DIMENSIONLESS TIME

Oilfield Units	SI Units	Preferred API Standard SI Units	cgs Units*	Groundwater Units
$t_D = \dfrac{0.000263679\,kt}{\phi\mu c_t r_w^2}$	$t_D = \dfrac{kt}{\phi\mu c_t r_w^2}$	$t_D = 3.6\times10^{-6}\dfrac{kt}{\phi\mu c_t r_w^2}$	$t_D = \dfrac{kt}{\phi\mu c_t r_w^2}$	$\alpha = 0.1336805\dfrac{Tt}{Sr_w^2}$

DARCY'S LAW FOR INCOMPRESSIBLE, RADIAL FLOW

Oilfield Units	SI Units	Preferred API Standard SI Units	cgs Units*	Groundwater Units
$q = \dfrac{0.00708188\,kh(p_e-p_w)}{B\mu\ln(r_e/r_w)}$	$q = 2\pi\dfrac{kh(p_e-p_w)}{B\mu\ln(r_e/r_w)}$	$q = \dfrac{2\pi\times10^{-6}\,kh(p_e-p_w)}{B\mu\ln(r_e/r_w)}$	$q = 2\pi\dfrac{kh(p_e-p_w)}{B\mu\ln(r_e/r_w)}$	$Q = \dfrac{0.00436332\,T(h_e-h_w)}{\ln(r_e/r_w)}$

DIFFUSIVITY EQUATION

Oilfield Units	SI Units	Preferred API Standard SI Units	cgs Units*	Groundwater Units
$\dfrac{\partial^2 p}{\partial r^2}+\dfrac{1}{r}\dfrac{\partial p}{\partial r} = \dfrac{1}{0.000263679}\dfrac{\phi\mu c_t}{k}\dfrac{\partial p}{\partial t}$	$\dfrac{\partial^2 p}{\partial r^2}+\dfrac{1}{r}\dfrac{\partial p}{\partial r} = \dfrac{\phi\mu c_t}{k}\dfrac{\partial p}{\partial t}$	$\dfrac{\partial^2 p}{\partial r^2}+\dfrac{1}{r}\dfrac{\partial p}{\partial r} = \dfrac{1}{3.6\times10^{-6}}\dfrac{\phi\mu c_t}{k}\dfrac{\partial p}{\partial t}$	$\dfrac{\partial^2 p}{\partial r^2}+\dfrac{1}{r}\dfrac{\partial p}{\partial r} = \dfrac{\phi\mu c_t}{k}\dfrac{\partial p}{\partial t}$	$\dfrac{\partial^2 h}{\partial r^2}+\dfrac{1}{r}\dfrac{\partial h}{\partial r} = \dfrac{1}{0.1336805}\dfrac{S}{T}\dfrac{\partial h}{\partial t}$

GENERALIZED TRANSIENT FLOW EQUATION

Oilfield Units	SI Units	Preferred API Standard SI Units	cgs Units*	Groundwater Units
$\Delta p = \dfrac{141.205\,qB\mu\,p_D(t_D)}{kh}$	$\Delta p = \dfrac{qB\mu\,p_D(t_D)}{2\pi kh}$	$\Delta p = 10^6\dfrac{qB\mu\,p_D(t_D)}{2\pi kh}$	$\Delta p = \dfrac{1}{2\pi}\dfrac{qB\mu}{kh}p_D(t_D)$	$s = 229.183\dfrac{Q}{T}p_D(\alpha)$

SLOPE OF SEMILOG STRAIGHT LINE

Oilfield Units	SI Units	Preferred API Standard SI Units	cgs Units*	Groundwater Units
$m = 162.568\dfrac{qB\mu}{kh}$	$m = 0.183234\dfrac{qB\mu}{kh}$	$m = 1.83234\times10^5\dfrac{qB\mu}{kh}$	$m = 0.183234\dfrac{qB\mu}{kh}$	$M = 263.857\dfrac{Q}{T}$

GENERALIZED SKIN-FACTOR EQUATION

Oilfield Units	SI Units	Preferred API Standard SI Units	cgs Units*	Groundwater Units
$s = 1.15129\left[\dfrac{p_{1hr}-p(\Delta t=0)}{m} - \log\left(\dfrac{k}{\phi\mu c_t r_w^2}\right) + 3.227546\right]$	$s = 1.15129\left[\dfrac{p_{1hr}-p(\Delta t=0)}{m} - \log\left(\dfrac{k}{\phi\mu c_t r_w^2}\right) - 0.351378\right]$	$s = 1.15129\left[\dfrac{p_{1hr}-p(\Delta t=0)}{m} - \log\left(\dfrac{k}{\phi\mu c_t r_w^2}\right) + 5.092319\right]$	$s = 1.15129\left[\dfrac{p_{1hr}-p(\Delta t=0)}{m} - \log\left(\dfrac{k}{\phi\mu c_t r_w^2}\right) - 0.351378\right]$	$\text{skin} = 1.15129\left[\dfrac{s_{1hr}-s(\Delta t=0)}{M} - \log\left(\dfrac{T}{Sr_w^2}\right) + 0.522555\right]$

*The cgs system is considered to be obsolete and is replaced by SI; cgs units are included only for comparison with published material. SI is a coherent system, so equations do not contain units conversion factors.

**See Table A.8.

Appendix B

Application of Superposition To Generate Dimensionless Pressures

B.1 Introduction

As indicated in Section 2.9 and by several authors,[1-9] the principle of superposition may be used to develop dimensionless pressure data for many finite and bounded systems. This appendix shows how to use superposition to form no-flow and constant-pressure boundaries, and closed systems. A method of "desuperposition" for changing existing dimensionless-pressures solutions to solutions for different systems is explained. Finally, a general equation for calculating pressures owing to variable production rates is derived.

B.2 Dimensionless Pressure Used

When using the principle of superposition, we must choose a dimensionless pressure, p_D, that applies for the system. Normally, superposition calculations are performed using dimensionless pressures for infinite-acting systems (even when the goal is to generate a closed system[4,6]), so the exponential-integral p_D, Eq. 2.5, is used. The exponential-integral p_D may be used when $r_D \geqslant 20$ and $t_D/r_D^2 \geqslant 0.5$ or when $t_D/r_D^2 \geqslant 25$. If neither of those restrictions is met, then Fig. C.1 must be used for the applicable r_D.

If superposition calculations are performed for a system that is not infinite-acting, the appropriate p_D from Appendix C must be used. In such a situation, there is no conceptual modification to the application of the superposition principle. However, p_D tables and figures for such systems often do not provide data for points other than at the well. Some useful p_D data are given in Appendix C and in Ref. 10.

B.3 Generating No-Flow and Constant-Pressure Boundaries

Fig. B.1 illustrates the method of images[3-6] when used to create a no-flow boundary in an infinite system. Well 1 operates at constant flow rate q, at distance L from a single impermeable boundary, represented by the y axis in Fig. B.1. The image well, Well 2 in Fig. B.1, at a distance $-L$ from the y axis mathematically generates the boundary. By applying superposition we can calculate the pressure at any point in the x-y plane of Fig. B.1:

$$p(t, x, y) = p_i - \frac{141.2\, qB\mu}{kh}\left[p_D(t_D, a_{D1}) + p_D(t_D, a_{D2})\right]. \quad \text{(B.1)}$$

In Eq. B.1, the dimensionless distances a_{D1} and a_{D2} are calculated from

$$a_{D1} = \frac{a_1}{r_w} = \frac{1}{r_w}\sqrt{(x-L)^2 + y^2}, \quad \text{(B.2a)}$$

$$a_{D2} = \frac{a_2}{r_w} = \frac{1}{r_w}\sqrt{(x+L)^2 + y^2}, \quad \text{(B.2b)}$$

where r_w is the same for both wells. For an infinite system with $a_D > 20$, Eq. 2.5a applies.

$$p_D(t_D, a_D) = -\frac{1}{2}\,\mathrm{Ei}\left(-\frac{a_D^2}{4t_D}\right), \quad \text{(B.3a)}$$

$$= -\frac{1}{2}\,\mathrm{Ei}\left(-\frac{(x \pm L)^2 + y^2}{4r_w^2\, t_D}\right), \quad \text{(B.3b)}$$

$$= \frac{1}{2}\int_{\left[\frac{(x \pm L)^2 + y^2}{4r_w^2\, t_D}\right]}^{\infty} \frac{e^{-u}}{u}\, du. \quad \text{(B.3c)}$$

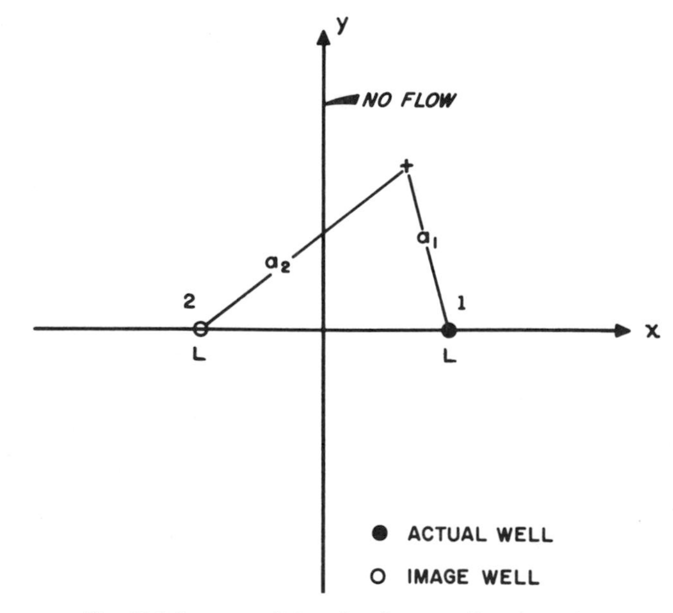

Fig. B.1 Image-well location for a no-flow boundary (sealing fault).

We wish to verify that no fluid flows across the impermeable barrier, the y axis. That is true if $(\partial p/\partial x)_{x=0} = 0$ at all points along the y axis. Differentiating Eq. B.1,

$$\frac{\partial p(t, x, y)}{\partial x} = \frac{-141.2\, qB\mu}{kh}\left[\frac{\partial p_D\,(t_D, a_{D1})}{\partial x} + \frac{\partial p_D\,(t_D, a_{D2})}{\partial x}\right]. \quad \ldots \text{(B.4)}$$

We wish to evaluate Eq. B.4 at $x = 0$ for arbitrary values of y. To do this we differentiate Eq. B.3c using Leibnitz's rule:[11]

$$\frac{\partial p_D\,(t_D, a_D)}{\partial x} = -\frac{(x \pm L)}{(x \pm L)^2 + y^2} \exp\left(-\frac{(x \pm L)^2 + y^2}{4r_w^2\, t_D}\right). \quad \ldots \text{(B.5)}$$

Eq. B.5 is substituted into Eq. B.4:

$$\frac{\partial p(t, x, y)}{\partial x} = \frac{141.2\, qB\mu}{kh}\left\{\left[\frac{x-L}{(x-L)^2 + y^2}\right] \exp\left(-\frac{(x-L)^2 + y^2}{4r_w^2\, t_D}\right) + \left[\frac{(x+L)}{(x+L)^2 + y^2}\right] \exp\left(-\frac{(x+L)^2 + y^2}{4r_w^2\, t_D}\right)\right\},$$

$$\ldots \text{(B.6)}$$

which, when evaluated at $x = 0$, becomes

$$\frac{\partial p(t, 0, y)}{\partial x} = \frac{141.2\, qB\mu}{kh}\left(\frac{1}{L^2 + y^2}\right)\left\{- L + L\right\} \exp\left(-\frac{L^2 + y^2}{4r_w^2\, t_D}\right) = 0 . \quad \ldots \text{(B.7)}$$

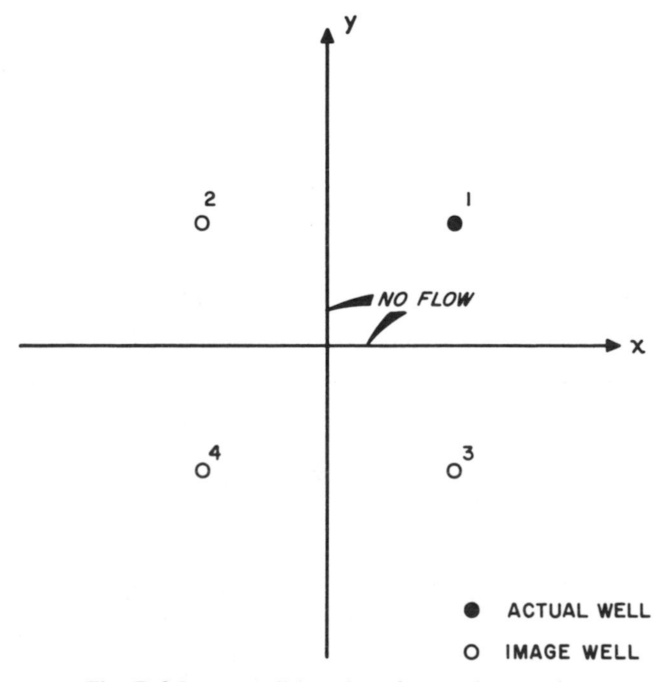

Fig. B.2 Image-well locations for two intersecting no-flow boundaries.

Since $(\partial p/\partial x)_{x=0} = 0$, we can see that an image of the operating well in the boundary creates a no-flow boundary. This is always true for straight-line boundaries no matter how many boundaries or how many wells there are. Some specific examples are given in Section B.4.

If the y axis in Fig. B.1 is to be a constant-pressure boundary, then the image well is an injection well with the same rate as the production well. In this case, superposition gives

$$p(t, x, y) = p_i - \frac{141.2\, qB\mu}{kh}\left[p_D(t_D, a_{D1}) - p_D(t_D, a_{D2})\right]. \quad \ldots \text{(B.8)}$$

Since

$$a_{D1} = a_{D2}, \quad \ldots \text{(B.9)}$$

at $x = 0$ for all y, then

$$p_D(t_D, a_{D1}) = p_D(t_D, a_{D2}), \quad \ldots \text{(B.10)}$$

and the pressure at all points along the boundary (y axis) is

$$p(t, 0, y) = p_i. \quad \ldots \text{(B.11)}$$

By using the method of images to form either no-flow or constant-pressure boundaries, one can generate dimensionless pressure solutions for many important situations.

B.4 Use of Method of Images To Generate Multiple Boundary and Closed Systems

Clearly, the method of images extends to systems with more than one no-flow barrier. Fig. B.2 shows a system with two no-flow barriers intersecting at right angles. Well 1 is located near the intersection of the two barriers. Well 2, the image of Well 1 in the y axis, prevents flow across that axis resulting from Well 1. Well 3, the image of Well 1 in the x axis, prevents flow across that axis owing to Well 1. Well 4 prevents flow across the x axis resulting from Well 2 and flow across the y axis resulting from Well 3. The method of images considers barriers to be of infinite length, so flow across any barrier caused by an *image* well must be prevented by other image wells, as demonstrated in Fig. B.2.

When barriers are formed by using images, pressure may be calculated at any point by superposition. Rather than write the equation for pressure change, we may write an equation for the dimensionless pressure at any point in the two-barrier system:

$$p_D(t_D, x_D, y_D) = p_D(t_D, a_{D1}) + p_D(t_D, a_{D2}) + p_D(t_D, a_{D3}) + p_D(t_D, a_{D4}),$$

$$\ldots \text{(B.12)}$$

where a_{D1} means the dimensionless distances from the point where the pressure is being calculated to Well 1, and so on, and x_D and y_D are dimensionless Cartesian coordinates. Their precise definition varies with the application; Refs. 6 and 10 use two practical definitions of those quantities. It is easy to verify that the four wells in Fig. B.2 do create the two intersecting no-flow barriers indicated.

In general, if there are more wells or more boundaries, the dimensionless pressure may be written

$$p_D(t_D, x_D, y_D) = \sum_{i=1}^{n} p_D(t_D, a_{Di}), \quad \ldots\ldots\ldots (B.13)$$

where the number of wells, n, includes all actual wells and all image wells.

Fig. B.3 shows a single well located between two parallel no-flow barriers and the image wells required to produce those barriers. Well a is the image of the actual well in Boundary A. Well b is the image of the actual well in Boundary B. Each image well would cause flow across the other boundary so additional image wells are required. Well (a)b is the image of Well a in Boundary B and is required to prevent Well a from causing fluid to flow across Boundary B. Well (b)a is required to prevent flow across Boundary A resulting from Well b. Some of the other image wells in Fig. B.3 are marked with a similar nomenclature to indicate the reason for each well. The line of image wells goes to infinity in both directions. The dimensionless pressure function for this system can be written

$$p_D(t_D, x_D, y_D) = \sum_{i=1}^{\infty} p_D(t_D, a_{Di}), \quad \ldots\ldots\ldots (B.14)$$

where a_{Di} is the dimensionless distance from Well i to x_D, y_D.

To obtain a closed system with a single well in it, add two horizontal no-flow boundaries to Fig. B.3 and image all the wells shown in Fig. B.3 in those two boundaries. Fig. B.4 shows the results — a single well in a closed rectangular drainage area. The image wells extend to infinity in all directions; the dimensionless pressure for the closed rectangular system is given by Eq. B.14. Although an infinite number of image wells is indicated, for calculation purposes it is usually only necessary to include several rows and columns of image wells before their contribution becomes negligible.

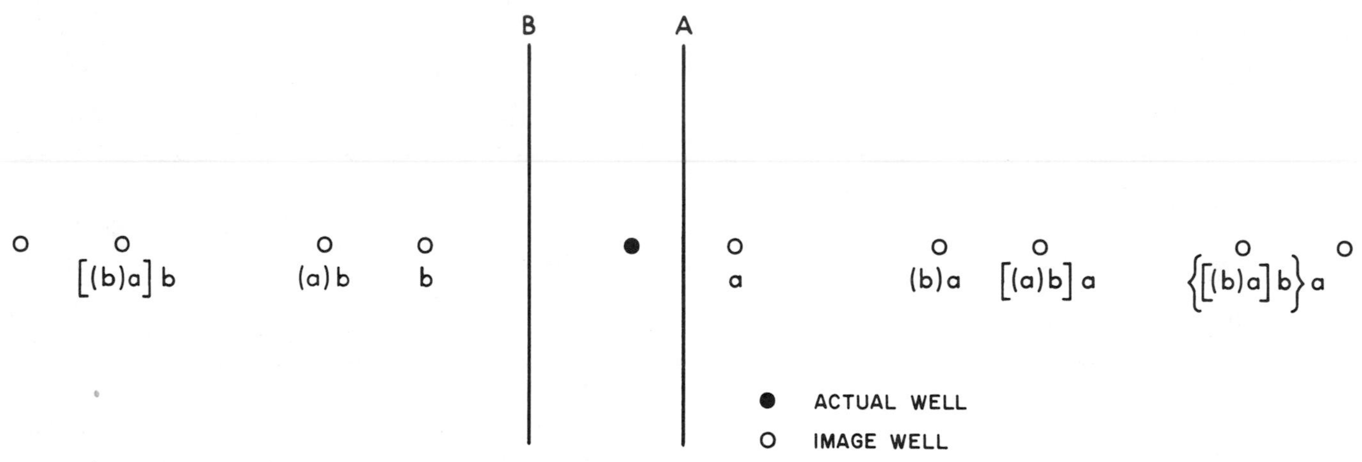

Fig. B.3 Image-well locations for two parallel no-flow boundaries.

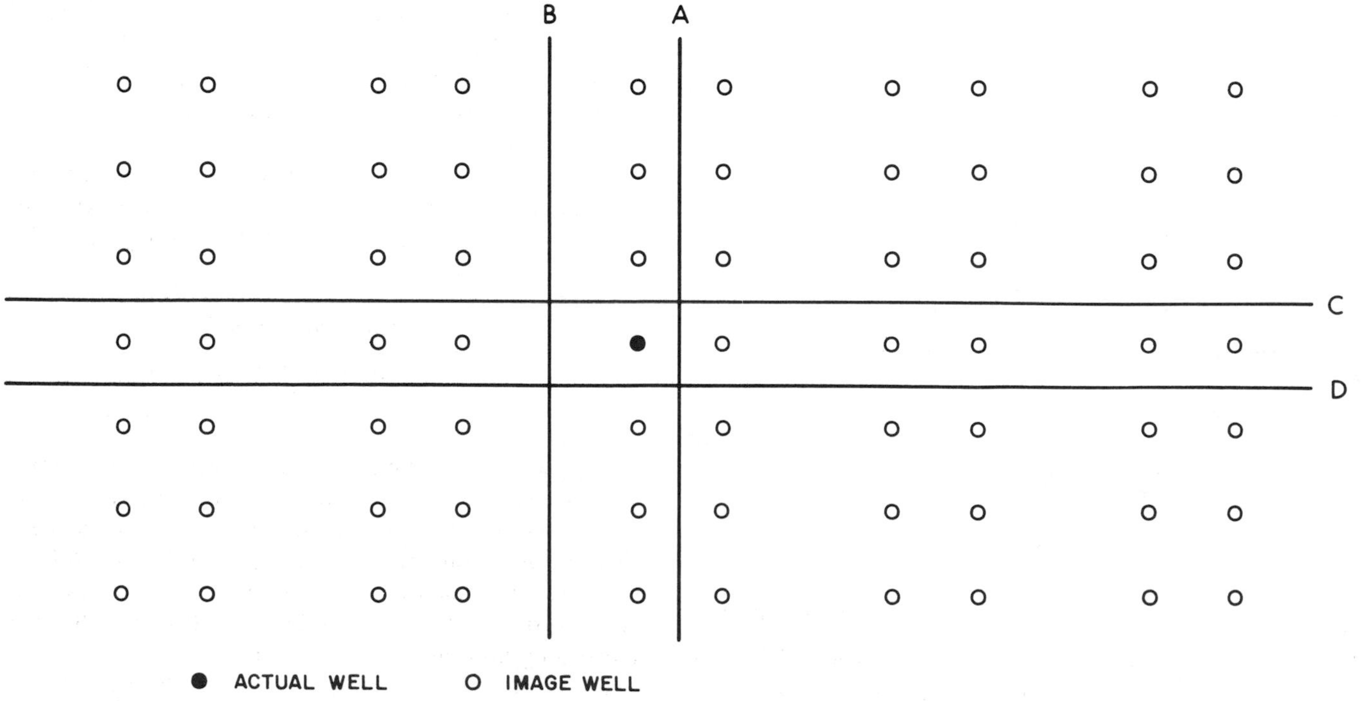

ACTUAL WELL IMAGE WELL

Fig. B.4 Image-well location for one well in a closed rectangle.

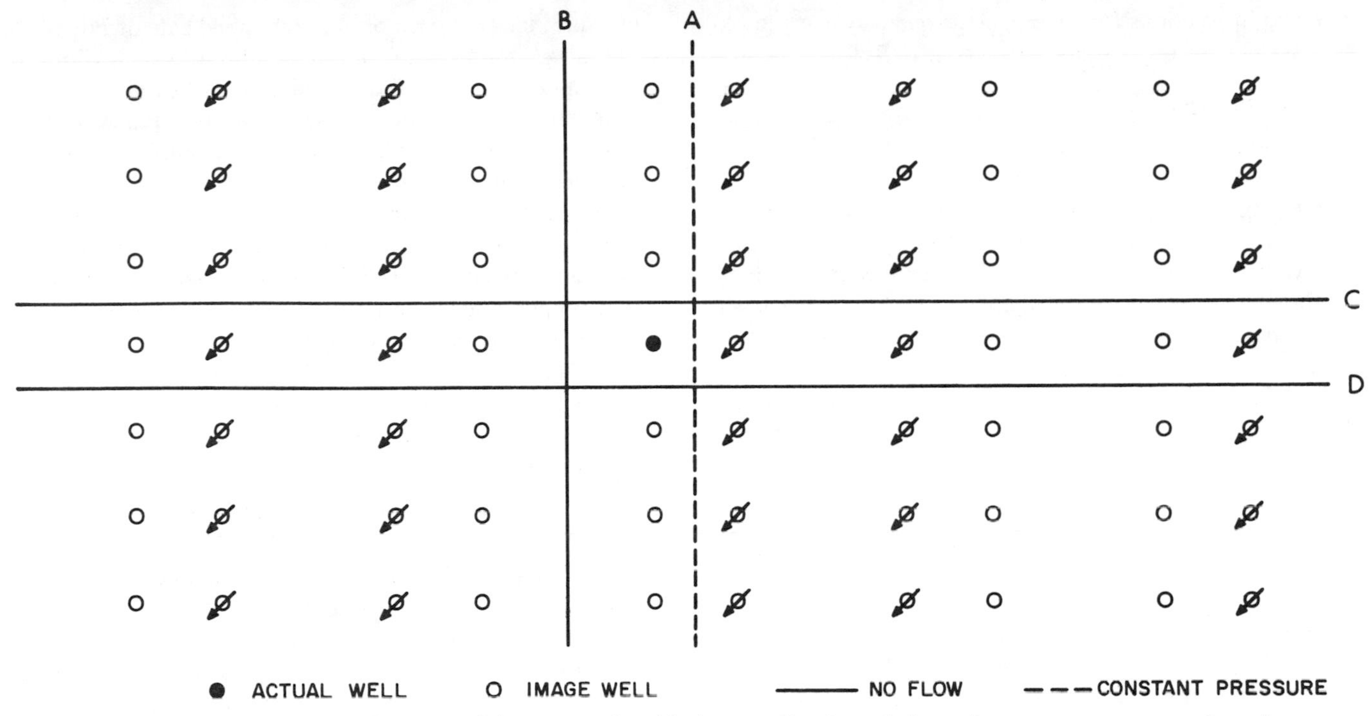

Fig. B.5 Image-well location for one well in a rectangle with three no-flow boundaries and one constant-pressure boundary.

To make one or more of the boundaries constant-pressure rather than no-flow, change some image production wells to injection wells.[6,9] Fig. B.5 shows the image-well array for the system of Fig. B.4 but with Boundary A at constant-pressure. Whether there are one, two, three, or four constant-pressure boundaries in a rectangular system, exactly one-half the image wells will be injectors and one-half will be producers. The location of each type of well depends on which boundaries are constant-pressure and which are no-flow. Ramey, Kumar, and Gulati[9] give several examples of systems with one or more constant-pressure boundaries and present a computer program for calculating p_D in such systems.

B.5 Superposition of Square Drainage Systems

Earlougher *et al.*[6] showed that a useful unit of superposition is a closed-square system with a single well located at its center (Fig. B.6). The dimensionless pressures at the well and at other locations within the square may be added together to obtain dimensionless pressures for systems with different well locations and different shapes. Fig. B.7 illustrates how two square systems are added to obtain a 2:1 rectangular system with the well at the center.[6] The open circles represent the well array for one square system; the closed circles show the well array for the second square system. The resulting well array creates a 2:1 rectangular drainage area with a well in the center as illustrated in Fig. B.7. As pointed out in Ref. 6, it is important to recognize that such superposition changes the area of the drainage system. That affects both the dimensionless time, t_{DA}, and the dimensionless pressure value *at the well* since p_D at a well in a closed drainage area depends on the value of $\sqrt{A}/r_w$. If p_D is desired at a well point for a system of different $\sqrt{A}/r_w$, it is necessary to make the correction

$$p_D\,(\sqrt{A}/r_w) = p_D\left(\left[\sqrt{A}/r_w\right]_{\text{table}}\right) + \ln\left[\frac{(\sqrt{A}/r_w)}{(\sqrt{A}/r_w)_{\text{table}}}\right], \quad \text{(B.15)}$$

where the p_D on the left hand side of the equation is for the desired value of $\sqrt{A}/r_w$ and on the right-hand side is for the value given in a table or figure (such as Table C.2 or Fig. C.12) with a different value of $\sqrt{A}/r_w$. Additional informa-

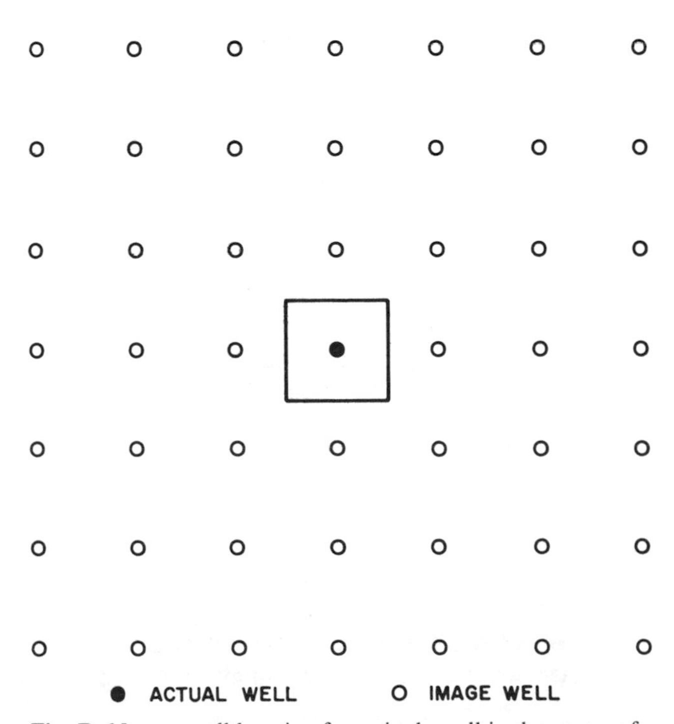

Fig. B.6 Image-well location for a single well in the center of a closed square.

tion and instructions for superposing square systems to obtain systems of other shapes are given in Ref. 6.

B.6 Desuperposition

Gringarten, Ramey, and Raghavan[7] and Chen and Brigham[8] have illustrated the concept of desuperposition for modifying known p_D values to p_D's for somewhat different systems. The approach may be used for any drainage shape and well location, although we illustrate it here only for a well in the center of a closed square. Suppose that we desire to compute p_D at a well in the center of a closed square for $C_D > 0$ and $s \neq 0$. Most p_D data for a closed-square system are for $C_D = 0$, $s = 0$, so those data are not what we need. However, we may use them to get the desired results by using

$$p_D(C_D, s, \boxdot) = p_D(C_D = 0, s = 0, \boxdot) - p_D(C_D = 0, s = 0, \infty) + p_D(C_D, s, \infty), \qquad \text{(B.16)}$$

as illustrated in Fig. B.8. We start with p_D for the closed-square system with zero skin and zero wellbore storage as indicated in Part a of Fig. B.8. From this dimensionless pressure, the first term on the right-hand side of Eq. B.16, subtract p_D for a single well in an infinite system with zero skin and zero storage, the second term on the right-hand side of Eq. B.16. The result, shown in Part b of Fig. B.8, is an infinite array of wells with the well in the center of the square removed. Finally, add p_D for a single well in an infinite system with the desired wellbore storage coefficient and skin factor, the last term on the right-hand side of Eq. B.16, to obtain Part c of Fig. B.8. The theoretically correct dimensionless pressure is given by the right-hand side of Part d in Fig. B.8, where all the image wells have the desired skin factor and wellbore storage coefficient. However, since skin factors and wellbore storage coefficients have only a small influence[7] at points away from the well, the approximation is a good one.[8]

Gringarten, Ramey, and Raghavan[7] use this approach to estimate dimensionless pressure for closed fractured systems. Chen and Brigham[8] use the approach to generate dimensionless pressures and then pressure buildup curves for a single well with wellbore storage and skin in the center of a closed square. Dimensionless pressure for many other systems can be computed with the same approach.

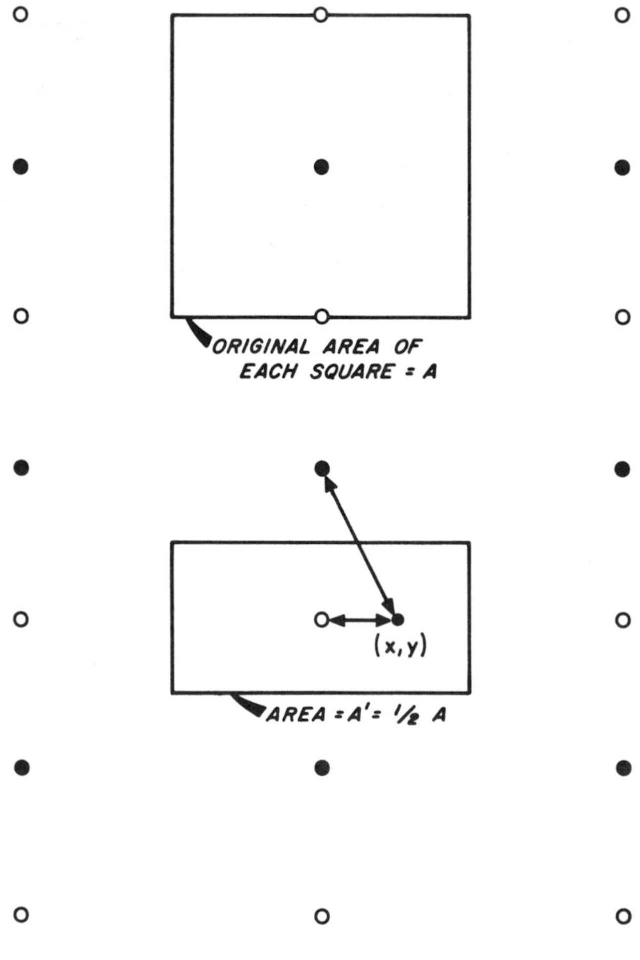

Fig. B.7 Superposition of two square arrays to form a 2:1 rectangle. After Earlougher *et al.*[6]

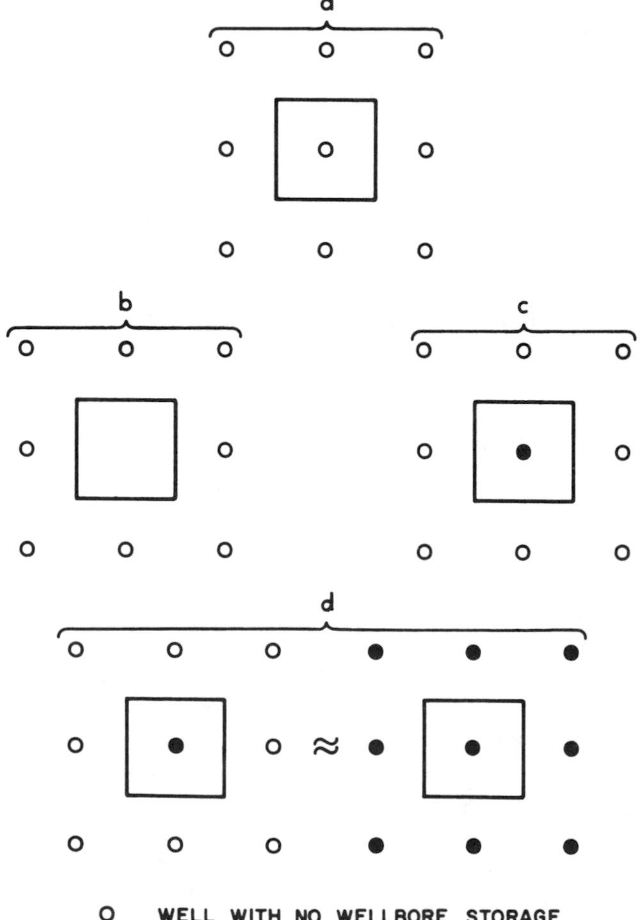

Fig. B.8 Desuperposition to approximate a single well with wellbore storage and skin in the center of a closed square. After Chen and Brigham.[8]

B.7 Superposition for Variable Rate

Fig. B.9 schematically shows a variable-rate history. In the nomenclature of that figure, production (or injection) starts at time 0; the rate remains constant at q_1 until time t_1, then changes to q_2 until time t_2, etc. Note that rate q_j ends at time t_j and that $t_0 = 0$. The last (and current rate) is always q_N. We may calculate the pressure at the well (or at any other point for which we know p_D) at any time during rate q_N by using the principle of superposition as indicated by Eq. 2.31. The pressure at the well is

$$p_{wf}(t) = p_i - \frac{141.2B\mu}{kh}\Big\{q_1\big[p_D(t_D) + s\big] + (q_2-q_1)\big[p_D([t-t_1]_D) + s\big] + (q_3-q_2)\big[p_D([t-t_2]_D) + s\big] + \ldots + (q_N-q_{N-1})\big[p_D([t-t_{N-1}]_D) + s\big]\Big\}. \quad \ldots (B.17)$$

This may be rearranged to

$$p_{wf}(t) = p_i - \frac{141.2B\mu}{kh}\Big\{q_1\big[p_D(t_D) - p_D([t-t_1]_D)\big] + q_2\big[p_D([t-t_1]_D) - p_D([t-t_2]_D)\big] + \ldots + q_{N-1}\big[p_D([t-t_{N-2}]_D) - p_D([t-t_{N-1}]_D)\big] + q_N\big[p_D([t-t_{N-1}]_D) + s\big]\Big\}. \quad \ldots\ldots\ldots (B.18)$$

If the system is infinite-acting and if the logarithmic approximation of the exponential integral, Eq. 2.5b, applies, Eq. B.18 may be written

$$p_{wf}(t) = p_i - \frac{70.60B\mu}{kh}\Big\{q_1 \ln\Big(\frac{t}{t-t_1}\Big) + q_2 \ln\Big(\frac{t-t_1}{t-t_2}\Big) + \ldots + q_{N-1}\ln\Big(\frac{t-t_{N-2}}{t-t_{N-1}}\Big) + q_N\Big[\ln(t-t_{N-1}) + \ln\Big(\frac{k}{\phi\mu c_t r_w^2}\Big) - 7.4316 + 2s\Big]\Big\}, \quad \ldots (B.19)$$

or

$$p_{wf}(t) = p_i - \frac{162.6B\mu}{kh}\Big\{\sum_{j=1}^{N-1} q_j \log\Big(\frac{t-t_{j-1}}{t-t_j}\Big) + q_N\Big[\log(t-t_{N-1}) + \log\Big(\frac{k}{\phi\mu c_t r_w^2}\Big) - 3.2275 + 0.86859s\Big]\Big\}. \quad \ldots\ldots\ldots\ldots (B.20)$$

Eqs. B.17 through B.20 are convenient for estimating pressures resulting from multiple-rate histories. However, the form

$$\frac{p_i - p_{wf}(t)}{q_N} = \frac{162.6B\mu}{kh}\Big\{\sum_{j=1}^{N}\Big[\Big(\frac{q_j-q_{j-1}}{q_N}\Big) \times \log(t-t_{j-1})\Big] + \log\Big(\frac{k}{\phi\mu c_t r_w^2}\Big) - 3.2275 + 0.86859s\Big\}, \quad \ldots\ldots\ldots (B.21)$$

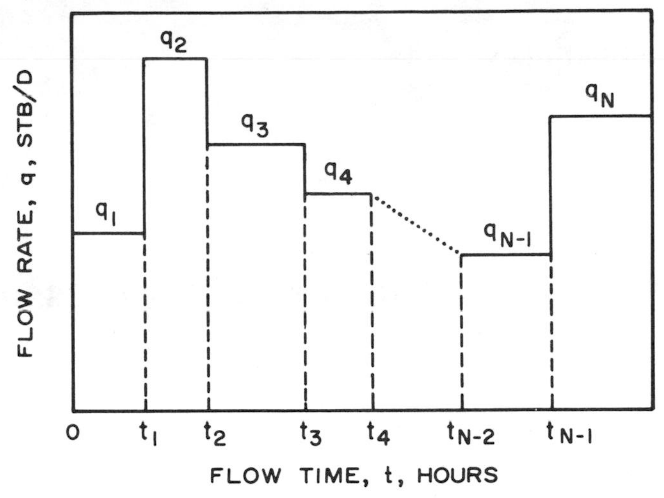

Fig. B.9 Schematic representation of a variable production-rate schedule.

which results from combining Eqs. B.17 and 2.5b, is more convenient for analyzing multiple-rate test data, as indicated in Section 4.2.

References

1. Matthews, C. S. and Russell, D. G.: *Pressure Buildup and Flow Tests in Wells*, Monograph Series, Society of Petroleum Engineers of AIME, Dallas (1967) **1.**
2. van Everdingen, A. F. and Hurst, W.: "The Application of the Laplace Transformation to Flow Problems in Reservoirs," *Trans.*, AIME (1949) **186,** 304-324.
3. Horner, D. R.: "Pressure Build-Up in Wells," *Proc.*, Third World Pet. Cong., The Hague (1951) Sec. II, 503-523. Also *Reprint Series, No. 9 — Pressure Analysis Methods*, Society of Petroleum Engineers of AIME, Dallas (1967) 25-43.
4. Matthews, C. S., Brons, F., and Hazebroek, P.: "A Method for Determination of Average Pressure in a Bounded Reservoir," *Trans.*, AIME (1954) **201,** 182-191. Also *Reprint Series, No. 9 — Pressure Analysis Methods*, Society of Petroleum Engineers of AIME, Dallas (1967) 51-60.
5. Collins, Royal Eugene: *Flow of Fluids Through Porous Materials*, Reinhold Publishing Corp., New York (1961) 109-113, 116-118.
6. Earlougher, Robert C., Jr., Ramey, H. J., Jr., Miller, F. G., and Mueller, T. D.: "Pressure Distributions in Rectangular Reservoirs," *J. Pet. Tech.* (Feb. 1968) 199-208; *Trans.*, AIME, **243.**
7. Gringarten, Alain C., Ramey, Henry J., Jr., and Raghavan, R.: "Unsteady-State Pressure Distributions Created by a Well With a Single Infinite-Conductivity Vertical Fracture," *Soc. Pet. Eng. J.* (Aug. 1974) 347-360; *Trans.*, AIME, **257.**
8. Chen, Hsiu-Kuo and Brigham, W. E.: "Pressure Buildup for a Well With Storage and Skin in a Closed Square," paper SPE 4890 presented at the SPE-AIME 44th Annual California Regional Meeting, San Francisco, April 4-5, 1974.
9. Ramey, Henry J., Jr., Kumar, Anil, and Gulati, Mohinder S.: *Gas Well Test Analysis Under Water-Drive Conditions*, AGA, Arlington, Va. (1973).
10. Earlougher, R. C., Jr., and Ramey, H. J., Jr.: "Interference Analysis in Bounded Systems," *J. Cdn. Pet. Tech.* (Oct.-Dec. 1973) 35-45.
11. Kaplan, Wilfred: *Advanced Calculus*, Addison Wesley Publishing Co., Inc., Reading, Mass. (1952) 220.

Appendix C

Dimensionless Pressure Solutions

C.1 Introduction

This appendix presents correlations of dimensionless pressure with dimensionless time for single-well systems producing at constant rate. Some data from the literature have been modified to be consistent with the nomenclature used in this monograph. We retain the definition of dimensionless pressure corresponding to

$$p_i - p = \Delta p = \frac{141.2\,qB\mu}{kh}\,p_D(t_D, \ldots) \;\ldots\ldots \text{(C.1)}$$

Dimensionless time is usually based on wellbore radius,

$$t_D = \frac{0.0002637\,kt}{\phi\mu c_t r_w^2}, \;\ldots\ldots\ldots\ldots\ldots\ldots \text{(C.2a)}$$

or based on drainage area,

$$t_{DA} = \frac{0.0002637\,kt}{\phi\mu c_t A} = t_D\left(\frac{r_w^2}{A}\right). \;\ldots\ldots \text{(C.2b)}$$

We clearly indicate when it is convenient to base the dimensionless time on some other characteristic dimension.

C.2 Infinite Systems

No Wellbore Storage, No Skin

After wellbore storage effects are no longer important, the dimensionless pressure for infinite and infinite-acting systems is given by[1-4]

$$p_D = -\frac{1}{2}\,\text{Ei}\left(-\frac{r_D^2}{4t_D}\right), \;\ldots\ldots\ldots\ldots\ldots\ldots \text{(C.3)}$$

when ($r_D \geqslant 20$ and $t_D/r_D^2 \geqslant 0.5$) or when $t_D/r_D^2 \geqslant 25$. Dimensionless pressure values for smaller t_D and r_D are given in Fig. C.1 for a range of r_D and t_D; tabulated values for $r_D = 1$ are given by van Everdingen and Hurst.[5] The lowermost curve in Fig. C.1 ($r_D > 20$), which is the exponential-integral solution (Eq. C.3), is shown on an expanded scale in Fig. C.2.*

A simplification of Eq. C.3 and Figs. C.1 and C.2 applies when $t_D/r_D^2 > 100$ (or with less than 1-percent error when $t_D/r_D^2 > 10$):

$$p_D = \frac{1}{2}\left[\ln(t_D/r_D^2) + 0.80907\right] \;\ldots\ldots\ldots\ldots \text{(C.4)}$$

*See footnote on Page 24.

These dimensionless pressure solutions apply for a single undamaged well in an infinite-acting system with no wellbore storage. Damage or improvement may be included as indicated in Eq. 2.2.

Single Vertical Fracture, No Wellbore Storage

Figs. C.3* and C.4 give dimensionless pressure data for a vertically fractured well in an infinite-acting system. A single fracture of half-length x_f intersects the well. Two situations are shown:

1. The *uniform-flux fracture* is a first approximation to the behavior of a vertically fractured well.[6,7] Fluid enters the fracture at a uniform flow rate per unit area of fracture face so that there is a pressure drop in the fracture. The dimensionless pressure at the well for the uniform-flux fracture case is computed from[6,7]

$$p_D = \sqrt{\pi t_{Dxf}}\,\text{erf}\left(\frac{1}{2\sqrt{t_{Dxf}}}\right) - \frac{1}{2}\,\text{Ei}\left(\frac{-1}{4t_{Dxf}}\right), \;\ldots \text{(C.5)}$$

where dimensionless time based on the half-fracture length is defined as

$$t_{Dxf} = t_D(r_w/x_f)^2 \;\ldots\ldots\ldots\ldots\ldots\ldots\ldots\ldots \text{(C.6)}$$

When $t_{Dxf} > 10$, Eq. C.5 becomes[6,7]

$$p_D = \frac{1}{2}\left[\ln t_{Dxf} + 2.80907\right], \;\ldots\ldots\ldots\ldots\ldots \text{(C.7)}$$

with less than 1-percent error. For $t_{Dxf} < 0.1$, Eq. C.5 becomes[6,7]

$$p_D = \sqrt{\pi t_{Dxf}}, \;\ldots\ldots\ldots\ldots\ldots\ldots\ldots\ldots \text{(C.8)}$$

indicating that at short times flow into the fracture is linear.

2. The *infinite-conductivity fracture* has infinite permeability and, therefore, uniform pressure throughout. The dimensionless well pressure for that case is given in Figs. C.3 and C.4, and may be computed from[6,7]

$$p_D = \frac{1}{2}\sqrt{\pi t_{Dxf}}\left[\text{erf}\left(\frac{0.134}{\sqrt{t_{Dxf}}}\right) + \text{erf}\left(\frac{0.866}{\sqrt{t_{Dxf}}}\right)\right] - 0.067\,\text{Ei}\left(-\frac{0.018}{t_{Dxf}}\right) - 0.433\,\text{Ei}\left(-\frac{0.750}{t_{Dxf}}\right). \;\ldots\ldots \text{(C.9)}$$

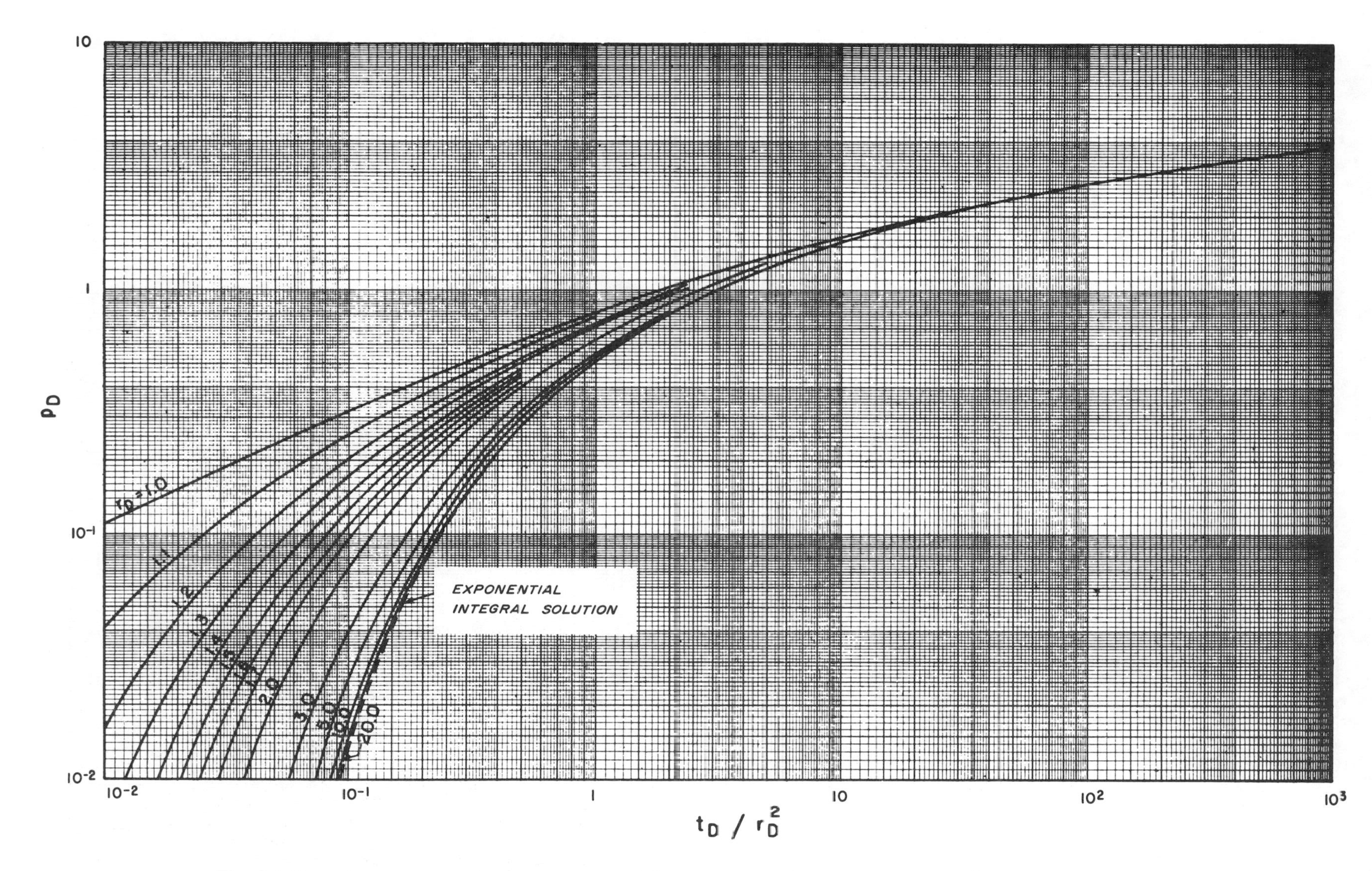

Fig. C.1 Dimensionless pressure for single well in an infinite system, small r_D, short time, no wellbore storage, no skin. After Mueller and Witherspoon.[4]

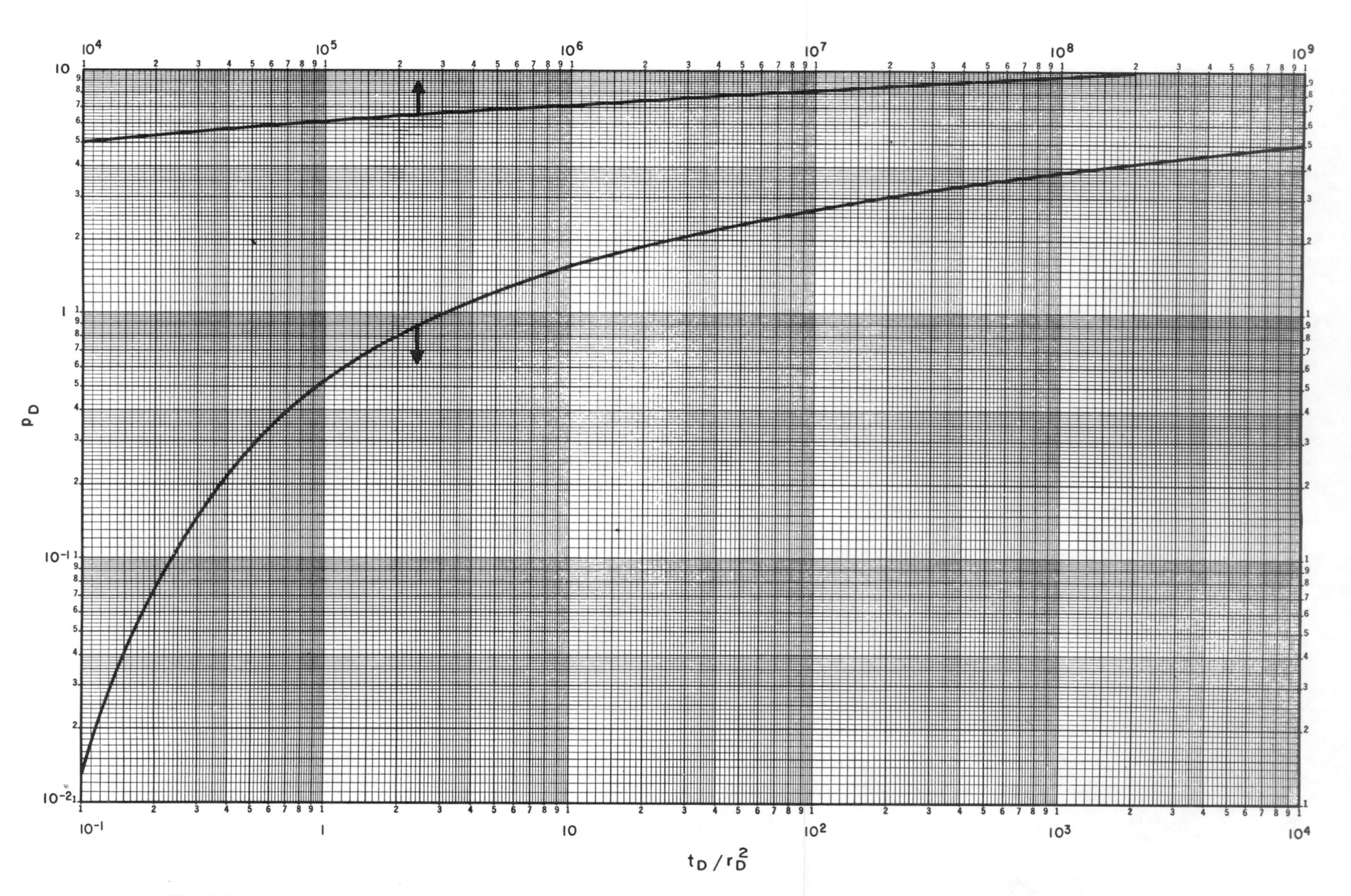

Fig. C.2 Dimensionless pressure for a single well in an infinite system, no wellbore storage, no skin. Exponential-integral solution.

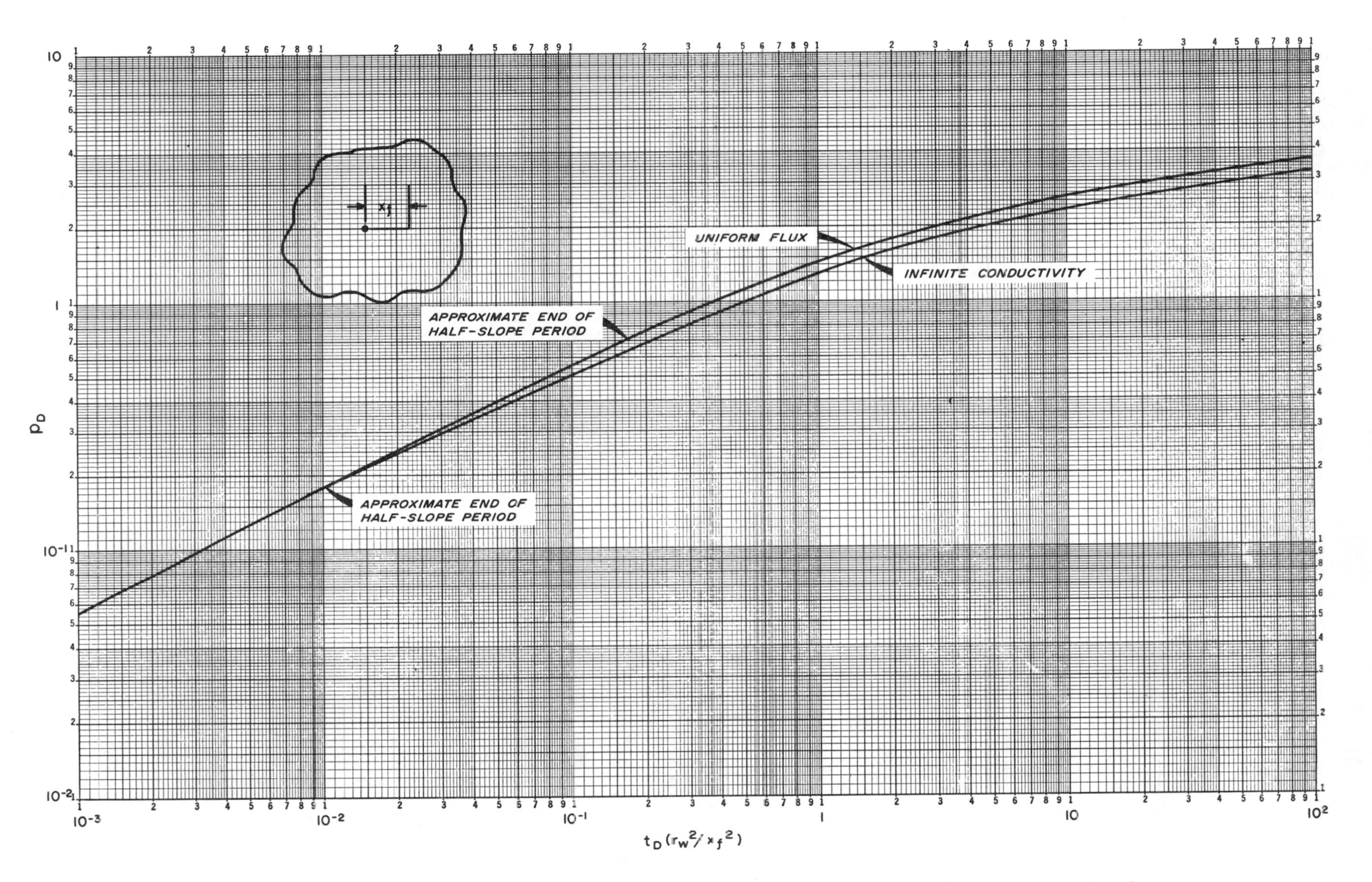

Fig. C.3 Dimensionless pressure for single, vertically fractured well in an infinite system, no wellbore storage. Log-log plot. Data of Gringarten, Ramey, and Raghavan.[6,7]

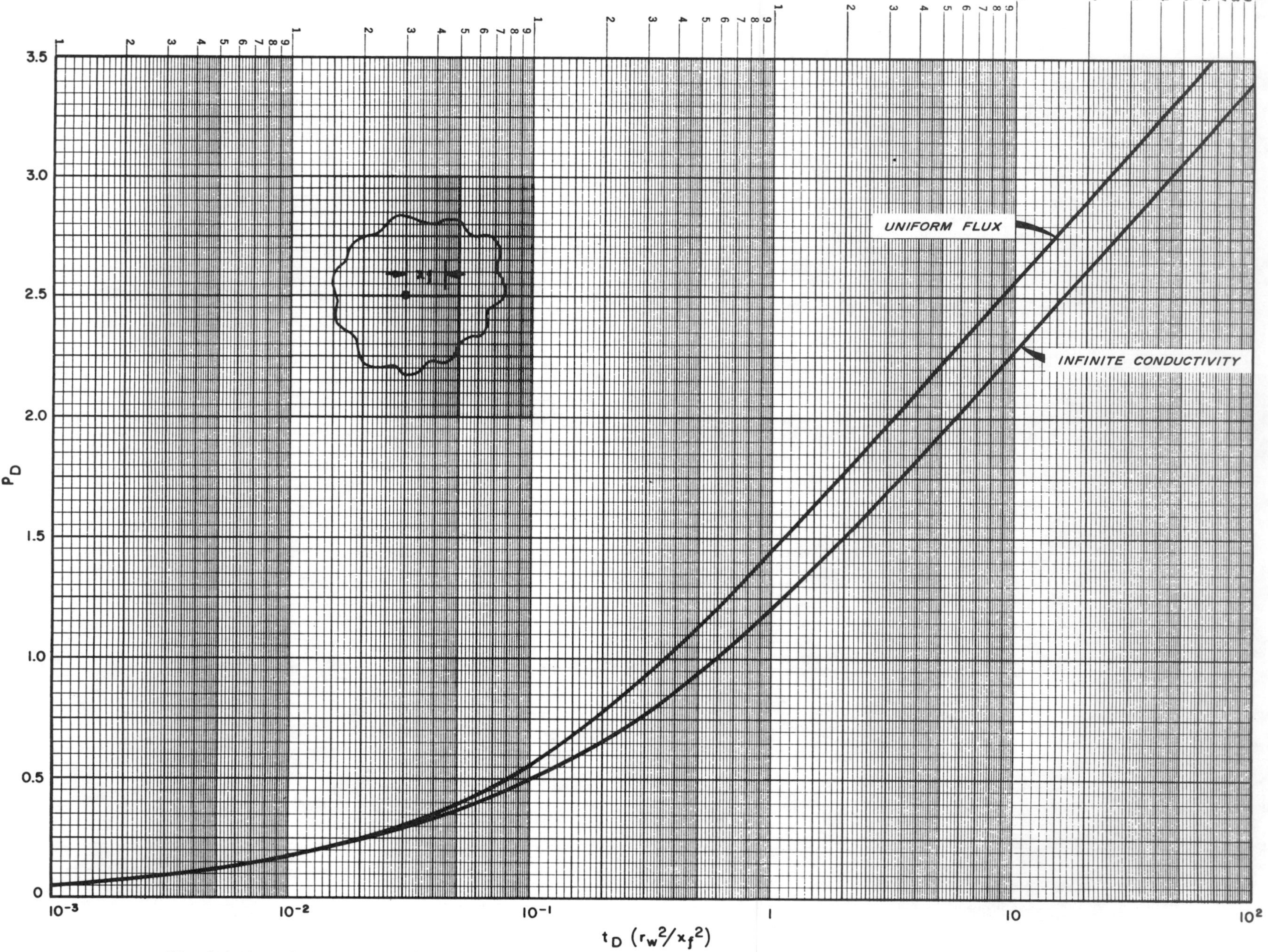

Fig. C.4 Dimensionless pressure for single, vertically fractured well in an infinite system, no wellbore storage. Semilog plot. Data of Gringarten, Ramey, and Raghavan.[6,7]

When $t_{Dxf} > 10$, Eq. C.9 becomes[6,7]

$$p_D = \frac{1}{2}\left[\ln t_{Dxf} + 2.2000\right], \quad \text{(C.10)}$$

with less than 1-percent error. When t_{Dxf} is less than 0.01, Eq. C.8 applies.

An important feature of Fig. C.3 is the early-time slope of one-half cycle in pressure per cycle in time. Such a half-slope straight line on the log-log plot is often diagnostic of a vertically fractured well.

Gringarten, Ramey, and Raghavan[6,7] tabulate p_D values for the two types of vertical fractures.

Single Horizontal Fracture, No Wellbore Storage

Fig. C.5* shows p_D data at the well for an infinite system with a single horizontal fracture located at the formation midpoint. The dimensionless time used,

$$t_{Drf} = \frac{0.0002637\,kt}{\phi\mu c_t r_f^2} = t_D\left(\frac{r_w^2}{r_f^2}\right), \quad \text{(C.11)}$$

is based on the horizontal fracture radius, r_f. In Fig. C.5, dimensionless pressure is normalized by the parameter on the curves,

$$h_D = \frac{h}{r_f}\sqrt{\frac{k_r}{k_z}}, \quad \text{(C.12)}$$

where k_r and k_z are radial and vertical permeabilities, respectively. At short times and for large h_D, the curves in Fig. C.5 have a half-slope portion. It is apparent, however, that many horizontally fractured systems would not exhibit a half-slope straight line on the log-log plot. At low values of h_D, the curves in Fig. C.5 have a unit slope, like the unit slope caused by wellbore storage effects. There is *no wellbore storage effect* included in Fig. C.5, so the unit slope there is a result of the *fracture*, not the wellbore. Gringarten[8] and Ramey[9] tabulate p_D values for the horizontal-fracture case.

Wellbore Storage and Thin Skin Included

Fig. C.6* shows dimensionless pressure data for a single well in an infinite system with wellbore storage and skin effect included.[10] The dimensionless wellbore storage coefficient is

$$C_D = \frac{5.6146\,C}{2\pi\phi c_t h r_w^2}. \quad \text{(C.13)}$$

When $C_D > 0$, Fig. C.6 shows that the log-log plot has an early-time unit slope. At later times, the curves approach those for zero wellbore storage. Tabulated dimensionless pressure data are given by Agarwal, Al-Hussainy, and Ramey.[10] Although t_D in Fig. C.6 is based on r_w, generation of the negative skin solutions involved use of an apparent larger wellbore radius as defined by Eq. 2.11.

Wellbore Storage and Finite Skin Included

Fig. C.7* gives dimensionless pressure data for a single well in an infinite reservoir with wellbore storage and a finite skin effect.[11] Fig. 2.6 schematically illustrates the finite skin. The skin factor is calculated from

$$s = \left(\frac{k}{k_s} - 1\right)\ln(r_{sD}), \quad \text{(C.14)}$$

where

$$r_{sD} = r_s/r_w. \quad \text{(C.15)}$$

Wattenbarger and Ramey[11] provide tables of p_D vs t_D for the conditions of Fig. C.7.

Other Useful Type Curves

Fig. C.8* shows another relation between pressure and time for a single well with wellbore storage and skin effect in an infinite system.[12] The graph can be changed to a dimensionless pressure-dimensionless time basis by using equations given in Ref. 12. This type curve is particularly useful for curve matching and is not recommended for calculating pressure response. Its use is illustrated in Section 3.3.

Fig. C.9* is a type curve presented by McKinley[13] for a single well with wellbore storage, but no skin factor, in an infinite system. Fig. C.9 assumes

$$\frac{k}{\phi\mu c_t r_w^2} = 9.728\times 10^6. \quad \text{(C.16)}$$

Although the figure is plotted on the basis of actual variables, it may be reduced to a dimensionless graph by using the definitions of C_D, t_D, and p_D with Eq. C.16. The main utility of Fig. C.9 is for type-curve matching of test data, not for calculating pressure response.

C.3 Closed Systems

All closed reservoir systems (that is, those with no-flow outer boundaries) have the transient behavior illustrated in Fig. 2.1. Within 1 percent,

$$p_D = \frac{1}{2}\left[\ln(t_{DA}) + \ln\left(\frac{A}{r_w^2}\right) + 0.80907\right], \quad \text{(C.17)}$$

if $0.000025 < t_{DA}$ and t_{DA} is less than the value in the "Use Infinite System Solution With Less Than 1% Error for $t_{DA} <$" column of Table C.1. At long times the system reaches pseudosteady state and[14]

$$p_D = 2\pi t_{DA} + \frac{1}{2}\ln\left(\frac{A}{r_w^2}\right) + \frac{1}{2}\ln\left(\frac{2.2458}{C_A}\right). \quad \text{(C.18)}$$

Eq. C.18 applies when t_{DA} exceeds the value in the "Less Than 1% Error for $t_{DA} >$" column of Table C.1. Values of C_A and of the last term on the right-hand side of Eq. C.18 are given in Table C.1 for many closed drainage shapes. Values of C_A are also given in Refs. 15 through 18 and in Table C.2.

Dimensionless pressure data *at the well* in closed reservoir systems are always given for a specific $\sqrt{A}/r_w$. If p_D is desired for a system of similar shape and geometry but with a different value of this parameter, it may be computed from[17]

$$(p_D)_{\text{desired}} = (p_D)_{\text{table}} + \ln\left[(\sqrt{A}/r_w)_{\text{desired}}/(\sqrt{A}/r_w)_{\text{table}}\right], \quad \text{(C.19)}$$

*See footnote on Page 24.

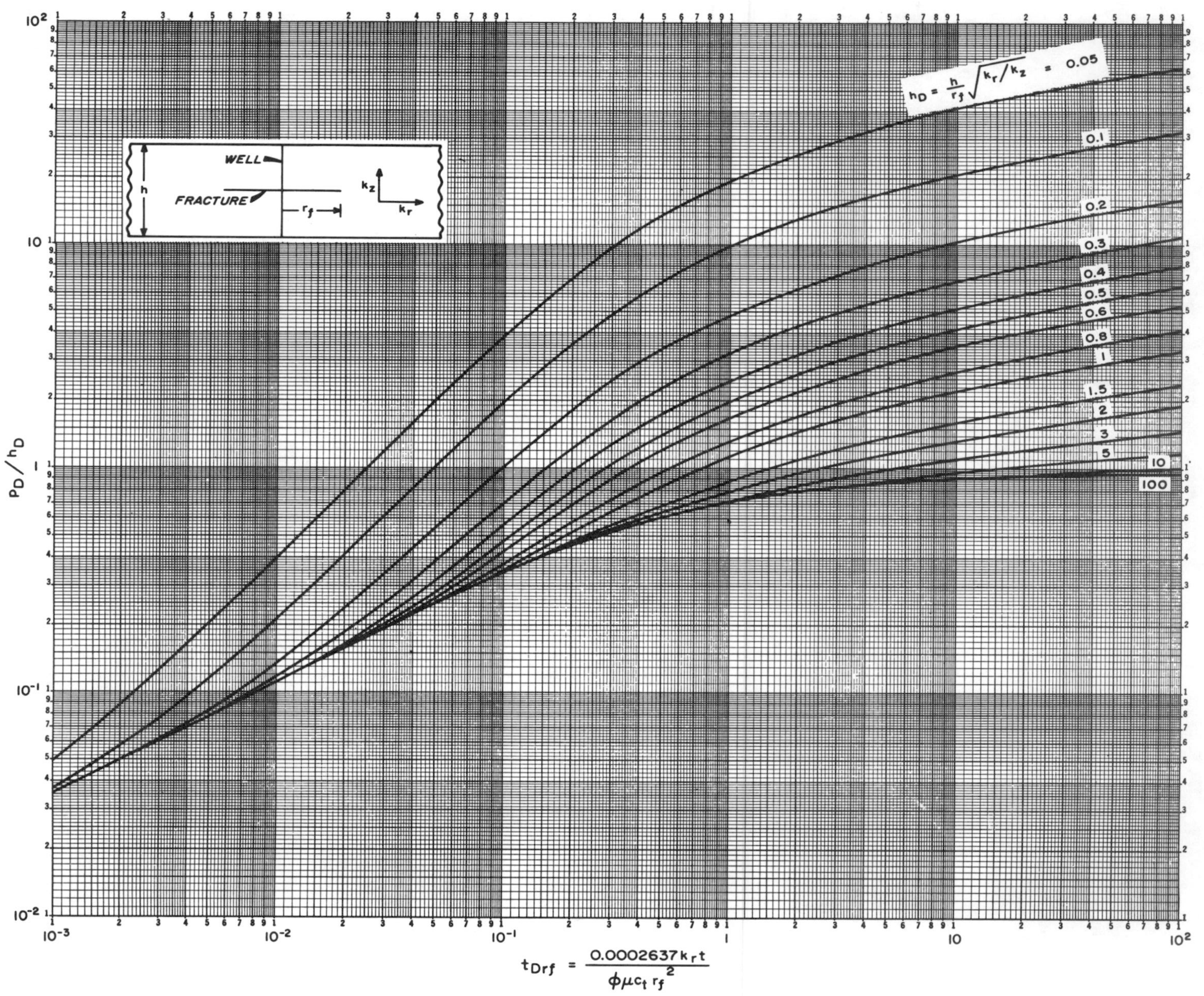

Fig. C.5 Dimensionless pressure for a single, horizontally fractured (uniform-flux) well in an infinite system, no wellbore storage. Fracture located in the center of the interval. After Gringarten, Ramey, and Raghavan.[6]

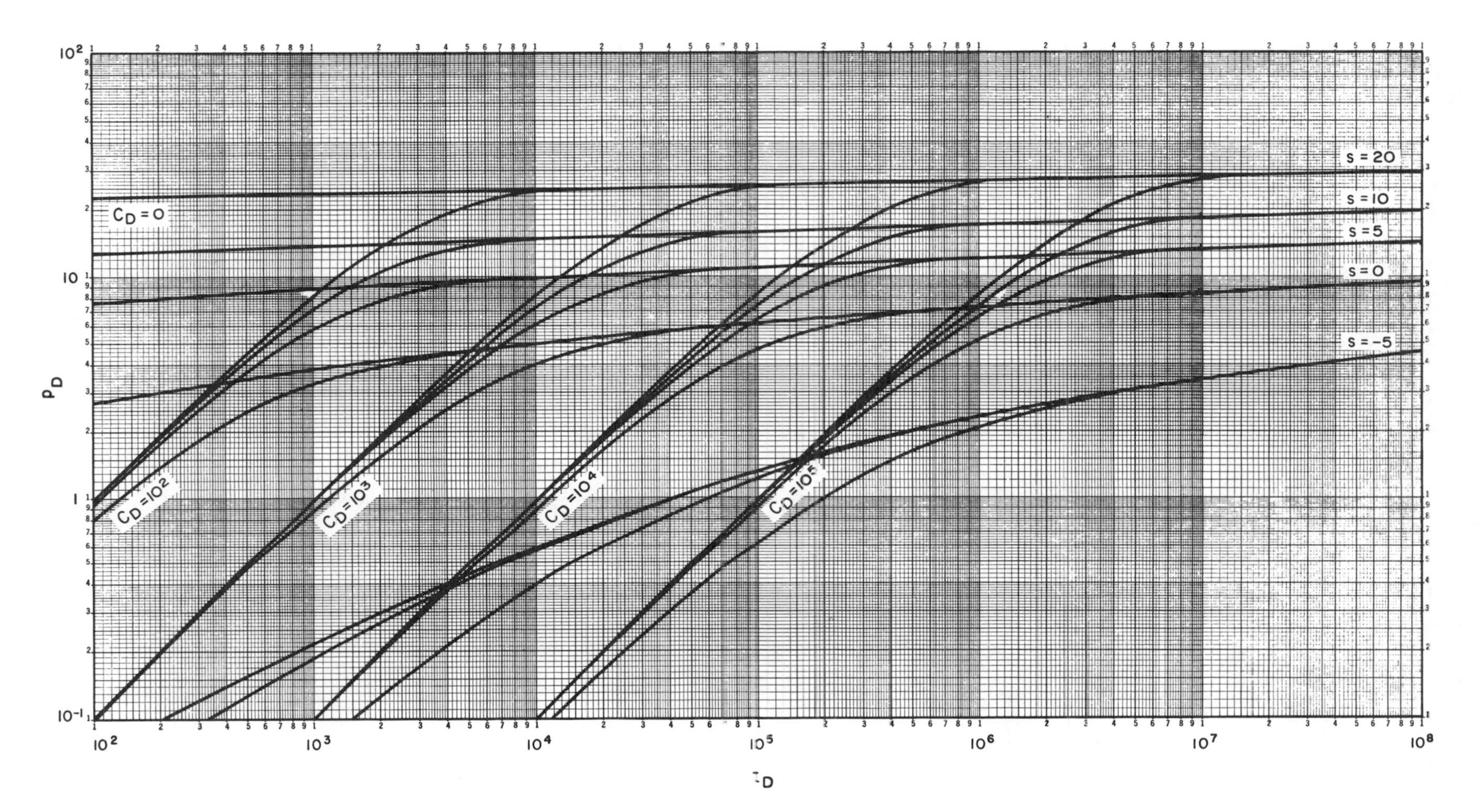

Fig. C.6 Dimensionless pressure for a single well in an infinite system, wellbore storage and skin included. After Agarwal, Al-Hussainy, and Ramey.[10] Graph courtesy H. J. Ramey, Jr.

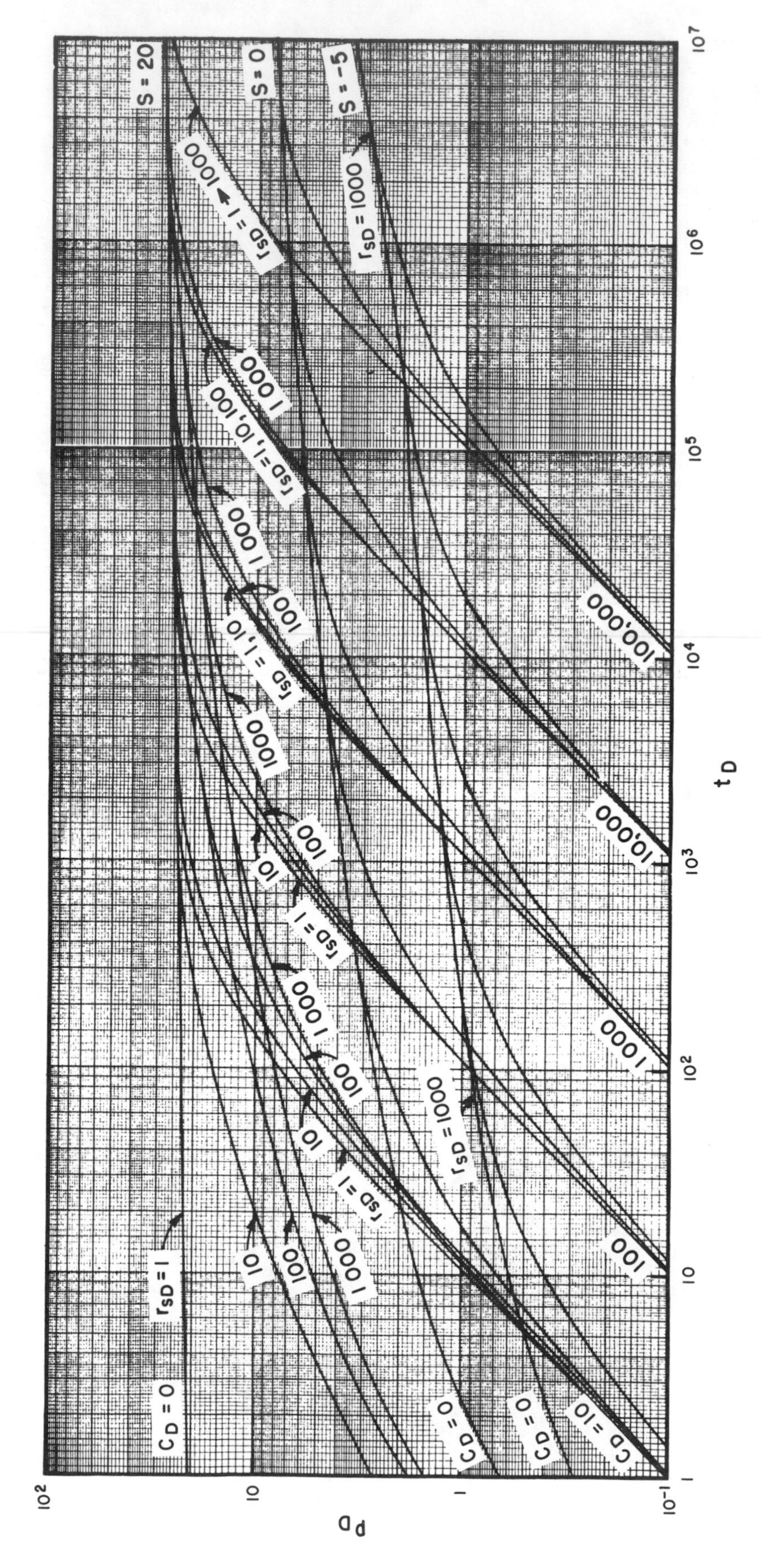

Fig. C.7 Dimensionless pressure for a single well in an infinite reservoir including wellbore storage and a finite skin (composite reservoir). After Wattenbarger and Ramey.[11] Graph courtesy H. J. Ramey, Jr.

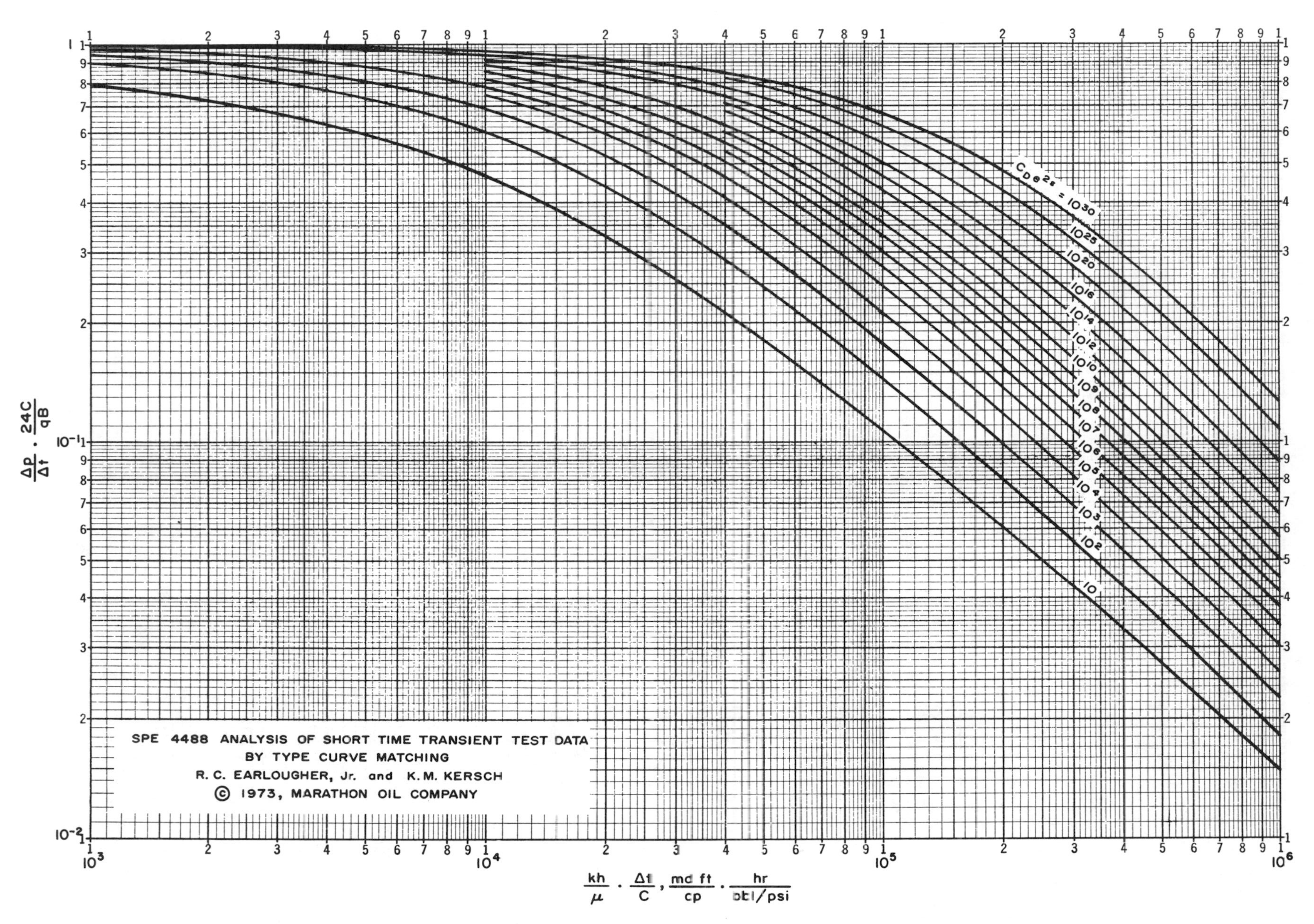

Fig. C.8 Type curve for a single well in an infinite system, wellbore storage and skin effects included. After Earlougher and Kersch.[12] Reprinted by permission of Marathon Oil Co.

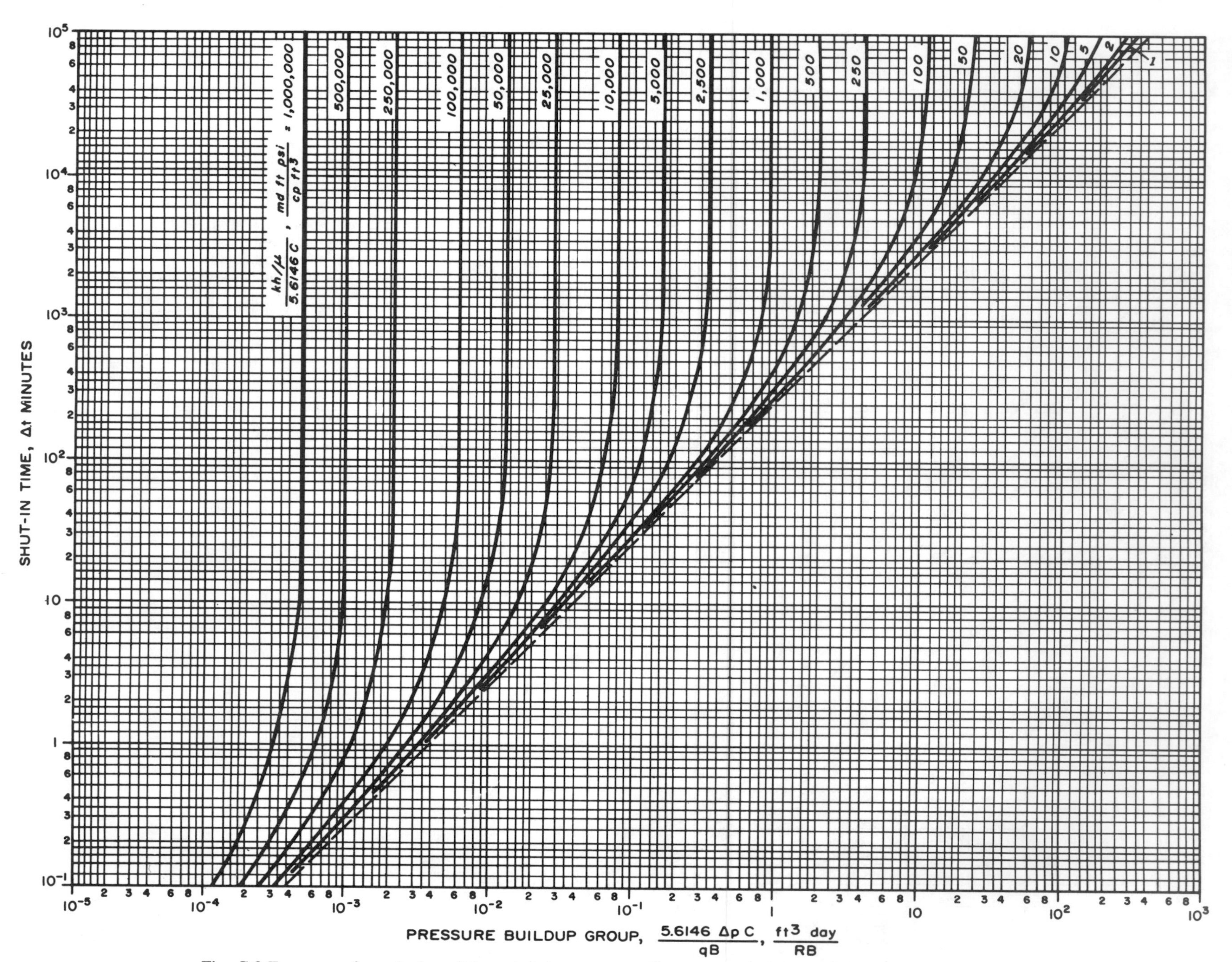

Fig. C.9 Type curve for a single well in an infinite system, wellbore storage included, no skin. After McKinley.[13]

TABLE C.1—SHAPE FACTORS FOR VARIOUS CLOSED SINGLE-WELL DRAINAGE AREAS.

IN BOUNDED RESERVOIRS	C_A	$\ln C_A$	$\frac{1}{2}\ln\left(\frac{2.2458}{C_A}\right)$	EXACT FOR t_{DA} >	LESS THAN 1% ERROR FOR t_{DA} >	USE INFINITE SYSTEM SOLUTION WITH LESS THAN 1% ERROR FOR t_{DA} <
	31.62	3.4538	−1.3224	0.1	0.06	0.10
	31.6	3.4532	−1.3220	0.1	0.06	0.10
	27.6	3.3178	−1.2544	0.2	0.07	0.09
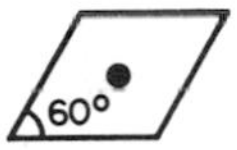 60°	27.1	3.2995	−1.2452	0.2	0.07	0.09
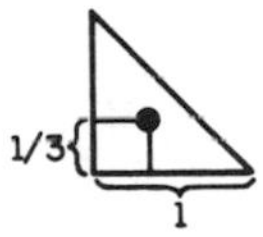 1/3, 1	21.9	3.0865	−1.1387	0.4	0.12	0.08
 3, 4	0.098	-2.3227	+1.5659	0.9	0.60	0.015
	30.8828	3.4302	−1.3106	0.1	0.05	0.09
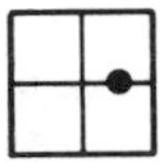	12.9851	2.5638	−0.8774	0.7	0.25	0.03
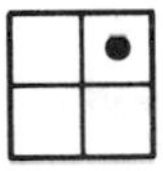	4.5132	1.5070	−0.3490	0.6	0.30	0.025
	3.3351	1.2045	−0.1977	0.7	0.25	0.01
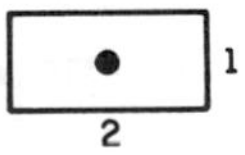 1, 2	21.8369	3.0836	−1.1373	0.3	0.15	0.025
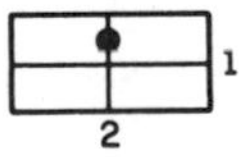 1, 2	10.8374	2.3830	−0.7870	0.4	0.15	0.025
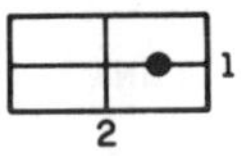 1, 2	4.5141	1.5072	−0.3491	1.5	0.50	0.06
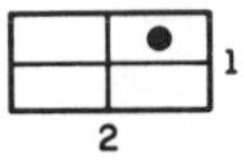 1, 2	2.0769	0.7309	+0.0391	1.7	0.50	0.02
 1, 2	3.1573	1.1497	−0.1703	0.4	0.15	0.005

TABLE C.1—CONT'D.

	C_A	$\ln C_A$	$1/2 \ln \left(\frac{2.2458}{C_A}\right)$	EXACT FOR t_{DA} >	LESS THAN 1% ERROR FOR t_{DA} >	USE INFINITE SYSTEM SOLUTION WITH LESS THAN 1% ERROR FOR t_{DA} <
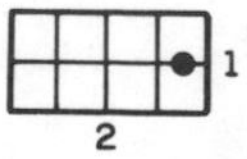 1, 2	0.5813	-0.5425	+0.6758	2.0	0.60	0.02
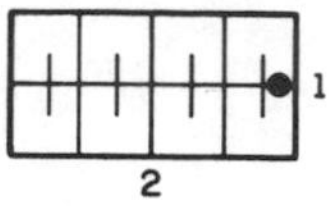 1, 2	0.1109	-2.1991	+1.5041	3.0	0.60	0.005
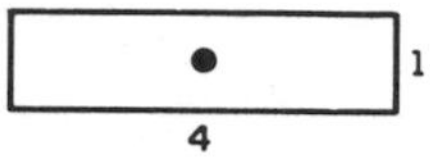 1, 4	5.3790	1.6825	-0.4367	0.8	0.30	0.01
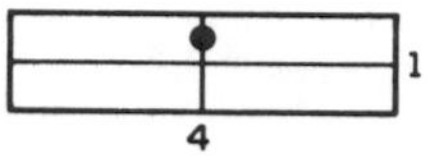 1, 4	2.6896	0.9894	-0.0902	0.8	0.30	0.01
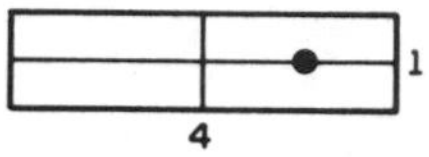 1, 4	0.2318	-1.4619	+1.1355	4.0	2.00	0.03
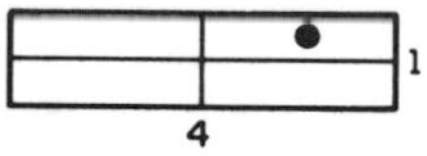 1, 4	0.1155	-2.1585	+1.4838	4.0	2.00	0.01
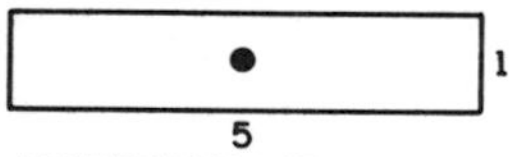 1, 5	2.3606	0.8589	-0.0249	1.0	0.40	0.025
IN VERTICALLY-FRACTURED RESERVOIRS	USE $(x_e/x_f)^2$ IN PLACE OF A/r_w^2 FOR FRACTURED SYSTEMS					
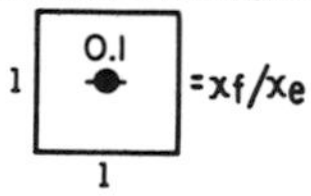 1, 1, 0.1 $= x_f/x_e$	2.6541	0.9761	-0.0835	0.175	0.08	CANNOT USE
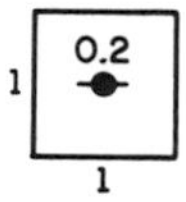 1, 1, 0.2	2.0348	0.7104	+0.0493	0.175	0.09	CANNOT USE
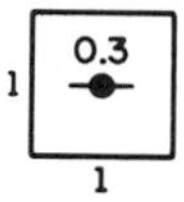 1, 1, 0.3	1.9986	0.6924	+0.0583	0.175	0.09	CANNOT USE
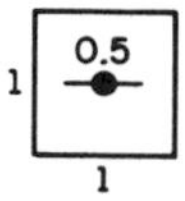 1, 1, 0.5	1.6620	0.5080	+0.1505	0.175	0.09	CANNOT USE
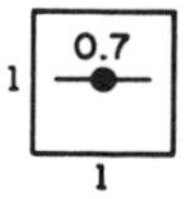 1, 1, 0.7	1.3127	0.2721	+0.2685	0.175	0.09	CANNOT USE
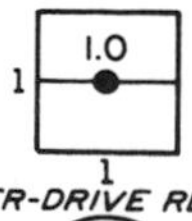 1, 1, 1.0	0.7887	-0.2374	+0.5232	0.175	0.09	CANNOT USE
IN WATER-DRIVE RESERVOIRS						
	19.1	2.95	-1.07	—	—	—
IN RESERVOIRS OF UNKNOWN PRODUCTION CHARACTER						
	25.0	3.22	-1.20	—	—	—

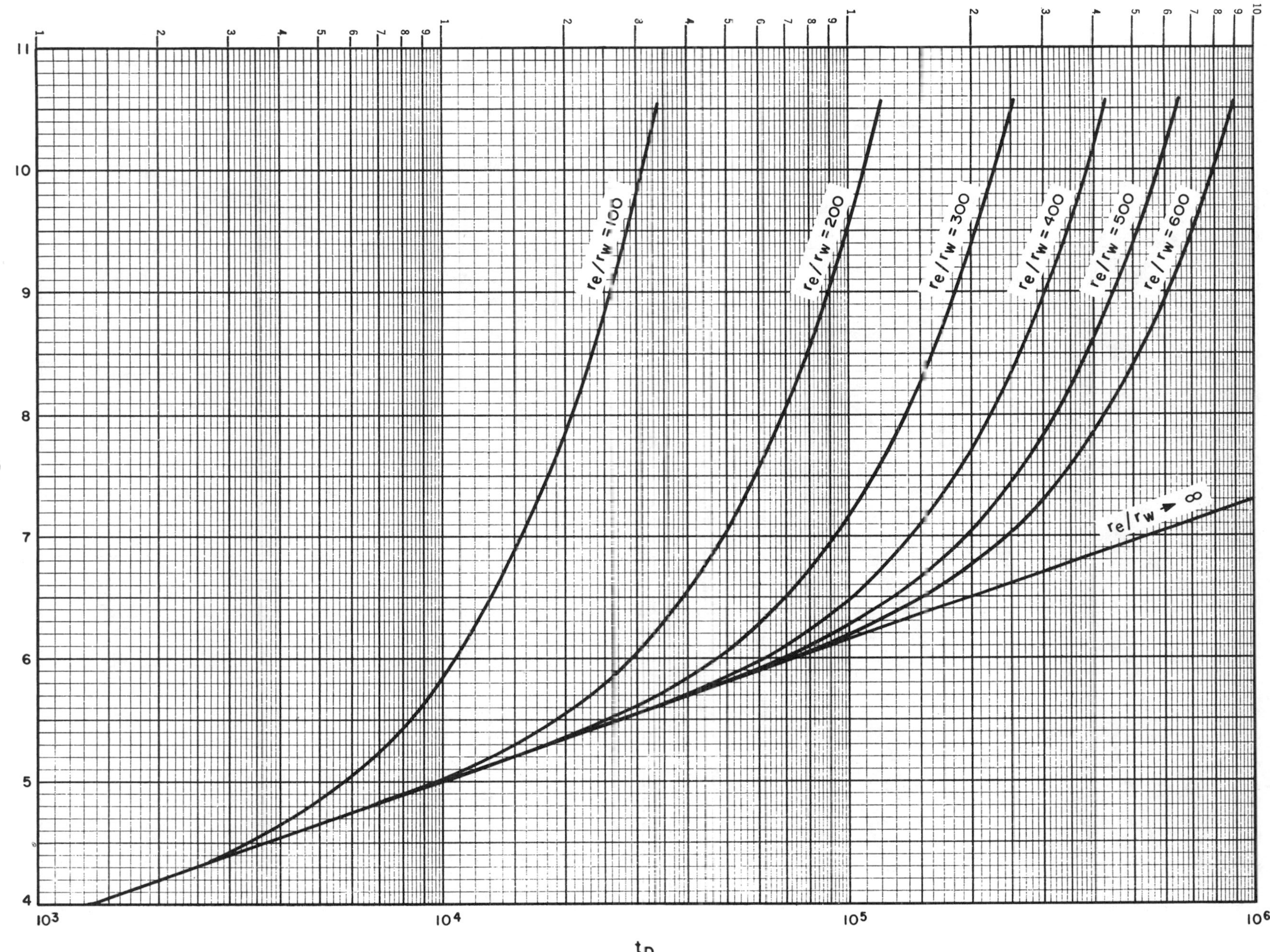

Fig. C.10A Dimensionless pressure for a well in the center of a closed circular reservoir, no wellbore storage, no skin. Calculated from Eq. C.20.

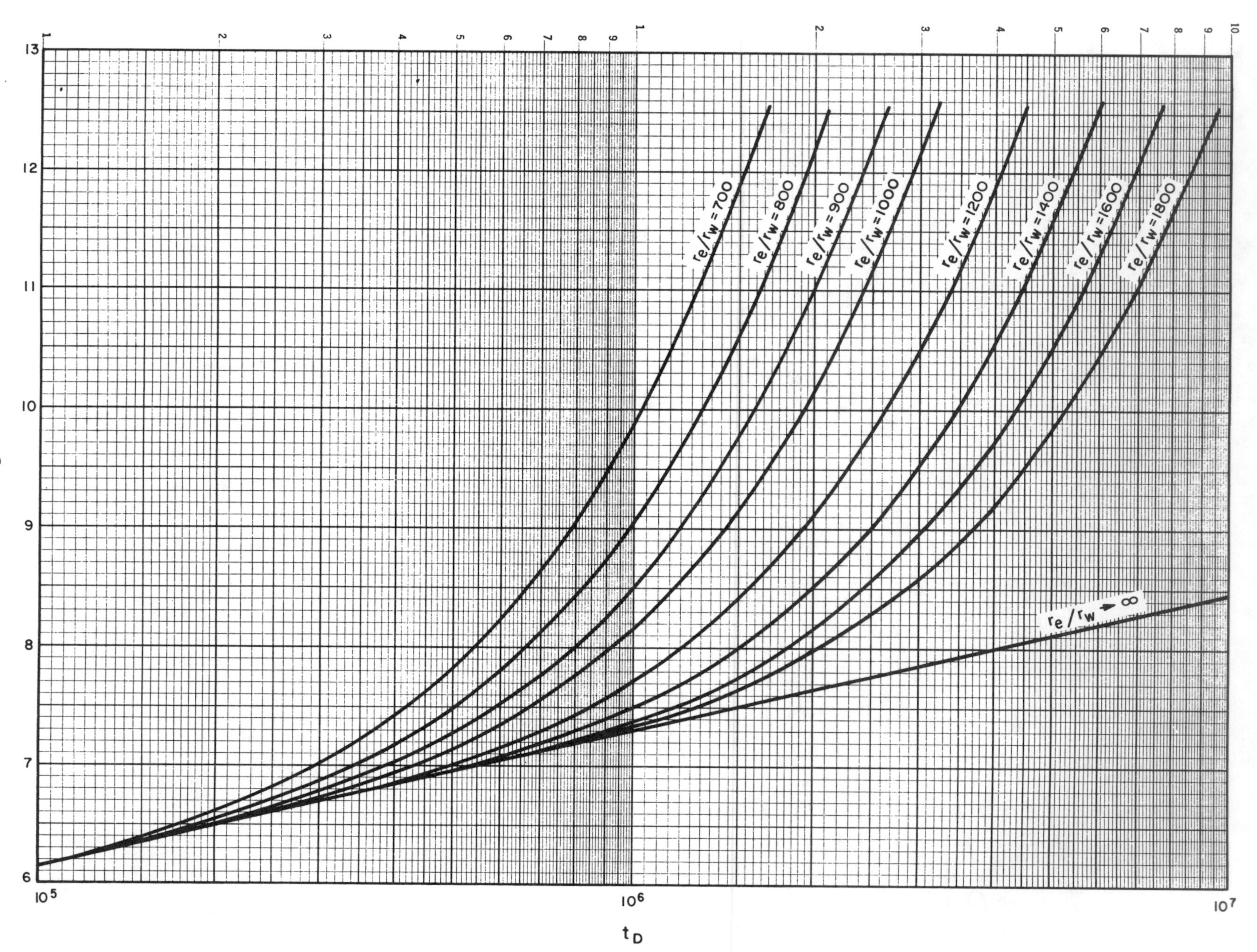

Fig. C.10B Dimensionless pressure for a well in the center of a closed circular reservoir, no wellbore storage, no skin. Calculated from Eq. C.20.

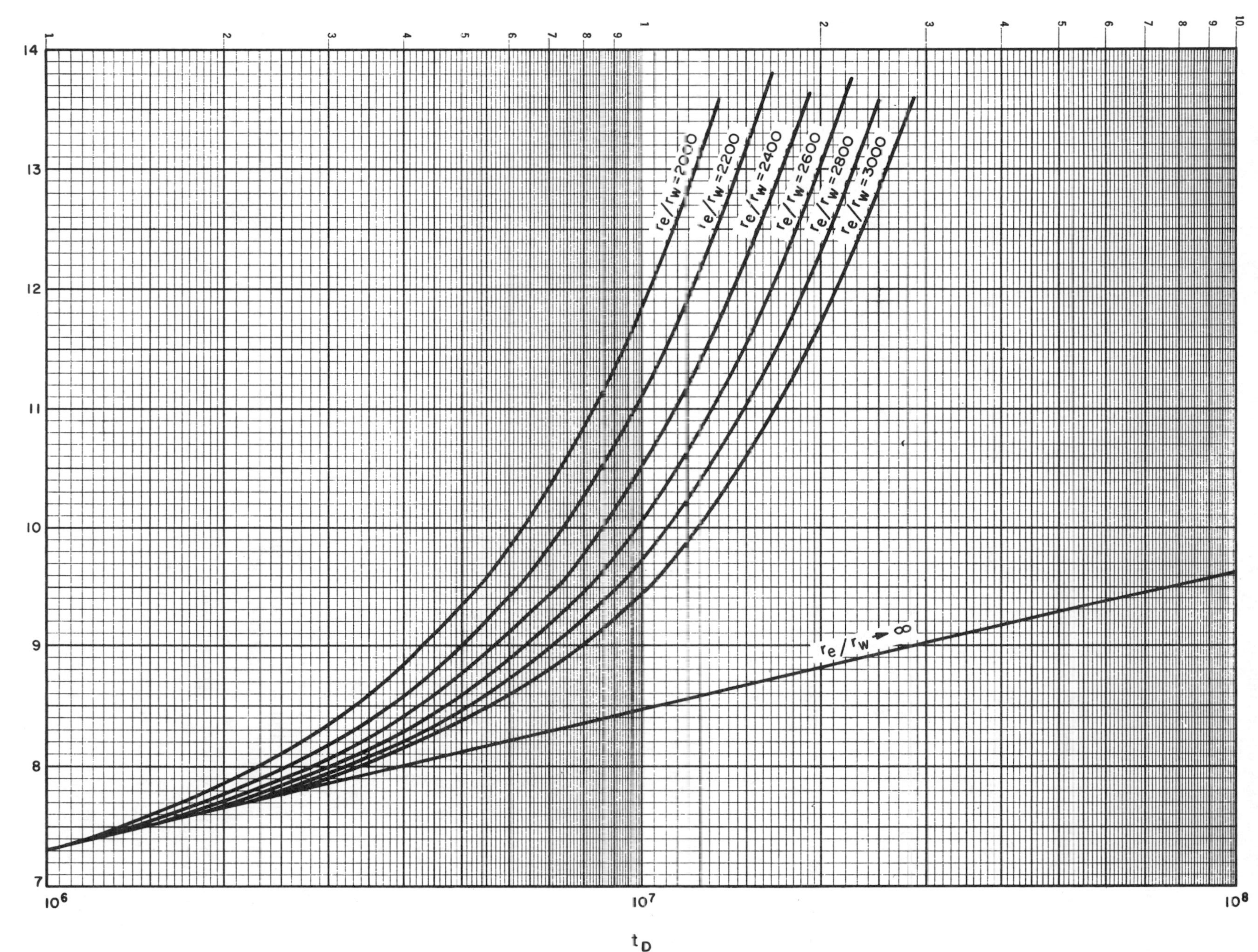

Fig. C.10C Dimensionless pressure for a well in the center of a closed circular reservoir, no wellbore storage, no skin. Calculated from Eq. C.20.

where

$(p_D)_{\text{desired}} =$ value of p_D at desired value of $\sqrt{A}/r_w$, and

$(p_D)_{\text{table}} =$ tabulated or plotted value of p_D using table or graph value of $\sqrt{A}/r_w$.

The quantity $\sqrt{A}/r_w$ affects the dimensionless pressure only at the well; values given at points away from the well are independent of this parameter.

Closed Circular Reservoir, No Wellbore Storage, No Skin

Figs. C.10A, C.10B, and C.10C show dimensionless pressure data for a well in the center of a closed circular reservoir with no wellbore storage and no skin. Skin effect may be included by using Eq. 2.2. The data in Fig. C.10 have been computed from[3]

$$p_D = -\frac{1}{2}\left\{\text{Ei}\left(-\frac{1}{4t_D}\right) - \text{Ei}\left(-\frac{1}{4t_{De}}\right) - 4t_{De}\exp\left(-1/4t_{De}\right)\right\}, \quad \text{(C.20)}$$

where the dimensionless time based on the external radius of the system is

$$t_{De} = \frac{0.0002637\,kt}{\phi\mu c_t r_e^2} = t_D\left(\frac{r_w^2}{r_e^2}\right). \quad \text{(C.21)}$$

Horner[3] points out that Eq. C.20 is "not even a mathematical solution of the basic flow equation," Eq. 2.1. However, it is an excellent approximation to the exact solution[3] (Ref. 3, Eq. XII, and Ref. 1, Eq. 2.36).

Closed-Square Reservoir, No Wellbore Storage, No Skin

Earlougher, Ramey, Miller, and Mueller[17] give dimensionless pressure data at several points in a closed-square drainage area with the well at the center of the system. Wellbore storage and skin effect are not included. Fig. C.11 schematically illustrates system geometry and the points for which p_D data are given. Figs. C.12A and C.12B show p_D at several points in the system. Table C.2 presents the data for this system.

Single-Well Rectangular Systems, No Wellbore Storage, No Skin

Figs. C.13 through C.16 present dimensionless pressure data at the well for a single well at various locations in various closed rectangular systems. Wellbore storage and skin factors are not included. Earlougher and Ramey[18] give tabular data for these figures; they also present p_D data for points away from the well.

The data in Figs. C.13 through C.16 are related to the data presented by Matthews, Brons, and Hazebroek[19] (Figs. 6.2 through 6.5) by[17]

$$p_D(t_{DA}) = 2\pi t_{DA} + \frac{1}{2}\left[\ln\left(t_{DA}\frac{A}{r_w^2}\right) + 0.80907 - p_{D\text{MBH}}(t_{DA})\right], \quad \text{(C.22)}$$

where

$$p_{D\text{MBH}} = \frac{kh\,(p^* - \bar{p})}{70.6\,qB\mu}. \quad \text{(C.23)}$$

Closed-Square Reservoir, Vertically Fractured Well, No Wellbore Storage

Fig. C.17* gives dimensionless pressure data for a single vertically fractured well (infinite-conductivity fracture case) in the center of a closed-square drainage region. Wellbore storage effects are not included. The data in Fig. C.17 are from Gringarten, Ramey, and Raghavan[6,7] and are considered to be slightly more accurate than other similar data.[20,21] As for the infinite, vertically fractured system, there is an initial half-slope straight line on the log-log plot; the duration of this line depends on the fracture length.

Fig. C.18* shows additional p_D data for a single, verti-

*See footnote on Page 24.

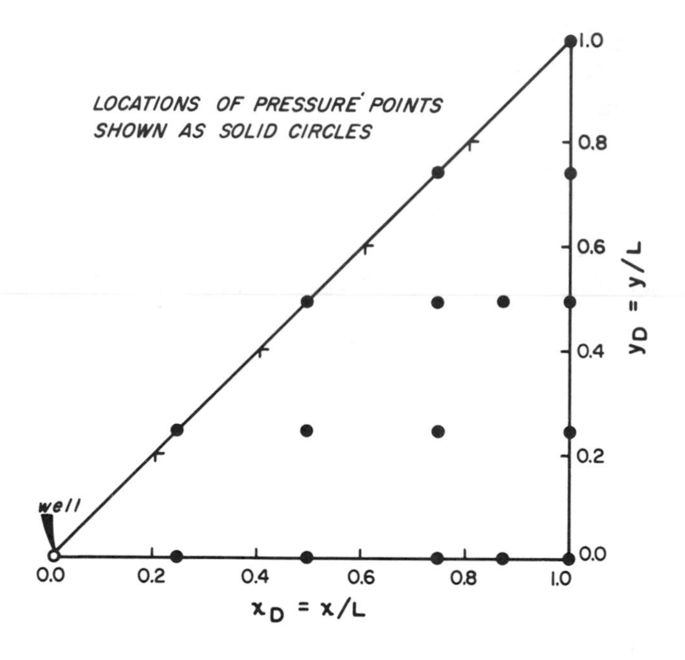

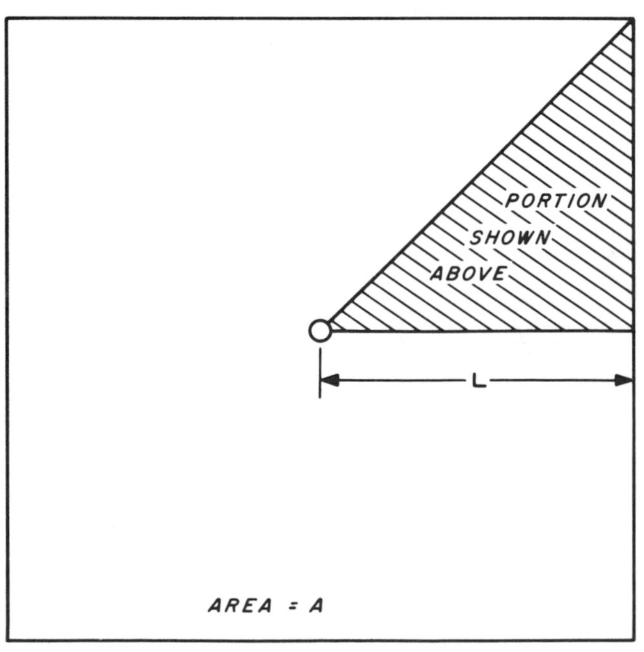

Fig. C.11 Well in the center of a closed-square system, well and pressure point location. After Earlougher, Ramey, Miller, and Mueller.[17]

cally fractured (infinite-conductivity) well in a closed-square drainage area. Wellbore storage is not included. Fig. C.18 may be more useful for type-curve matching than Fig. C.17 under some circumstances. Gringarten, Ramey, and Raghavan[6] give tabular p_D data for this case.

Fig. C.19* gives dimensionless pressure data for the systems of Figs. C.17 and C.18 but for a *uniform-flux* vertical fracture. Tabular data appear in Ref. 6. It is generally believed[6,7] that the uniform-flux fracture solution more closely approximates actual fractured systems than does the infinite-conductivity fracture solution.

*See footnote on Page 24.

A vertically fractured well in a closed system has the same general transient behavior as an unfractured well in a closed system. For the fractured well, the dimensionless pressure during the infinite-acting period is given by Eqs. C.5 and C.7 through C.10, depending on the fracture solution (infinite-conductivity or uniform-flux) and the time. Vertically fractured systems also exhibit pseudosteady-state behavior:

$$p_D = 2\pi t_{DA} + \frac{1}{2}\ln\left[\left(\frac{x_e}{x_f}\right)^2\right] + \frac{1}{2}\ln\left(\frac{2.2458}{C_A}\right) . \qquad \text{(C.24)}$$

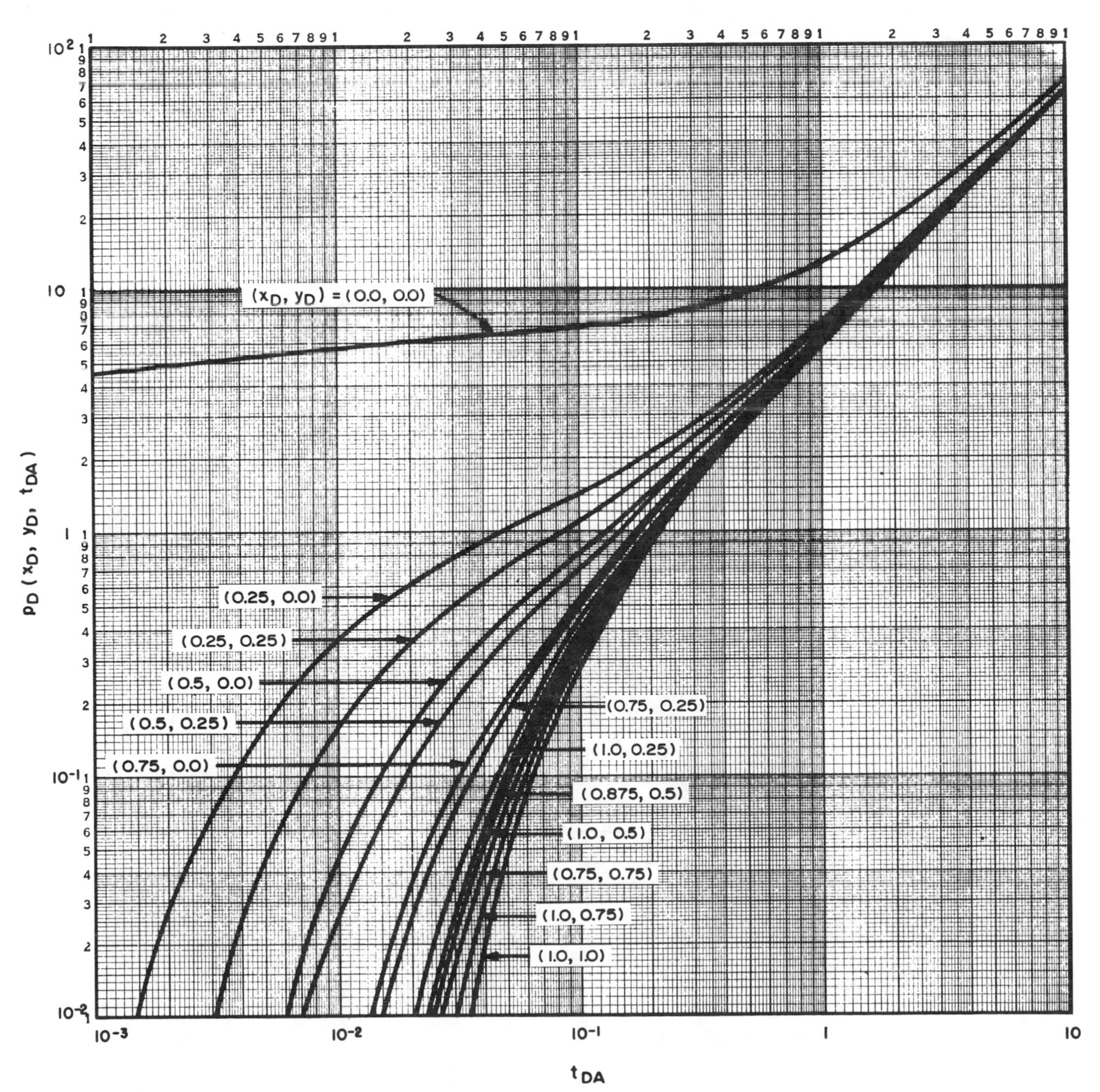

Fig. C.12A Dimensionless pressure at various points in a closed-square system with the well at the center, no wellbore storage, no skin, $\sqrt{A}/r_w = 2{,}000$. Log-log plot. See Fig. C.11 for point locations. After Earlougher, Ramey, Miller, and Mueller.[17]

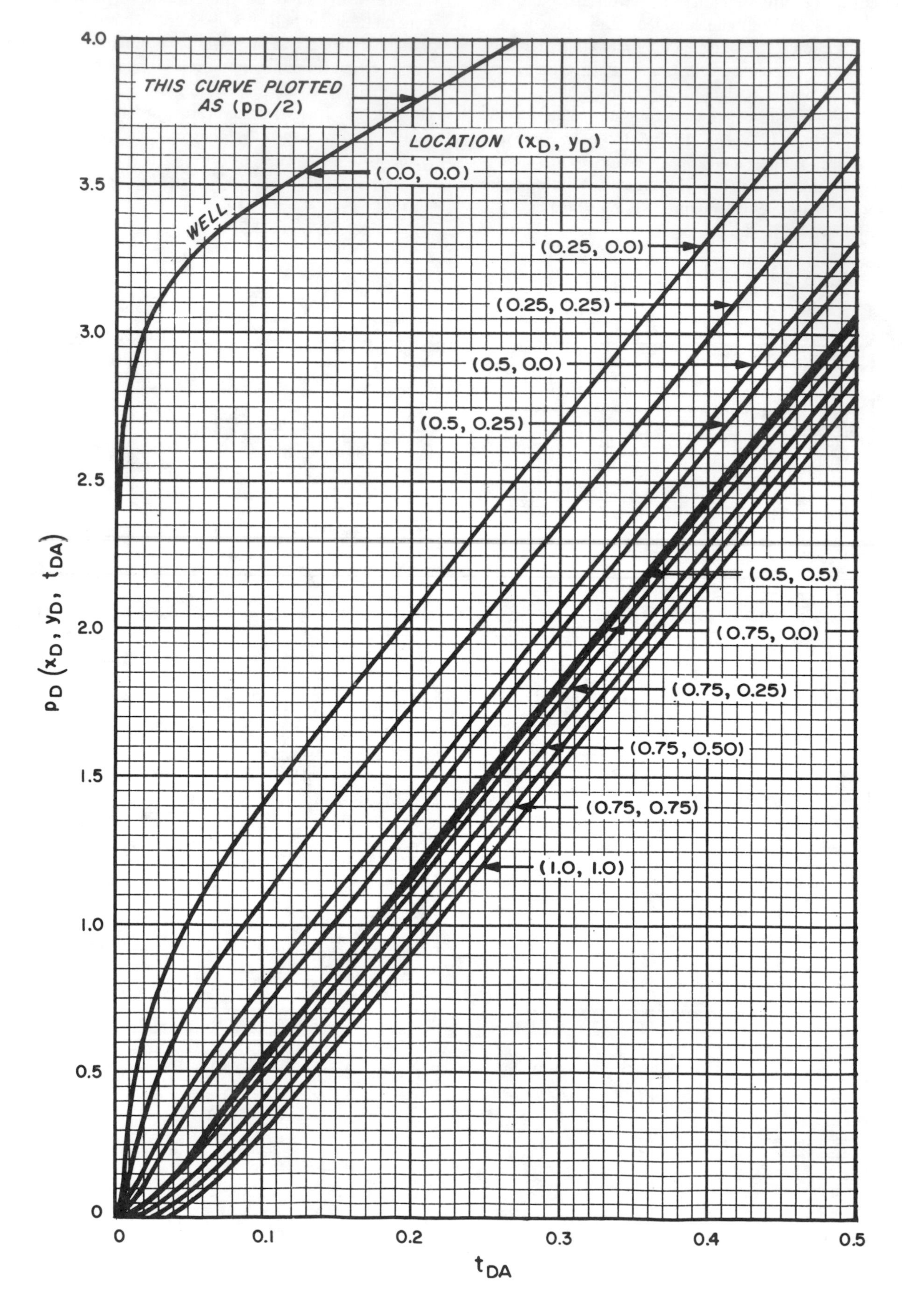

Fig. C.12B Dimensionless pressure at various points in a closed-square system with the well at the center, no wellbore storage, no skin, $\sqrt{A}/r_w = 2{,}000$. Coordinate plot. See Fig. C.11 for point locations. After Earlougher, Ramey, Miller, and Mueller.[17]

TABLE C.2—DIMENSIONLESS PRESSURE AT VARIOUS POINTS IN A CLOSED SQUARE WITH A WELL AT THE CENTER, NO WELLBORE STORAGE, NO SKIN. SEE FIG. C.11 FOR DEFINITION OF POINTS. $\sqrt{A}/r_w$ = 2,000. After Earlougher, Ramey, Miller, and Mueller.[17]

p_D

t_{DA}	XD=0.000 YD=0.000	XD=0.250 YD=0.000	XD=0.250 YD=0.250	XD=0.500 YD=0.000	XD=0.500 YD=0.250	XD=0.500 YD=0.500	XD=0.750 YD=0.000	XD=0.750 YD=0.250
0.0010	4.5516	0.0021	0.0000	0.0000	0.0000	0.0000	0.0000	0.0000
0.0015	4.7543	0.0109	0.0004	0.0000	0.0000	0.0000	0.0000	0.0000
0.0020	4.8981	0.0261	0.0021	0.0000	0.0000	0.0000	0.0000	0.0000
0.0025	5.0097	0.0456	0.0056	0.0001	0.0000	0.0000	0.0000	0.0000
0.0030	5.1009	0.0675	0.0109	0.0004	0.0001	0.0000	0.0000	0.0000
0.0040	5.2447	0.1141	0.0261	0.0021	0.0007	0.0000	0.0000	0.0000
0.0050	5.3563	0.1607	0.0456	0.0056	0.0021	0.0001	0.0001	0.0000
0.0060	5.4474	0.2053	0.0675	0.0109	0.0047	0.0004	0.0002	0.0001
0.0070	5.5245	0.2475	0.0906	0.0178	0.0085	0.0011	0.0006	0.0003
0.0080	5.5913	0.2871	0.1141	0.0261	0.0135	0.0021	0.0012	0.0007
0.0090	5.6502	0.3243	0.1376	0.0354	0.0194	0.0036	0.0021	0.0013
0.0100	5.7029	0.3592	0.1607	0.0456	0.0261	0.0056	0.0034	0.0021
0.0150	5.9056	0.5063	0.2676	0.1023	0.0675	0.0218	0.0154	0.0109
0.0200	6.0494	0.6211	0.3592	0.1607	0.1142	0.0456	0.0350	0.0266
0.0250	6.1610	0.7147	0.4379	0.2164	0.1609	0.0735	0.0597	0.0474
0.0300	6.2522	0.7939	0.5065	0.2685	0.2061	0.1032	0.0876	0.0716
0.0400	6.3965	0.9232	0.6224	0.3628	0.2906	0.1650	0.1485	0.1263
0.0500	6.5099	1.0279	0.7192	0.4470	0.3685	0.2276	0.2125	0.1854
0.0600	6.6050	1.1178	0.8041	0.5242	0.4415	0.2904	0.2772	0.2466
0.0700	6.6888	1.1983	0.8815	0.5968	0.5112	0.3532	0.3418	0.3086
0.0800	6.7654	1.2728	0.9539	0.6661	0.5786	0.4160	0.4061	0.3711
0.0900	6.8374	1.3434	1.0231	0.7334	0.6446	0.4788	0.4700	0.4338
0.1000	6.9063	1.4114	1.0902	0.7992	0.7095	0.5417	0.5336	0.4965
0.1500	7.2311	1.7347	1.4119	1.1186	1.0274	0.8558	0.8492	0.8106
0.2000	7.5468	2.0501	1.7271	1.4335	1.3421	1.1700	1.1636	1.1248
0.2500	7.8611	2.3644	2.0414	1.7478	1.6563	1.4841	1.4778	1.4390
0.3000	8.1753	2.6786	2.3556	2.0620	1.9705	1.7983	1.7919	1.7531
0.4000	8.8036	3.3069	2.9839	2.6903	2.5988	2.4266	2.4202	2.3814
0.5000	9.4320	3.9352	3.6122	3.3186	3.2271	3.0549	3.0486	3.0098
0.6000	10.0603	4.5636	4.2406	3.9469	3.8555	3.6833	3.6769	3.6381
0.7000	10.6886	5.1919	4.8689	4.5752	4.4838	4.3116	4.3052	4.2664
0.8000	11.3169	5.8202	5.4972	5.2036	5.1121	4.9399	4.9335	4.8947
0.9000	11.9452	6.4485	6.1255	5.8319	5.7404	5.5682	5.5618	5.5230
1.0000	12.5735	7.0768	6.7530	6.4602	6.3687	6.1965	6.1902	6.1513
2.0000	18.8567	13.3600	13.0370	12.7433	12.6519	12.4797	12.4733	12.4345
4.0000	31.4230	25.9263	25.6033	25.3097	25.2182	25.0460	25.0397	25.0009
8.0000	56.5557	51.0590	50.7360	50.4423	50.3509	50.1787	50.1723	50.1335
10.0000	69.1220	63.6253	63.3023	63.0087	62.9172	62.7450	62.7386	62.6999
C_A	30.8828	1.837×10^6	3.504×10^6	6.303×10^6	7.570×10^6	10.68×10^6	10.82×10^6	11.69×10^6
$\frac{1}{2} \ln \left(\frac{2.2458}{C_A}\right)$	-1.3106	-6.807	-7.130	-7.424	-7.515	-7.688	-7.694	-7.733

p_D

t_{DA}	XD=0.750 YD=0.500	XD=0.750 YD=0.750	XD=0.875 YD=0.000	XD=0.875 YD=0.500	XD=1.000 YD=0.000	XD=1.000 YD=0.250	XD=1.000 YD=0.500	XD=1.000 YD=0.750	XD=1.000 YD=1.000
0.0010	0.0000	0.0000	0.0000	0.0000	0.0000	0.0000	0.0000	0.0000	0.0000
0.0015	0.0000	0.0000	0.0000	0.0000	0.0000	0.0000	0.0000	0.0000	0.0000
0.0020	0.0000	0.0000	0.0000	0.0000	0.0000	0.0000	0.0000	0.0000	0.0000
0.0025	0.0000	0.0000	0.0000	0.0000	0.0000	0.0000	0.0000	0.0000	0.0000
0.0030	0.0000	0.0000	0.0000	0.0000	0.0000	0.0000	0.0000	0.0000	0.0000
0.0040	0.0000	0.0000	0.0000	0.0000	0.0000	0.0000	0.0000	0.0000	0.0000
0.0050	0.0000	0.0000	0.0000	0.0000	0.0000	0.0000	0.0000	0.0000	0.0000
0.0060	0.0000	0.0000	0.0000	0.0000	0.0000	0.0000	0.0000	0.0000	0.0000
0.0070	0.0000	0.0000	0.0001	0.0000	0.0000	0.0000	0.0000	0.0000	0.0000
0.0080	0.0001	0.0000	0.0002	0.0000	0.0000	0.0000	0.0000	0.0000	0.0000
0.0090	0.0003	0.0000	0.0004	0.0001	0.0001	0.0001	0.0000	0.0000	0.0000
0.0100	0.0005	0.0001	0.0008	0.0001	0.0003	0.0002	0.0000	0.0000	0.0000
0.0150	0.0040	0.0008	0.0055	0.0016	0.0031	0.0023	0.0009	0.0002	0.0001
0.0200	0.0121	0.0036	0.0164	0.0060	0.0111	0.0087	0.0042	0.0014	0.0005
0.0250	0.0245	0.0091	0.0329	0.0143	0.0249	0.0203	0.0112	0.0045	0.0023
0.0300	0.0404	0.0177	0.0539	0.0264	0.0436	0.0365	0.0219	0.0102	0.0062
0.0400	0.0805	0.0437	0.1050	0.0600	0.0913	0.0793	0.0532	0.0307	0.0223
0.0500	0.1281	0.0800	0.1628	0.1030	0.1469	0.1308	0.0947	0.0623	0.0498
0.0600	0.1807	0.1241	0.2237	0.1525	0.2065	0.1871	0.1431	0.1029	0.0872
0.0700	0.2366	0.1740	0.2859	0.2064	0.2678	0.2460	0.1962	0.1502	0.1321
0.0800	0.2948	0.2279	0.3486	0.2632	0.3299	0.3064	0.2525	0.2023	0.1826
0.0900	0.3546	0.2846	0.4114	0.3219	0.3925	0.3677	0.3109	0.2579	0.2369
0.1000	0.4153	0.3433	0.4744	0.3820	0.4551	0.4296	0.3708	0.3157	0.2939
0.1500	0.7257	0.6500	0.7888	0.6913	0.7692	0.7421	0.6797	0.6209	0.5976
0.2000	1.0393	0.9632	1.1030	1.0047	1.0834	1.0561	0.9931	0.9338	0.9103
0.2500	1.3534	1.2772	1.4172	1.3188	1.3975	1.3702	1.3071	1.2478	1.2243
0.3000	1.6676	1.5913	1.7313	1.6330	1.7117	1.6843	1.6213	1.5620	1.5384
0.4000	2.2959	2.2196	2.3597	2.2613	2.3400	2.3127	2.2496	2.1903	2.1667
0.5000	2.9242	2.8479	2.9880	2.8896	2.9683	2.9410	2.8779	2.8186	2.7950
0.6000	3.5525	3.4763	3.6163	3.5179	3.5966	3.5693	3.5062	3.4469	3.4233
0.7000	4.1808	4.1046	4.2446	4.1462	4.2249	4.1976	4.1346	4.0752	4.0517
0.8000	4.8092	4.7329	4.8729	4.7745	4.8533	4.8259	4.7629	4.7036	4.6800
0.9000	5.4375	5.3612	5.5012	5.4029	5.4816	5.4542	5.3912	5.3319	5.3083
1.0000	6.0658	5.9895	6.1296	6.0312	6.1099	6.0826	6.0195	5.9602	5.9366
2.0000	12.3490	12.2727	12.4127	12.3144	12.3930	12.3657	12.3027	12.2434	12.2198
4.0000	24.9153	24.8391	24.9791	24.8807	24.9594	24.9321	24.8690	24.8097	24.7861
8.0000	50.0480	49.9717	50.1117	50.0134	50.0921	50.0647	50.0017	49.9424	49.9188
10.0000	62.6143	62.5381	62.6781	62.5797	62.6584	62.6311	62.5680	62.5087	62.4851
C_A	13.87×10^6	16.16×10^6	12.21×10^6	14.87×10^6	12.70×10^6	13.41×10^6	15.22×10^6	17.14×10^6	18.96×10^6
$\frac{1}{2} \ln \left(\frac{2.2458}{C_A}\right)$	-7.818	-7.894	-7.754	-7.853	-7.774	-7.801	-7.865	-7.924	-7.974

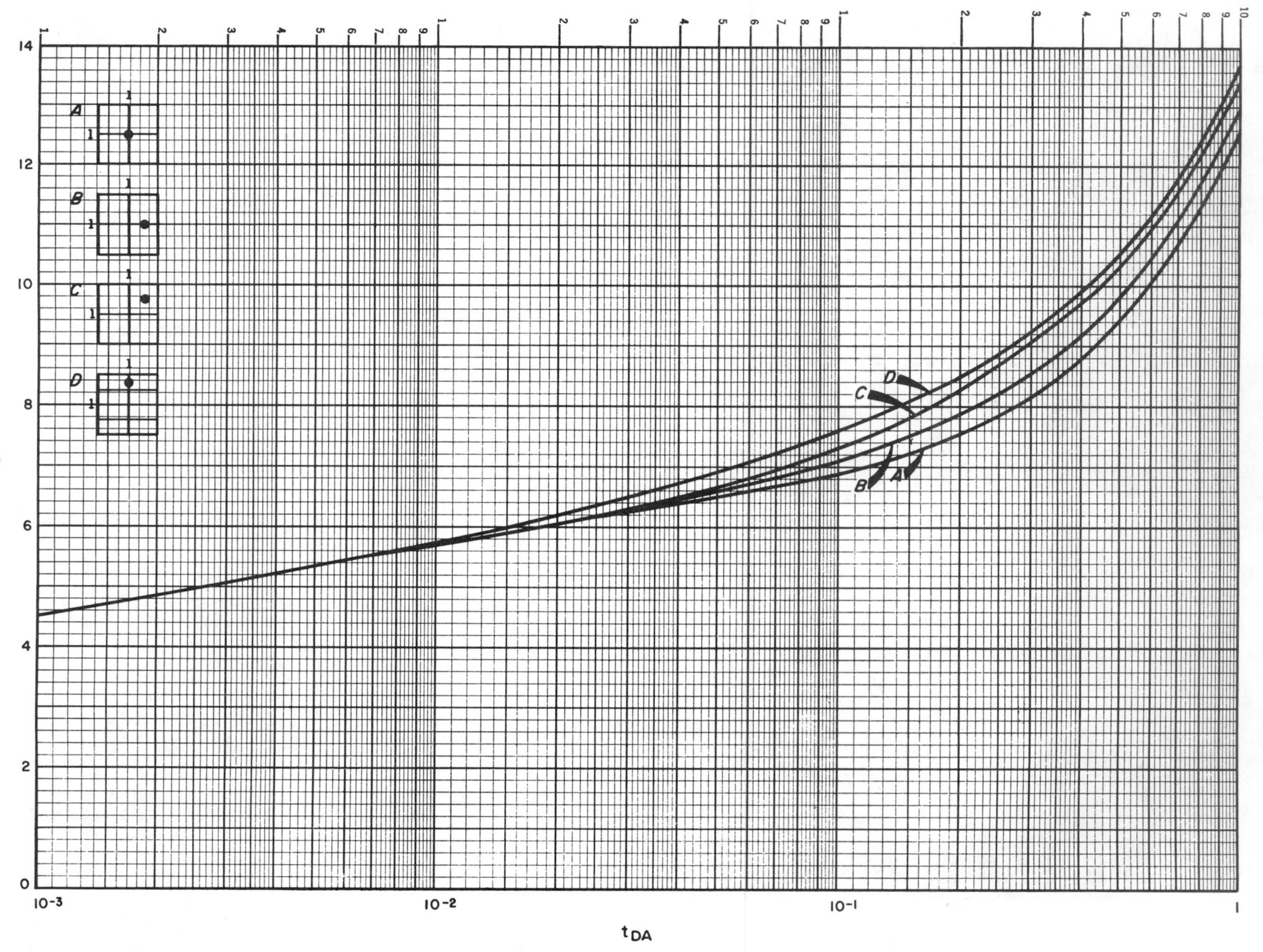

Fig. C.13 Dimensionless pressure for a single well in various closed rectangular systems, no wellbore storage, no skin, $\sqrt{A}/r_w = 2{,}000$. Data of Earlougher and Ramey.[18]

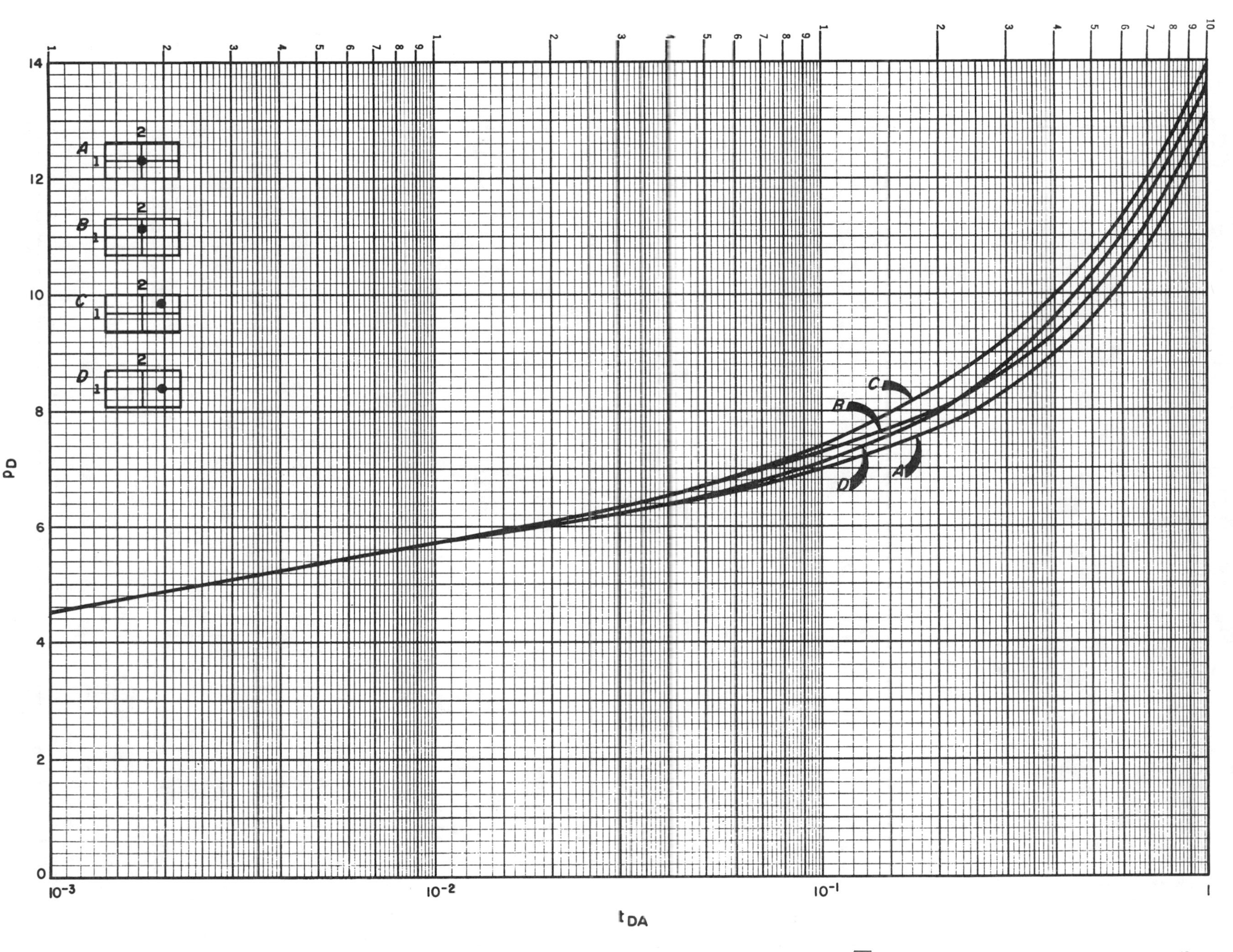

Fig. C.14 Dimensionless pressure for a single well in various closed rectangular systems, no wellbore storage, no skin, $\sqrt{A}/r_w = 2{,}000$. Data of Earlougher and Ramey.[18]

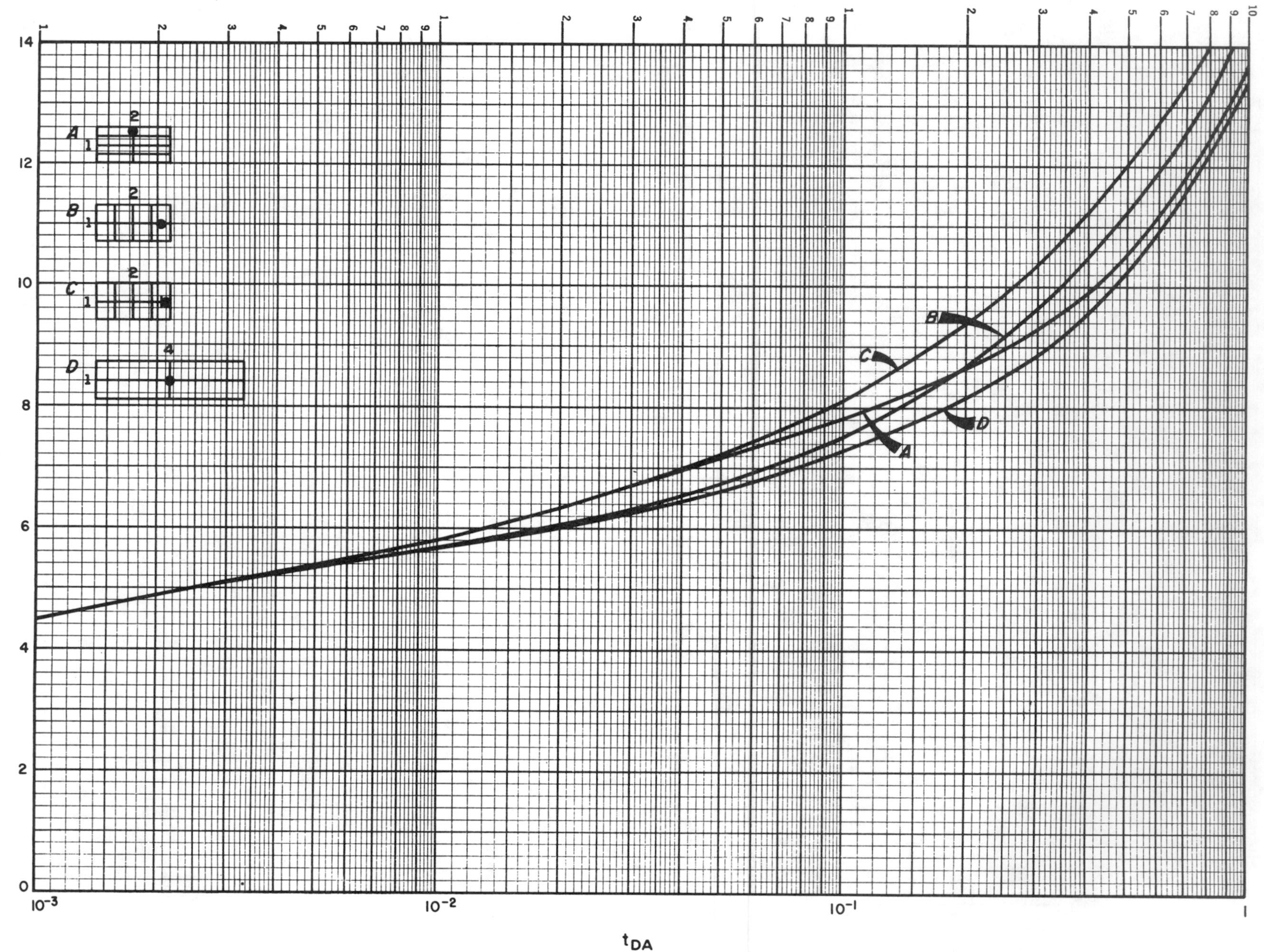

Fig. C.15 Dimensionless pressure for a single well in various closed rectangular systems, no wellbore storage, no skin, $\sqrt{A}/r_w = 2{,}000$. Data of Earlougher and Ramey.[18]

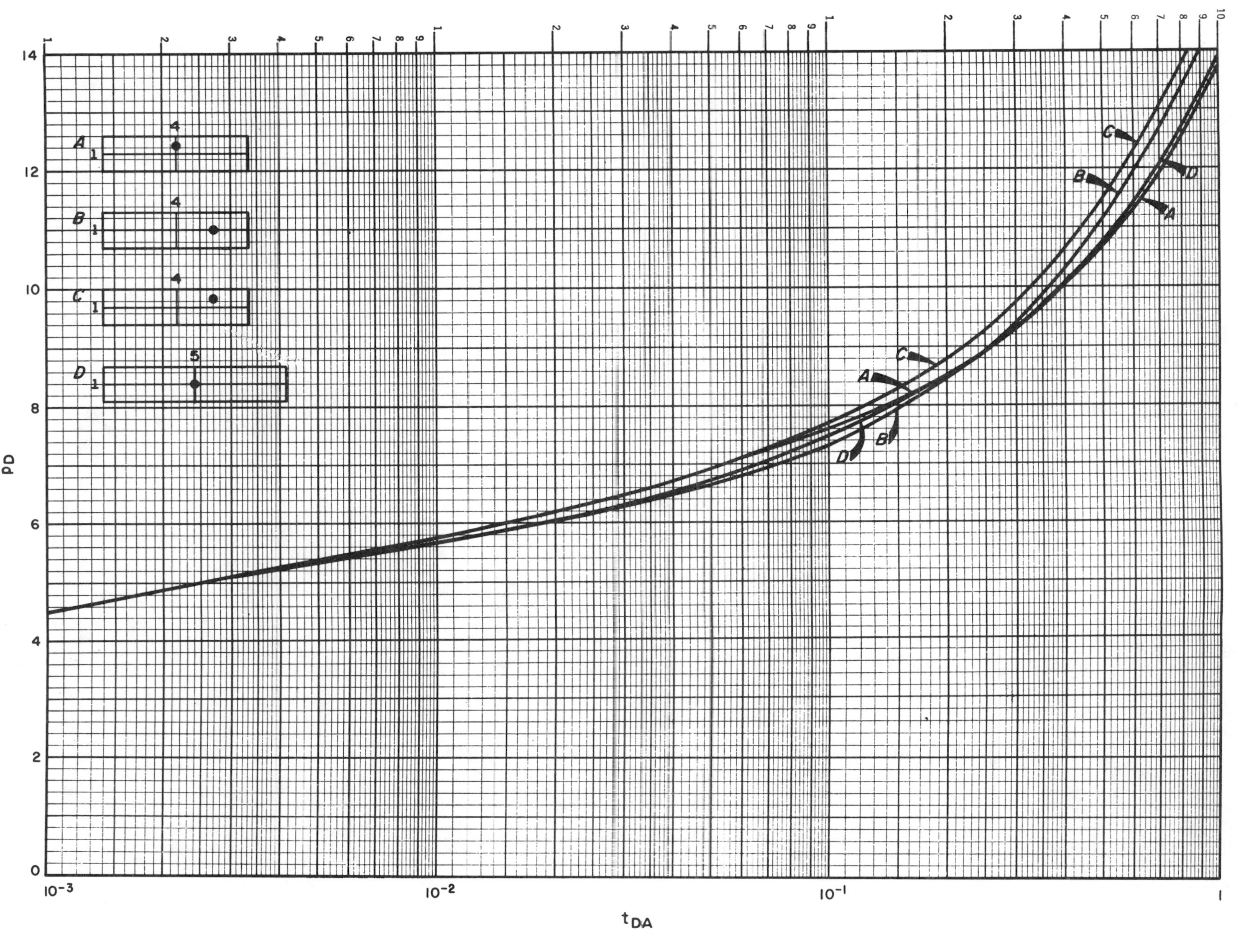

Fig. C.16 Dimensionless pressure for a single well in various closed rectangular systems, no wellbore storage, no skin, $\sqrt{A}/r_w = 2{,}000$. Data of Earlougher and Ramey.[18]

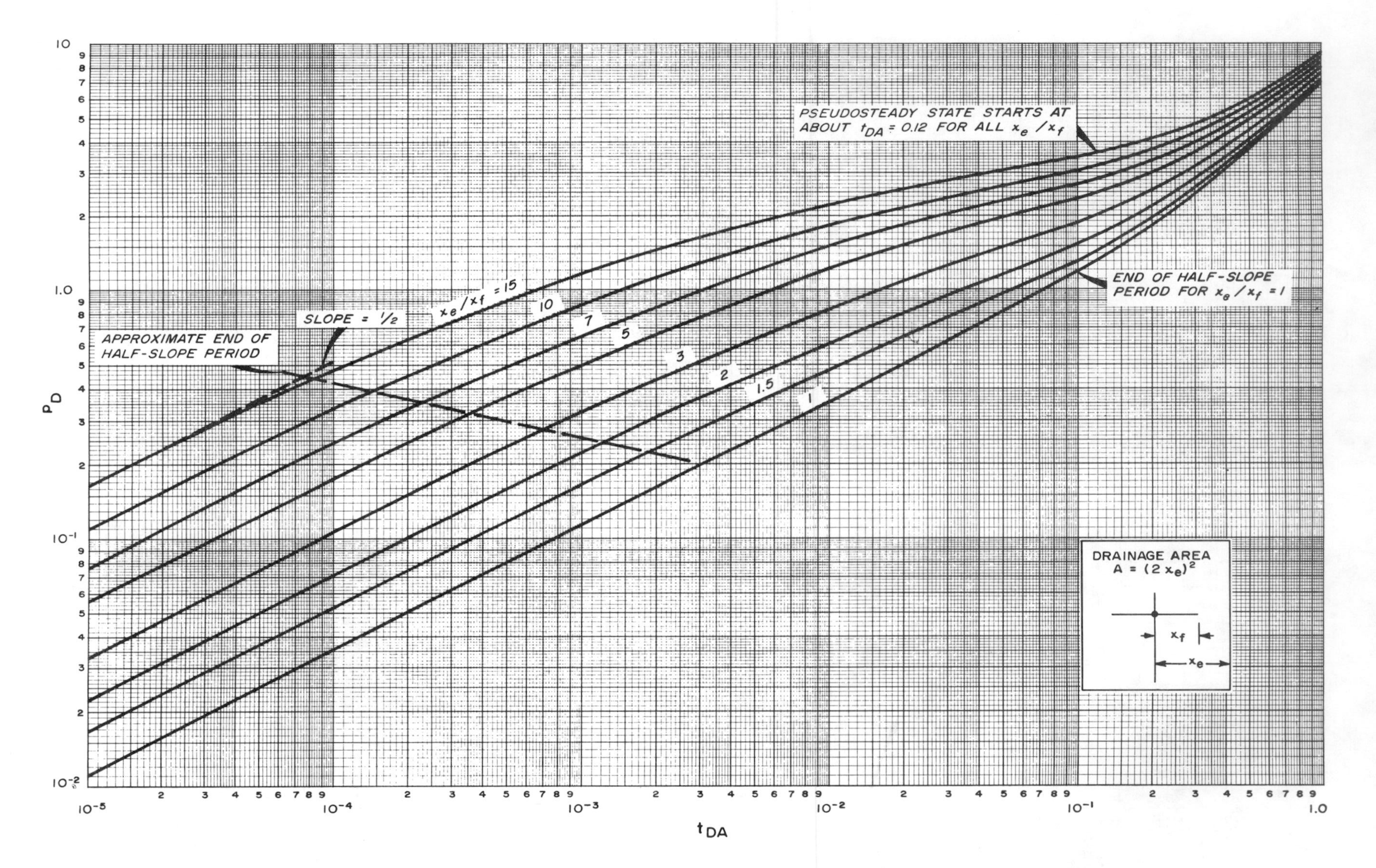

Fig. C.17 Dimensionless pressure for vertically fractured well in the center of a closed system, no wellbore storage, infinite-conductivity fracture. Data of Gringarten, Ramey, and Raghavan.[6,7] Graph courtesy H. J. Ramey, Jr.

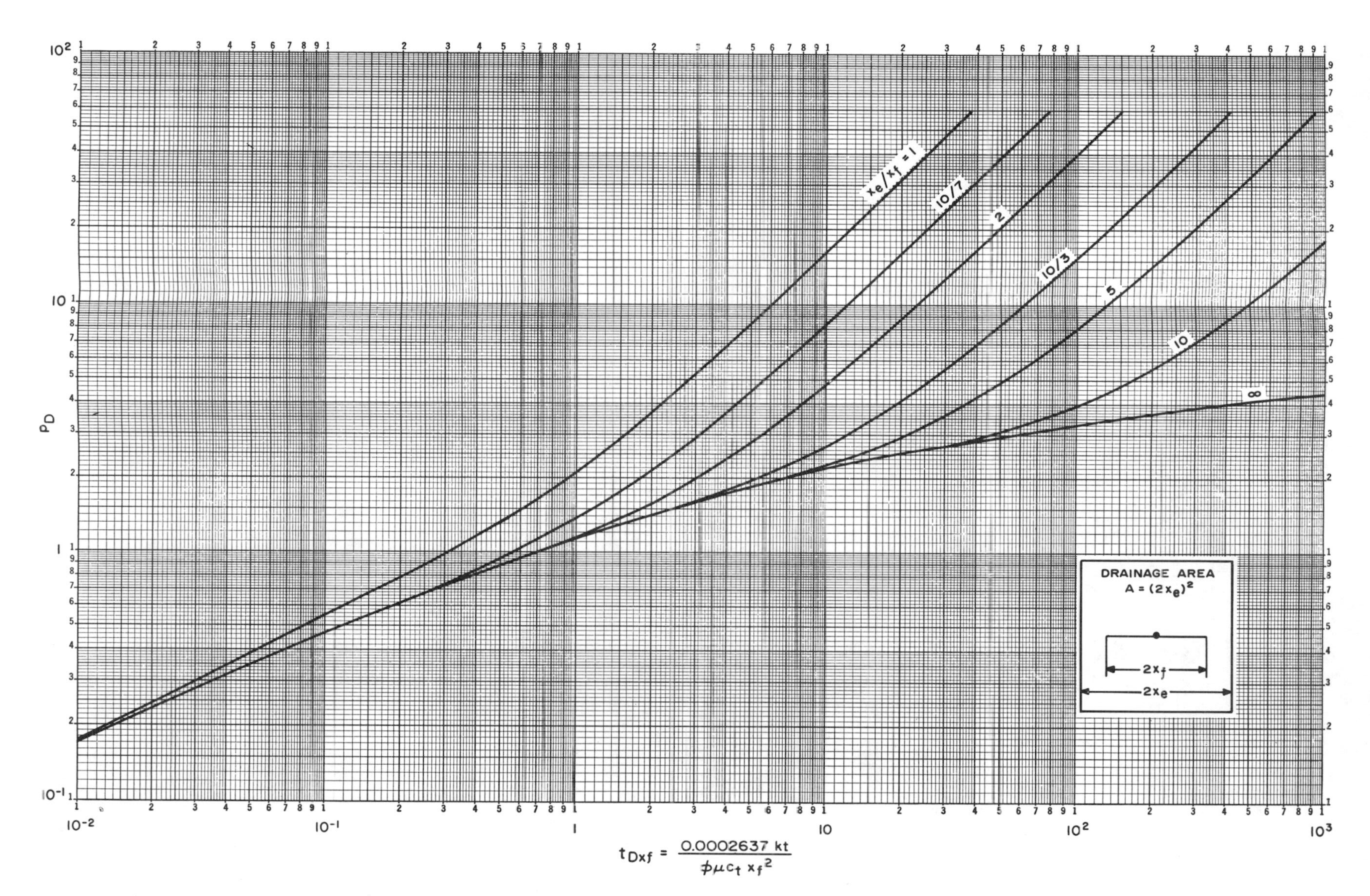

Fig. C.18 Dimensionless pressure for vertically fractured well in the center of a closed square, no wellbore storage, infinite-conductivity fracture. After Gringarten, Ramey, and Raghavan.[6,7]

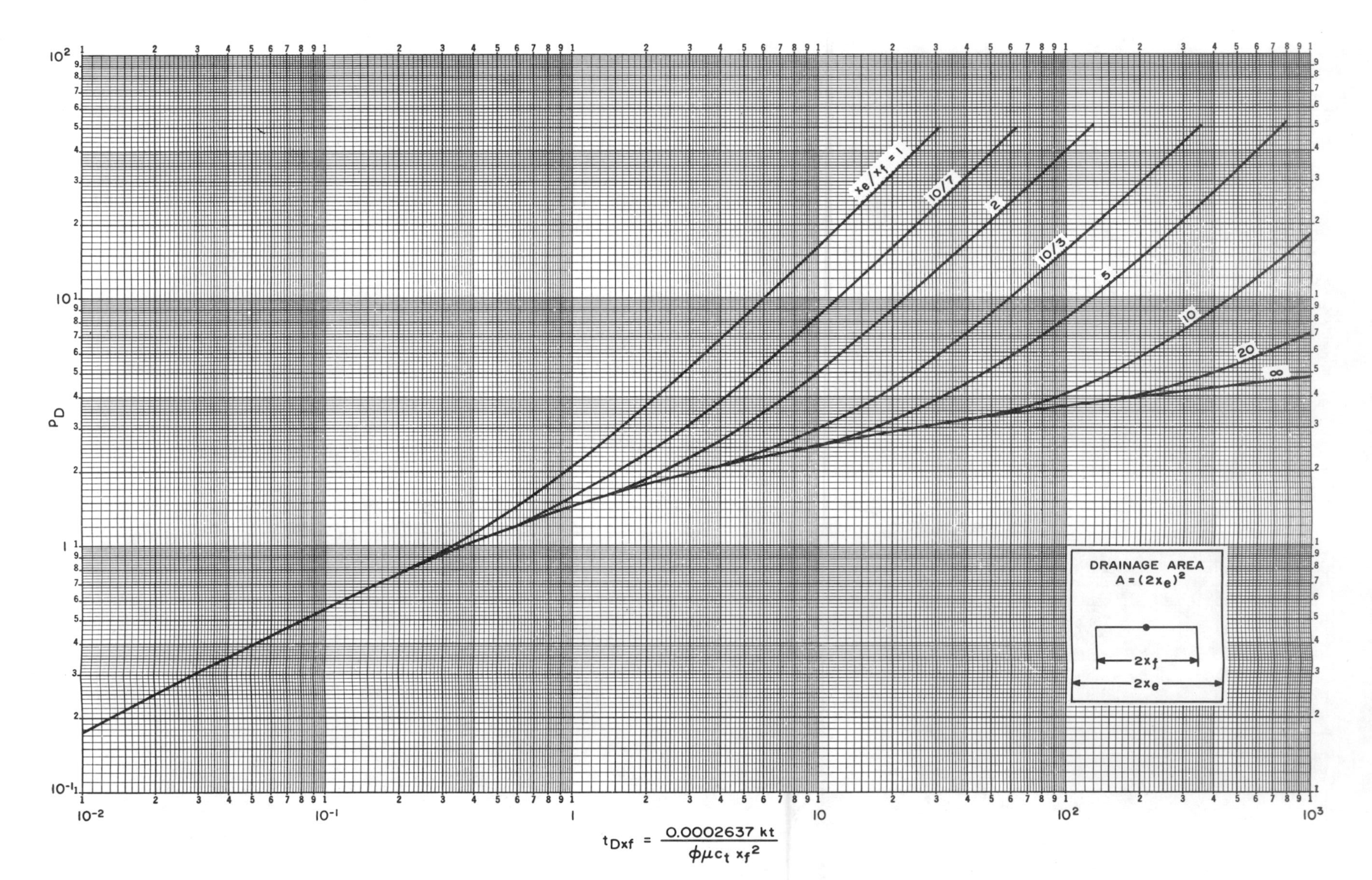

Fig. C.19 Dimensionless pressure for vertically fractured well in the center of a closed square, no wellbore storage, uniform-flux fracture. After Gringarten, Ramey and Raghavan.[6,7]

Eq. C.24 is Eq. C.18 with x_e/x_f substituted for $\sqrt{A}/r_w$. Shape factors for fractured systems based on this dimensional parameter are given in Table C.1.

C.4 Constant-Pressure Systems

Circular Reservoir, No Wellbore Storage, No Skin

Fig. C.20 shows p_D for a single well in the center of a circular reservoir with constant external pressure, no wellbore storage, and no skin. The system reaches true steady state when

$$t_D > 1.25 \left(\frac{r_e^2}{r_w^2}\right), \quad \text{(C.25a)}$$

or

$$t_{DA} > 0.40 \quad \text{(C.25b)}$$

After that time the dimensionless pressure is given by Eq. 2.26a. Tabular data are presented by van Everdingen and Hurst.[5]

Rectangular Reservoirs, No Wellbore Storage, No Skin

Fig. C.21 gives dimensionless pressures for several square and rectangular systems with a single well and one or more boundaries at constant pressure. No wellbore storage or skin effects are included in Fig. C.21. Each system reaches steady state at some time. Ramey, Kumar, and Gulati[22] give additional information about steady state and

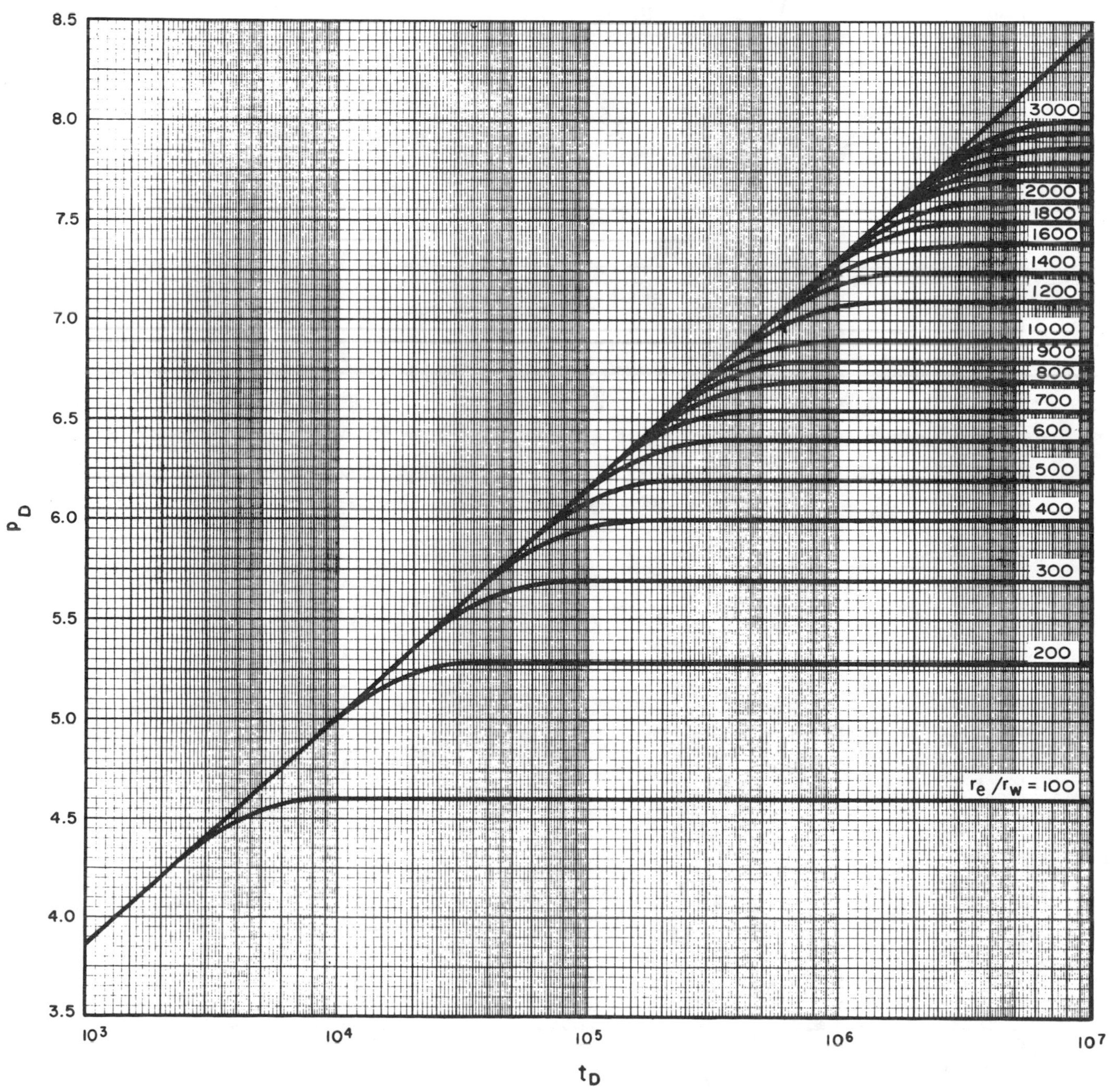

Fig. C.20 Dimensionless pressure for a well in the center of a closed circular reservoir with constant external pressure, no wellbore storage, no skin. After van Everdingen and Hurst.[5]

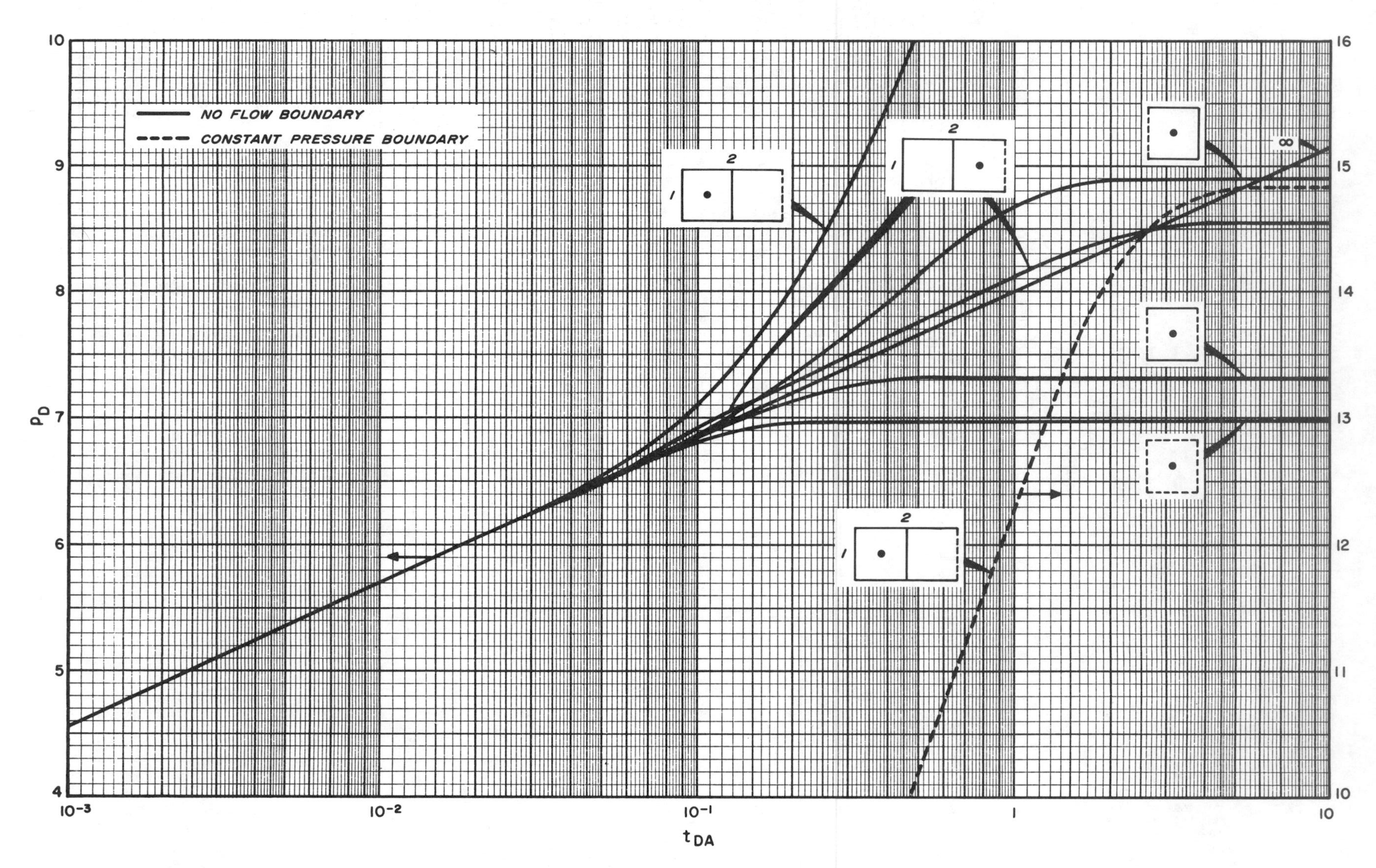

Fig. C.21 Dimensionless pressure for single wells in various rectangular shapes with one or more constant-pressure boundaries, no wellbore storage, no skin. Data of Ramey, Kumar, and Gulati.[22]

the time required to reach it. They also tabulate the p_D data of Fig. C.21.

In Fig. C.21, some systems show closed-system effects (even to the point of exhibiting pseudosteady-state behavior) before the effect of the constant-pressure boundary is felt. That is particularly clear for the 2:1 rectangle with the well located three-fourths of the length from the constant-pressure boundary.

References

1. Matthews, C. S. and Russell, D. G.: *Pressure Buildup and Flow Tests in Wells,* Monograph Series, Society of Petroleum Engineers of AIME, Dallas (1967) **1,** Chap. 2.
2. Theis, Charles V.: "The Relation Between the Lowering of the Piezometric Surface and the Rate and Duration of Discharge of a Well Using Ground-Water Storage," *Trans.,* AGU (1935) 519-524.
3. Horner, D. R.: "Pressure Build-Up in Wells," *Proc.,* Third World Pet. Cong., The Hague (1951) Sec. II, 503-523. Also *Reprint Series, No. 9 — Pressure Analysis Methods,* Society of Petroleum Engineers of AIME, Dallas (1967) 25-43.
4. Mueller, Thomas D. and Witherspoon, Paul A.: "Pressure Interference Effects Within Reservoirs and Aquifers," *J. Pet. Tech.* (April 1965) 471-474; *Trans.,* AIME, **234.**
5. van Everdingen, A. F. and Hurst, W.: "The Application of the Laplace Transformation to Flow Problems in Reservoirs," *Trans.,* AIME (1949) **186,** 305-324.
6. Gringarten, Alain C., Ramey, Henry J., Jr., and Raghavan, R.: "Pressure Analysis for Fractured Wells," paper SPE 4051 presented at the SPE-AIME 47th Annual Fall Meeting, San Antonio, Tex., Oct. 8-11, 1972.
7. Gringarten, Alain C., Ramey, Henry J., Jr., and Raghavan, R.: "Unsteady-State Pressure Distributions Created by a Well With a Single Infinite-Conductivity Vertical Fracture," *Soc. Pet. Eng. J.* (Aug. 1974) 347-360; *Trans.,* AIME, **257.**
8. Gringarten, Alain C.: "Unsteady-State Pressure Distributions Created by a Well With a Single Horizontal Fracture, Partial Penetration, or Restricted Flow Entry," PhD dissertation, Stanford U., Stanford, Calif. (1971) 106. (Order No. 71-23,512, University Microfilms, P.O. Box 1764, Ann Arbor, Mich. 48106.)
9. Gringarten, Alain C. and Ramey, Henry J., Jr.: "Unsteady-State Pressure Distributions Created by a Well With a Single Horizontal Fracture, Partial Penetration, or Restricted Entry," *Soc. Pet. Eng. J.* (Aug. 1974) 413-426; *Trans.,* AIME, **257.**
10. Agarwal, Ram G., Al-Hussainy, Rafi, and Ramey, H. J., Jr.: "An Investigation of Wellbore Storage and Skin Effect in Unsteady Liquid Flow: I. Analytical Treatment," *Soc. Pet. Eng. J.* (Sept. 1970) 279-290; *Trans.,* AIME, **249.**
11. Wattenbarger, Robert A. and Ramey, H. J., Jr.: "An Investigation of Wellbore Storage and Skin Effect in Unsteady Liquid Flow: II. Finite Difference Treatment," *Soc. Pet. Eng. J.* (Sept. 1970) 291-297; *Trans.,* AIME, **249.**
12. Earlougher, Robert C., Jr., and Kersch, Keith M.: "Analysis of Short-Time Transient Test Data by Type-Curve Matching," *J. Pet. Tech.* (July 1974) 793-800; *Trans.,* AIME, **257.**
13. McKinley, R. M.: "Wellbore Transmissibility From Afterflow-Dominated Pressure Buildup Data," *J. Pet. Tech.* (July 1971) 863-872; *Trans.,* AIME, **251.**
14. Ramey, H. J., Jr., and Cobb, William M.: "A General Buildup Theory for a Well in a Closed Drainage Area," *J. Pet. Tech.* (Dec. 1971) 1493-1505; *Trans.,* AIME, **251.**
15. Brons, F. and Miller, W. C.: "A Simple Method for Correcting Spot Pressure Readings," *J. Pet. Tech.* (Aug. 1961) 803-805; *Trans.,* AIME, **222.**
16. Dietz, D. N.: "Determination of Average Reservoir Pressure From Build-Up Surveys," *J. Pet. Tech.* (Aug. 1965) 955-959; *Trans.,* AIME, **234.**
17. Earlougher, Robert C., Jr., Ramey, H. J., Jr., Miller, F. G., and Mueller, T. D.: "Pressure Distributions in Rectangular Reservoirs," *J. Pet. Tech.* (Feb. 1968) 199-208; *Trans.,* AIME, **243.**
18. Earlougher, R. C., Jr., and Ramey, H. J., Jr.: "Interference Analysis in Bounded Systems," *J. Cdn. Pet. Tech.* (Oct.-Dec. 1973) 33-45.
19. Matthews, C. S., Brons, F., and Hazebroek, P.: "A Method for Determination of Average Pressure in a Bounded Reservoir," *Trans.,* AIME (1954) **201,** 182-191. Also *Reprint Series, No. 9 — Pressure Analysis Methods,* Society of Petroleum Engineers of AIME, Dallas (1967) 51-60.
20. Russell, D. G. and Truitt, N. E.: "Transient Pressure Behavior in Vertically Fractured Reservoirs," *J. Pet. Tech.* (Oct. 1964) 1159-1170; *Trans.,* AIME, **231.** Also *Reprint Series, No. 9 — Pressure Analysis Methods,* Society of Petroleum Engineers of AIME, Dallas (1967) 149-160.
21. Raghavan, R., Cady, Gilbert V., and Ramey, Henry J., Jr.: "Well-Test Analysis for Vertically Fractured Wells," *J. Pet. Tech.* (Aug. 1972) 1014-1020; *Trans.,* AIME, **253.**
22. Ramey, Henry J., Jr., Kumar, Anil, and Gulati, Mohinder S.: *Gas Well Test Analysis Under Water-Drive Conditions,* AGA, Arlington, Va. (1973).

Appendix D

Rock and Fluid Property Correlations

D.1 Introduction

This appendix provides information useful for estimating fluid and rock properties needed when analyzing transient-test pressure data. We believe that the correlations presented are among the most reliable presently available, although they are only a small sample of those available. The correlations may be used when necessary, but laboratory data measured on representative samples taken from the reservoir are always superior to general correlations and should be used whenever possible.

D.2 PVT Properties

This section presents correlations of pressure-volume-temperature (PVT) relations for reservoir fluids. The information can be used when laboratory data are not available. However, to ensure the best possible reservoir engineering and transient-test results, laboratory data should be obtained and used. It is both poor economics and poor engineering to resist obtaining good laboratory data simply because correlations are available.

Table D.1 gives physical properties of methane through decane and some other compounds commonly associated with petroleum reservoirs. More complete data are given in Ref. 1. Such information can be used to estimate some of the properties of hydrocarbon mixtures.

The pseudocritical temperature, T_{pc}, and pressure, p_{pc}, of a mixture are used in many correlations and equations in this appendix. If mixture composition is known, those quantities may be estimated from

$$T_{pc} = \sum_{i=1}^{N} y_i T_{ci}, \qquad \text{(D.1)}$$

and

$$p_{pc} = \sum_{i=1}^{N} y_i p_{ci}, \qquad \text{(D.2)}$$

where

N = number of components in the mixture
y_i = mole fraction of Component i
T_{ci} = critical temperature of Component i, °R
p_{ci} = critical pressure of Component i, psia.

If the system composition is not known, Figs. D.1 through D.3 may be used to estimate T_{pc} and p_{pc}. Fig. D.1 provides a way to estimate those quantities for undersaturated oil at reservoir pressure; the oil specific gravity corrected to 60 °F (the value normally reported) is used. If the API gravity is reported at other than 60 °F, it may be corrected to 60 °F

TABLE D.1—PHYSICAL PROPERTIES OF HYDROCARBONS AND ASSOCIATED COMPOUNDS.

Constituent	Molecular Weight	Normal Boiling Point °F	Normal Boiling Point °R	Liquid Density (lb_m/cu ft)	Gas Density at 60 °F, 1 atm (lb_m/cu ft)	Critical Temperature (°R)	Critical Pressure (psia)
Methane, CH_4	16.04	−258.7	201	18.72*	0.04235	344	673
Ethane, C_2H_6	30.07	−127.5	332	23.34*	0.07986	550	712
Propane, C_3H_8	44.09	−43.8	416	31.68**	0.1180	666	617
iso-butane, C_4H_{10}	58.12	10.9	471	35.14**	0.1577	735	528
n-butane, C_4H_{10}	58.12	31.1	491	36.47**	0.1581	766	551
iso-pentane, C_5H_{12}	72.15	82.1	542	38.99	—	830	483
n-pentane, C_5H_{12}	72.15	96.9	557	39.39	—	847	485
n-hexane, C_6H_{14}	86.17	155.7	615	41.43	—	914	435
n-heptane, C_7H_{16}	100.20	209.2	669	42.94	—	972	397
n-octane, C_8H_{18}	114.22	258.1	718	44.10	—	1,025	362
n-nonane, C_9H_{20}	128.25	303.3	763	45.03	—	1,073	335
n-decane, $C_{10}H_{22}$	142.28	345.2	805	45.81	—	1,115	313
Nitrogen, N_2	28.02	−320.4	140	—	0.0739	227	492
Air (O_2 + N_2)	29	−317.7	142	—	0.0764	239	547
Carbon dioxide, CO_2	44.01	−109.3	351	68.70	0.117	548	1,073
Hydrogen sulfide, H_2S	34.08	−76.5	383	87.73	0.0904	673	1,306
Water	18.02	212	672	62.40	—	1,365	3,206

*Apparent density in liquid phase.
**Density at saturation pressure.

using the technique described in Ref. 4. (In Ref. 4, Table 5 is used for hydrometer measurements at other than 60 °F; Table 7 allows correction of volume at a given temperature to volume at 60 °F.) Fig. D.2 applies to bubble-point liquids, again using the specific gravity corrected to 60 °F. The bubble-point pressure at 60 °F should be determined in the laboratory.* Fig. D.3 applies to condensate well fluids and natural gases; knowledge of the gas gravity is required to use Fig. D.3.

T_{pc} and p_{pc} are normally used to estimate the pseudoreduced temperature and pressure:

$$T_{pr} = \frac{T}{T_{pc}}, \quad \text{(D.3)}$$

and

$$p_{pr} = \frac{p}{p_{pc}}, \quad \text{(D.4)}$$

where

T = temperature of interest, °R
p = pressure of interest, psia.

Note in Eqs. D.1 through D.4 that temperature must be absolute temperature and pressure must be absolute pressure.

Since many correlations in this appendix use specific gravity or API gravity, it is worthwhile to restate the relationship between those two quantities:

$$°\text{API} = \frac{141.5}{\gamma} - 131.5. \quad \text{(D.5)}$$

In Eq. D.5, the specific gravity, γ, must be corrected to 60 °F and atmospheric pressure.

Figs. D.4 through D.6 are Standing's[5] correlations for properties of mixtures of hydrocarbon gases and liquids. Examples of their use are shown in the figures. Standing's correlations are based mainly on the properties of California crude oils. Cronquist[6] gives correlations that may be useful for Gulf Coast oils.

Fig. D.7 is the well known chart of real gas deviation factor for natural gases. Pseudoreduced properties may be estimated from Eqs. D.1 through D.4.

*If only the value at reservoir temperature is known, Fig. D.5 may be used to estimate the 60 °F value by going vertically upward from the bubble-point pressure to reservoir temperature, horizontally left to 60 °F, and vertically downward to the estimated bubble-point pressure.

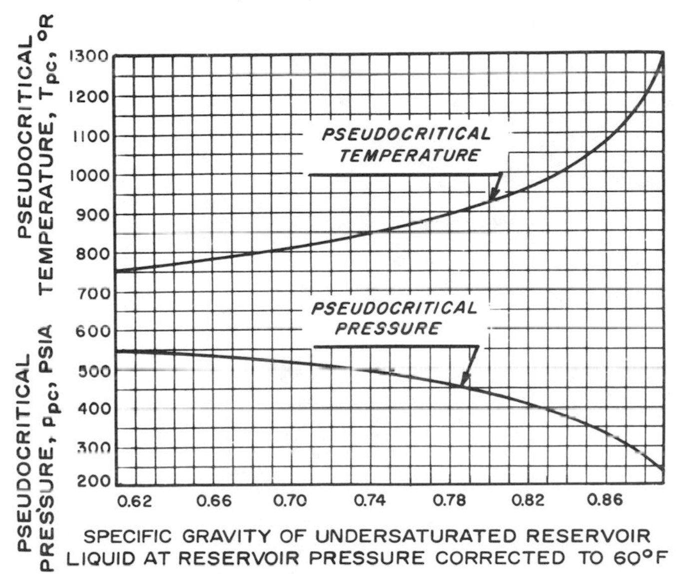

Fig. D.1 Approximate correlation of liquid pseudocritical pressure and temperature with specific gravity. After Trube.[2]

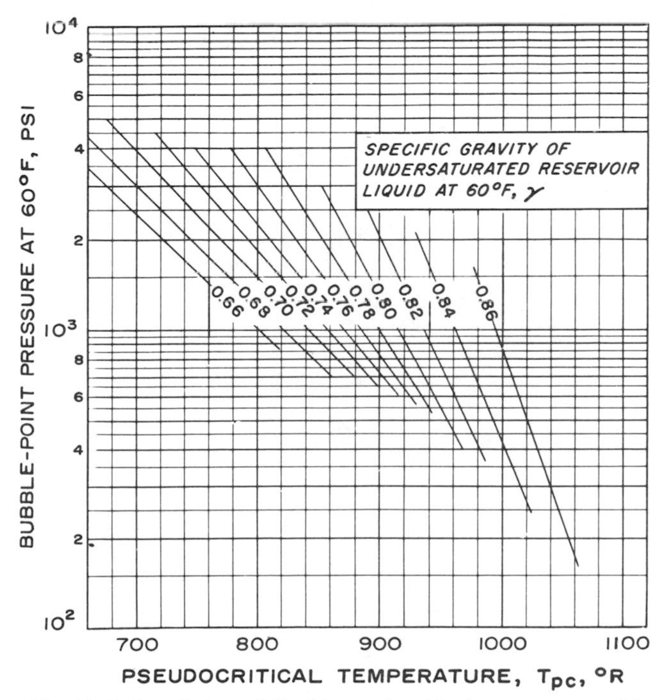

Fig. D.2 Correlation of liquid pseudocritical temperature with specific gravity and bubble point. After Trube.[2]

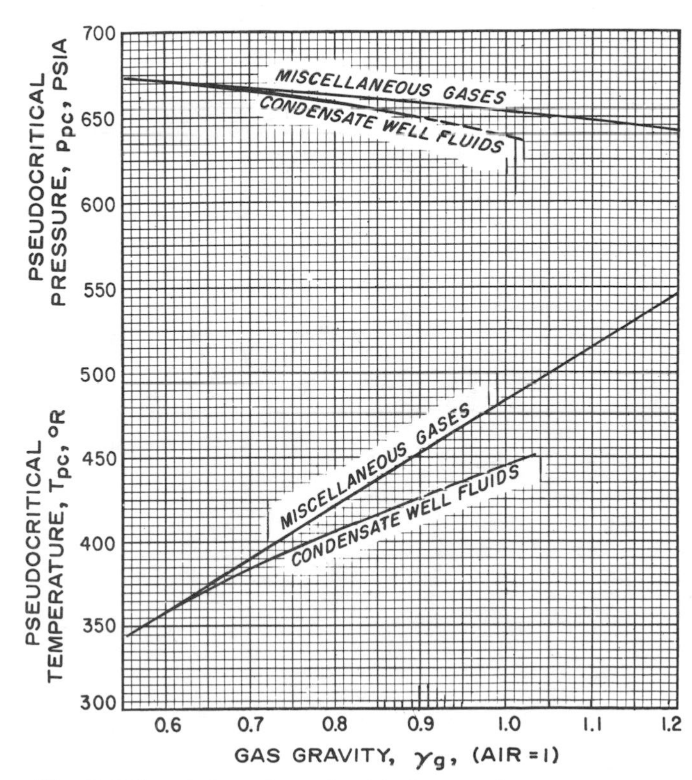

Fig. D.3 Correlation of pseudocritical properties of condensate well fluids and miscellaneous natural gas with fluid gravity. After Brown *et al.*[3]

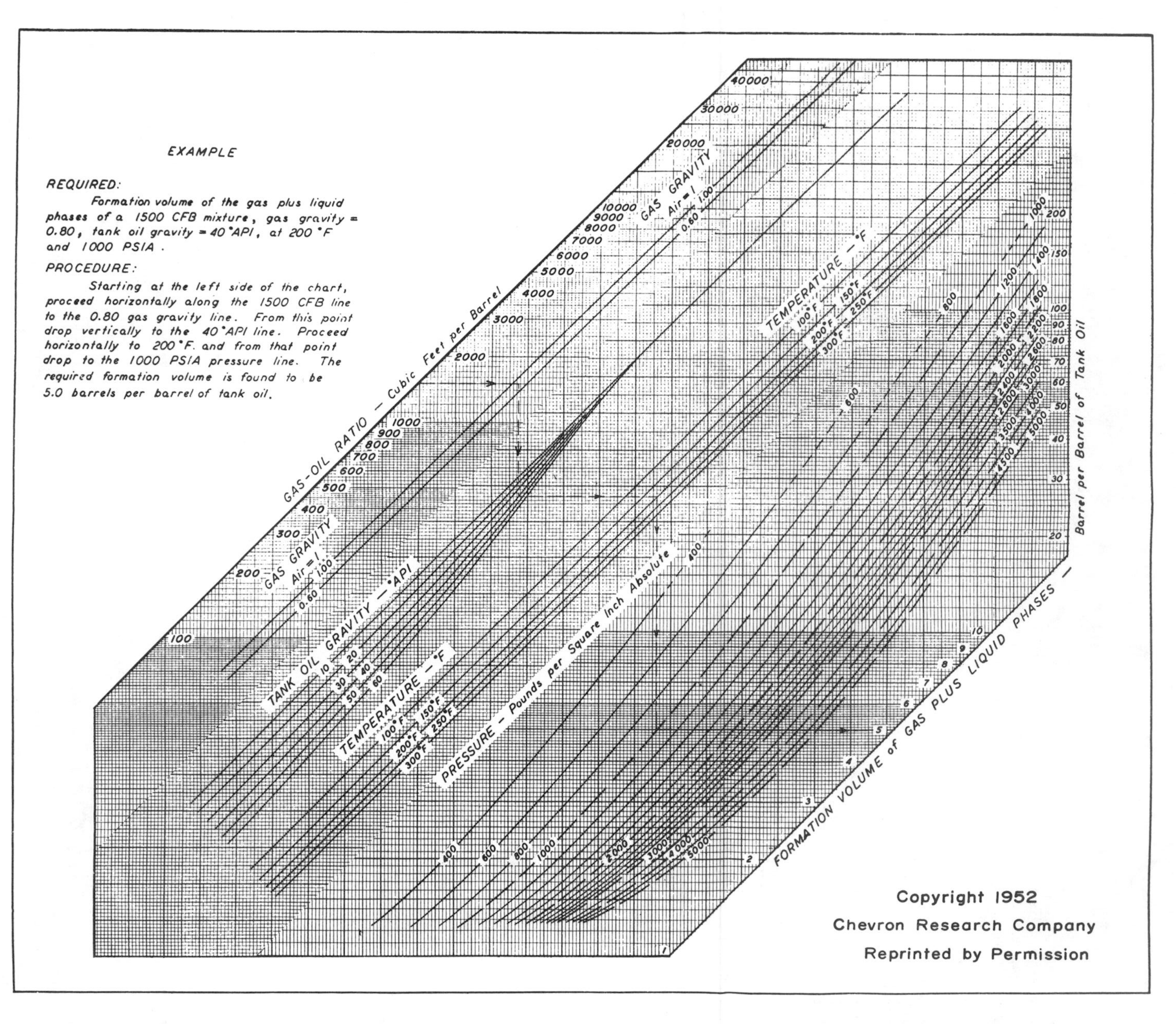

Fig. D.4 Properties of natural mixtures of hydrocarbon gas and liquids, formation volume of gas plus liquid phase. After Standing.[5]

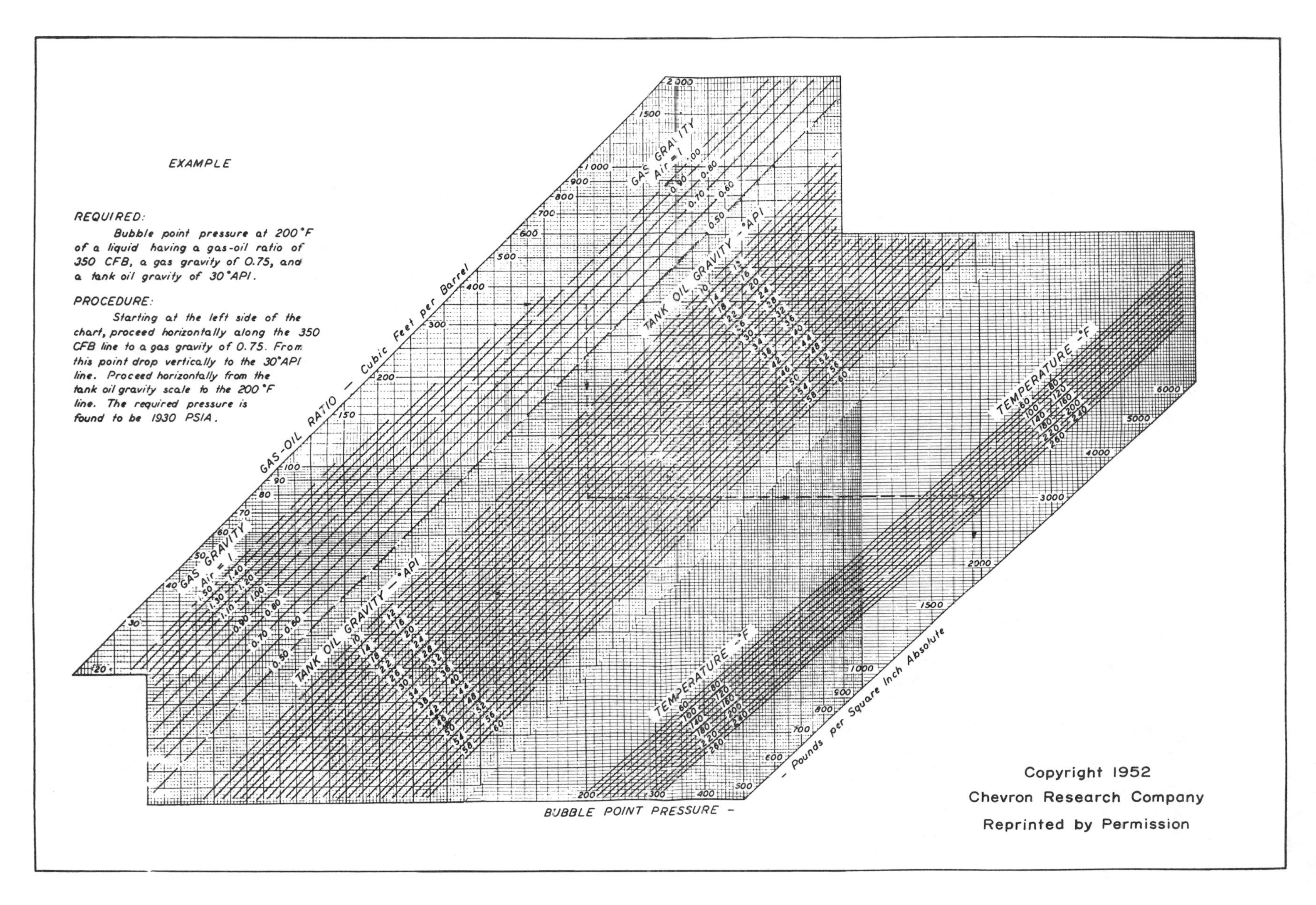

Fig. D.5 Properties of natural mixtures of hydrocarbon gas and liquids, bubble-point pressure. After Standing.[5]

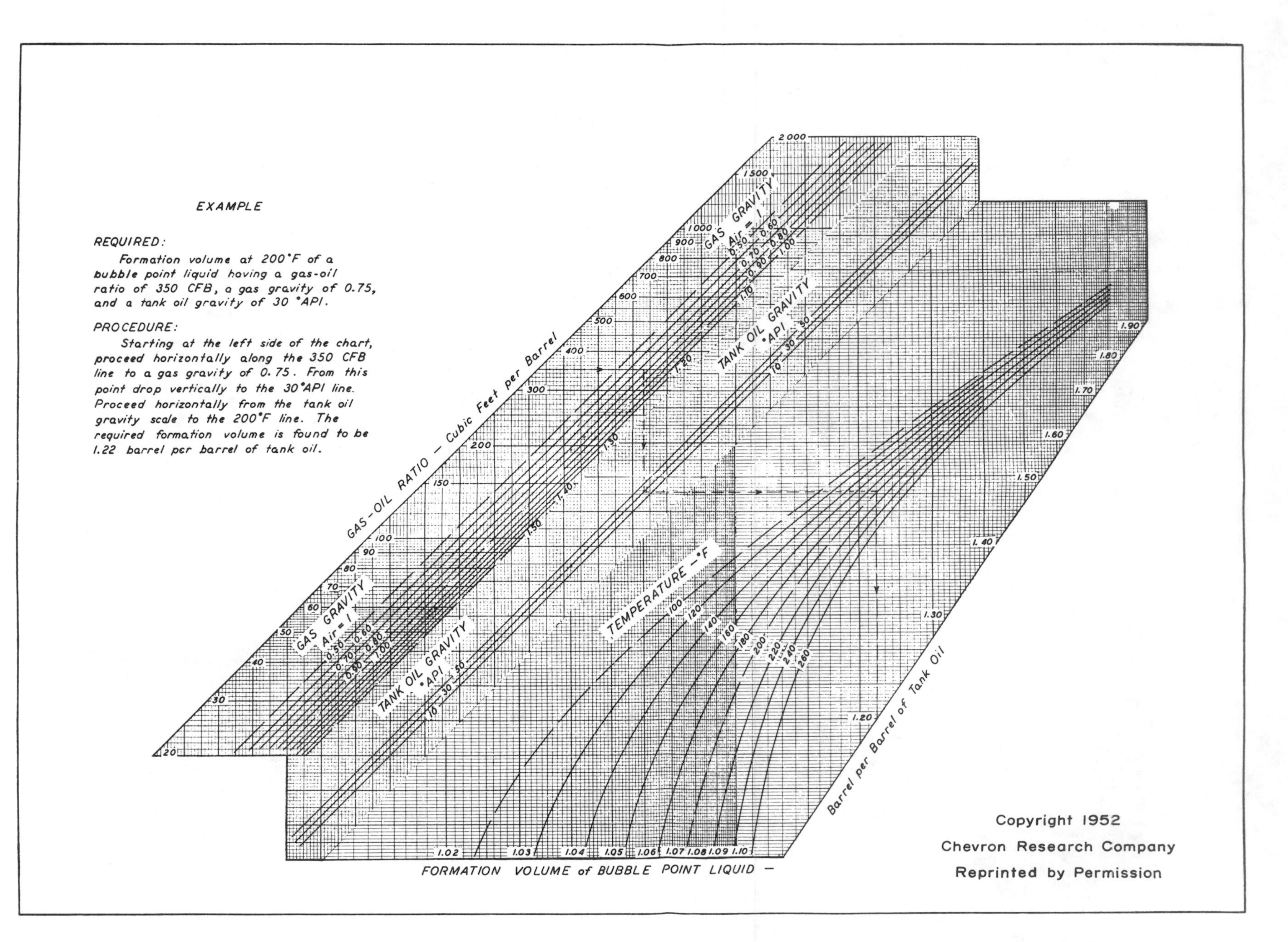

Fig. D.6 Properties of natural mixtures of hydrocarbon gas and liquids, formation volume of bubble-point liquids. After Standing.[5]

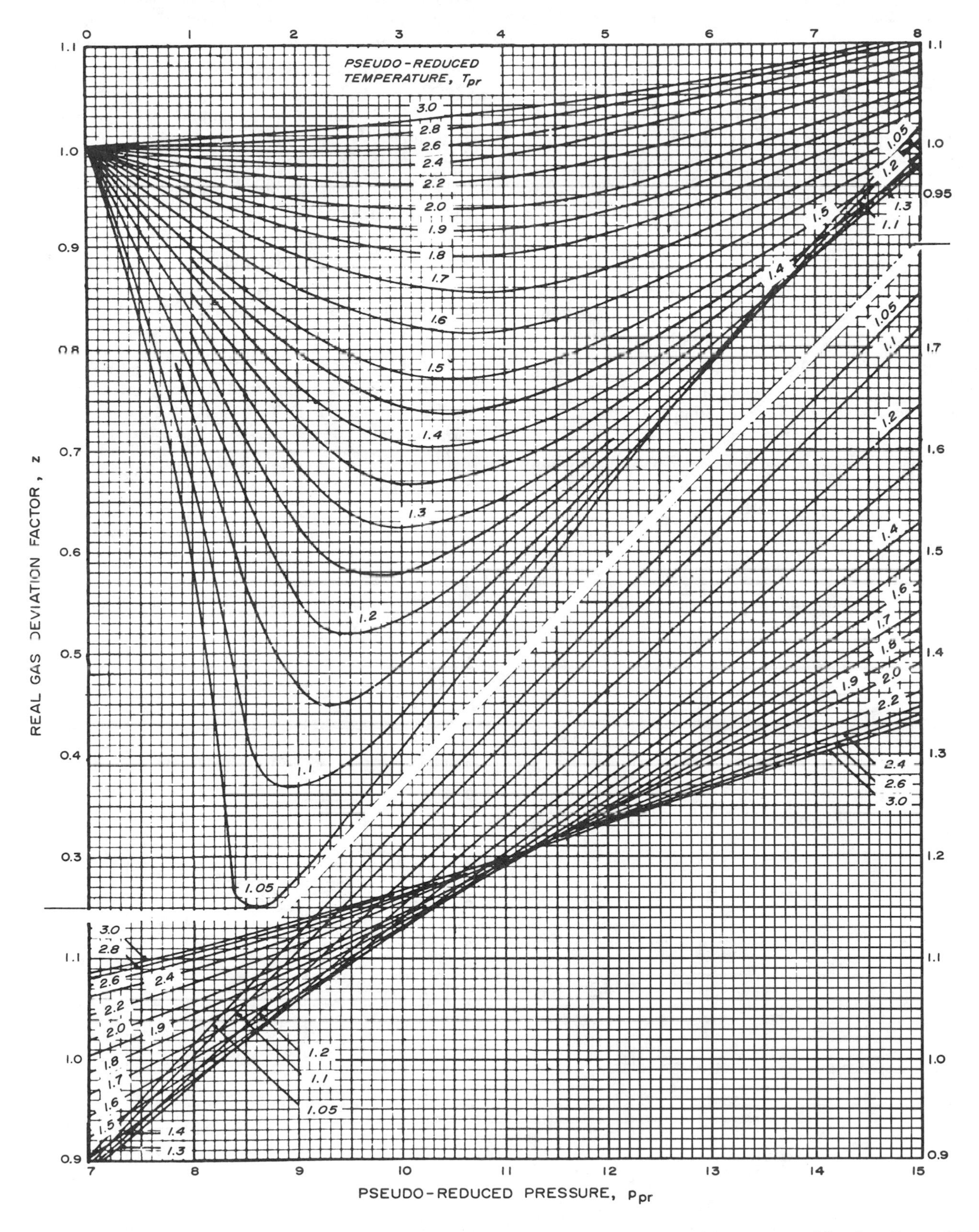

Fig. D.7 Real gas deviation factor for natural gases as a function of pseudoreduced pressure and temperature. After Standing and Katz.[7]

The gas formation volume factor may be estimated from

$$B_g = 5.039 \times 10^{-3} \frac{zT}{p}, \quad \text{(D.6)}$$

where z is from Fig. D.7.

The water formation volume factor, B_w, may be estimated from Fig. D.8.

D.3 Rock Pore-Volume Compressibility

The isothermal formation (rock, pore volume) compressibility is generally defined as

$$c_f = \frac{1}{V_p}\left(\frac{\partial V_p}{\partial p}\right)_T . \quad \text{(D.7)}$$

The subscript T indicates that the partial derivative is taken at constant temperature. All compressibilities used in this monograph are isothermal compressibilities; the subscript is frequently omitted. Formation compressibility is defined so that it is a positive number. Thus, Eq. D.7 indicates that as fluid pressure decreases, the pore volume decreases. That occurs because the confining lithostatic pressure is essentially constant while the reservoir is depleted, thus causing compression of the rock.

Several authors[9-11] have attempted to correlate formation compressibility with various physical parameters. The correlations of Hall[10] and van der Knaap[11] have been used extensively in the petroleum literature. Recently, Newman[9] has shown that those correlations do not apply to a very wide range of reservoir rocks. Fig. D.9 shows data for compressibility of limestone samples superimposed on both van der Knaap's and Hall's limestone correlations. In Fig. D.9, and in other figures in this section, the lithostatic pressure is defined as the pressure obtained by multiplying reservoir depth by 1 psi/ft.

Figs. D.10 through D.12 compare Newman's and other data with Hall's sandstone correlation. In preparing the three figures, Newman used the following definitions:[9]

1. Consolidated samples consisted of hard rocks (thin edges could not be broken off by hand).
2. Friable samples could be cut into cylinders but the edges could be broken off by hand.
3. Unconsolidated samples could fall apart under their own weight unless they had undergone special treatment, such as freezing.

As can be seen in the three figures, no correlation would provide a good description of the large suite of samples studied. It is apparent from Fig. D.11 that there is no correlation at all for the friable samples. Fig. D.12 indicates that if there is any correlation for unconsolidated samples, the trend may be opposite the trend for consolidated samples (Fig. D.10).

Unfortunately, Figs. D.9 through D.12 lead to only one conclusion: formation compressibility should be measured for the reservoir being studied. At best, correlations can be expected to give only order-of-magnitude estimates.

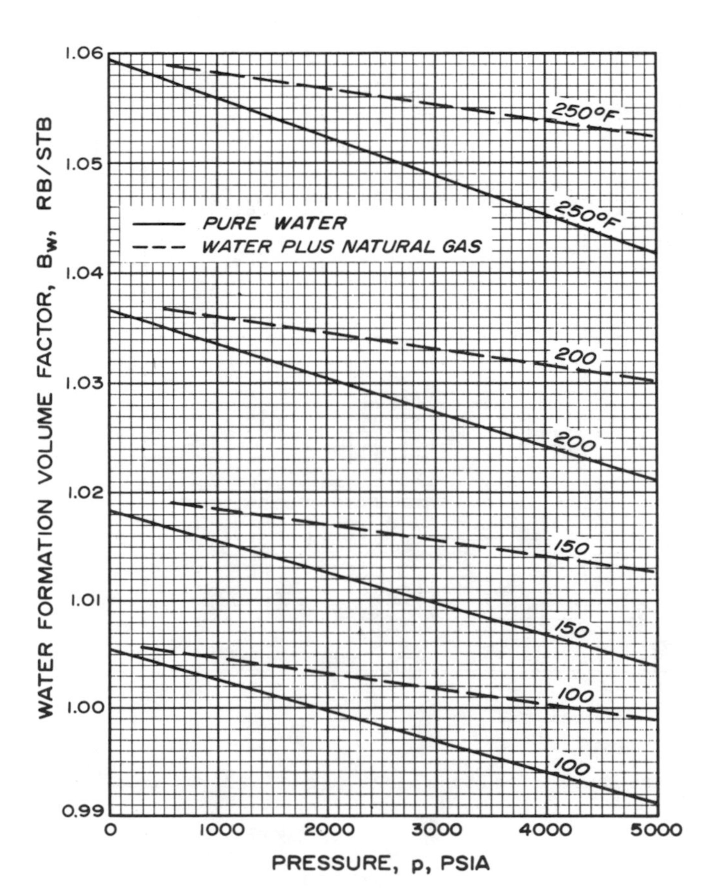

Fig. D.8 Formation volume factor of pure water and a mixture of natural gas and water. Data of Dodson and Standing.[8]

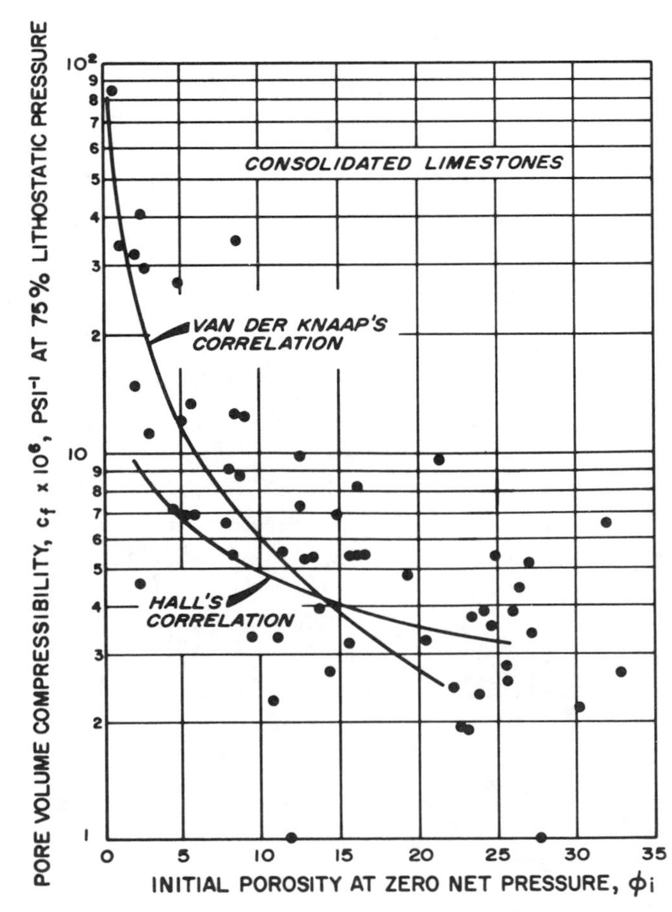

Fig. D.9 Pore-volume compressibility at 75-percent lithostatic pressure vs initial sample porosity for limestones. After Newman.[9]

D.4 Oil Compressibility

The isothermal compressibility of an undersaturated oil (oil above the bubble point) is defined as

$$c_o = -\frac{1}{V_o}\left(\frac{\partial V_o}{\partial p}\right)_T = \frac{1}{\rho_o}\left(\frac{\partial \rho_o}{\partial p}\right)_T = -\frac{1}{B_o}\left(\frac{\partial B_o}{\partial p}\right)_T . \quad \text{(D.8)}$$

Since the volume of an undersaturated liquid decreases as the pressure increases, c_o is positive.

Generally, oil compressibility should be computed from laboratory PVT data for the oil existing in the reservoir being studied. The final equality in Eq. D.8 is useful for calculating c_o from such data. In some reservoirs, c_o is essentially constant above the bubble point, while in others it varies with pressure.

If laboratory data are not available, Trube's[2] correlation for compressibility of an undersaturated oil (Fig. D.13) may be used. It is necessary to estimate T_{pr} and p_{pr} from Fig. D.1 or Fig. D.2. The pseudoreduced compressibility, c_{pr}, is read from Fig. D.13 and the oil compressibility is estimated from

$$c_o = \frac{c_{pr}}{p_{pc}} . \quad \text{(D.9)}$$

Below the bubble point, dissolved gas must be considered in computing compressibilities used in transient test and reservoir analysis. Thus, we define an *apparent* oil compressibility for the region below the bubble point where oil volume increases with pressure as a result of gas going into solution:

$$c_{oa} = -\frac{1}{B_o}\frac{\partial B_o}{\partial p} + \frac{B_g}{B_o}\frac{\partial R_s}{\partial p} . \quad \text{(D.10)}$$

Note that Eq. D.10 reduces to Eq. D.8 above the bubble point when R_s is constant with pressure. If available, laboratory data should be used to estimate c_{oa}; otherwise, correlations may be used with caution. When using correlations, the $\partial R_s/\partial p$ term in Eq. D.10 may be estimated from Fig. D.14 or from

$$\frac{\partial R_s}{\partial p} \simeq \frac{R_s}{(0.83p + 21.75)} . \quad \text{(D.11)}$$

Eq. D.11 and Fig. D.14 are from Ramey[12] and are based on Standing's data.[5] The gas formation volume factor may be estimated from Eq. D.6, where the z factor is estimated from Fig. D.7. The term $\partial B_o/\partial p$ in Eq. D.10 may be estimated from

$$\frac{\partial B_o}{\partial p} \simeq \frac{\partial R_s}{\partial p} \cdot \frac{\partial B_o}{\partial R_s} , \quad \text{(D.12)}$$

where the first term on the right-hand side is from Eq. D.11 or Fig. D.14 and the second term on the right-hand side is from Fig. D.15. Oil and gas gravities must be known to use

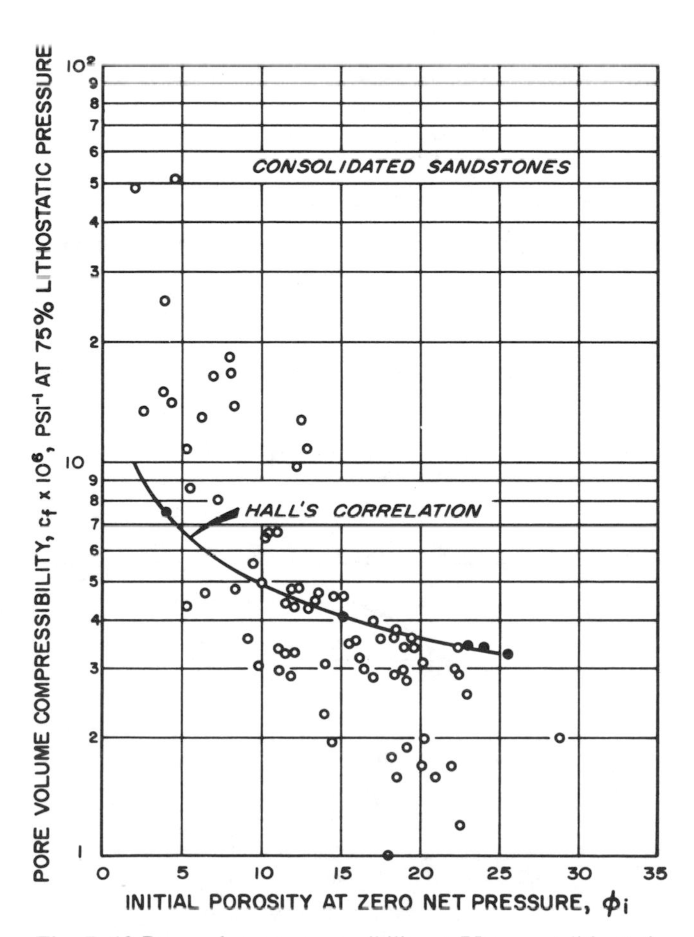

Fig. D.10 Pore-volume compressibility at 75-percent lithostatic pressure vs initial sample porosity for consolidated sandstones. After Newman.[9]

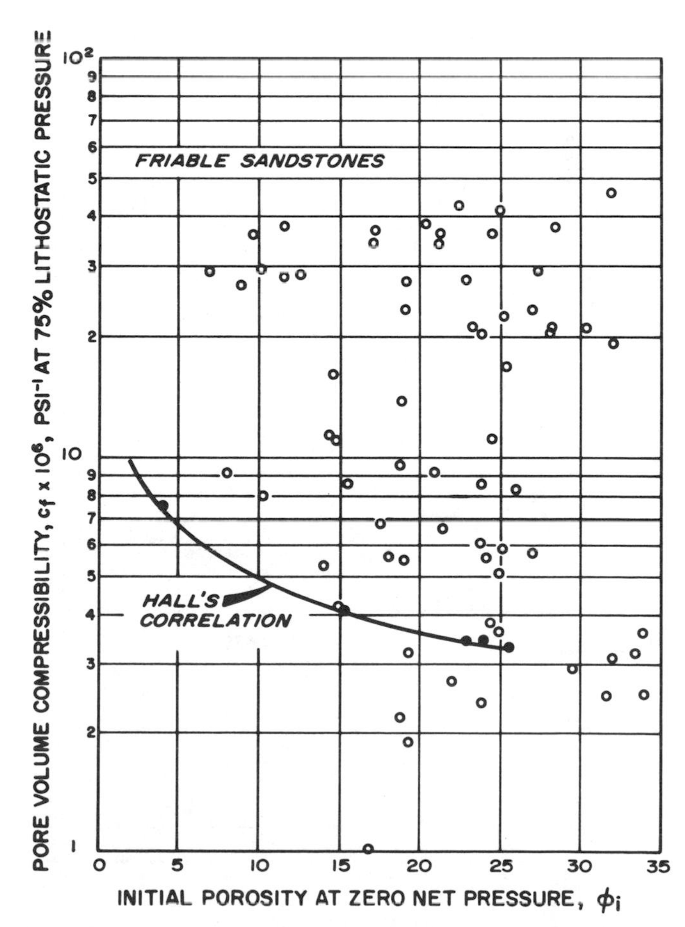

Fig. D.11 Pore-volume compressibility at 75-percent lithostatic pressure vs initial sample porosity for friable sandstones. After Newman.[9]

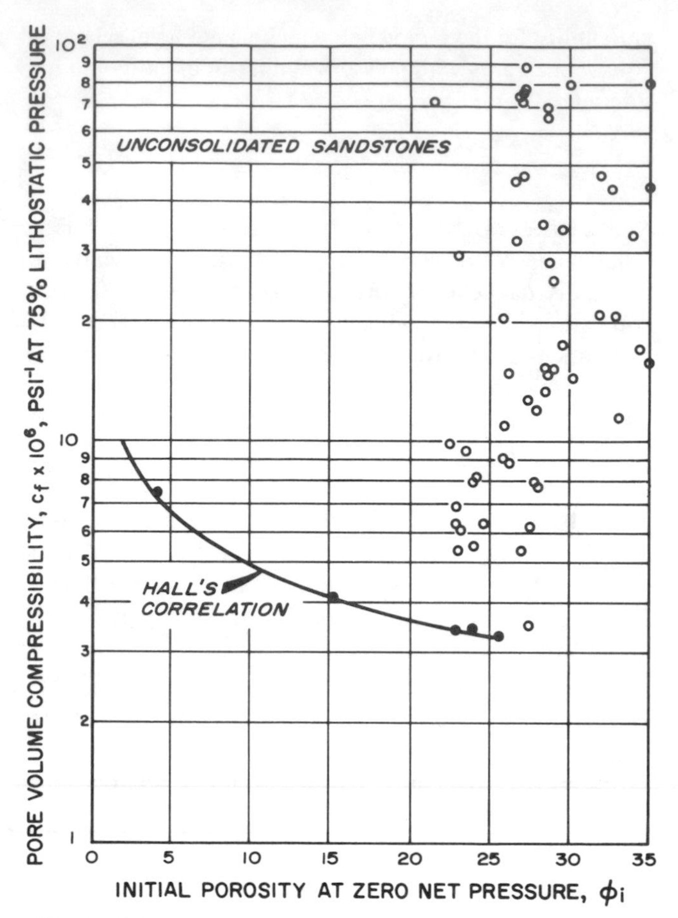

Fig. D.12 Pore-volume compressibility at 75-percent lithostatic pressure vs initial sample porosity for unconsolidated sandstones. After Newman.[9]

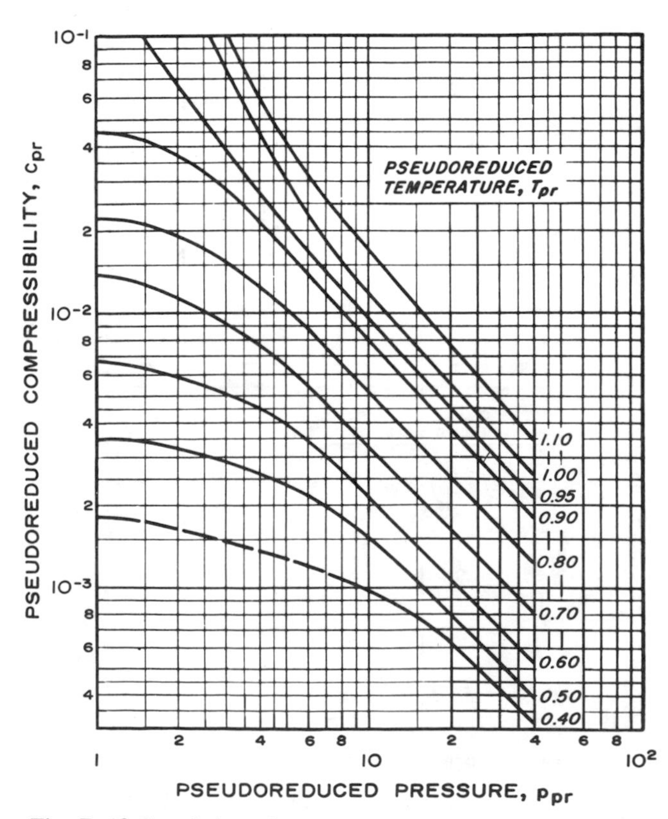

Fig. D.13 Correlation of pseudoreduced compressibility for an undersaturated oil. After Trube.[2]

Fig. D.15. The oil formation volume factor, B_o, may be estimated from Standing's correlation (Fig. D.6).

D.5 Water Compressibility

The water compressibility is defined analogously to the oil compressibility (Eq. D.8). The compressibility of water or brine *without* any solution gas is estimated from Figs. D.16 through D.19. Linear interpolation may be used for intermediate pressures and salinities.

To estimate the compressibility of undersaturated water or brine (that is, with solution gas), Long and Chierici[13] recommend using

$$c_w = (c_w)_{0,n}\left[1 + 0.0088 \times 10^{-Kn}(R_{sw})\right], \quad \ldots . \text{(D.13)}$$

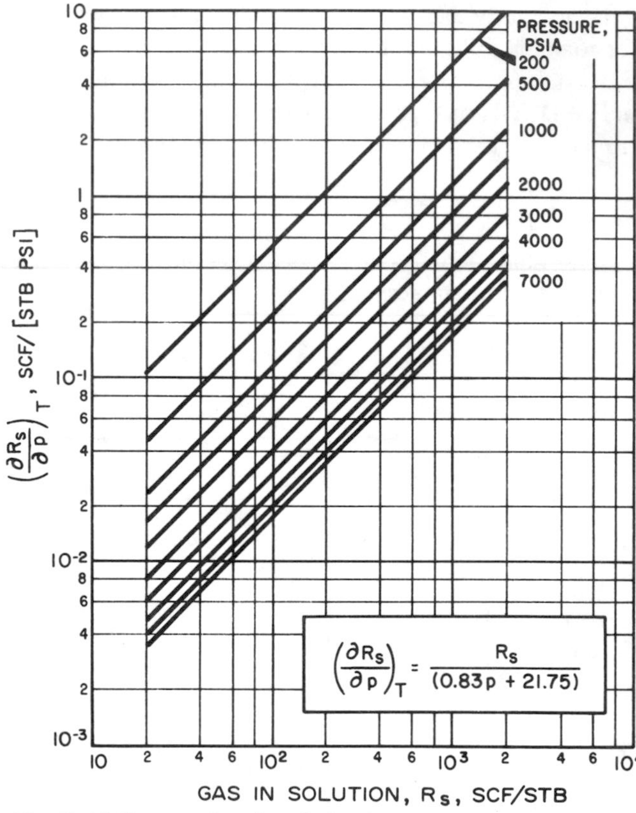

Fig. D.14 Change of gas in solution in oil with pressure vs gas in solution. After Ramey,[12] data of Standing.[5]

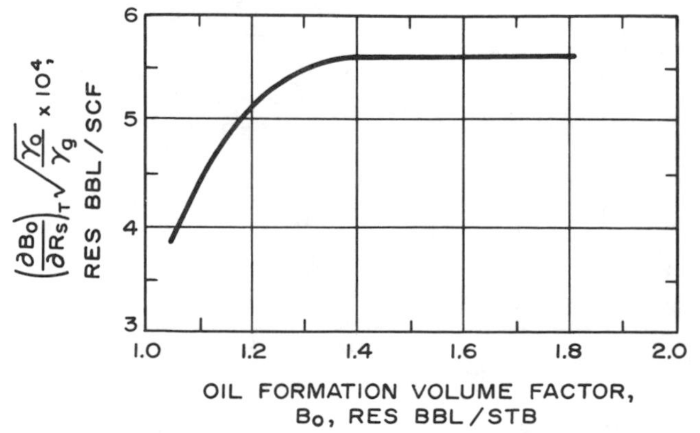

Fig. D.15 Change of oil formation volume factor with gas in solution vs oil formation volume factor. After Ramey.[12]

where

c_w = compressibility of an undersaturated brine containing solution gas and n gram-equivalents of dissolved solids, psi^{-1}

$(c_w)_{0,n}$ = compressibility of a *gas-free* brine containing n gram-equivalents of dissolved solids, psi^{-1}, from Figs. D.16 through D.19

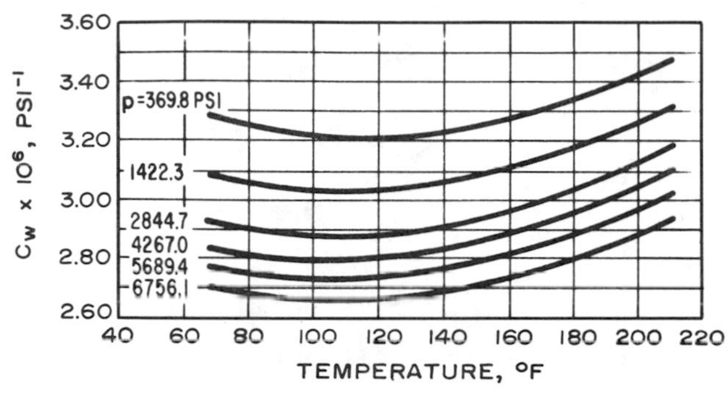

Fig. D.16 Average compressibility of distilled water. After Long and Chierici.[13]

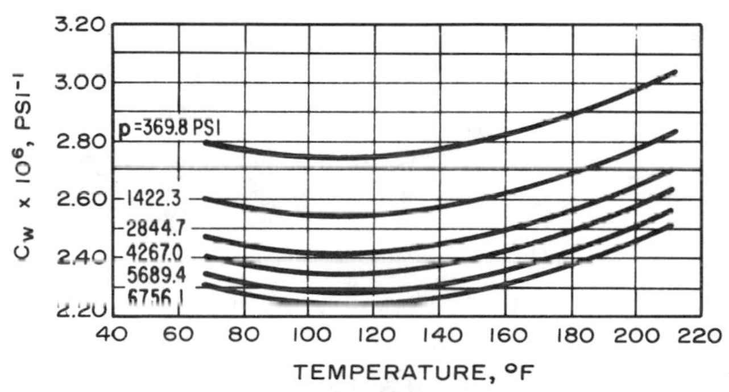

Fig. D.17 Average compressibility of 100,000-ppm NaCl in distilled water. After Long and Chierici.[13]

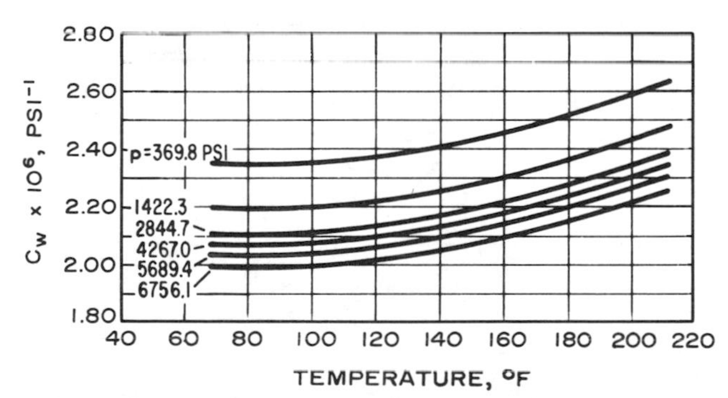

Fig. D.18 Average compressibility of 200,000-ppm NaCl in distilled water. After Long and Chierici.[13]

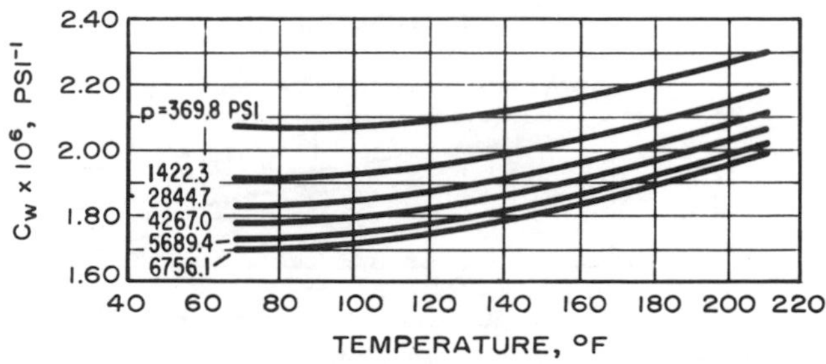

Fig. D.19 Average compressibility of 300,000-ppm NaCl in distilled water. After Long and Chierici.[13]

n = dissolved solids (ppm) ÷ 58,443, concentration of dissolved solids, gram-equivalents/liter

K = Secenov's coefficient, obtained at reservoir temperature from Fig. D.20

R_{sw} = gas solubility in distilled water at the required pressure and temperature, from Fig. D.21, scf/bbl.

An alternate approach to estimating the compressibility of undersaturated water is to use Fig. D.22 to estimate water compressibility at reservoir temperature, pressure, and solution gas-oil ratio. Fig. D.23 is used to estimate the solution gas-water ratio as a function of temperature, pressure, and salinity.

The apparent compressibility of water below the bubble point is given by

$$c_{wa} = -\frac{1}{B_w}\frac{\partial B_w}{\partial p} + \frac{B_g}{B_w}\frac{\partial R_{sw}}{\partial p} \quad \text{. . . . . (D.14)}$$

Again, it is best to compute c_{wa} from PVT analyses if they are available. However, since they seldom are, the use of

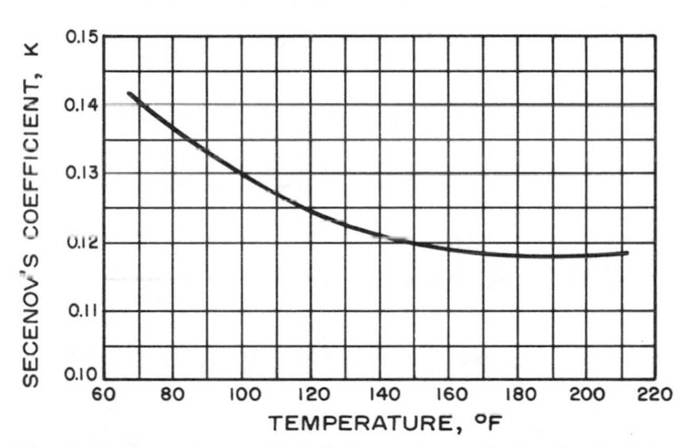

Fig. D.20 Secenov's coefficient for methane, for Eq. D.13. After Long and Chierici.[13]

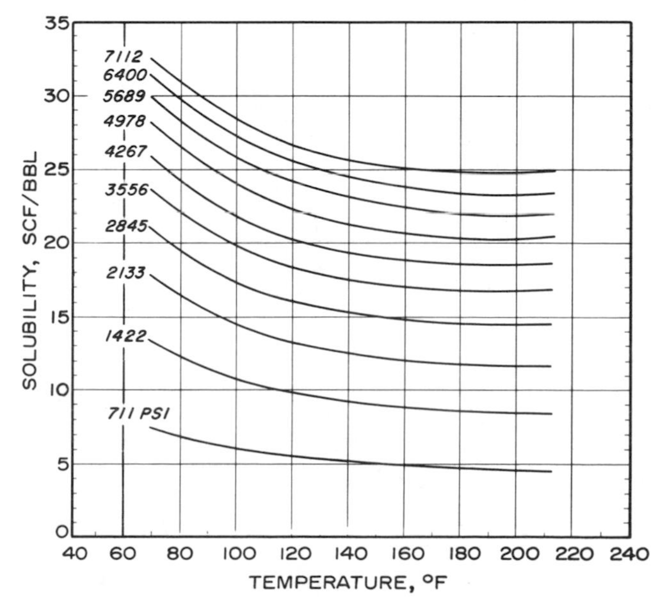

Fig. D.21 Solubility of methane in distilled water. After Long and Chierici.[13]

correlations is often required. The $\partial R_{sw}/\partial p$ term may be approximated from Fig. D.24, while B_w may be approximated from Fig. D.8. B_g is estimated with Eq. D.6. The first term on the right-hand side of Eq. D.14 must be estimated from Fig. D.22 or Eq. D.13.

D.6 Gas Compressibility

Isothermal gas compressibility is defined analogously to the oil compressibility (Eq. D.8). The gas equivalent of Eq. D.8 may be written using the real gas deviation factor, z:

$$c_g = \frac{1}{p} - \frac{1}{z}\left(\frac{\partial z}{\partial p}\right)_T . \qquad \text{(D.15a)}$$

If pseudoreduced pressures and temperatures are introduced into Eq. D.15a, the isothermal gas compressibility may be written as

$$c_g = \frac{1}{p_{pc}}\left[\frac{1}{p_{pr}} - \frac{1}{z}\left(\frac{\partial z}{\partial p_{pr}}\right)_{T_{pr}}\right] . \qquad \text{(D.15b)}$$

The z-factor chart, Fig. D.7, may be used directly to estimate the derivative term for Eq. D.15b.

Gas compressibility also may be estimated from the pseudoreduced-compressibility correlation shown in Figs. D.25 and D.26. The pseudoreduced compressibility is read

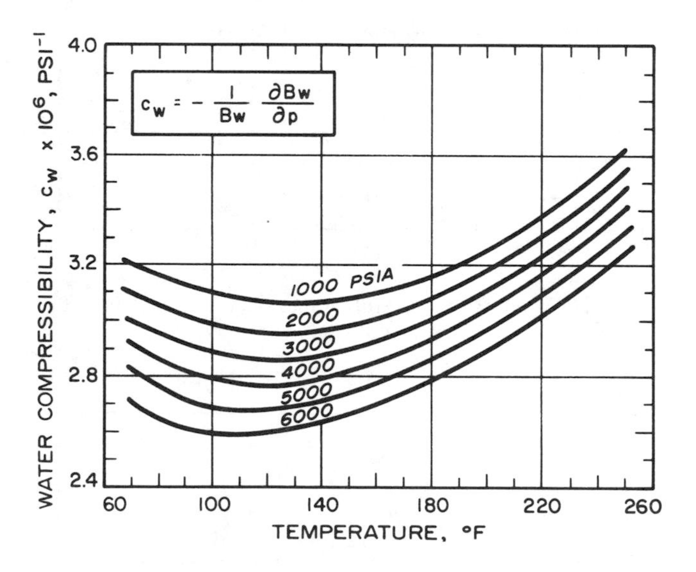

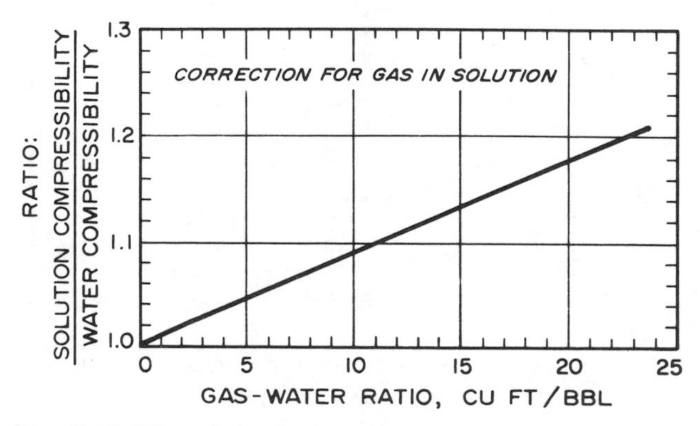

Fig. D.22 Effect of dissolved gas on water compressibility. After Dodson and Standing.[8]

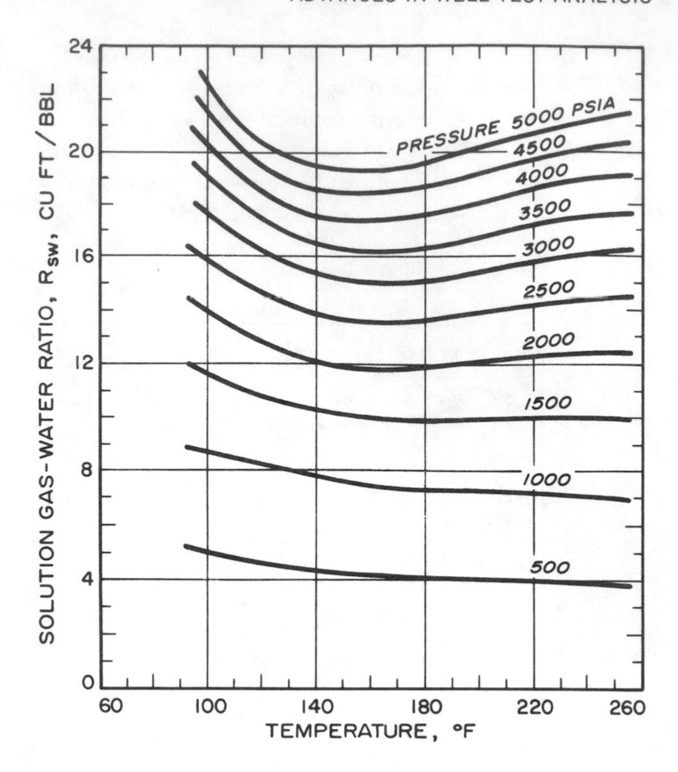

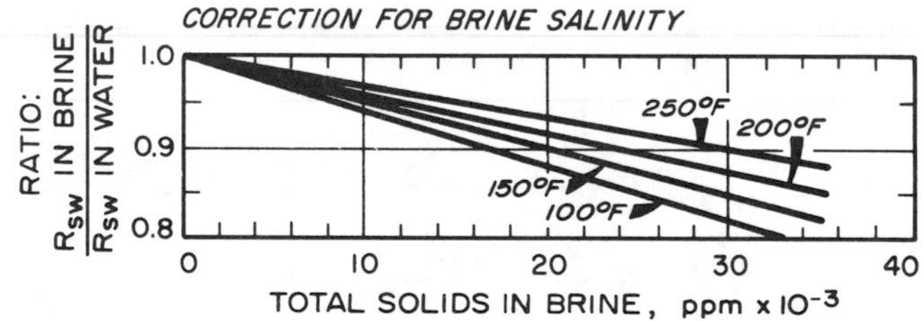

Fig. D.23 Solubility of natural gas in water. After Dodson and Standing.[8]

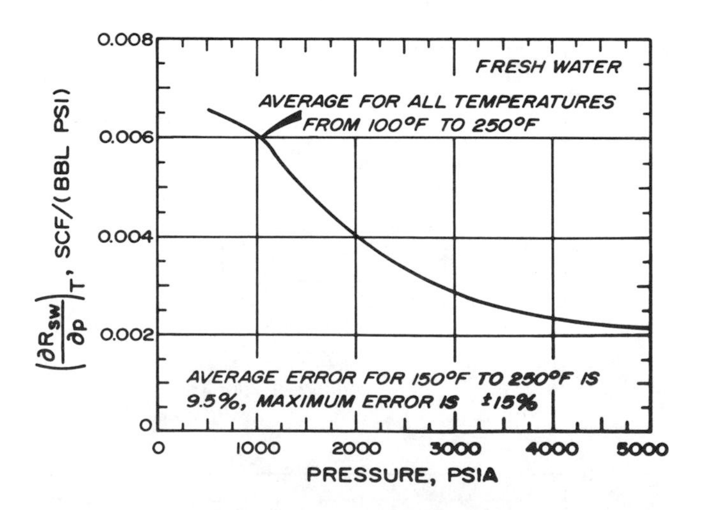

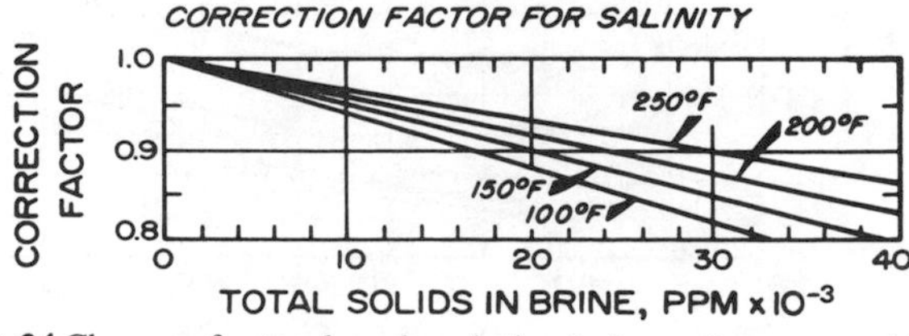

Fig. D.24 Change of natural gas in solution in formation water with pressure vs pressure. Multiply $(\partial R_{sw}/\partial p)_T$ by the correction factor to get result for brine. After Ramey,[12] data of Standing.[5]

from one of those figures and the gas compressibility is computed from

$$c_g = \frac{c_{pr}}{p_{pc}} . \quad \text{(D.16)}$$

D.7 Gas Viscosity

Fig. D.27 is one of the simplest hydrocarbon gas viscosity correlations available.[1] That figure gives gas viscosity as a function of gas gravity, pressure, and temperature. Its use is illustrated by the arrows. For a 0.7-gravity gas at 750 psia and 220 °F, viscosity is 0.0158 cp.

Carr, Kobayashi, and Burrows[15] present a method for estimating natural gas viscosity that is widely used. That method requires knowledge of the gas composition and of the viscosity of each component at atmospheric pressure and reservoir temperature. The viscosity of the mixture at atmospheric pressure is estimated from

$$\mu_{ga} = \frac{\sum_{i=1}^{N} y_i \mu_i \sqrt{M_i}}{\sum_{i=1}^{N} y_i \sqrt{M_i}} , \quad \text{(D.17)}$$

where

μ_{ga} = viscosity of the gas mixture at the desired temperature and *atmospheric* pressure, cp
y_i = mole fraction of the ith component
μ_i = viscosity of ith component at the desired temperature and *atmospheric* pressure, obtained from Fig. D.28
M_i = molecular weight of ith component (Table D.1)
N = number of components in the gas.

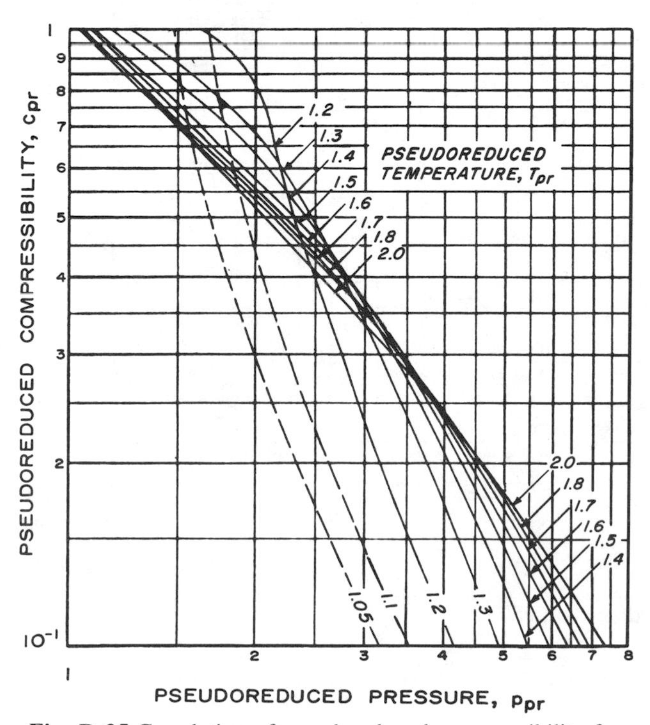

Fig. D.25 Correlation of pseudoreduced compressibility for natural gases. After Trube.[14]

Viscosity for many gaseous components is shown in Fig. D.28 at 1 atm and various temperatures. If the gas composition is not known, Fig. D.29 may be used with gas molecular weight to estimate the gas viscosity at reservoir temperature and atmospheric pressure. Molecular weight is related to gas gravity by

$$M \simeq 29\gamma . \quad \text{(D.18)}$$

The gas viscosity at reservoir pressure is estimated by determining the ratio μ_g/μ_{ga} at the appropriate temperature and pressure from Fig. D.30 or Fig. D.31. Then, that ratio is applied to μ_{ga} computed from Eq. D.17 or Fig. D.29. The pseudoreduced temperatures and pressures for use in Figs. D.30 and D.31 are estimated from Eqs. D.1 through D.4 or from Fig. D.3.

D.8 Oil Viscosity

Whenever possible, oil viscosity should be determined by laboratory measurements at reservoir temperature and pressure. Oil viscosity is usually reported in standard PVT analyses. If such laboratory data are not available, the Chew and Connally[16] correlation for viscosity of gas-saturated oil, Fig. D.32, may be used. Both solution gas-oil ratio and oil viscosity at reservoir temperature and atmospheric pressure must be known to use Fig. D.32. If the dead oil viscosity is not determined from laboratory data, it may be estimated from Fig. D.33.

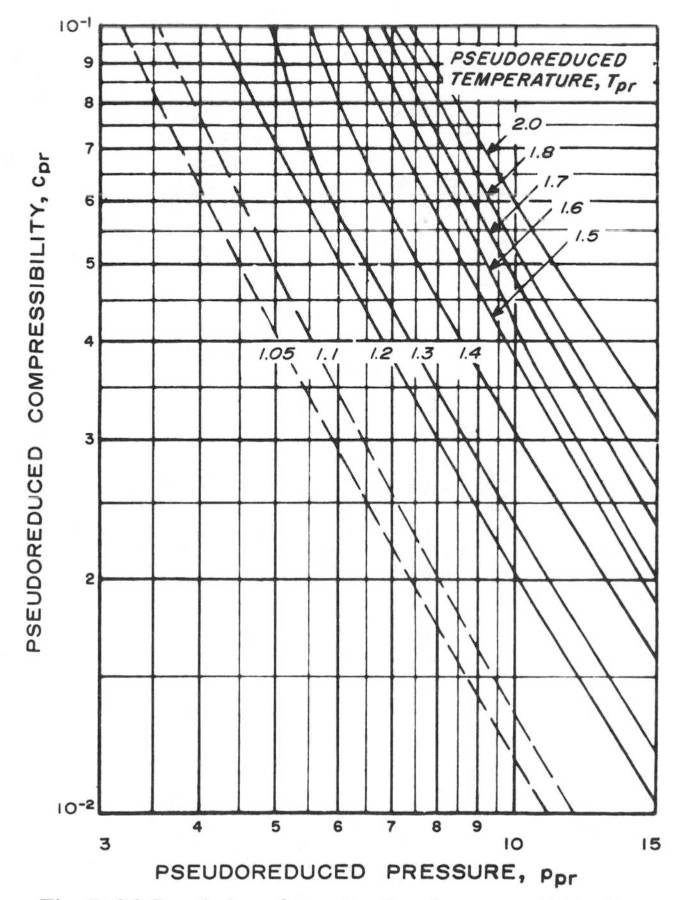

Fig. D.26 Correlation of pseudoreduced compressibility for natural gases. After Trube.[14]

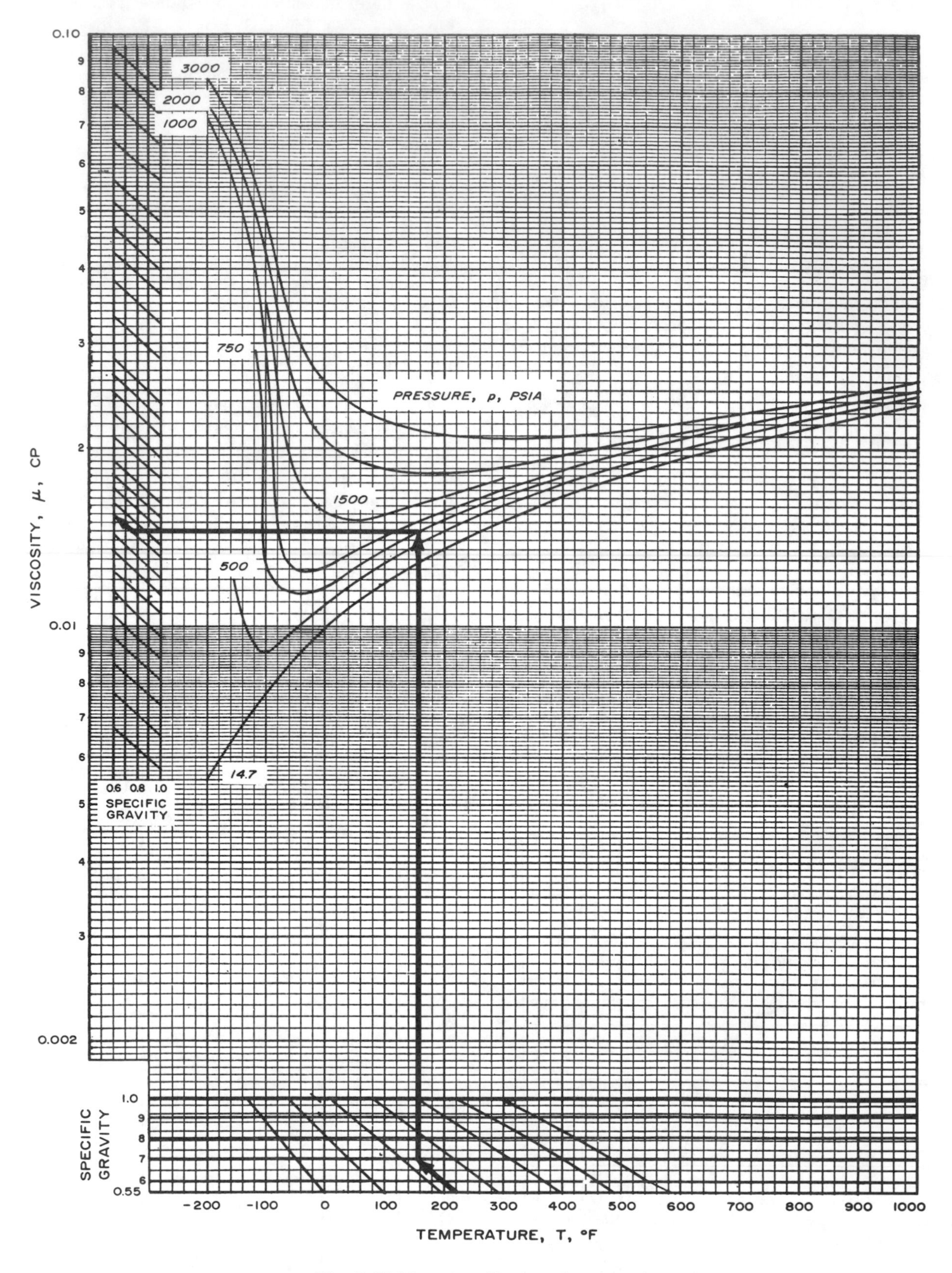

Fig. D.27 Viscosity of hydrocarbon gases.[1]

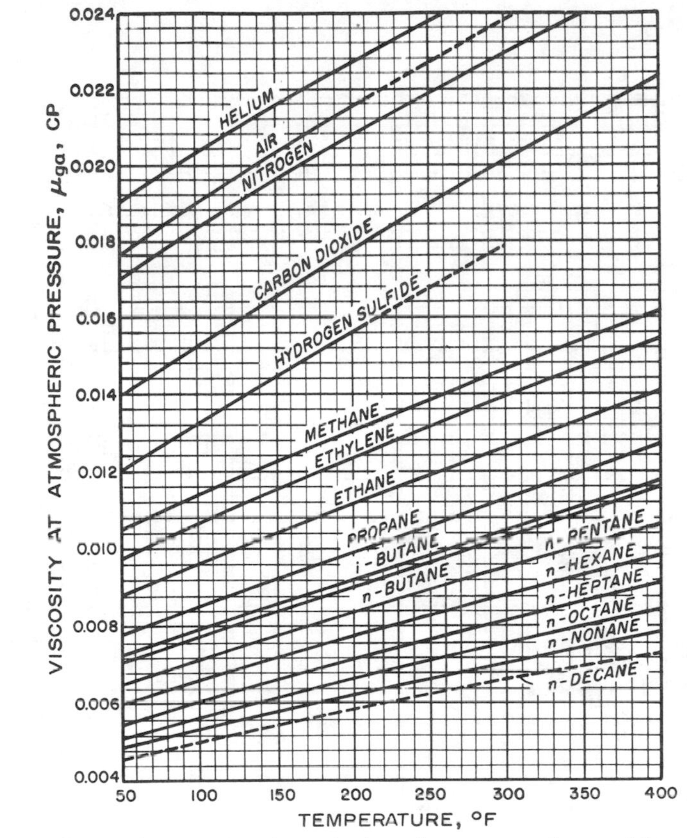

Fig. D.28 Viscosity of pure hydrocarbon gases at 1 atm. After Carr, Kobayashi, and Burrows.[15]

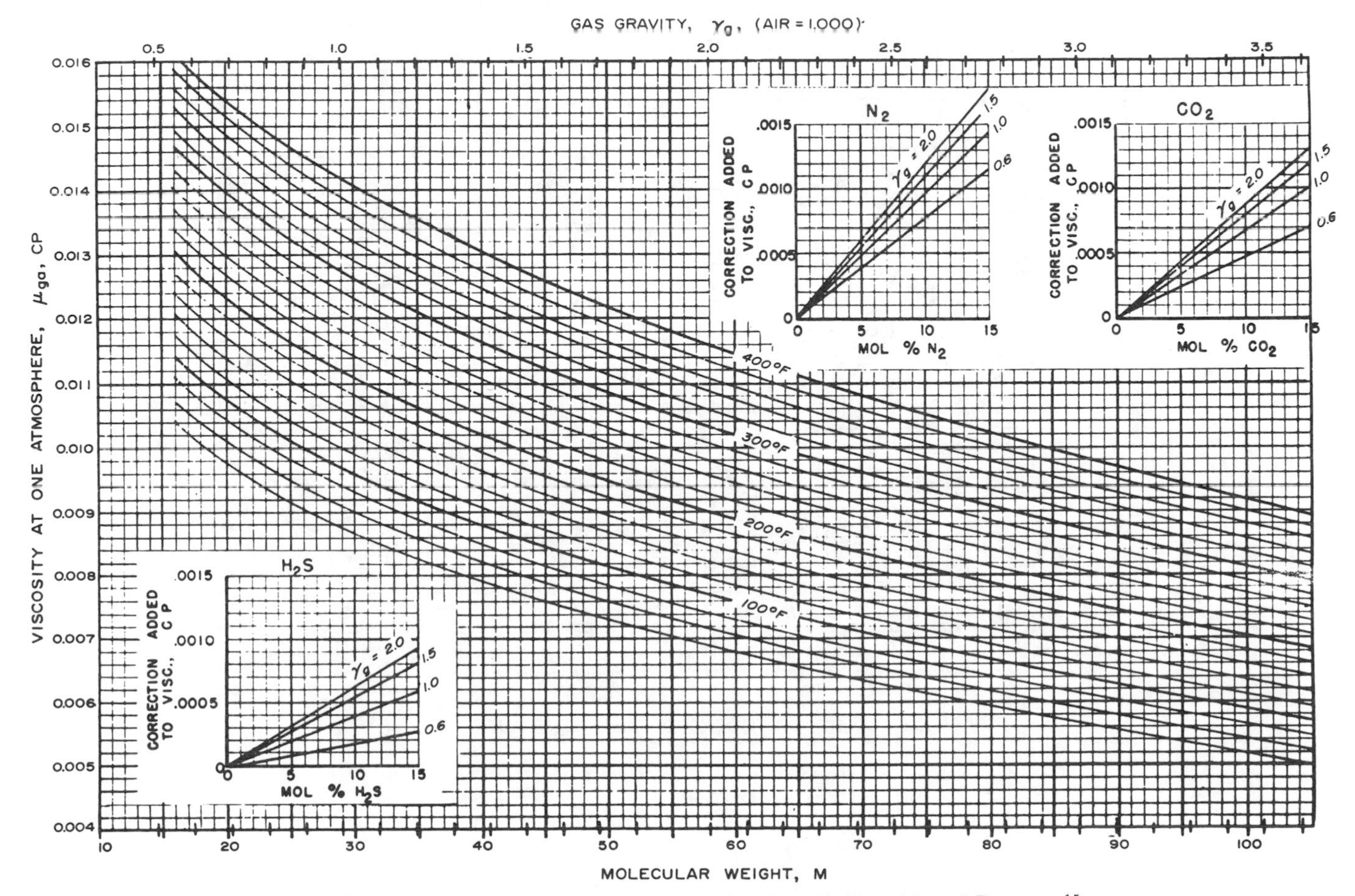

Fig. D.29 Viscosity of natural gases at 1 atm. After Carr, Kobayashi, and Burrows.[15]

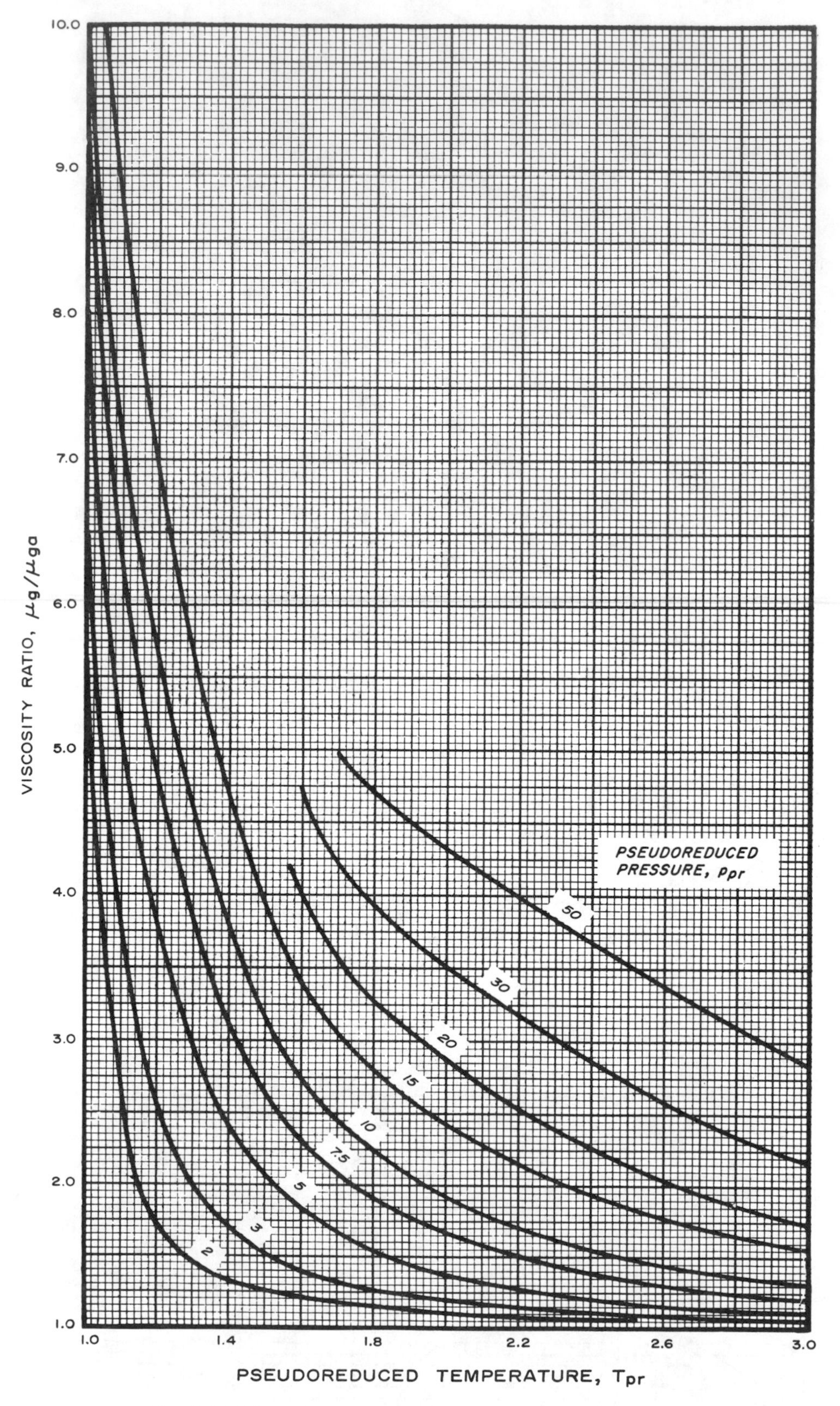

Fig. D.30 Effect of temperature and pressure on gas viscosity; μ_{ga} is estimated from Eq. D.17 or Fig. D.29. After Carr, Kobayashi, and Burrows.[15]

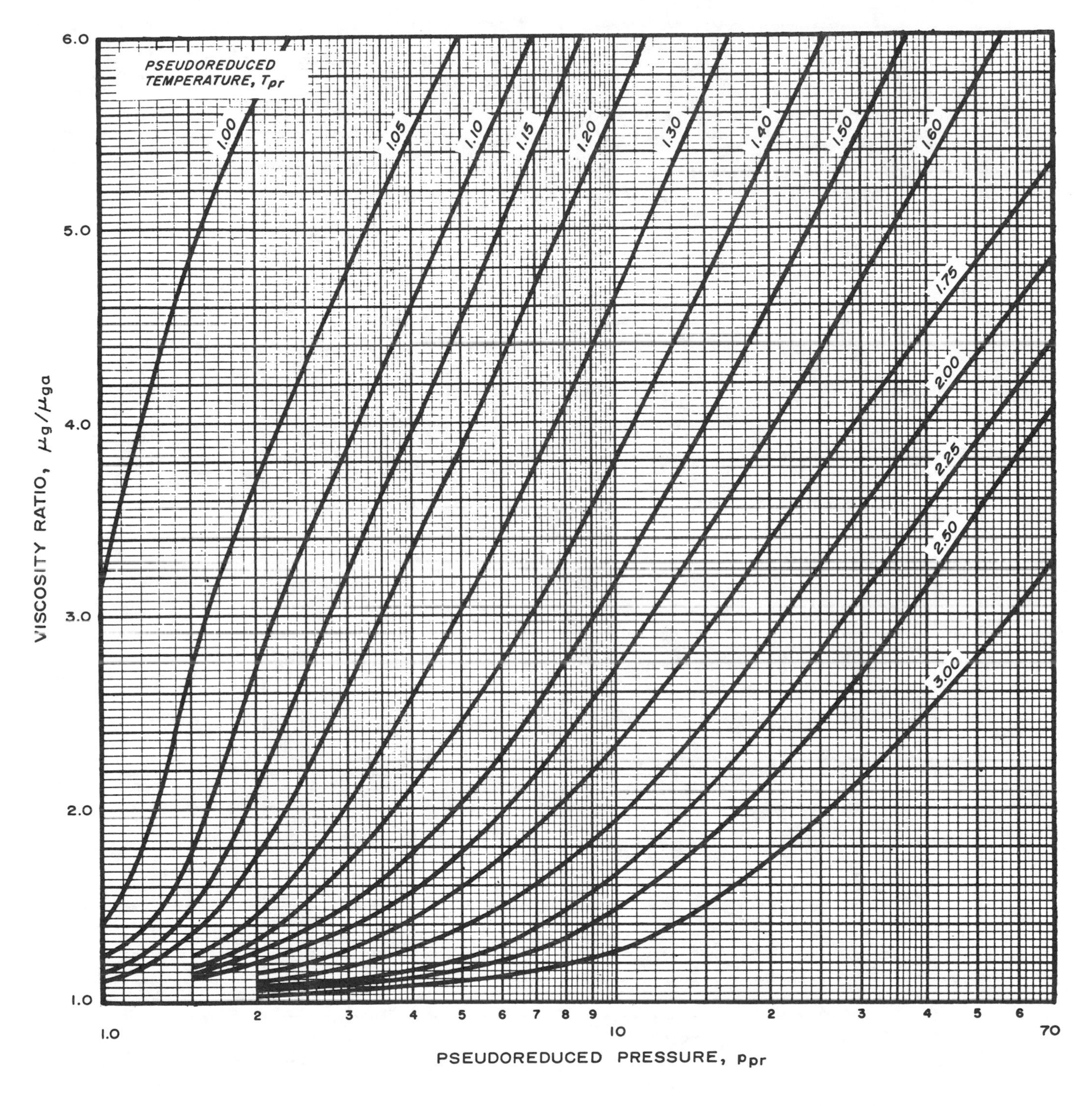

Fig. D.31 Effect of pressure and temperature on gas viscosity; μ_{ga} is estimated from Eq. D.17 or Fig. D.29. After Carr, Kobayashi, and Burrows.[15]

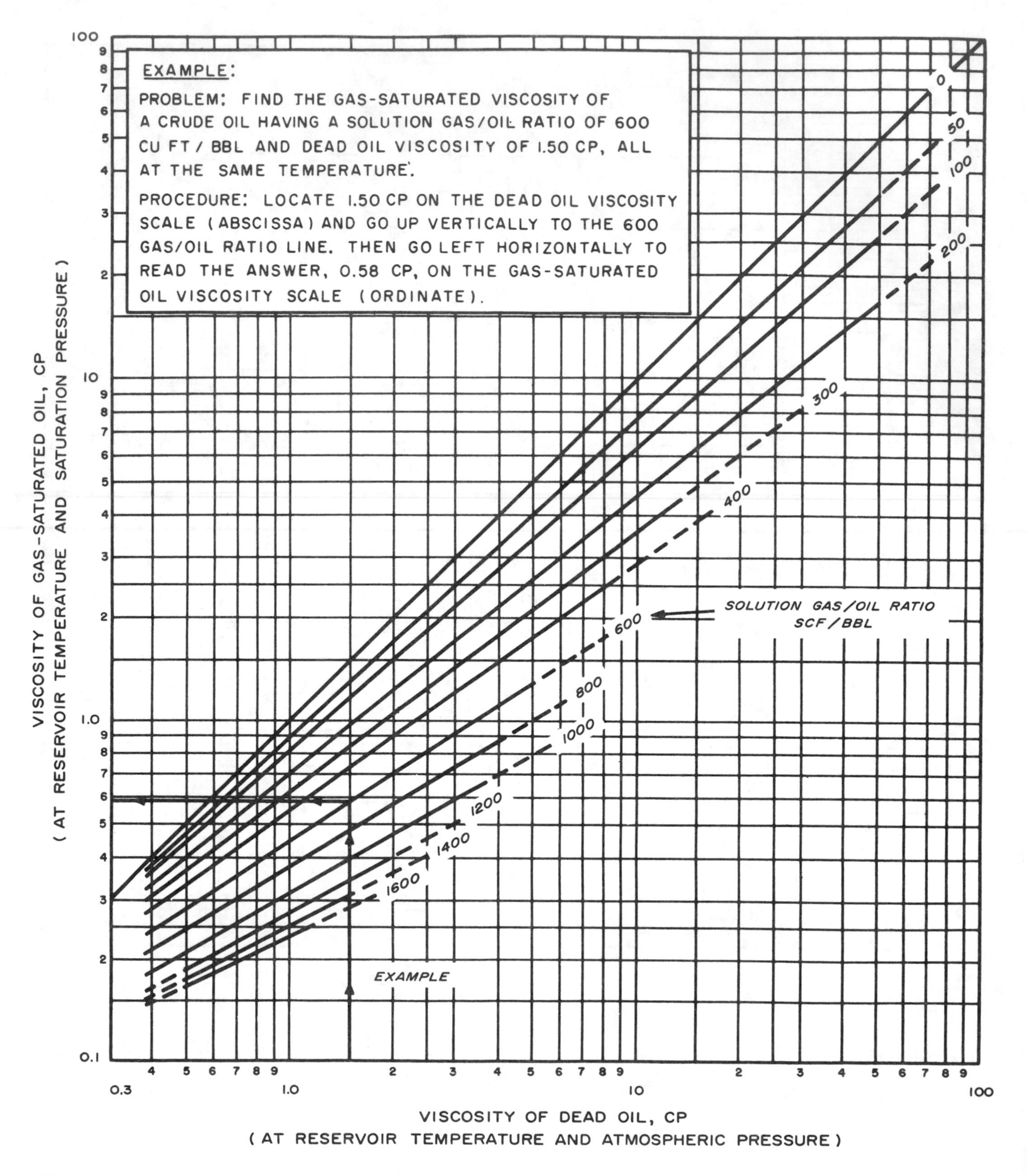

Fig. D.32 Viscosity of gas-saturated crude oil at reservoir temperature and pressure. Dead oil viscosity from laboratory data, or from Fig. D.33. After Chew and Connally.[16]

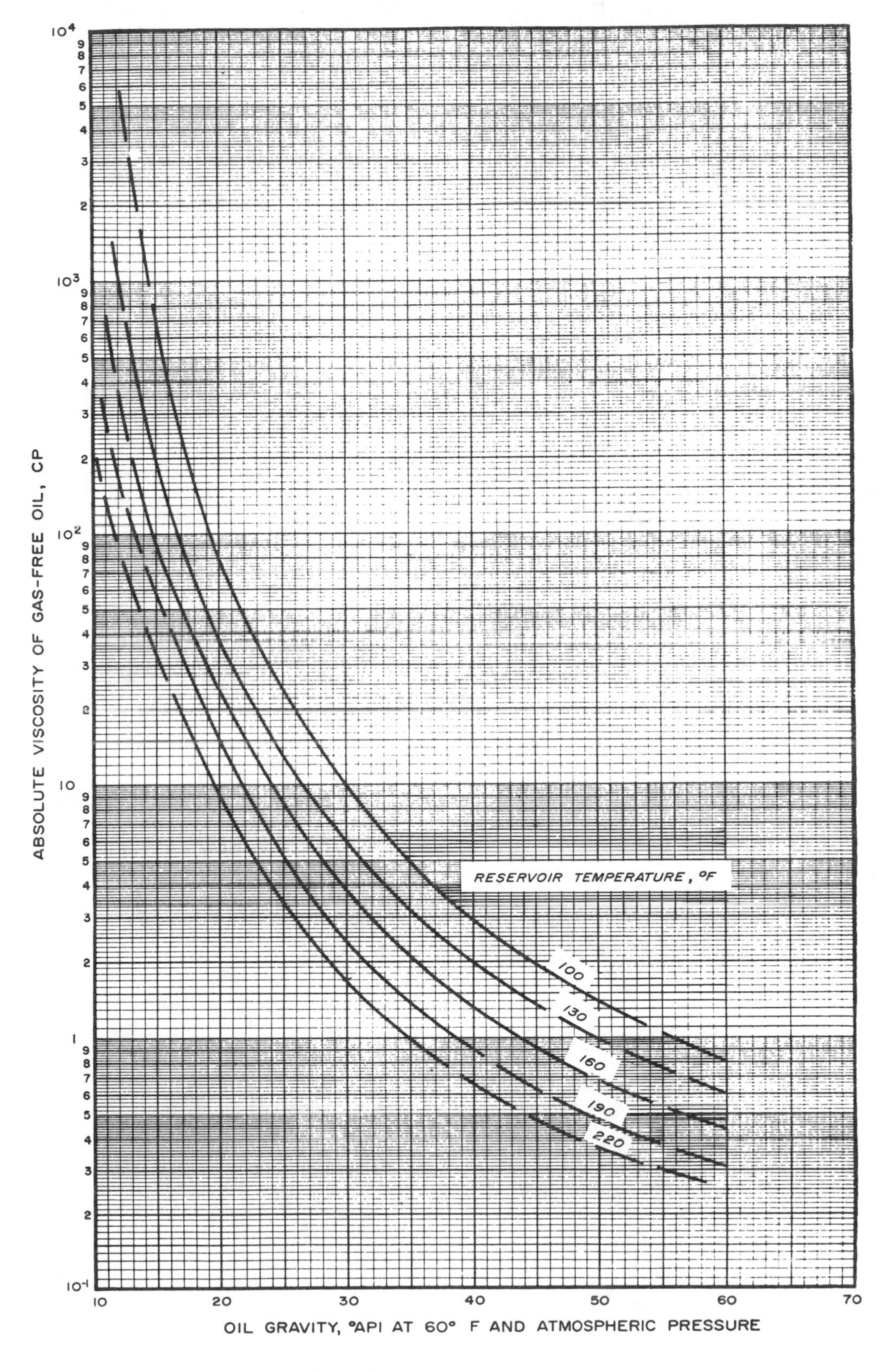

Fig. D.33 Dead oil viscosity at reservoir temperature and atmospheric pressure. After Beal.[17]

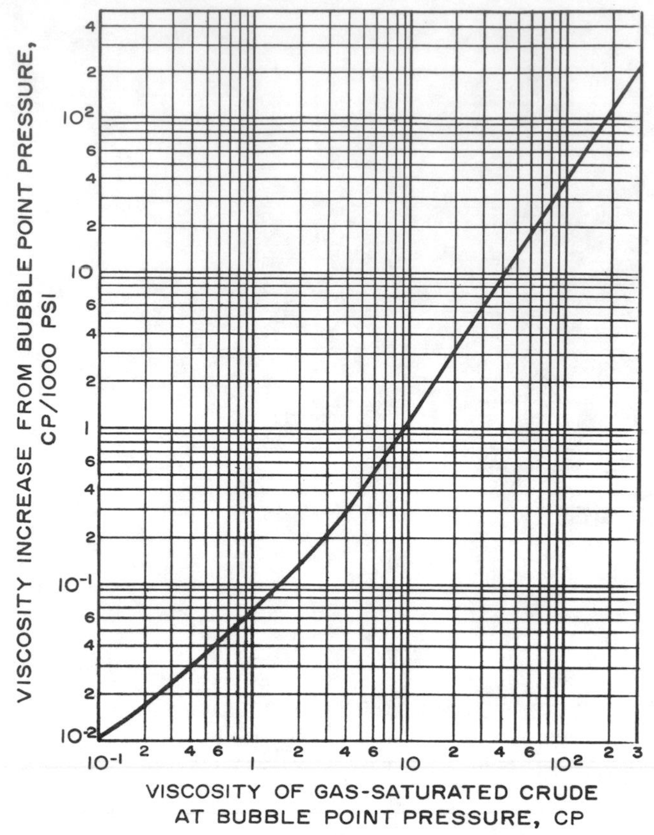

Fig. D.34 Rate of increase of oil viscosity above bubble-point pressure. After Beal.[17]

To estimate oil viscosity above the bubble-point pressure, use Fig. D.34. The figure shows the increase in viscosity above the bubble-point viscosity per 1,000 psi. It is based on a small amount of data, and so should be used only as a rough guide.

D.9 Water Viscosity

Fig. D.35 provides a means for estimating water viscosity as a function of salinity and temperature; a pressure correction is included. There are no provisions in Fig. D.35 for modifying the viscosity of water as a function of gas saturation. As for oil, it is best to measure water viscosity as a function of pressure at reservoir temperature. The water should have the gas saturation and salinity existing in the reservoir.

References

1. *Engineering Data Book,* 9th ed., Gas Processors Suppliers Assn., Tulsa (1972) Sec. 16.
2. Trube, Albert S.: "Compressibility of Undersaturated Hydrocarbon Reservoir Fluids," *Trans.,* AIME (1957) **210,** 341-344.
3. Brown, George G., Katz, Donald L., Oberfell, George G., and Alden, Richard C.: *Natural Gasoline and the Volatile Hydrocarbons,* Natural Gasoline Assn. of America, Tulsa (1948).
4. "Measuring, Sampling, and Testing Crude Oil," API Standard 2500, American Petroleum Institute. Reproduced in Frick, Thomas C. and Taylor, R. William: *Petroleum Production Handbook,* McGraw-Hill Book Co., Inc., New York (1962) **1,** Chap. 16.
5. Standing, M. B.: *Volumetric and Phase Behavior of Oil Field Hydrocarbon Systems,* Reinhold Publishing Corp., New York (1952).
6. Cronquist, Chapman: "Dimensionless PVT Behavior of Gulf Coast Reservoir Oils," *J. Pet. Tech.* (May 1973) 538-542.
7. Standing, Marshall B. and Katz, Donald L.: "Density of Natural Gases," *Trans.,* AIME (1942) **146,** 140-149.
8. Dodson, C. R. and Standing, M. B.: "Pressure-Volume-Temperature and Solubility Relations for Natural-Gas-Water Mixtures," *Drill. and Prod. Prac.,* API (1944) 173-179.
9. Newman, G. H.: "Pore-Volume Compressibility of Consolidated, Friable, and Unconsolidated Reservoir Rocks Under Hydrostatic Loading," *J. Pet. Tech.* (Feb. 1973) 129-134.
10. Hall, Howard N.: "Compressibility of Reservoir Rocks," *Trans.,* AIME (1953) **198,** 309-311.
11. van der Knaap, W.: "Nonlinear Behavior of Elastic Porous Media," *Trans.,* AIME (1959) **216,** 179-187.
12. Ramey, H. J., Jr.: "Rapid Method of Estimating Reservoir Compressibility," *J. Pet. Tech.* (April 1964) 447-454; *Trans.,* AIME, **231.**
13. Long, Giordano and Chierici, Gianluigi: "Salt Content Changes Compressibility of Reservoir Brines," *Pet. Eng.* (July 1961) B-25 to B-31.
14. Trube, Albert S.: "Compressibility of Natural Gases," *Trans.,* AIME (1957) **210,** 355-357.
15. Carr, Norman L., Kobayashi, Riki, and Burrows, David B.: "Viscosity of Hydrocarbon Gases Under Pressure," *Trans.,* AIME (1954) **201,** 264-272.
16. Chew, Ju-Nam and Connally, Carl A., Jr.: "A Viscosity Correlation for Gas-Saturated Crude Oils," *Trans.,* AIME (1959) **216,** 23-25.
17. Beal, Carlton: "The Viscosity of Air, Water, Natural Gas, Crude Oil and Its Associated Gases at Oil-Field Temperatures and Pressures," *Trans.,* AIME (1946) **165,** 94-115.
18. Matthews, C. S. and Russell, D. G.: *Pressure Buildup and Flow Tests in Wells,* Monograph Series, Society of Petroleum Engineers of AIME, Dallas (1967) **1,** Appendix G.

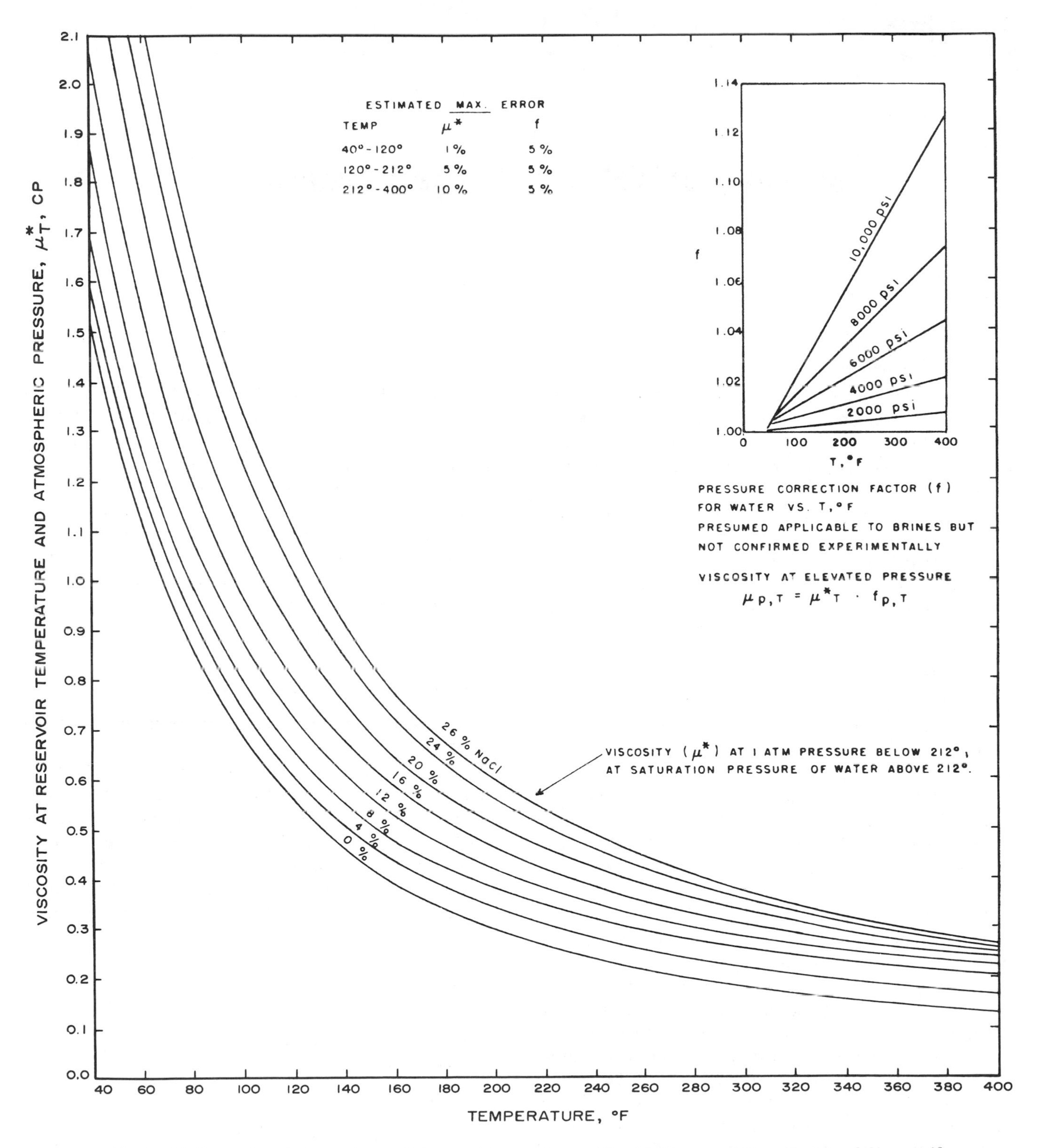

Fig. D.35 Water viscosity at various salinities and temperatures. After Matthews and Russell, data of Chesnut.[18]

Appendix E

Summary of Well Test Analysis Methods

E.1 Introduction

This monograph considers many types of well tests and discusses the influence of many factors on well-test response. Although some analysis techniques are unique, most have similarities. This appendix summarizes the equations and data plots for a variety of single-well-test analysis methods. Although the monograph is consistent in nomenclature and sign convention, and although many of the equations used are the same, there are some slight variations from one test type to another that are most clearly described in the tabular form presented here. This appendix also shows buildup-curve shapes resulting from various influences.

E.2 Pressure-Buildup Curve Shapes

Conceptually, graphs of pressure buildup, drawdown, injectivity, or falloff behavior in individual wells can be divided into three areas: (1) front-end effects (wellbore storage, fractures, damage); (2) the semilog straight line for which most analysis techniques apply; and (3) boundary effects. Those three portions are illustrated for a pressure buildup case in Fig. E.1. Throughout the text, criteria have been given for when each effect can be expected to be significant.

Wellbore storage effects always take priority at early time and can completely obscure early-time formation response. Thus, afterflow can be devastating to test analysis methods that depend heavily on early-time data. Fig. E.1 shows that fracture cases and large-negative-skin cases approach the semilog straight line from above when wellbore storage is small. As also shown in that figure, that behavior can be hidden by large wellbore storage effects, so the buildup curve may have the characteristic shape associated with wellbore storage only, or with a positive skin. There is no guarantee that a fractured-well pressure buildup curve will approach the semilog straight line from above.

Analysis methods that use late-time pressure data can be highly sensitive to variations in boundary conditions. Attempting to find a semilog straight line in late-time data affected by boundary or interference effects from offset wells can give highly misleading results. Generally, it is advisable to make time checks to estimate the end of the semilog straight line. Application of late-time analysis methods, such as the Muskat method, to middle-time data can also give misleading results. In some situations, as when wellbore storage effects are extremely severe or induced fractures are deep ($x_e/x_f < 10$), a classic semilog straight line may never develop, or its slope may be incorrect so that correction factors must be applied to apparent semilog slopes. (Correction factors are given for vertically fractured wells in Section 11.3.) Sometimes type-curve matching techniques may be used for those situations as well as for situations with severe wellbore storage. Curve matching is described in detail in Section 3.3 and is illustrated in other places in the monograph.

E.3 Well-Test Analysis Equations

Table E.1 summarizes analysis equations for unfractured, single-well drawdown, buildup, injectivity, and falloff tests. The equation numbers given in that table refer to the equations listed below. Also shown are the sections containing thorough discussions of the tests, cautions, and alternate analysis techniques. Table E.1 is presented to provide a quick reference and summary only, and should not be used as a replacement for material presented elsewhere in the monograph. To do so blindly will lead to incorrect analysis results.

Time Axis

$$\sum_{j=1}^{N} \frac{(q_j - q_{j-1})}{q_N} \log(t - t_{j-1}). \quad \text{(E.1)}$$

$$\sum_{j=1}^{N} \frac{q_j}{q_N} \log\left(\frac{t_N - t_{j-1} + \Delta t}{t_N - t_j + \Delta t}\right). \quad \text{(E.2)}$$

Permeability

$$k = \frac{-162.6\, qB\mu}{mh}. \quad \text{(E.3)}$$

$$k = \frac{162.6\, B\mu}{m'h}. \quad \text{(E.4)}$$

$$k = \frac{162.6 B\mu}{m_q (p_i - p_{wf}) h} \quad \text{(E.5)}$$

$$k = \frac{162.6 qB\mu}{mh} \quad \text{(E.6)}$$

$$k = \frac{141.2 qB\mu}{h(\bar{p} - p_{ws})_{int}} p_{D\,\mathrm{Mint}}(t_{pDA}). \quad \text{Use Fig. 5.10.} \quad \text{(E.7)}$$

Skin Factor

$$s = 1.1513 \left[\frac{p_{1hr} - p_i}{m} - \log\left(\frac{k}{\phi\mu c_t r_w^2}\right) + 3.2275 \right]. \quad \text{(E.8)}$$

$$s = 1.1513 \left[\frac{\Delta p_{1hr} - p_{ws}(\Delta t{=}0)}{m} - \log\left(\frac{k}{\phi\mu c_t r_w^2}\right) + 3.2275 \right]. \quad \text{(E.9)}$$

$$s = 1.1513 \left[\frac{b'}{m'} - \log\left(\frac{k}{\phi\mu c_t r_w^2}\right) + 3.2275 \right]. \quad \text{(E.10)}$$

$$s = 1.1513 \left[\frac{(1/q)_{1hr}}{m_q} - \log\left(\frac{k}{\phi\mu c_t r_w^2}\right) + 3.2275 \right]. \quad \text{(E.11)}$$

$$s = 1.1513 \left[\frac{p_{1hr} - p_{wf}(\Delta t{=}0)}{m} - \log\left(\frac{k}{\phi\mu c_t r_w^2}\right) + 3.2275 \right]. \quad \text{(E.12)}$$

$$s = 1.1513 \left[\frac{\Delta p_{1hr}}{m} - \log\left(\frac{k}{\phi\mu c_t r_w^2}\right) + 3.2275 \right]. \quad \text{(E.13)}$$

Connected Pore Volume

$$\phi hA = -\frac{0.23395 qB}{c_t m^*} \quad \text{(E.14)}$$

$$\phi hA = -\frac{0.00471 kh}{\mu c_t m_M} \quad \text{Closed square.} \quad \text{(E.15a)}$$

$$\phi hA = -\frac{0.00233 kh}{\mu c_t m_M} \quad \text{Square with constant-pressure boundaries.} \quad \text{(E.15b)}$$

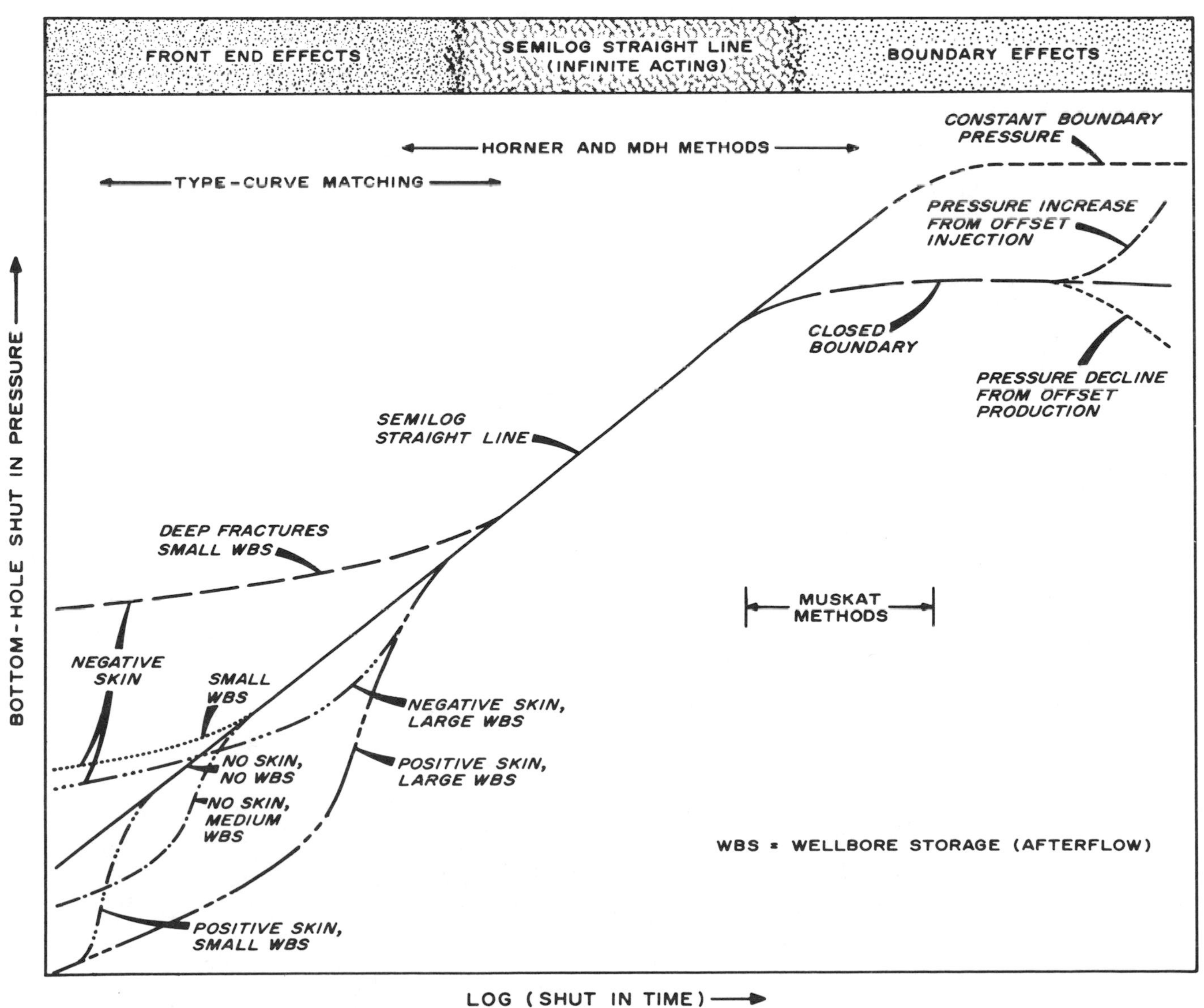

Fig. E.1 Typical bottom-hole pressure-buildup curve shapes. For production at pseudosteady state before shut-in.

TABLE E.1—SUMMARY OF COMMON WELL-TEST ANALYSIS EQUATIONS. EQUATIONS ARE IN SECTION E.3 OR REFERENCED CHAPTER.

Test Type	Analysis Method	Analysis Graph Axes: Pressure	Analysis Graph Axes: Time	Slope Data: Sign of Slope	Slope Data: Relation to m[9]	Permeability Equation	Skin-Factor Equation	Connected Pore-Volume Equation	Limits of Straight-Line Equation: Beginning	Limits of Straight-Line Equation: End	Section Number in Text	Comments
Drawdown	MDH	p_{wf}	$\log t$	−	$+m$	E.3	E.8	N/A	E.16	E.21	3.2	
	Developed system	[1]$p_{wf}-m^*\Delta t$	$\log \Delta t$	−	$+m$	E.3	E.9	N/A	E.16	E.21	3.4	
	Reservoir limit	p_{wf}	t	−	$+m^*$	N/A	N/A	E.14	E.17	—	3.5	
	Variable rate	$(p_i-p_{wf})/q_N$	Eq. E.1	+	$+m'$	E.4	E.10	N/A	E.16	E.21	4.2	Many specialized forms. Line limits approximate.
	Constant pressure	$1/q$	$\log t$	+	$+m_q$	E.5	E.11	N/A	E.18	—	4.6	Also curve matching with Fig. 4.12.
Buildup	Horner	p_{ws}	[2]$\log[(t_p+\Delta t)/\Delta t]$	−	$-m$	E.6	E.12	N/A	E.19	[3]E.22	5.2, 5.3	
	MDH	p_{ws}	$\log \Delta t$	+	$+m$	E.6	E.12	N/A	E.19	[4]E.22	5.3	
	Muskat	$\log(\bar{p}-p_{ws})$	Δt	−	$+m_M$	[5]E.7	N/A	E.15	[6]E.20	[6]E.20	5.3	Square system only.
	Developed system	[7]$\Delta p_{\Delta t}$	$\log \Delta t$	+	$+m$	E.6	E.13	N/A	E.19	[4]E.22	5.3	Line limits approximate.
	Developed system	[1]$p_{ws}-m^*\Delta t$	$\log \Delta t$	+	$+m$	E.6	E.9	N/A	E.19	[4]E.22	5.3	Line limits approximate.
	Variable production rate	p_{ws}	Eq. E.2	−	$-m$	E.6	E.12	N/A	—	—	5.4	Short production time only; see Section 5.4.
Injectivity	MDH	p_{wf}	$\log t$	+	$+m$	E.3	E.8	N/A	E.16	E.21	7.2	[8]See footnotes.
	Developed system	[1]$p_{wf}-m^*\Delta t$	$\log \Delta t$	+	$+m$	E.3	E.9	N/A	E.16	E.21	3.4	[8]See footnotes.
	Reservoir limit	p_{wf}	t	+	$+m^*$	N/A	N/A	E.14	E.17	—	3.5	[8]See footnotes.
	Variable rate	$(p_i-p_{wf})/q_N$	Eq. E.1	+	$+m'$	E.4	E.10	N/A	E.16	E.21	4.2	[8]Many specialized forms.
	Constant pressure	$1/q$	$\log t$	−	$+m_q$	E.5	E.11	N/A	E.18	—	4.6	[8]Also curve matching with Fig. 4.12.
Falloff	Horner	p_{ws}	[2]$\log[(t_p+\Delta t)/\Delta t]$	+	$-m$	E.6	E.12	N/A	E.19	[3]E.22	7.3	[8]See footnotes.
	MDH	p_{ws}	$\log \Delta t$	−	$+m$	E.6	E.12	N/A	E.19	[4]E.22	5.3	[8]See footnotes.
	Muskat	$\log(\bar{p}-p_{ws})$	Δt	+	$+m_M$	[5]E.7	N/A	E.15	[6]E.20	[6]E.20	5.3	[8]Square system only.
	Developed system	[7]$\Delta p_{\Delta t}$	$\log \Delta t$	−	$+m$	E.6	E.13	N/A	E.19	[4]E.22	5.3	[8]Line limits approximate.
	Developed system	[1]$p_{ws}-m^*\Delta t$	$\log \Delta t$	−	$+m$	E.6	E.9	N/A	E.19	[4]E.22	5.3	[8]Line limits approximate.
	Variable injection rate	p_{ws}	Eq. E.2	+	$-m$	E.6	E.12	N/A	—	—	5.4	[8]Short injection time only; see Section 5.4.

1. Requires linear pressure change before testing.
2. t_p from Eq. 5.9
3. Use Fig. 5.6
4. Use Fig. 5.7.
5. Use Fig. 5.10.
6. Use Fig. 5.11.
7. Extrapolate prior pressure trend to get $\Delta p_{\Delta t}$.
8. Liquid filled, unit mobility ratio assumed.
9. Use as an equation, (slope value) = $\pm m$.

Beginning Time for Analyzed Straight Line

$$t \simeq \frac{(200{,}000 + 12{,}000\, s)\, C}{(kh/\mu)} \quad \text{. . . . . (E.16)}$$

$$t \simeq \frac{\phi \mu c_t A}{0.0002637\, k} (t_{DA})_{pss} \quad \text{. . . . . (E.17)}$$

$$t \simeq \frac{19 \times 10^6\, \phi \mu c_t r_w^{\,2}}{k} \quad \text{. . . . . (E.18)}$$

$$\Delta t \simeq \frac{170{,}000\, C e^{0.14s}}{(kh/\mu)} \quad \text{. . . . . (E.19)}$$

$$\Delta t = \frac{\phi \mu c_t A}{0.0002637\, k} (\Delta t_{DA})_{sl}. \quad \text{Use with Fig. 5.11.} \quad \text{. . . . . (E.20)}$$

End Time for Semilog Straight Line

$$t \simeq \frac{\phi \mu c_t A}{0.0002637\, k} (t_{DA})_{eia}. \quad \text{. . . . . (E.21)}$$

$$\Delta t \simeq \frac{\phi \mu c_t A}{0.0002637\, k} (\Delta t_{DA})_{esl}. \quad \text{Use with Figs. 5.6 and 5.7.} \quad \text{. . . . . (E.22)}$$

Nomenclature

a = distance to an image well, Appendix B, ft
a_D = dimensionless distance to an image well, Appendix B
A = area, sq ft
b = intercept on Cartesian plot of transient-test pressure data, psi
b' = intercept on semilog plot of transient-test pressure data normalized by rate, psi/(STB/D)
B = formation volume factor, RB/STB
B_g = gas formation volume factor, RB/scf
B_o = oil formation volume factor, RB/STB
B_w = water formation volume factor, RB/STB
c = compressibility, psi^{-1}
c_f = formation (rock, pore volume) compressibility, psi^{-1}
c_g = gas compressibility, psi^{-1}
c_o = oil compressibility, psi^{-1}
c_{oa} = apparent oil-phase compressibility, including effects of dissolved gas, psi^{-1}
c_t = system total compressibility, psi^{-1}, Eq. 2.38
c_w = water compressibility, psi^{-1}
c_{wa} = apparent water-phase compressibility, including effects of dissolved gas, psi^{-1}
C = wellbore storage constant (coefficent, factor) RB/psi
C_A = shape constant or factor
C_D = dimensionless wellbore storage constant (coefficient, factor)
D = non-Darcy flow coefficient, D/Mcf
E_k = error in permeability estimated by simplified two-rate test analysis, fraction
E_s = error in skin factor estimated by simplified two-rate test analysis, dimensionless skin units
Ei = exponential integral, Eq. 2.7
e = 2.7182 . . .
erf = error function
exp = e
F_{cor} = correction factor when calculating permeability for a vertically fractured well, Section 11.3
F_{ft} = ratio of porosity-compressibility product of fracture to total porosity-compressibility product of reservoir rock
F_{HL} = Higgins-Leighton shape factor
F' = ratio of pulse length to total cycle length, Eq. 9.13
g = acceleration of gravity, ft/sec²
g_c = units conversion factor, 32.17 lb_m ft/(lb_f sec²)
G_P = primal geometric fraction for vertical pulse testing
G_R = reciprocal geometric fraction for vertical pulse testing
G^* = geometric fraction for vertical interference testing
h = formation thickness, ft
h_D = dimensionless thickness for horizontal fracture cases, Appendix C and Section 11.3
J = productivity index, (STB/D)/psi
J' = modified productivity index for a deliverability test
J^* = productivity index for a deliverability test
k = permeability, md
k_f = fracture permeability, md
k_{ma} = matrix permeability, md
k_{max} = maximum directional permeability, md
k_{min} = minimum directional permeability, md
k_o = permeability to oil, md
k_r = permeability in the radial (horizontal) direction, md
k_{rg} = relative permeability to gas, fraction
k_{ro} = relative permeability to oil, fraction
k_{rw} = relative permeability to water, fraction
k_s = permeability in the skin zone
k_z = permeability in the vertical direction, md
$\bar{k}$ = average permeability for anisotropic system, md
K = Secenov's coefficient, litre/gram-equivalent
log = logarithm, base 10
ln = logarithm, base e
L = length or distance, ft
m = ± slope of linear portion of semilog plot of pressure transient data, psi/cycle
$m(p)$ = real gas "potential" or pseudo pressure, Eq. 2.32, psi²/cp
m_H = slope of a Hall plot, psi/(STB/D)
m_M = slope of the straight-line portion of a Muskat plot of pressure buildup data, cycle/hour
m_q = slope of the $1/q$ vs $\log t$ plot for a constant-pressure test, (D/STB)/cycle
m_{Hf} = slope of p_{ws} vs $\sqrt{t}$ plot for horizontal-fracture well test data, psi/$\sqrt{\text{hours}}$
m_{Vf} = slope of p_{ws} vs $\sqrt{t}$ plot for vertical-fracture well test data, psi/$\sqrt{\text{hours}}$
m' = slope of the data plot for a multiple-rate test, psi/(cycle STB/D)
m'_1 = slope (based on q_1) of the data plot for a two-rate test, psi/cycle
m'_3 = slope (based on q_3) of the data plot for a drawdown after a shut-in period, psi/cycle
m'' = slope of simplified or special data plot for a multiple-rate test, psi/cycle
m^* = slope of the straight line on a linear plot of p_w vs t, psi/hour
M = mobility ratio
M = molecular weight, lb_m/mole
n = concentration of dissolved solids, gram-equivalents/litre, Appendix D

n = power in productivity-index formula
p = pressure, psi
p_c = critical pressure, psia
p_D = dimensionless pressure
$(p_D)_M$ = dimensionless pressure at the match point for type-curve analysis
p_{DMBH} = Matthews-Brons-Hazebroek-type dimensionless pressure
$\bar{p}_{DMBH}$ = Matthews-Brons-Hazebroek dimensionless pressure for a square, water-drive system based on average pressure
p_{DMBHe} = Matthews-Brons-Hazebroek dimensionless pressure for a square, water-drive system based on boundary pressure
p_{DMDH} = Miller-Dyes-Hutchinson-type dimensionless pressure
p_{DMint} = dimensionless pressure of extrapolated straight line at intercept of a Muskat plot
p_{DR} = dimensionless pressure ratio used in type-curve matching DST flow-period data
$\bar{p}_D$ = dimensionless average reservoir pressure for water-drive reservoir
p_e = external pressure, psi
p_{ext} = pressure correctly extrapolated from past behavior, psi
p_{ff} = final flowing pressure in a DST (subscript 1 or 2 indicates flow period), psi
p_{fhm} = final hydrostatic mud pressure in a DST, psi
p_{fsi} = final shut-in pressure in a DST, psi
p_i = initial pressure, psi
p_{if} = initial flowing pressure in a DST (subscript 1 or 2 indicates flow period), psi
p_{ihm} = initial hydrostatic mud pressure in a DST, psi
p_{int} = pressure at intercept (abscissa value = 0) of various kinds of $f(p_w)$ vs $f(t)$ plots, psi
p_{isi} = initial shut-in pressure in a DST, psi
p_o = pressure in drillstring just before a flow period of a DST, psi
p_{pc} = pseudocritical pressure, psia
p_{pr} = pseudoreduced pressure
p_{sc} = pressure at standard conditions, psi
p_{tf} = tubing or wellhead flowing pressure, psi
p_{ts} = tubing or wellhead shut-in pressure, psi
p_w = bottom-hole pressure, psi
$p_w(\Delta t = 0)$ = bottom-hole pressure just before starting a transient well test, psi
p_{wext} = bottom-hole pressure correctly extrapolated from past behavior, psi
p_{wf} = flowing bottom-hole pressure, psi
p_{ws} = shut-in bottom-hole pressure, psi
p_{1hr} = pressure on straight-line portion of semilog plot 1 hour after beginning a transient test; usually a special kind of p_{int}, psi
$\bar{p}$ = average reservoir pressure, psi
p^* = false pressure, pressure obtained when linear portion of the plot of p_{ws} vs $\log[(t_p + \Delta t)/\Delta t]$ is extrapolated to $(t_p + \Delta t)/\Delta t = 1$, psi
δp = pressure offset between two semilog straight lines in transient-test data plot for a naturally fractured system, psi
Δp = pressure change (or pulse response amplitude in pulse testing), psi
Δp_{bsl} = Δp at beginning of semilog straight line, psi
Δp_{DV} = dimensionless response amplitude for vertical pulse testing
$(\Delta p_{DV})_\infty$ = dimensionless response amplitude for vertical pulse testing in an infinite-acting system
Δp_{el} = Δp at end of linear flow period (half-slope log-log line, or $\sqrt{t}$ straight line) for a vertical fracture, psi
Δp_M = pressure change from transient test data at the match point for type-curve analysis, psi
Δp_{ow} = pressure drop at a well owing to operation of other wells in the reservoir, psi
Δp_s = pressure drop across skin, psi
Δp_{tw} = pressure difference between wellhead and bottom hole, psi
Δp_{1hr} = pressure difference on straight-line portion of semilog plot 1 hour after beginning a transient test; used in any kind of Δp vs $\log \Delta t$ plot, psi
$\Delta p_{\Delta t}$ = difference between observed and extrapolated pressure at time Δt, Eq. 5.22, psi
q = flow rate, > 0 for production, < 0 for injection, STB/D for liquid, Mscf/D for gas
q_D = dimensionless flow rate
$(q_D)_M$ = dimensionless flow rate at match point for type-curve matching
q_g = gas flow rate, Mcf/D
q_o = oil flow rate, STB/D
q_M = flow rate at match point for type-curve matching, STB/D
q_N = flow rate during Nth rate period in a variable-rate test, STB/D or Mcf/D
q_{sf} = sand-face flow rate expressed at standard conditions, STB/D
q_w = water flow rate, STB/D
$\bar{q}$ = average flow rate, STB/D
q^* = modified flow rate for pressure buildup analysis with variable rate before shut-in, STB/D
$(1/q)_{1hr}$ = ordinate value at 1 hour on straight-line plot of $(1/q)$ vs $\log t$, D/STB
r = radius, ft
r_d = radius of drainage as defined in Section 2.12, ft
r_D = dimensionless radial distance
r_e = external radius, ft
r_f = horizontal fracture radius, ft
r_{f1} = radial distance to fluid front number 1, ft
r_{inf} = influence radius for interference testing, ft
r_s = radius of skin zone, ft
r_w = wellbore radius, ft
r_{wa} = apparent or effective wellbore radius (includes effects of wellbore damage or improvement), ft
r_{wb} = radius to a water bank, ft
R_s = solution gas-oil ratio, scf/STB
R_{sw} = solution gas-water ratio, scf/STB
s = van Everdingen-Hurst skin factor
s_{cp} = pseudo skin factor resulting from sand consolidation
s_p = pseudo skin factor resulting from partial completion or restricted flow entry
s_{swp} = pseudo skin factor resulting from slanted well
s'' = additional skin factor resulting from anisotropic effects
S_g = gas saturation, fraction
S_o = oil saturation, fraction
S_w = water saturation, fraction
t = time, hours
t_{bsl} = time to beginning of the semilog straight line, hours
t_D = dimensionless time
$(t_D/r_D{}^2)_M$ = dimensionless time parameter from type curve at the match point for type-curve analysis
t_{DA} = dimensionless time based on drainage area
t_{De} = dimensionless time based on external radius, r_e
$(t_{DA})_{pss}$ = dimensionless time at the beginning of pseudosteady-state flow
t_{Drf} = dimensionless time based on horizontal fracture radius
t_{Dxf} = dimensionless time based on half-fracture length of a vertical fracture

t_{eia} = time at the end of the infinite-acting period, hours
t_{esl} = time to end of the semilog straight line, hours
t_L = time lag used in pulse testing, hours
$(t_L)_D$ = dimensionless time lag used in pulse testing
$(t_L)_\infty$ = time lag in vertical pulse testing for an infinite-acting system, hours
t_M = time value from transient-test data at the match point for type-curve analysis, hours
t_p = equivalent time well was on production or injection before shut-in, hours
t_p* = modified production time for pressure buildup analysis with variable rate before shut-in, hours
t_{pDA} = dimensionless production time based on drainage area
t_{pss} = time at the beginning of pseudosteady-state flow, hours
t_R = readjustment time, hours
t_s = stabilization time, hours
t_x = intersection time of two semilog straight-line segments on transient-test data plot, hours
t_1 = any time in a transient test, hours
t_2 = any time in a transient test, hours
$t*$ = time that transient-test data start deviating from semilog straight line, hours
Δt = running testing time, hours
Δt_C = total cycle length in pulse testing, hours
$(\Delta t_{DA})_{esl}$ = dimensionless time at end of Horner or Miller-Dyes-Hutchinson straight line for pressure buildup test analysis
$(\Delta t_{DA})_{sl}$ = dimensionless time at beginning or end of Muskat straight line for pressure buildup analysis
Δt_{Dfx} = dimensionless intersection time of two semilog straight lines for falloff test in a composite system
$\Delta t_{Df1}*$ = dimensionless time for deviation of data from first semilog straight line for falloff test in a composite system
Δt_{dyn} = time for reading dynamic pressure (used in reservoir simulation) from straight line of a buildup plot, hours
Δt_{fx} = time of intersection of two semilog straight lines for falloff test in a composite system, hours
$\Delta t_{f1}*$ = time of deviation of data from first semilog straight line for falloff test in a composite system, hours
$(\Delta t)_M$ = time at match point for type-curve matching, hours
$\Delta t_{\bar{p}}$ = shut-in time corresponding to Dietz's average reservoir pressure, hours
Δt_P = pulse length used in pulse testing, hours
Δt_{PDV} = dimensionless pulse length used in vertical pulse testing
Δt_{si} = shut-in time before drawdown test, hours
T = temperature, °R
T_c = critical temperature, °R
T_{pc} = pseudocritical temperature, °R
T_{pr} = pseudoreduced temperature
T_{sc} = temperature at standard conditions, °R
V = volume, bbl
V_p = pore volume, bbl
V_{pi} = drainage pore volume of Well i, bbl
V_{pt} = total system pore volume, bbl
V_P = volume produced, bbl
V_u = wellbore volume per unit length, bbl/ft
V_w = wellbore volume, bbl
ΔV = change in volume, bbl
W_i = cumulative water injection, bbl
x = x coordinate, ft
x_D = dimensionless x coordinate, Fig. 2.15 and Appendix B
x_e = x distance from a centered well to the edge of its square drainage region (half-length of the side of a square), ft
x_f = x distance from a well in the center of a square drainage region to the end of a vertical fracture that is parallel to the x axis (half-length of a vertical fracture), ft
x' = transformed x coordinate for an anisotropic system, ft
Δx = x length of a grid in a reservoir simulator, ft
y = y coordinate, ft
y_D = dimensionless y coordinate, Fig. 2.15 and Appendix B
y_i = mole fraction of Component i in the gas phase
y' = transformed y coordinate in an anisotropic system, ft
Δy = y length of a grid in a reservoir simulator, ft
z = real gas deviation factor
z_i = real gas deviation factor at initial conditions
ΔZ_P = vertical distance from upper formation boundary to center of upper perforations; for vertical well testing; Fig. 10.25; ft
ΔZ_R = vertical (response) distance between upper and lower perforations; for vertical pulse testing; Fig. 10.25; ft
ΔZ_{wf} = vertical distance from lower formation boundary to flow perforations; for vertical well testing; Fig. 10.25; ft
ΔZ_{ws} = vertical distance from lower formation boundary to observation (static) perforations; for vertical well testing; Fig. 10.25; ft
Δ = difference
γ = specific gravity; referenced to water for liquids, to air for gases
ϵ = interporosity flow parameter
θ = angle between positive x axis and direction of $k_{\max}$ in an anisotropic reservoir, degrees
λ = mobility, md/cp
λ_g = mobility of gas phase, md/cp
λ_o = mobility of oil phase, md/cp
λ_t = total flowing mobility, md/cp
λ_w = mobility of water phase, md/cp
μ = viscosity, cp
μ_g = gas viscosity, cp
μ_{ga} = gas viscosity at atmospheric pressure and reservoir temperature, cp
μ_{gi} = gas viscosity at initial conditions, cp
μ_o = oil viscosity, cp
μ_w = water viscosity, cp
ρ = density, lb_m/cu ft
ρ_w = water density, lb_m/cu ft
ϕ = porosity, fraction

Subscripts

a = apparent
b = base
bsl = beginning of semilog straight line
C = calculated
dyn = dynamic pressure value for use in reservoir simulation
D = dimensionless
e = external
eia = end of infinite-acting period
el = end of linear flow period
esl = end of straight-line portion
ext = on extrapolated pressure trend
E = estimated
f = flowing
f = in fracture

F = future
g = gas
i = initial, index, component number
int = intercept value, value of ordinate at zero abscissa value
j = index
ma = in formation matrix
M = match point in type-curve matching
n = total in summation
N = last rate interval in a multiple-rate flow test
N = total in summation
N = number of components in a mixture
o = oil
OB = observed value
s or si = shut-in or static
s = skin zone
sl = beginning or end of straight-line portion
t = total
tr = true
w = water
w = well
wb = wellbore
x = intersection point of two semilog straight-line segments on transient-test data plot
1hr = data from straight-line portion of semilog plot at 1 hour of test time, extrapolated if necessary
1, 2 = layer, zone numbers, or time numbers
∞ = infinite-acting system

Bibliography

A

Abramowitz, Milton and Stegun, Irene A. (ed.): *Handbook of Mathematical Functions With Formulas, Graphs and Mathematical Tables,* National Bureau of Standards Applied Mathematics Series-55 (June 1964) 227-253.

Adams, A. R., Ramey, H. J., Jr., and Burgess, R. J.: "Gas Well Testing in a Fractured Carbonate Reservoir," *J. Pet. Tech.* (Oct. 1968) 1187-1194; *Trans.,* AIME, **243.**

Agarwal, Ram G., Al-Hussainy, Rafi, and Ramey, H. J., Jr.: "An Investigation of Wellbore Storage and Skin Effect in Unsteady Liquid Flow: I. Analytical Treatment," *Soc. Pet. Eng. J.* (Sept. 1970) 279-290; *Trans.,* AIME, **249.**

Agostini, M. D.: "Wireline Formation-Tester Performance on the North West Shelf," *Australian Pet. Expl. Assn. J.* (1975) **15,** Part 1, 127-132.

Al-Hussainy, R. and Ramey, H. J., Jr.: "Application of Real Gas Flow Theory to Well Testing and Deliverability Forecasting," *J. Pet. Tech.* (May 1966) 637-642; *Trans.,* AIME, **237.**

Al-Hussainy, R., Ramey, H. J., Jr., and Crawford, P. B.: "The Flow of Real Gases Through Porous Media," *J. Pet. Tech.* (May 1966) 624-636; *Trans.,* AIME, **237.**

Ammann, Charles B.: "Case Histories of Analyses of Characteristics of Reservoir Rock From Drill-Stem Tests," *J. Pet. Tech.* (May 1960) 27-36.

Amyx, James W., Bass, Daniel M., Jr., and Whiting, Robert L.: *Petroleum Reservoir Engineering: Physical Properties,* McGraw-Hill Book Co., Inc., New York (1960).

Aronofsky, J. A. and Jenkins, R.: "A Simplified Analysis of Unsteady Radial Gas Flow," *Trans.,* AIME (1954) **201,** 149-154.

Arps, J. J. and Smith, A. E.: "Practical Use of Bottom-Hole Pressure Buildup Curves," *Drill. and Prod. Prac.,* API (1949) 155-165.

B

Baldwin, David E., Jr.: "A Monte Carlo Model for Pressure Transient Analysis," paper SPE 2568 presented at the SPE-AIME 44th Annual Fall Meeting, Denver, Sept. 28-Oct. 1, 1969.

Banks, K. M.: "Recent Achievements With the Formation Tester in Canada," *J. Cdn. Pet. Tech.* (July-Sept. 1963) 84-94.

Barbe, J. A. and Boyd, B. L.: "Short-Term Buildup Testing," *J. Pet. Tech.* (July 1971) 800-804.

Beal, Carlton: "The Viscosity of Air, Water, Natural Gas, Crude Oil and Its Associated Gases at Oil-Field Temperatures and Pressures," *Trans.,* AIME (1946) **165,** 94-115.

Bixel, H. C., Larkin, B. K., and van Poollen, H. K.: "Effect of Linear Discontinuities on Pressure Build-Up and Drawdown Behavior," *J. Pet. Tech.* (Aug. 1963) 885-895; *Trans.,* AIME, **228.**

Bixel, H. C. and van Poollen, H. K.: "Pressure Drawdown and Buildup in the Presence of Radial Discontinuities," *Soc. Pet. Eng. J.* (Sept. 1967) 301-309; *Trans.,* AIME, **240.**

Breitenbach, E. A., Thurnau, D. H., and van Poollen, H. K.: "Solution of the Immiscible Fluid Flow Simulation Equations," *Soc. Pet. Eng. J.* (June 1969) 155-169.

Brigham, W. E.: "Planning and Analysis of Pulse-Tests," *J. Pet. Tech.* (May 1970) 618-624; *Trans.,* AIME, **249.**

Brill, J. P., Bourgoyne, A. T., and Dixon, T. N.: "Numerical Simulation of Drillstem Tests as an Interpretation Technique," *J. Pet. Tech.* (Nov. 1969) 1413-1420.

Brons, F. and Marting, V. E.: "The Effect of Restricted Fluid Entry on Well Productivity," *J. Pet. Tech.* (Feb. 1961) 172-174; *Trans.,* AIME, **222.**

Brons, F. and Miller, W. C.: "A Simple Method for Correcting Spot Pressure Readings," *J. Pet. Tech.* (Aug. 1961) 803-805; *Trans.,* AIME, **222.**

Brown, George G., Katz, Donald L., Oberfell, George G., and Alden, Richard C.: *Natural Gasoline and the Volatile Hydrocarbons,* Natural Gasoline Assn. of America, Tulsa (1948).

Brownscombe, E. R.: "A Field Calibration Technique for Bottom-Hole Pressure Measurement," *Pet. Eng.* (Aug. 1947) 84-88.

Brownscombe, E. R. and Conlon, D. R.: "Precision in Bottom-Hole Pressure Measurements," *Trans.,* AIME (1946) **165,** 159-174.

Bruce, G. H., Peaceman, D. W., Rachford, H. H., Jr., and Rice, J. D.: "Calculations of Unsteady-State Gas Flow Through Porous Media," *Trans.,* AIME (1953) **198,** 79-92.

Burnett, O. W. and Mixa, E.: "Application of the Formation Interval Tester in the Rocky Mountain Area," *Drill. and Prod. Prac.,* API (1964) 131-140.

Burns, William A., Jr.: "New Single-Well Test for Determining Vertical Permeability," *J. Pet. Tech.* (June 1969) 743-752; *Trans.,* AIME, **246.**

C

Cannon, John R. and Dogru, Ali H.: "Estimation of Permeability and Porosity From Well Test Data," paper SPE 5345 presented at the SPE-AIME 45th Annual California Regional Meeting, Ventura, April 2-4, 1975.

Carr, Norman L., Kobayashi, Riki, and Burrows, David B.: "Viscosity of Hydrocarbon Gases Under Pressure," *Trans.,* AIME (1954) **201,** 264-272.

Carter, R. D.: "Pressure Behavior of a Limited Circular Composite Reservoir," *Soc. Pet. Eng. J.* (Dec. 1966) 328-334; *Trans.,* AIME, **237.**

Chatas, Angelos T.: "A Practical Treatment of Non-Steady State Flow Problems in Reservoir Systems," *Pet. Eng.,* Part 1 (May 1953) B-42 through B-50; Part 2 (June 1953) B-38 through B-50; Part 3 (Aug. 1953) B-44 through B-56.

Chen, Hsiu-Kuo and Brigham, W. E.: "Pressure Buildup for a Well With Storage and Skin in a Closed Square," paper SPE 4890 presented at the SPE-AIME 44th Annual California Regional Meeting, San Francisco, April 4-5, 1974.

Chew, Ju-Nam and Connally, Carl A., Jr.: "A Viscosity Correlation for Gas-Saturated Crude Oils," *Trans.*, AIME (1959) **216,** 23-25.

Cinco, H., Miller, F. G., and Ramey, H. J., Jr.: "Unsteady-State Pressure Distribution Created by a Directionally Drilled Well," *J. Pet. Tech.* (Nov. 1975) 1392-1400; *Trans.*, AIME, **259.**

Cinco-Ley, Heber, Ramey, H. J., Jr., and Miller, Frank G.: "Pseudo-Skin Factors for Partially Penetrating Directionally Drilled Wells," paper SPE 5589 presented at the SPE-AIME 50th Annual Fall Technical Conference and Exhibition, Dallas, Sept. 28-Oct. 1, 1975.

Cinco-Ley, Heber, Ramey, Henry J., Jr., and Miller, Frank G.: "Unsteady-State Pressure Distribution Created by a Well With an Inclined Fracture," paper SPE 5591 presented at the SPE-AIME 50th Annual Fall Technical Conference and Exhibition, Dallas, Sept. 28-Oct. 1, 1975.

Cinco-L., Heber, Samaniego-V., F., and Dominguez-A., N.: "Transient Pressure Behavior for a Well With a Finite Conductivity Vertical Fracture," paper SPE 6014 presented at the SPE-AIME 51st Annual Fall Technical Conference and Exhibition, New Orleans, Oct. 3-6, 1976.

Cinco-L., Heber, Samaniego-V., F., and Dominguez-A., N.: "Unsteady-State Flow Behavior for a Well Near a Natural Fracture," paper SPE 6019 presented at the SPE-AIME 51st Annual Fall Technical Conference and Exhibition, New Orleans, Oct. 3-6, 1976.

Clark, K. K.: "Transient Pressure Testing of Fractured Water Injection Wells," *J. Pet. Tech.* (June 1968) 639-643; *Trans.*, AIME, **243.**

Coats, K. H., Dempsey, J. R., and Henderson, J. H.: "A New Technique for Determining Reservoir Description From Field Performance Data," *Soc. Pet. Eng. J.* (March 1970) 66-74; *Trans.*, AIME, **249.**

Cobb, William M. and Dowdle, Walter L.: "A Simple Method for Determining Well Pressure in Closed Rectangular Reservoirs," *J. Pet. Tech.* (Nov. 1973) 1305-1306.

Cobb, William M., Ramey, H. J., Jr., and Miller, Frank G.: "Well-Test Analysis for Wells Producing Commingled Zones," *J. Pet. Tech.* (Jan. 1972) 27-37; *Trans.*, AIME, **253.**

Cobb, William M. and Smith, James T.: "An Investigation of Pressure Buildup Tests in Bounded Reservoirs," paper SPE 5133 presented at the SPE-AIME 49th Annual Fall Meeting, Houston, Oct. 6-9, 1974. An abridged version appears in *J. Pet. Tech.* (Aug. 1975) 991-996; *Trans.*, AIME, **259.**

Collins, Royal Eugene: *Flow of Fluids Through Porous Materials,* Reinhold Publishing Corp., New York (1961) 108-123.

"Conversion of Operational and Process Measurement Units to the Metric (SI) System," *Manual of Petroleum Measurement Standards,* Pub. API 2564, API (March 1974) Chap. 15, Sec. 2.

Cooper, Hilton, H., Jr., Bredehoeft, John D., and Papadopulos, Istavros S.: "Response of a Finite-Diameter Well to an Instantaneous Charge of Water," *Water Resources Res.* (1967) **3,** No. 1, 263-269.

Crawford, G. E., Hagedorn, A. R., and Pierce, A. E.: "Analysis of Pressure Buildup Tests in a Naturally Fractured Reservoir," *J. Pet. Tech.* (Nov. 1976) 1295-1300.

Cronquist, Chapman: "Dimensionless PVT Behavior of Gulf Coast Reservoir Oils," *J. Pet. Tech.* (May 1973) 538-542.

Culham, W. E.: "Amplification of Pulse-Testing Theory," *J. Pet. Tech.* (Oct. 1969) 1245-1247.

Culham, W. E.: "Pressure Buildup Equations for Spherical Flow Regime Problems," *Soc. Pet. Eng. J.* (Dec. 1974) 545-555.

Cullender, M. H.: "The Isochronal Performance Method of Determining the Flow Characteristics of Gas Wells," *Trans.*, AIME (1955) **204,** 137-142.

D

Denson, A. H., Smith, J. T., and Cobb, W. M.: "Determining Well Drainage Pore Volume and Porosity From Pressure Buildup Tests," *Soc. Pet. Eng. J.* (Aug. 1976) 209-216; *Trans.*, AIME, **261.**

de Swaan O., A.: "Analytic Solutions for Determining Naturally Fractured Reservoir Parameters by Well Testing," *Soc. Pet. Eng. J.* (June 1976) 117-122; *Trans.*, AIME, **261.**

Dietz, D. N.: "Determination of Average Reservoir Pressure From Build-Up Surveys," *J. Pet. Tech.* (Aug. 1965) 955-959; *Trans.*, AIME, **234.**

Dixon, Thomas N., Seinfeld, John H., Startzman, Richard A., and Chen, W. H.: "Reliability of Reservoir Parameters From History Matched Drill Stem Tests," paper SPE 4282 presented at the SPE-AIME Third Symposium on Numerical Simulation of Reservoir Performance, Houston, Jan. 10-12, 1973.

Dodson, C. R. and Standing, M. B.: "Pressure-Volume-Temperature and Solubility Relations for Natural-Gas-Water Mixtures," *Drill. and Prod. Prac.*, API (1944) 173-179.

Dolan, John P., Einarsen, Charles A., and Hill, Gilman A.: "Special Application of Drill-Stem Test Pressure Data," *Trans.*, AIME (1957) **210,** 318-324.

Dowdle, Walter L.: "Discussion of Pressure Falloff Analysis in Reservoirs With Fluid Banks," *J. Pet. Tech.* (July 1974) 818; *Trans.*, AIME, **257.**

Doyle, R. E. and Sayegh, E. F.: "Real Gas Transient Analysis of Three-Rate Flow Tests," *J. Pet. Tech.* (Nov. 1970) 1347-1356.

Driscoll, Vance J.: "Use of Well Interference and Build Up Data for Early Quantitative Determination of Reserves, Permeability and Water Influx," *J. Pet. Tech.* (Oct. 1963) 1127-1136; *Trans.*, AIME, **228.**

Duvaut, G.: "Drainage des Systèmes Hétérogènes," *Revue IFP* (Oct. 1961) 1164-1181.

E

Earlougher, Robert C., Jr.: "Comparing Single-Point Pressure Buildup Data With Reservoir Simulator Results," *J. Pet. Tech.* (June 1972) 711-712.

Earlougher, Robert C., Jr.: "Discussion of Interference Analysis for Anisotropic Formations — A Case History," *J. Pet. Tech.* (Dec. 1975) 1525; *Trans.*, AIME, **259.**

Earlougher, R. C., Jr.: "Estimating Drainage Shapes From Reservoir Limit Tests," *J. Pet. Tech.* (Oct. 1971) 1266-1268; *Trans.*, AIME, **251.**

Earlougher, R. C., Jr.: "Estimating Errors When Analyzing Two-Rate Flow Tests," *J. Pet. Tech.* (May 1973) 545-547.

Earlougher, Robert C., Jr.: "Variable Flow Rate Reservoir Limit Testing," *J. Pet. Tech.* (Dec. 1972) 1423-1429.

Earlougher, Robert C., Jr., and Kersch, Keith M.: "Analysis of Short-Time Transient Test Data by Type-Curve Matching," *J. Pet. Tech.* (July 1974) 793-800; *Trans.*, AIME, **257.**

Earlougher, Robert C., Jr., and Kersch, Keith M.: "Field Examples of Automatic Transient Test Analysis," *J. Pet. Tech.* (Oct. 1972) 1271-1277.

Earlougher, Robert C., Jr., Kersch, K. M., and Kunzman, W. J.: "Some Characteristics of Pressure Buildup Behavior in Bounded Multiple-Layer Reservoirs Without Crossflow," *J. Pet. Tech.* (Oct. 1974) 1178-1186; *Trans.*, AIME, **257.**

Earlougher, Robert C., Jr., Kersch, K. M., and Ramey, H. J., Jr.: "Wellbore Effects in Injection Well Testing," *J. Pet. Tech.* (Nov. 1973) 1244-1250.

Earlougher, R. C., Jr., Miller, F. G., and Mueller, T. D.: "Pressure Buildup Behavior in a Two-Well Gas-Oil System," *Soc. Pet. Eng. J.* (June 1967) 195-204; *Trans.*, AIME, **240.**

Earlougher, R. C., Jr., and Ramey, H. J., Jr.: "Interference Analysis in Bounded Systems," *J. Cdn. Pet. Tech.* (Oct.-Dec. 1973) 33-45.

Earlougher, Robert C., Jr., and Ramey, H. J., Jr.: "The Use of Interpolation to Obtain Shape Factors for Pressure Buildup Calculations," *J. Pet. Tech.* (May 1968) 449-450.

Earlougher, Robert C., Jr., Ramey, H. J., Jr., Miller, F. G., and Mueller, T. D.: "Pressure Distributions in Rectangular Reservoirs," *J. Pet. Tech.* (Feb. 1968) 199-208; *Trans.*, AIME, **243.**

Edwards, A. G. and Shryock, S. H.: "New Generation Drill Stem Testing Tools/Technology," *Pet. Eng.* (July 1974) 46, 51, 56, 58, 61.

Edwards, A. G. and Winn, R. H.: "A Summary of Modern Tools and Techniques Used in Drill Stem Testing," Pub. T-4069, Halliburton Co., Duncan, Okla. (Sept. 1973).

Eilerts, C. Kenneth: "Methods for Estimating Deliverability After Massive Fracture Completions in Tight Formations," paper SPE 5112 presented at the SPE-AIME Deep Drilling and Production Symposium, Amarillo, Tex., Sept. 8-10, 1974.

El-Hadidi, Samir M. and Ritter, A. W.: "Interpretation of Bottom-Hole Pressure Build-Up Tests on Fractured Oil Reservoirs," paper SPE 6020 presented at the SPE-AIME 51st Annual Fall Technical Conference and Exhibition, New Orleans, Oct. 3-6, 1976.

Elkins, Lincoln F. and Skov, Arlie M.: "Determination of Fracture Orientation From Pressure Interference," *Trans.*, AIME (1960) **219,** 301-304.

Engineering Data Book, 9th ed., Gas Processors Suppliers Assn., Tulsa (1972) Sec. 1.

Ershaghi, Iraj, Rhee, Shie-Woo, and Yang, Hsun-Tiao: "Analysis of Pressure Transient Data in Naturally Fractured Reservoirs With Spherical Flow," paper SPE 6018 presented at the SPE-AIME 51st Annual Fall Technical Conference and Exhibition, New Orleans, Oct. 3-6, 1976.

Evans, John G.: "The Use of Pressure Buildup Information to Analyze Non-Respondent Vertically Fractured Oil Wells," paper SPE 3345 presented at the SPE-AIME Rocky Mountain Regional Meeting, Billings, Mont., June 2-4, 1971.

Evers, John F. and Soeiinah, Edy: "Transient Tests and Long-Range Performance Predictions in Stress-Sensitive Reservoirs," paper SPE 5423 presented at the SPE-AIME Northern Plains Section Regional Meeting, Omaha, May 15-16, 1975.

F

Falade, Gabriel K. and Brigham, William E.: "The Analysis of Single-Well Pulse Tests in a Finite-Acting Slab Reservoir," paper SPE 5055B presented at the SPE-AIME 49th Annual Fall Meeting, Houston, Oct. 6-9, 1974.

Falade, Gabriel K. and Brigham, William E.: "The Dynamics of Vertical Pulse Testing in a Slab Reservoir," paper SPE 5055A presented at the SPE-AIME 49th Annual Fall Meeting, Houston, Oct. 6-9, 1974.

Felsenthal, Martin: "Step-Rate Tests Determine Safe Injection Pressures in Floods," *Oil and Gas J.* (Oct. 28, 1974) 49-54.

Fetkovich, M. J.: "The Isochronal Testing of Oil Wells," paper SPE 4529 presented at the SPE-AIME 48th Annual Fall Meeting, Las Vegas, Sept. 30-Oct. 3, 1973.

G

Gibson, J. A. and Campbell, A. T., Jr.: "Calculating the Distance to a Discontinuity From D.S.T. Data," paper SPE 3016 presented at the SPE-AIME 45th Annual Fall Meeting, Houston, Oct. 4-7, 1970.

Gladfelter, R. E., Tracy, G. W., and Wilsey, L. E.: "Selecting Wells Which Will Respond to Production-Stimulation Treatment," *Drill. and Prod. Prac.*, API (1955) 117-129.

Gogarty, W. B., Kinney, W. L., and Kirk, W. B.: "Injection Well Stimulation With Micellar Solutions," *J. Pet. Tech.* (Dec. 1970) 1577-1584.

Gray, K. E.: "Approximating Well-to-Fault Distance From Pressure Buildup Tests," *J. Pet. Tech.* (July 1965) 761-767.

Greenkorn, R. A. and Johnson, C. R.: "Method for Defining Reservoir Heterogeneities," U.S. Patent No. 3,285,064 (Nov. 15, 1966).

Griffith, H. D. and Collins, T.: "Determining Average Reservoir Properties From Gathering-Line Transient Analysis for a Multiwell Reservoir," *J. Pet. Tech.* (July 1975) 835-842.

Gringarten, Alain C.: "Unsteady-State Pressure Distributions Created by a Well With a Single Horizontal Fracture, Partial Penetration, or Restricted Flow Entry," PhD dissertation, Stanford U. (1971) 106. (Order No. 71-23,512 University Microfilms, P. O. Box 1764, Ann Arbor, Mich. 48106.)

Gringarten, Alain C. and Ramey, Henry J., Jr.: "An Approximate Infinite Conductivity Solution for a Partially Penetrating Line-Source Well," *Soc. Pet. Eng. J.* (April 1975) 140-148; *Trans.*, AIME, **259.**

Gringarten, Alain C. and Ramey, Henry J., Jr.: "Unsteady-State Pressure Distributions Created by a Well With a Single Horizontal Fracture, Partial Penetration, or Restricted Entry," *Soc. Pet. Eng. J.* (Aug. 1974) 413-426; *Trans.*, AIME, **257.**

Gringarten, A. C., Ramey, H. J., Jr., and Raghavan, R.: "Applied Pressure Analysis for Fractured Wells," *J. Pet. Tech.* (July 1975) 887-892; *Trans.*, AIME, **259.**

Gringarten, Alain C., Ramey, Henry J., Jr., and Raghavan, R.: "Pressure Analysis for Fractured Wells," paper SPE 4051 presented at the SPE-AIME 47th Annual Fall Meeting, San Antonio, Tex., Oct. 8-11, 1972.

Gringarten, Alain C., Ramey, Henry J., Jr., and Raghavan, R.: "Unsteady-State Pressure Distributions Created by a Well With a Single Infinite-Conductivity Vertical Fracture," *Soc. Pet. Eng. J.* (Aug. 1974) 347-360.

Gringarten, A. C. and Witherspoon, P. A.: "A Method of Analyzing Pump Test Data From Fractured Aquifers," *Proc.*, Symposium on Percolation Through Fissured Rock, International Society for Rock Mechanics, Stuttgart (Sept. 18-19, 1972).

"Guide for Calculating Static Bottom-Hole Pressures Using Fluid-Level Recording Devices," ERCB Report 74-S, Energy Resources Conservation Board, Calgary, Alta., Canada (Nov. 1974).

"Guide for the Planning, Conducting, and Reporting of Subsurface Pressure Tests," ERCB Report 74-T, Energy Resources Conservation Board, Calgary, Alta., Canada (Nov. 1974).

Gutek, A. M. H. and Clark, K. K.: "Vertical Permeability Measurements, Swan Hills Reef Complex," paper SPE 4122 presented at the SPE-AIME 47th Annual Fall Meeting, San Antonio, Tex., Oct. 8-11, 1972.

H

Hall, H. N.: "How to Analyze Waterflood Injection Well Performance," *World Oil* (Oct. 1963) 128-130.

Hall, Howard N.: "Compressibility of Reservoir Rocks," *Trans.*, AIME (1953) **198,** 309-311.

Harrill, J. R.: "Determining Transmissivity From Water-Level Recovery of a Step-Drawdown Test," Prof. Paper 700-C, USGS (1970) C212 through C213.

Hartsock, J. H. and Warren, J. E.: "The Effect of Horizontal Hydraulic Fracturing on Well Performance," *J. Pet. Tech.* (Oct. 1961) 1050-1056; *Trans.*, AIME, **222.**

Hawkins, Murray F., Jr.: "A Note on the Skin Effect," *Trans.*, AIME (1956) **207,** 356-357.

Hazebroek, P., Rainbow, H., and Matthews, C. S.: "Pressure Fall-Off in Water Injection Wells," *Trans.*, AIME (1958) **213,** 250-260.

Hernandez, Victor M. and Swift, George W.: "A Method for Determining Reservoir Parameters From Early Drawdown Data," paper SPE 3982 presented at the SPE-AIME 47th Annual Fall Meeting, San Antonio, Oct. 8-11, 1972.

Higgins, R. V., Boley, D. W., and Leighton, A. J.: "Aids to Forecasting the Performance of Waterfloods," *J. Pet. Tech.* (Sept. 1964) 1076-1082; *Trans.*, AIME, **231.**

Higgins, R. V. and Leighton, A. J.: "A Method of Predicting Performance of Five-Spot Waterfloods in Stratified Reservoirs Using Streamlines," *Report of Investigations 5921,* USBM (1962).

Higgins, R. V. and Leighton, A. J.: "Quick Way to Find Reservoir Pressure Distribution," *Oil and Gas J.* (Jan. 6, 1969) 67-70.

Hirasaki, George J.: "Pulse Tests and Other Early Transient Pressure Analyses for In-Situ Estimation of Vertical Permeability," *Soc. Pet. Eng. J.* (Feb. 1974) 75-90; *Trans.*, AIME, **257.**

Hopkins, Robert A.: *The International (SI) Metric System and How It Works,* Polymetric Services, Inc., Tarzana, Calif. (1974).

Horner, D. R.: "Pressure Build-Up in Wells," *Proc.*, Third World Pet. Cong., The Hague (1951) Sec. II, 503-523.

Howard, G. C. and Fast, C. R.: *Hydraulic Fracturing,* Monograph Series, Society of Petroleum Engineers of AIME, Dallas (1970) **2.**

Hubbert, M. King: "The Theory of Ground-Water Motion," *J. of Geol.* (Nov.-Dec. 1940) **XLVIII,** 785-944.

Hurst, William: "Establishment of the Skin Effect and Its Impediment to Fluid Flow Into a Well Bore," *Pet. Eng.* (Oct. 1953) B-6 through B-16.

Hurst, William: "Interference Between Oil Fields," *Trans.*, AIME (1960) **219,** 175-192.

Hurst, William, Clark, J. Donald, and Brauer, E. Bernard: "The Skin Effect in Producing Wells," *J. Pet. Tech.* (Nov. 1969) 1483-1489; *Trans.*, AIME, **246.**

Huskey, William L. and Crawford, Paul B.: "Performance of Petroleum Reservoirs Containing Vertical Fractures in the Matrix," *Soc. Pet. Eng. J.* (June 1967) 221-228; *Trans.*, AIME, **240.**

J

Jacob, C. E. and Lohman, S. W.: "Nonsteady Flow to a Well of Constant Drawdown in an Extensive Aquifer," *Trans.*, AGU (Aug. 1952) 559-569.

Jahns, Hans O.: "A Rapid Method for Obtaining a Two-Dimensional Reservoir Description From Well Pressure Response Data," *Soc. Pet. Eng. J.* (Dec. 1966) 315-327; *Trans.*, AIME, **237.**

Jargon, J. R.: "Effect of Wellbore Storage and Wellbore Damage at the Active Well on Interference Test Analysis," *J. Pet. Tech.* (Aug. 1976) 851-858.

Jargon, J. R. and van Poollen, H. K.: "Unit Response Function From Varying-Rate Data," *J. Pet. Tech.* (Aug. 1965) 965-969; *Trans.*, AIME, **234.**

Johnson, C. R.: "Portable 'Radar' for Testing Reservoirs Developed by Esso Production Research," *Oil and Gas J.* (Nov. 20, 1967) 162-164.

Johnson, C. R., Greenkorn, R. A., and Woods, E. G.: "Pulse-Testing: A New Method for Describing Reservoir Flow Properties Between Wells," *J. Pet. Tech.* (Dec. 1966) 1599-1604; *Trans.*, AIME, **237.**

Johnson, C. R. and Raynor, R.: "System for Measuring Low Level Pressure Differential," U.S. Patent No. 3,247,712 (April 26, 1966).

Jones, L. G.: "Reservoir Reserve Tests," *J. Pet. Tech.* (March 1963) 333-337; *Trans.*, AIME, **228.**

Jones, L. G. and Watts, J. W.: "Estimating Skin Effect in a Partially Completed Damaged Well," *J. Pet. Tech.* (Feb. 1971) 249-252; *Trans.*, AIME, **251.**

Jones, Park: "Drawdown Exploration Reservoir Limit, Well and Formation Evaluation," paper 824-G presented at the SPE-AIME Permian Basin Oil Recovery Conference, Midland, April 18-19, 1957.

Jones, Park: "Reservoir Limit Test," *Oil and Gas J.* (June 18, 1956) 184-196.

K

Kamal, M. and Brigham, W. E.: "Design and Analysis of Pulse Tests With Unequal Pulse and Shut-In Periods," *J. Pet. Tech.* (Feb. 1976) 205-212; *Trans.*, AIME, **261.**

Kamal, M. and Brigham, W. E.: "The Effect of Linear Pressure Trends on Interference Tests," *J. Pet. Tech.* (Nov. 1975) 1383-1384.

Kamal, Medhat and Brigham, William E.: "Pulse-Testing Response for Unequal Pulse and Shut-In Periods," *Soc. Pet. Eng. J.* (Oct. 1975) 399-410; *Trans.*, AIME, **259.**

Kaplan, Wilfred: *Advanced Calculus,* Addison Wesley Publishing Co., Inc., Reading, Mass. (1952) 220.

Katz, Donald L., Cornell, David, Kobayashi, Riki L., Poettmann, Fred H., Vary, John A., Elenbaas, John R., and Weinaug, Charles F.: *Handbook of Natural Gas Engineering,* McGraw-Hill Book Co., Inc., New York (1959) Chap. 11.

Kazemi, H.: "A Reservoir Simulator for Studying Productivity Variation and Transient Behavior of a Well in a Reservoir Undergoing Gas Evolution," *J. Pet. Tech.* (Nov. 1975) 1401-1412; *Trans.*, AIME, **259.**

Kazemi, Hossein: "Damage Ratio From Drill-Stem Tests With Variable Back Pressure," paper SPE 1458 presented at the SPE-AIME 36th Annual California Regional Meeting, Santa Barbara, Nov. 17-18, 1966.

Kazemi, Hossein: "Determining Average Reservoir Pressure From Pressure Buildup Tests," *Soc. Pet. Eng. J.* (Feb. 1974) 55-62; *Trans.*, AIME, **257.**

Kazemi, Hossein: "Discussion of Variable Flow Rate Reservoir Limit Testing," *J. Pet. Tech.* (Dec. 1972) 1429-1430.

Kazemi, Hossein: "Locating a Burning Front by Pressure Transient Measurements," *J. Pet. Tech.* (Feb. 1966) 227-232; *Trans.*, AIME, **237.**

Kazemi, Hossein: "Pressure Buildup in Reservoir Limit Testing of Stratified Systems," *J. Pet. Tech.* (April 1970) 503-511; *Trans.*, AIME, **249.**

Kazemi, H.: "Pressure Transient Analysis of Naturally Fractured Reservoirs With Uniform Fracture Distribution," *Soc. Pet. Eng. J.* (Dec. 1969) 451-462; *Trans.*, AIME, **246.**

Kazemi, Hossein, Merrill, L. S., and Jargon, J. R.: "Problems in Interpretation of Pressure Fall-Off Tests in Reservoirs With and Without Fluid Banks," *J. Pet. Tech.* (Sept. 1972) 1147-1156.

Kazemi, Hossein and Seth, Mohan S.: "Effect of Anisotropy and Stratification on Pressure Transient Analysis of Wells With Restricted Flow Entry," *J. Pet. Tech.* (May 1969) 639-647; *Trans.*, AIME, **246.**

Kazemi, H., Seth, M. S., and Thomas, G. W. "The Interpretation of Interference Tests in Naturally Fractured Reservoirs With Uniform Fracture Distribution," *Soc. Pet. Eng. J.* (Dec. 1969) 463-472; *Trans.*, AIME, **246.**

Khurana, A. K.: "Influence of Tidal Phenomena on Interpretation of Pressure Build-Up and Pulse Tests," *Australian Pet. Expl. Assn. J.* (1976) **16,** Part 1, 99-105.

Knutson, G. C.: "A Computer-Oriented Method of Pressure Build-Up Analysis," *J. Cdn. Pet. Tech.* (July-Sept. 1967) 111-114.

Kohlhaas, Charles A.: "A Method for Analyzing Pressures Measured During Drillstem-Test Flow Periods," *J. Pet. Tech.* (Oct. 1972) 1278-1282; *Trans.*, AIME, **253.**

Kohlhaas, C. A. and Miller, F. G.: "Rock-Compaction and Pressure-Transient Analysis With Pressure-Dependent Rock Properties," paper SPE 2563 presented at the SPE-AIME 44th Annual Fall Meeting, Denver, Sept. 28-Oct. 1, 1969.

Kolb, R. H.: "Two Bottom-Hole Pressure Instruments Providing Automatic Surface Recording," *Trans.*, AIME (1960) **219,** 346-349.

Kumar, Anil and Ramey, Henry J., Jr.: "Well-Test Analysis for a Well in a Constant-Pressure Square," paper SPE 4054 presented at the SPE-AIME 47th Annual Fall Meeting, San Antonio, Tex., Oct. 8-11, 1972. An abridged version appears in *Soc. Pet. Eng. J.* (April 1974) 107-116.

L

Laird, A. and Birks, J.: "Performance and Accuracy of Amerada Bottom-Hole Pressure Recorder With Special Reference to Use in Drill Stem Formation Tests and Repeatability of Reservoir Pressures Obtained Therein," *J. Inst. Pet.* (1951) **37,** 678-695.

Langston, E. P.: "Field Application of Pressure Buildup Tests, Jay-Little Escambia Creek Fields," paper SPE 6199 presented at the SPE-AIME 51st Annual Fall Technical Conference and Exhibition, New Orleans, Oct. 3-6, 1976.

Larkin, Bert K.: "Solutions to the Diffusion Equation for a Region Bounded by a Circular Discontinuity," *Soc. Pet. Eng. J.* (June 1963) 113-115; *Trans.*, AIME, **228.**

Lee, W. J., Jr.: "Analysis of Hydraulically Fractured Wells With Pressure Buildup Tests," paper SPE 1820 presented at the SPE-AIME 42nd Annual Fall Meeting, Houston, Oct. 1-4, 1967.

Lee, W. John: "Wellbore Storage: How It Affects Pressure Buildup and Pressure Drawdown Tests," paper presented at the SPWLA 12th Annual Logging Symposium, Dallas, May 2-5, 1971.

Lee, W. John, Harrell, Robert R., and McCain, William D., Jr.: "Evaluation of a Gas Well Testing Method," paper SPE 3872 presented at the SPE-AIME Northern Plains Section Regional Meeting, Omaha, May 18-19, 1972.

Lefkovits, H. C., Hazebroek, P., Allen, E. E., and Matthews, C. S.: "A Study of the Behavior of Bounded Reservoirs Composed of Stratified Layers," *Soc. Pet. Eng. J.* (March 1961) 43-58; *Trans.*, AIME, **222.**

Lescarboura, Jaime A.: "New Downhole Shut-in Tool Boosts BHP Test Accuracy," *World Oil* (Nov. 1974) 71-73.

"Letter Symbols for Petroleum Reservoir Engineering, Natural Gas Engineering, and Well Logging Quantities," Society of Petroleum Engineers of AIME, Dallas (1965); *Trans.*, AIME (1965) **234,** 1463-1496.

Levorsen, A. I.: *Geology of Petroleum,* 2nd ed., W. H. Freedman and Co., San Francisco (1967) 125.

Locke, C. D. and Sawyer, W. K.: "Constant Pressure Injection Test in a Fractured Reservoir — History Match Using Numerical Simulation and Type-Curve Analysis," paper SPE 5594 presented at the SPE-AIME 50th Annual Fall Technical Conference and Exhibition, Dallas, Sept. 28-Oct. 1, 1975.

Long, Giordano and Chierici, Gianluigi: "Salt Content Changes Compressibility of Reservoir Brines," *Pet. Eng.* (July, 1961) B-25 through B-31.

Loucks, T. L. and Guerrero, E. T.: "Pressure Drop in a Composite Reservoir," *Soc. Pet. Eng. J.* (Sept. 1961) 170-176; *Trans.*, AIME, **222.**

M

Maer, N. K., Jr.: "Type Curves for Analysis of Afterflow-Dominated Gas Well Buildup Data," *J. Pet. Tech.* (Aug. 1976) 915-924.

Martin, John C.: "Simplified Equations of Flow in Gas Drive Reservoirs and the Theoretical Foundation of Multiphase Pressure Buildup Analyses," *Trans.*, AIME (1959) **216,** 309-311.

Mathur, Shri B.: "Determination of Gas Well Stabilization Factors in the Hugoton Field," *J. Pet. Tech.* (Sept. 1969) 1101-1106.

Matthews, C. S.: "Analysis of Pressure Build-Up and Flow Test Data," *J. Pet. Tech.* (Sept. 1961) 862-870.

Matthews, C. S., Brons, F., and Hazebroek, P.: "A Method for Determination of Average Pressure in a Bounded Reservoir," *Trans.*, AIME (1954) **201,** 182-191.

Matthews, C. S. and Lefkovits, H. C.: "Studies on Pressure Distribution in Bounded Reservoirs at Steady State," *Trans.*, AIME (1955) **204,** 182-189.

Matthews, C. S. and Russell, D. G.: *Pressure Buildup and Flow Tests in Wells,* Monograph Series, Society of Petroleum Engineers of AIME, Dallas (1967) **1.**

Matthies, E. Peter: "Practical Application of Interference Tests," *J. Pet. Tech.* (March 1964) 249-252.

McAlister, J. A., Nutter, B. P., and Lebourg, M.: "A New System of Tools for Better Control and Interpretation of Drill-Stem Tests," *J. Pet. Tech.* (Feb. 1965) 207-214; *Trans.*, AIME, **234.**

McKinley, R. M.: "Estimating Flow Efficiency From Afterflow-Distorted Pressure Buildup Data," *J. Pet. Tech.* (June 1974) 696-697.

McKinley, R. M.: "Wellbore Transmissibility From Afterflow-Dominated Pressure Buildup Data," *J. Pet. Tech.* (July 1971) 863-872; *Trans.*, AIME, **251.**

McKinley, R. M., Vela, Saul, and Carlton, L. A.: "A Field Application of Pulse-Testing for Detailed Reservoir Description," *J. Pet. Tech.* (March 1968) 313-321; *Trans.*, AIME, **243.**

McLeod, H. O., Jr., and Coulter, A. W., Jr.: "The Stimulation Treatment Pressure Record — An Overlooked Formation Evaluation Tool," *J. Pet. Tech.* (Aug. 1969) 951-960.

"Measuring, Sampling, and Testing Crude Oil," API Standard 2500, American Petroleum Institute. Reproduced in Frick, Thomas C. and Taylor, R. William: *Petroleum Production Handbook,* McGraw-Hill Book Co., Inc., New York (1962) **1,** Chap. 16.

Merrill, L. S., Jr., Kazemi, Hossein, and Gogarty, W. Barney: "Pressure Falloff Analysis in Reservoirs With Fluid Banks," *J. Pet. Tech.* (July 1974) 809-818; *Trans.*, AIME, **257.**

Miller, C. C., Dyes, A. B., and Hutchinson, C. A., Jr.: "The Estimation of Permeability and Reservoir Pressure From Bottom Hole Pressure Build-Up Characteristics," *Trans.*, AIME (1950) **189,** 91-104.

Miller, G. B., Seeds, R. W. S., and Shira, H. W.: "A New, Surface-Recording, Down-Hole Pressure Gauge," paper SPE 4125 presented at the SPE-AIME 47th Annual Fall Meeting, San Antonio, Tex., Oct. 8-11, 1972.

Moran, J. H. and Finklea, E. E.: "Theoretical Analysis of Pressure Phenomena Associated With the Wireline Formation Tester," *J. Pet. Tech.* (Aug. 1962) 899-908; *Trans.*, AIME, **225.**

Morris, Earl E. and Tracy, G. W.: "Determination of Pore Volume in a Naturally Fractured Reservoir," paper SPE 1185 presented at the SPE-AIME 40th Annual Fall Meeting, Denver, Oct. 3-6, 1965.

Morse, J. V. and Ott, Frank III: "Field Application of Unsteady-State Pressure Analysis in Reservoir Diagnosis," *J. Pet. Tech.* (July 1967) 869-876.

Mueller, Thomas D. and Witherspoon, Paul A.: "Pressure Interference Effects Within Reservoirs and Aquifers," *J. Pet. Tech.* (April 1965) 471-474; *Trans.*, AIME, **234.**

Murphy, W. C.: "The Interpretation and Calculation of Formation Characteristics From Formation Test Data," Pamphlet T-101, Halliburton Co., Duncan, Okla. (1970).

Muskat, Morris: *Physical Principles of Oil Production*, McGraw-Hill Book Co., Inc., New York (1949) Chap. 12.

Muskat, Morris: "Use of Data on the Build-Up of Bottom-Hole Pressures," *Trans.*, AIME (1937) **123,** 44-48.

N

Najurieta, Humberto L.: "A Theory for the Pressure Transient Analysis in Naturally Fractured Reservoirs," paper SPE 6017 presented at the SPE-AIME 51st Annual Fall Technical Conference and Exhibition, New Orleans, Oct. 3-6, 1976.

Nestlerode, W. A.: "Permanently Installed Bottom-Hole Pressure Gauge," paper 875-16-L presented at the API Div. of Production Meeting, Denver, April 11-13, 1962.

Newman, G. H.: "Pore-Volume Compressibility of Consolidated, Friable, and Unconsolidated Reservoir Rocks Under Hydrostatic Loading," *J. Pet. Tech.* (Feb. 1973) 129-134.

Nowak, T. J. and Lester, G. W.: "Analysis of Pressure Fall-Off Curves Obtained in Water Injection Wells to Determine Injective Capacity and Formation Damage," *Trans.*, AIME (1955) **204,** 96-102.

O

Odeh, A. S.: "Flow Test Analysis for a Well With Radial Discontinuity," *J. Pet. Tech.* (Feb. 1969) 207-210; *Trans.*, AIME, **246.**

Odeh, A. S.: "Pseudo Steady-State Flow Capacity of Oil Wells With Limited Entry, and With an Altered Zone Around the Wellbore," paper SPE 6132 presented at the SPE-AIME 51st Annual Fall Technical Conference and Exhibition, New Orleans, Oct. 3-6, 1976.

Odeh, A. S.: "Steady-State Flow Capacity of Wells With Limited Entry to Flow," *Soc. Pet. Eng. J.* (March 1968) 43-51; *Trans.*, AIME, **243.**

Odeh, A. S.: "Unsteady-State Behavior of Naturally Fractured Reservoirs," *Soc. Pet. Eng. J.* (March 1965) 60-64; *Trans.*, AIME, **234.**

Odeh, A. S. and Al-Hussainy, R.: "A Method for Determining the Static Pressure of a Well From Buildup Data," *J. Pet. Tech.* (May 1971) 621-624; *Trans.*, AIME, **251.**

Odeh, A. S. and Jones, L. G.: "Pressure Drawdown Analysis, Variable-Rate Case," *J. Pet. Tech.* (Aug. 1965) 960-964; *Trans.*, AIME, **234.**

Odeh, A. S. and Jones, L. G.: "Two-Rate Flow Test, Variable-Rate Case — Application to Gas-Lift and Pumping Wells," *J. Pet. Tech.* (Jan. 1974) 93-99; *Trans.*, AIME, **257.**

Odeh, A. S., Moreland, E. E., and Schueler, S.: "Characterization of a Gas Well From One Flow-Test Sequence," *J. Pet. Tech.* (Dec. 1975) 1500-1504; *Trans.*, AIME, **259.**

Odeh, A. S. and Selig, F.: "Pressure Build-Up Analysis, Variable-Rate Case," *J. Pet. Tech.* (July 1963) 790-794; *Trans.*, AIME, **228.**

Otoumagie, Robert and Menzie, D. E.: "Analysis of Pressure Buildup in an Infinite Two-Layered Oil Reservoir Without Crossflow," paper SPE 6130 presented at the SPE-AIME 51st Annual Fall Technical Conference and Exhibition, New Orleans, Oct. 3-6, 1976.

P

Papadopulos, Istavros S.: "Nonsteady Flow to a Well in an Infinite Anisotropic Aquifer," *Proc.*, 1965 Dubrovnik Symposium on Hydrology of Fractured Rocks, Inter. Assoc. of Sci. Hydrology (1965) **I,** 21-31.

Papadopulos, Istavros S., Bredehoeft, John D., and Cooper, Hilton H., Jr.: "On the Analysis of 'Slug Test' Data," *Water Resources Res.* (Aug. 1973) **9,** No. 4, 1087-1089.

Papadopulos, Istavros S. and Cooper, Hilton H., Jr.: "Drawdown in a Well of Large Diameter," *Water Resources Res.* (1967) **3,** No. 1, 241-244.

Perrine, R. L.: "Analysis of Pressure Buildup Curves," *Drill. and Prod. Prac.*, API (1956) 482-509.

Pierce, Aaron E.: "Case History: Waterflood Performance Predicted by Pulse Testing," paper SPE 6196 presented at the SPE-AIME 51st Annual Fall Technical Conference and Exhibition, New Orleans, Oct. 3-6, 1976.

Pierce, A. E., Vela, Saul, and Koonce, K. T.: "Determination of the Compass Orientation and Length of Hydraulic Fractures by Pulse Testing," *J. Pet. Tech.* (Dec. 1975) 1433-1438.

Pinson, A. E., Jr.: "Concerning the Value of Producing Time Used in Average Pressure Determinations From Pressure Buildup Analysis," *J. Pet. Tech.* (Nov. 1972) 1369-1370.

Pinson, A. E., Jr.: "Conveniences in Analyzing Two-Rate Flow Tests," *J. Pet. Tech.* (Sept. 1972) 1139-1141.

Pirson, Richard S. and Pirson, Sylvain J.: "An Extension of the Pollard Analysis Method of Well Pressure Build-Up and Drawdown Tests," paper SPE 101 presented at the SPE-AIME 36th Annual Fall Meeting, Dallas, Oct. 8-11, 1961.

Pitzer, Sidney C., Rice, John D., and Thomas, Clifford E.: "A Comparison of Theoretical Pressure Build-Up Curves With Field Curves Obtained From Bottom-Hole Shut-In Tests," *Trans.*, AIME (1959) **216,** 416-419.

Pollard, P.: "Evaluation of Acid Treatments From Pressure Build-Up Analysis," *Trans.*, AIME (1959) **216,** 38-43.

Polubarinova-Kochina, P. Ya.: *Theory of Ground Water Movement*, Princeton U. Press, Princeton, N.J. (1962) 343-369.

Prasad, Raj K.: "A Practical Way to Find Minimum Drainage Area for a Well," *Oil and Gas J.* (July 30, 1973) 118-120.

Prasad, Raj K.: "Pressure Transient Analysis in the Presence of Two Intersecting Boundaries," *J. Pet. Tech.* (Jan. 1975) 89-96; *Trans.*, AIME, **259.**

Prats, Michael: "A Method for Determining the Net Vertical Permeability Near a Well From In-Situ Measurements," *J. Pet. Tech.* (May 1970) 637-643; *Trans.*, AIME, **249.**

Prats, M. and Scott, J. B.: "Effect of Wellbore Storage on Pulse-Test Pressure Response," *J. Pet. Tech.* (June 1975) 707-709.

R

Raghavan, R.: "Well Test Analysis: Wells Producing by Solution Gas Drive," *Soc. Pet. Eng. J.* (Aug. 1976) 196-208; *Trans.*, AIME, **261.**

Raghavan, R., Cady, Gilbert V., and Ramey, Henry J., Jr.: "Well-Test Analysis for Vertically Fractured Wells," *J. Pet. Tech.* (Aug. 1972) 1014-1020; *Trans.*, AIME, **253.**

Raghavan, R. and Clark, K. K.: "Vertical Permeability From Limited Entry Flow Tests in Thick Formations," *Soc. Pet. Eng. J.* (Feb. 1975) 65-73; *Trans.*, AIME, **259.**

Raghavan, R. and Hadinoto, Nico: "Analysis of Pressure Data for Fractured Wells: The Constant Pressure Outer Boundary," paper SPE 6015 presented at the SPE-AIME 51st Annual Fall Technical Conference and Exhibition, New Orleans, Oct. 3-6, 1976.

Raghavan, R., Scorer, J. D. T., and Miller, F. G.: "An Investigation by Numerical Methods of the Effect of Pressure-Dependent Rock and Fluid Properties on Well Flow Tests," *Soc. Pet. Eng. J.* (June 1972) 267-275; *Trans.*, AIME, **253.**

Raghavan, R., Topaloglu, H. N., Cobb, W. M., and Ramey, H. J., Jr.: "Well-Test Analysis for Wells Producing From Two Commingled Zones of Unequal Thickness," *J. Pet. Tech.* (Sept. 1974) 1035-1043; *Trans.*, AIME, **257.**

Raghavan, R., Uraiet, A., and Thomas, G. W.: "Vertical Fracture Weight: Effect on Transient Flow Behavior," paper SPE 6016 presented at the SPE-AIME 51st Annual Fall Technical Conference and Exhibition, New Orleans, Oct. 3-6, 1976.

Ramey, H. J., Jr.: "Application of the Line Source Solution to Flow in Porous Media — A Review," *Prod. Monthly* (May 1967) 4-7, 25-27.

Ramey, Henry J., Jr.: "Interference Analysis for Anisotropic Formations — A Case History," *J. Pet. Tech.* (Oct. 1975) 1290-1298; *Trans.*, AIME, **259.**

Ramey, H. J., Jr.: "Non-Darcy Flow and Wellbore Storage Effects in Pressure Build-Up and Drawdown of Gas Wells," *J. Pet. Tech.* (Feb. 1965) 223-233; *Trans.*, AIME, **234.**

Ramey, H. J., Jr.: "Rapid Method of Estimating Reservoir Compressibility," *J. Pet. Tech.* (April 1964) 447-454; *Trans.*, AIME, **231.**

Ramey, H. J., Jr.: "Short-Time Well Test Data Interpretation in the Presence of Skin Effect and Wellbore Storage," *J. Pet. Tech.* (Jan. 1970) 97-104; *Trans.*, AIME, **249.**

Ramey, H. J., Jr.: "Verification of the Gladfelter-Tracy-Wilsey Concept for Wellbore Storage Dominated Transient Pressures During Production," *J. Cdn. Pet. Tech.*, (April-June 1976) 84-85.

Ramey, Henry J., Jr., and Agarwal, Ram G.: "Annulus Unloading Rates as Influenced by Wellbore Storage and Skin Effect," *Soc. Pet. Eng. J.* (Oct. 1972) 453-462; *Trans.*, AIME, **253.**

Ramey, Henry J., Jr., Agarwal, Ram G., and Martin, Ian: "Analysis of 'Slug Test' or DST Flow Period Data," *J. Cdn. Pet. Tech.* (July-Sept. 1975) 37-47.

Ramey, H. J., Jr., and Cobb, William M.: "A General Buildup Theory for a Well in a Closed Drainage Area," *J. Pet. Tech.* (Dec. 1971) 1493-1505.

Ramey, H. J., Jr., and Earlougher, R. C., Jr.: "A Note on Pressure Buildup Curves," *J. Pet. Tech.* (Feb. 1968) 119-120.

Ramey, Henry J., Jr., Kumar, Anil, and Gulati, Mohinder S.: *Gas Well Test Analysis Under Water-Drive Conditions,* AGA, Arlington, Va. (1973).

Rawlins, E. L. and Schellhardt, M. A.: *Back-Pressure Data on Natural-Gas Wells and Their Application to Production Practices,* Monograph 7, USBM (1936).

Reprint Series No. 9 — Pressure Analysis Methods, Society of Petroleum Engineers of AIME, Dallas (1967).

Reprint Series No. 11 — Numerical Simulation, Society of Petroleum Engineers of AIME, Dallas (1973).

"Review of Basic Formation Evaluation," Form J-328, Johnston, Houston (1974).

Ridley, Trevor P.: "The Unified Analysis of Well Tests," paper SPE 5587 presented at the SPE-AIME 50th Annual Fall Technical Conference and Exhibition, Dallas, Sept. 28-Oct. 1, 1975.

Robertson, D. C. and Kelm, C. H.: "Injection-Well Testing To Optimize Waterflood Performance," *J. Pet. Tech.* (Nov. 1975) 1337-1342.

Russell, D. G.: "Determination of Formation Characteristics From Two-Rate Flow Tests," *J. Pet. Tech.* (Dec. 1963) 1347-1355; *Trans.*, AIME, **228.**

Russell, D. G.: "Extensions of Pressure Build-Up Analysis Methods," *J. Pet. Tech.* (Dec. 1966) 1624-1636; *Trans.*, AIME, **237.**

Russell, D. G., Goodrich, J. H., Perry, G. E., and Bruskotter, J. F.: "Methods for Predicting Gas Well Performance," *J. Pet. Tech.* (Jan. 1966) 99-108; *Trans.*, AIME, **237.**

Russell, D. G. and Prats, M.: "The Practical Aspects of Interlayer Crossflow," *J. Pet. Tech.* (June 1962) 589-594.

Russell, D. G. and Truitt, N. E.: "Transient Pressure Behavior in Vertically Fractured Reservoirs," *J. Pet. Tech.* (Oct. 1964) 1159-1170; *Trans.*, AIME, **231.**

S

Samaniego-V., F., Brigham, W. E., and Miller, F. G.: "A Performance Prediction Procedure for Transient Flow of Fluids Through Pressure Sensitive Formations," paper SPE 6051 presented at the SPE-AIME 51st Annual Fall Technical Conference and Exhibition, New Orleans, Oct. 3-6, 1976.

Samaniego-V., F., Brigham, W. E., and Miller, F. G.: "An Investigation of Transient Flow of Reservoir Fluids Considering Pressure-Dependent Rock and Fluid Properties," paper SPE 5593 presented at the SPE-AIME 50th Annual Fall Technical Conference and Exhibition, Dallas, Sept. 28-Oct. 1, 1975.

Schultz, A. L., Bell, W. T., and Urbanosky, H. J.: "Advances in Uncased-Hole, Wireline Formation-Tester Techniques," *J. Pet. Tech.* (Nov. 1975) 1331-1336.

Sinha, B. K., Sigmon, J. E., and Montgomery, J. M.: "Comprehensive Analysis of Drillstem Test Data With the Aid of Type Curves," paper SPE 6054 presented at the SPE-AIME 51st Annual Fall Technical Conference and Exhibition, New Orleans, Oct. 3-6, 1976.

Slater, G. E. and Hegeman, P. S. "Visual Studies of Well Interference," *J. Pet. Tech.* (Feb. 1976) 119-122.

Slider, H. C.: "A Simplified Method of Pressure Buildup Analysis for a Stabilized Well," *J. Pet. Tech.* (Sept. 1971) 1155-1160; *Trans.*, AIME, **251.**

Slider, H. C.: "Application of Pseudo-Steady-State Flow to Pressure-Buildup Analysis," paper SPE 1403 presented at the SPE-AIME Regional Symposium, Amarillo, Tex., Oct. 27-28, 1966.

Smith, R. V. and Dewees, E. J.: "Sources of Error in Subsurface-Pressure-Gage Calibration and Usage," *Oil and Gas J.* (Dec. 9, 1948) 85-98.

Standing, M. B.: "Concerning the Calculation of Inflow Performance of Wells Producing Solution Gas Drive Reservoirs," *J. Pet. Tech.* (Sept. 1971) 1141-1142.

Standing, M. B.: *Volumetric and Phase Behavior of Oil Field Hydrocarbon Systems,* Reinhold Publishing Corp., New York (1952).

Standing, Marshall B. and Katz, Donald L.: "Density of Natural Gases," *Trans.*, AIME (1942) **146,** 140-149.

Startzman, R. A.: "A Further Note on Pulse-Test Interpretation," *J. Pet. Tech.* (Sept. 1971) 1143-1144.

Stegemeier, G. L. and Matthews, C. S.: "A Study of Anomalous Pressure Build-Up Behavior," *Trans.*, AIME (1958) **213,** 44-50.

Strobel, C. J., Gulati, M. S., and Ramey, H. J., Jr.: "Reservoir Limit Tests in a Naturally Fractured Reservoir — A Field Case Study Using Type Curves," *J. Pet. Tech.* (Sept. 1976) 1097-1106; *Trans.*, AIME, **261.**

"Supplements to Letter Symbols and Computer Symbols for Petroleum Reservoir Engineering, Natural Gas Engineering, and Well Logging Quantities," Society of Petroleum Engineers of AIME, Dallas (1972); *Trans.*, AIME (1972) **253,** 556-574.

"Supplements to Letter Symbols and Computer Symbols for Petroleum Reservoir Engineering, Natural Gas Engineering, and Well Logging Quantities," Society of Petroleum Engineers of AIME, Dallas (1975); *Trans.*, AIME (1975) **259,** 517-537.

Swift, S. C. and Brown, L. P.: "Interference Testing for Reservoir Definition — The State of the Art," paper SPE 5809 presented at the SPE-AIME Fourth Symposium on Improved Oil Recovery, Tulsa, March 22-24, 1976.

T

Taylor, George S. and Luthin, James N.: "Computer Methods for Transient Analysis of Water-Table Aquifers," *Water Resources Res.* (Feb. 1969) **5,** No. 1, 144-152.

Theis, Charles V.: "The Relation Between the Lowering of the Piezometric Surface and the Rate and Duration of Discharge of a Well Using Ground-Water Storage," *Trans.*, AGU (1935) 519-524.

Theory and Practice of the Testing of Gas Wells, 3rd ed., Pub. ECRB-75-34, Energy Resources and Conservation Board, Calgary, Alta., Canada (1975).

Thomas, M. D. and Gupta, M. C.: "Can Your DST Results Be Improved?" *Oilweek* (Sept. 21, 1970) 35, 38, 42-44.

Thomas, Rex D. and Ward, Don C.: "Effect of Overburden Pressure and Water Saturation on the Gas Permeability of Tight Sandstone Cores," *J. Pet. Tech.* (Feb. 1972) 120-124.

Tiab, Djebbar and Kumar, Anil: "Application of p_D' Function to Interference Analysis," paper SPE 6053 presented at the SPE-AIME 51st Annual Fall Technical Conference and Exhibition, New Orleans, Oct. 3-6, 1976.

Timmerman, E. H. and van Poollen, H. K.: "Practical Use of Drill-Stem Tests," *J. Cdn. Pet. Tech.* (April-June 1972) 31-41.

Trube, Albert S.: "Compressibility of Natural Gases," *Trans.*, AIME (1957) **210,** 355-357.

Trube, Albert S.: "Compressibility of Undersaturated Hydrocarbon Reservoir Fluids," *Trans.*, AIME (1957) **210,** 341-344.

V

Vairogs, Juris, Hearn, C. L., Dareing, Donald W., and Rhoades, V. W.: "Effect of Rock Stress on Gas Production From Low-Permeability Reservoirs," *J. Pet. Tech.* (Sept. 1971) 1161-1167; *Trans.*, AIME, **251.**

Vairogs, Juris and Rhoades, Vaughan W.: "Pressure Transient Tests in Formations Having Stress-Sensitive Permeability," *J. Pet. Tech.* (Aug. 1973) 965-970; *Trans.*, AIME, **255.**

van der Knaap, W.: "Nonlinear Behavior of Elastic Porous Media," *Trans.*, AIME (1959) **216,** 179-187.

van Everdingen, A. F.: "The Skin Effect and Its Influence on the Productive Capacity of a Well," *Trans.*, AIME (1953) **198,** 171-176.

van Everdingen, A. F. and Hurst, W.: "The Application of the Laplace Transformation to Flow Problems in Reservoirs," *Trans.*, AIME (1949) **186,** 305-324.

van Everdingen, A. F. and Meyer, L. Joffre: "Analysis of Buildup Curves Obtained After Well Treatment," *J. Pet. Tech.* (April 1971) 513-524; *Trans.*, AIME, **251.**

van Poollen, H. K.: "Radius-of-Drainage and Stabilization-Time Equations," *Oil and Gas J.* (Sept. 14, 1964) 138-146.

van Poollen, H. K.: "Status of Drill-Stem Testing Techniques and Analysis," *J. Pet. Tech.* (April 1961) 333-339.

van Poollen, H. K.: "Transient Tests Find Fire Front in an In-Situ Combustion Project," *Oil and Gas J.* (Feb. 1, 1965) 78-80.

van Poollen, H. K., Bixel, H. C., and Jargon, J. R.: "Reservoir Modeling — 1: What It Is, What It Does," *Oil and Gas J.* (July 28, 1969) 158-160.

van Poollen, H. K., Bixel, H. C., and Jargon, J. R.: "Reservoir Modeling — 2: Single-Phase Fluid-Flow Equations," *Oil and Gas J.* (Aug. 18, 1969) 94-96.

van Poollen, H. K., Bixel, H. C., and Jargon, J. R.: "Reservoir Modeling — 3: Finite Differences," *Oil and Gas J.* (Sept. 15, 1969) 120-121.

van Poollen, H. K., Bixel, H. C., and Jargon, J. R.: "Reservoir Modeling — 4: Explicit Finite-Difference Technique," *Oil and Gas J.* (Nov. 3, 1969) 81-87.

van Poollen, H. K., Bixel, H. C., and Jargon, J. R.: "Reservoir Modeling — 5: Implicit Finite-Difference Approximation," *Oil and Gas J.* (Jan. 5, 1970) 88-92.

van Poollen, H. K., Bixel, H. C., and Jargon, J. R.: "Reservoir Modeling — 6: General Form of Finite-Difference Approximations," *Oil and Gas J.* (Jan. 19, 1970) 84-86.

van Poollen, H. K., Bixel, H. C., and Jargon, J. R.: "Reservoir Modeling — 7: Single-Phase Reservoir Models," *Oil and Gas J.* (March 2, 1970) 77-80.

van Poollen, H. K., Bixel, H. C., and Jargon, J. R.: "Reservoir Modeling — 8: Single-Phase Gas Flow," *Oil and Gas J.* (March 30, 1970) 106-107.

van Poollen, H. K., Bixel, H. C., and Jargon, J. R.: "Reservoir Modeling — 9: Here Are Fundamental Equations for Multiphase Fluid Flow," *Oil and Gas J.* (May 11, 1970) 72-78.

van Poollen, H. K., Bixel, H. C., and Jargon, J. R.: "Reservoir Modeling — 10: Applications of Multiphase Immiscible Fluid-Flow Simulator," *Oil and Gas J.* (June 29, 1970) 58-63.

van Poollen, H. K., Bixel, H. C., and Jargon, J. R.: "Reservoir Modeling — 11: Comparison of Multiphase Models," *Oil and Gas J.* (July 27, 1970) 124-130.

van Poollen, H. K., Bixel, H. C., and Jargon, J. R.: "Reservoir Modeling — 12: Individual Wells Pressures in Reservoir Modeling," *Oil and Gas J.* (Oct. 26, 1970) 78-80.

van Poollen, H. K., Bixel, H. C., and Jargon, J. R.: "Reservoir Modeling — 13: (Conclusion) A Review — and A Look Ahead," *Oil and Gas J.* (March 1, 1971) 78-79.

van Poollen, H. K., Breitenbach, E. A., and Thurnau, D. H.: "Treatment of Individual Wells and Grids in Reservoir Modeling," *Soc. Pet. Eng. J.* (Dec. 1968) 341-346.

van Poollen, H. K. and Jargon, J. R.: "Steady-State and Unsteady-State Flow of Non-Newtonian Fluids Through Porous Media," *Soc. Pet. Eng. J.* (March 1969) 80-88; *Trans.*, AIME, **246.**

Vela, Saul and McKinley, R. M.: "How Areal Heterogeneities Affect Pulse-Test Results," *Soc. Pet. Eng. J.* (June 1970) 181-191; *Trans.*, AIME, **249.**

Vogel, J. V.: "Inflow Performance Relationships for Solution-Gas Drive Wells," *J. Pet. Tech.* (Jan. 1968) 83-92; *Trans.*, AIME, **243.**

W

Warren, J. E. and Hartsock, J. H.: "Well Interference," *Trans.*, AIME (1960) **219,** 89-91.

Warren, J. E. and Root, P. J.: "Discussion of Unsteady-State Behavior of Naturally Fractured Reservoirs," *Soc. Pet. Eng. J.* (March 1965) 64-65; *Trans.*, AIME, **234.**

Warren, J. E. and Root, P. J.: "The Behavior of Naturally Fractured Reservoirs," *Soc. Pet. Eng. J.* (Sept. 1963) 245-255; *Trans.*, AIME, **228.**

Wattenbarger, Robert A. and Ramey, H. J., Jr.: "An Investigation of Wellbore Storage and Skin Effect in Unsteady Liquid Flow: II. Finite Difference Treatment," *Soc. Pet. Eng. J.* (Sept. 1970) 291-297; *Trans.*, AIME, **249.**

Wattenbarger, Robert A. and Ramey, H. J., Jr.: "Gas Well Testing With Turbulence, Damage and Wellbore Storage," *J. Pet. Tech.* (Aug. 1968) 877-887; *Trans.*, AIME, **243.**

Wattenbarger, Robert A. and Ramey, Henry J., Jr.: "Well Test Interpretation of Vertically Fractured Gas Wells," *J. Pet. Tech.* (May 1969) 625-632.

Weeks, Edwin P.: "Determining the Ratio of Horizontal to Vertical Permeability by Aquifer-Test Analysis," *Water Resources Res.* (Feb. 1969) **5,** No. 1, 196-214.

Weeks, Steve G. and Farris, Gerald F.: "Permagauge — A Permanent Surface-Recording Downhole Pressure Monitor — Through a Tube," paper SPE 5607 presented at the SPE-AIME 50th Annual Fall Technical Conference and Exhibition, Dallas, Sept. 28-Oct. 1, 1975.

West, W. J., Garvin, W. W., and Sheldon, J. W.: "Solution of the Equations of Unsteady-State Two-Phase Flow in Oil Reservoirs," *Trans.*, AIME (1954) **201,** 217-229.

Winestock, A. G. and Colpitts, G. P.: "Advances in Estimating Gas Well Deliverability," *J. Cdn. Pet. Tech.* (July-Sept. 1965) 111-119.

Witherspoon, Paul A., Narasimhan, T. N., and McEdwards, D. G.: "Results of Interference Tests From Two Geothermal Reservoirs," paper SPE 6052 presented at the SPE-AIME 51st Annual Fall Technical Conference and Exhibition, New Orleans, Oct. 3-6, 1976.

Woods, E. G.: "Pulse-Test Response of a Two-Zone Reservoir," *Soc. Pet. Eng. J.* (Sept. 1970) 245-256; *Trans.*, AIME, **249.**

Z

Zana, E. T. and Thomas, G. W.: "Some Effects of Contaminents on Real Gas Flow," *J. Pet. Tech.* (Sept. 1970) 1157-1168; *Trans.*, AIME, **249.**

Author Index

(List of authors and organizations referred to in the Monograph text, references, and the bibliography.)

Subject Index

N

O

P

Pulse testing: vertical Cont'd.

R

S